Fractional Differential Equations: Numerical Methods for Applications

Fractional Differential Equations: Numerical Methods for Applications

Edited by
Bernard Gibson

Willford Press
www.willfordpress.com

Published by Willford Press,
118-35 Queens Blvd., Suite 400,
Forest Hills, NY 11375, USA

ISBN: 978-1-64728-513-5

Cataloging-in-Publication Data

Fractional differential equations : numerical methods for applications / edited by Bernard Gibson.
p. cm.
Includes bibliographical references and index.
ISBN 978-1-64728-513-5
1. Fractional differential equations. 2. Differential equations. 3. Differential equations--Numerical solutions.
4. Fractional calculus. 5. Mathematics. I. Gibson, Bernard.
QA372 .F73 2023
515.35--dc23

For information on all Willford Press publications
visit our website at www.willfordpress.com

WILLFORD PRESS

Contents

Preface

This book has been a concerted effort by a group of academicians, researchers and scientists, who have contributed their research works for the realization of the book. This book has materialized in the wake of emerging advancements and innovations in this field. Therefore, the need of the hour was to compile all the required researches and disseminate the knowledge to a broad spectrum of people comprising of students, researchers and specialists of the field.

Fractional calculus is a branch of mathematical analysis that studies the different possibilities of defining real number powers or complex number powers of the differentiation operator and the integration operator. It also focuses on developing a calculus for such operators generalizing the classical one. Fractional differential equations are a generalization of differential equations through the application of fractional calculus. They are also referred to as extraordinary differential equations. They are widely used in various disciplines such as mathematics, physics, chemistry, biology, medicine, mechanics, control theory, signal and image processing, and environmental science. Most of the computational tools do not have in-built functions for solving fractional differential equations or differential equations having non-integer order. However, there are certain numerical methods for solving fractional-order problems such as certain MATLAB routines. This book provides a broad overview of the numerical methods used for solving fractional differential equations. It will serve as a valuable source of reference for students and mathematics researchers.

At the end of the preface, I would like to thank the authors for their brilliant chapters and the publisher for guiding us all-through the making of the book till its final stage. Also, I would like to thank my family for providing the support and encouragement throughout my academic career and research projects.

Editor

1

Existence and Stability Results for a Fractional Order Differential Equation with Non-Conjugate Riemann-Stieltjes Integro-Multipoint Boundary Conditions

Bashir Ahmad [1], Ymnah Alruwaily [1], Ahmed Alsaedi [1,*] and Sotiris K. Ntouyas [1,2]

1 Nonlinear Analysis and Applied Mathematics (NAAM)-Research Group, Department of Mathematics, Faculty of Science, King Abdulaziz University, P.O. Box 80203, Jeddah 21589, Saudi Arabia; bashirahmad_qau@yahoo.com (B.A.); ymnah@ju.edu.sa (Y.A.); sntouyas@uoi.gr (S.K.N.)
2 Department of Mathematics, University of Ioannina, Ioannina 45110, Greece
* Correspondence: aalsaedi@hotmail.com

Abstract: We discuss the existence and uniqueness of solutions for a Caputo-type fractional order boundary value problem equipped with non-conjugate Riemann-Stieltjes integro-multipoint boundary conditions on an arbitrary domain. Modern tools of functional analysis are applied to obtain the main results. Examples are constructed for the illustration of the derived results. We also investigate different kinds of Ulam stability, such as Ulam-Hyers stability, generalized Ulam-Hyers stability, and Ulam-Hyers-Rassias stability for the problem at hand.

Keywords: Caputo fractional derivative; nonlocal; integro-multipoint boundary conditions; existence; uniqueness; Ulam-Hyers stability

---

1. Introduction

Fractional calculus played a pivotal role in improving the mathematical modeling of many real-world problems. The extensive application of fractional order (differential and integral) operators indeed reflects the popularity of this branch of mathematical analysis. In contrast to the integer order operators, such operators are nonlocal in nature and do have the capacity to trace the history of the phenomenon under investigation. A detailed account of the use of fractional calculus tools can be found in several scientific disciplines such as, chaos and fractional dynamics [1], evolution in honeycomb lattice via fractional Schrödinger equation [2], financial economics [3], ecology [4], bio-engineering [5], etc. For theoretical development and further application of the topic, see the texts [6–9].

During the past two decades, the study of fractional order boundary value problems has been one of the hot topics of scientific research. Several researchers contributed to the development of this class of problems by producing a huge number of articles, special issues, monographs, etc. Now the literature on the topic contains a variety of existence and uniqueness results, and analytic and numerical methods of solutions for these problems. In particular, there has been shown a great interest in the formulation and investigation of fractional order boundary value problems involving non-classical (nonlocal and integral) boundary conditions. The nonlocal boundary conditions are found to be of great utility in modeling the changes happening within the domain of the given scientific phenomena, while the concept of integral boundary conditions is applied to model the physical problems, such as blood flow problems on arbitrary structures and ill-posed backward problems. For some recent works on fractional order differential equations involving Riemann-Liouville, Caputo, and Hadamard type fractional derivatives, equipped with classical, nonlocal, and integral boundary conditions, we refer the reader to a series of papers [10–28] and the references cited therein.

In this paper, we study the existence of solutions for a nonlinear Liouville-Caputo-type fractional differential equation on an arbitrary domain:

$$^{c}D^{q}x(t) = f(t, x(t)),\ 3 < q \leq 4,\ t \in [a, b], \tag{1}$$

supplemented with non-conjugate Riemann-Stieltjes integro-multipoint boundary conditions of the form:

$$x(a) = \sum_{i=1}^{n-2} \alpha_i x(\eta_i) + \int_a^b x(s)dA(s),\ x'(a) = 0,\ x(b) = 0,\ x'(b) = 0, \tag{2}$$

where $^{c}D^{q}$ denotes the Caputo fractional derivative of order q, $a < \eta_1 < \eta_2 < \cdots < \eta_{n-2} < b$, $f : [a, b] \times \mathbb{R} \longrightarrow \mathbb{R}$ is a given continuous function, A is a function of bounded variation, and $\alpha_i \in \mathbb{R}$, $i = 1, 2, \cdots, n-2$.

The main emphasis in the present work is to introduce non-conjugate Riemann-Stieltjes integro-multipoint boundary conditions and develop the existence theory for a Caputo-type fractional order boundary value problem equipped with these conditions on an arbitrary domain. Conjugate conditions on the body/fluid interface provide continuity of the thermal fields by specifying the equalities of temperatures and heat fluxes of a body and a flow at the vicinity of interface. The results obtained in this paper may have potential applications in diffraction-free and self-healing optoelectronic devices. Moreover, propagation properties for fractional Schrödinger equation similar to our results are well known theoretically [29].

The rest of the paper is organized as follows. An auxiliary result related to the linear variant of the problems (1) and (2), which plays a key role in the forthcoming analysis, is presented in Section 2. Some basic ideas of fractional calculus are also given in this section. In Section 3, we obtain some existence results for the given problem, while Section 4 contains a uniqueness result for the problem at hand. Ulam stability of different kinds for the problem (1) and (2) is studied in Section 5.

2. Preliminary Material

We begin this section with some basic definitions of fractional calculus [6]. Later we prove an auxiliary lemma, which plays a key role in defining a fixed-point problem associated with the given problem.

Definition 1. *Let g be a locally integrable real-valued function on $-\infty \leq a < t < b \leq +\infty$. The Riemann-Liouville fractional integral I_a^p of order $p \in \mathbb{R}$ $(p > 0)$ for the function g is defined as*

$$I_a^p g(t) = (g * K_p)(t) = \frac{1}{\Gamma(p)} \int_a^t (t-s)^{p-1} g(s) ds,$$

where $K_p(t) = \frac{t^{p-1}}{\Gamma(p)}$, Γ denotes the Euler gamma function.

Definition 2. *The Caputo derivative of fractional order p for an $(m-1)$-times absolutely continuous function $g : [a, \infty) \longrightarrow \mathbb{R}$ is defined as*

$$^{c}D^{p}g(t) = \frac{1}{\Gamma(m-p)} \int_a^t (t-\bar{t})^{m-p-1} g^{(m)}(\bar{t}) d\bar{t},\ m-1 < p \leq m,\ m = [p]+1,$$

where $[p]$ denotes the integer part of the real number p.

Lemma 1. *[6] The general solution of the fractional differential equation* ${}^cD^q x(t) = 0,\ m-1 < q < m,\ t \in [a,b]$ *is*

$$x(t) = \omega_0 + \omega_1(t-a) + \omega_2(t-a)^2 + \ldots + \omega_{m-1}(t-a)^{m-1},$$

where $\omega_i \in \mathbb{R},\ i = 0,1,\ldots,m-1$. *Furthermore,*

$$I^q\, {}^cD^q x(t) = x(t) + \sum_{i=0}^{m-1} \omega_i (t-a)^i.$$

Lemma 2. *Let*

$$\gamma_1 = \frac{-A_1(b-a)^2}{3} + \frac{2A_3}{3(b-a)} - A_2 \neq 0. \tag{3}$$

For $\widehat{f} \in C([a,b],\mathbb{R})$, *the unique solution of the linear equation*

$${}^cD^q x(t) = \hat{f}(t),\ 3 < q \leq 4,\ t \in [a,b], \tag{4}$$

supplemented with the boundary conditions (2) is given by

$$\begin{aligned} x(t) &= \int_a^t \frac{(t-s)^{q-1}}{\Gamma(q)}\widehat{f}(s)ds - g_1(t)\int_a^b \frac{(b-s)^{q-1}}{\Gamma(q)}\widehat{f}(s)ds - g_2(t)\int_a^b \frac{(b-s)^{q-2}}{\Gamma(q-1)}\widehat{f}(s)ds \\ &\quad + g_3(t)\int_a^b \Big(\int_a^s \frac{(s-u)^{q-1}}{\Gamma(q)}\widehat{f}(u)du\Big)dA(s), \end{aligned} \tag{5}$$

where

$$\begin{aligned} g_1(t) &= \lambda_1 - \frac{(t-a)^2 A_1}{\gamma_1} + (t-a)^3\lambda_4,\ g_2(t) = \lambda_2 + \frac{(t-a)^2\gamma_2}{\gamma_1} + (t-a)^3\lambda_5, \\ g_3(t) &= \lambda_3 + \frac{(t-a)^2}{\gamma_1} + (t-a)^3\lambda_6, \end{aligned} \tag{6}$$

$$\gamma_2 = \frac{A_3 + (b-a)^3 A_1}{3(b-a)^2}, \tag{7}$$

$$\begin{aligned} \lambda_1 &= 1 + \frac{(b-a)^2 A_1}{3\gamma_1},\ \lambda_2 = \frac{-(b-a)}{3} - \frac{(b-a)^2\gamma_2}{3\gamma_1},\ \lambda_3 = -\frac{(b-a)^2}{3\gamma_1}, \\ \lambda_4 &= \frac{2A_1}{3(b-a)\gamma_1},\ \lambda_5 = \frac{\gamma_1 - 2(b-a)\gamma_2}{3(b-a)^2\gamma_1},\ \lambda_6 = \frac{-2}{3(b-a)\gamma_1}, \end{aligned} \tag{8}$$

$$\begin{aligned} A_1 &= 1 - \sum_{i=1}^{n-2}\alpha_i - \int_a^b dA(s),\ A_2 = \sum_{i=1}^{n-2}\alpha_i(\eta_i - a)^2 + \int_a^b (s-a)^2 dA(s), \\ A_3 &= \sum_{i=1}^{n-2}\alpha_i(\eta_i - a)^3 + \int_a^b (s-a)^3 dA(s). \end{aligned} \tag{9}$$

Proof. Applying the integral operator I^q to both sides of (4) and using Lemma 1, we get

$$x(t) = \int_a^t \frac{(t-s)^{q-1}}{\Gamma(q)}\hat{f}(s)ds + c_0 + c_1(t-a) + c_2(t-a)^2 + c_3(t-a)^3, \tag{10}$$

where $c_i \in \mathbb{R},\ \ i = 0,1,2,3$ are unknown arbitrary constants. Differentiating (10) with respect to t, we have

$$x'(t) = \int_a^t \frac{(t-s)^{q-2}}{\Gamma(q-1)}\hat{f}(s)ds + c_1 + 2c_2(t-a) + 3c_3(t-a)^2. \tag{11}$$

Using the boundary conditions (2) in (10) and (11), we obtain $c_1 = 0$ and

$$c_0 + (b-a)^2 c_2 + (b-a)^3 c_3 = I_1, \tag{12}$$

$$2(b-a)c_2 + 3(b-a)^2 c_3 = I_2, \tag{13}$$

$$A_1 c_0 - A_2 c_2 - A_3 c_3 = I_3, \tag{14}$$

where A_i $(i = 1, 2, 3)$ are given by (9) and

$$\begin{aligned} I_1 &= -\int_a^b \frac{(b-s)^{q-1}}{\Gamma(q)} \widehat{f}(s)ds, \; I_2 = -\int_a^b \frac{(b-s)^{q-2}}{\Gamma(q-1)} \widehat{f}(s)ds, \\ I_3 &= \int_a^b \Big(\int_a^s \frac{(s-u)^{q-1}}{\Gamma(q)} \widehat{f}(u)du \Big) dA(s). \end{aligned} \tag{15}$$

Solving (12) and (13), for c_0 and c_3 in terms of c_2, we get

$$c_0 = I_1 - \frac{(b-a)}{3} I_2 - \frac{(b-a)^2}{3} c_2, \tag{16}$$

$$c_3 = \frac{1}{3(b-a)^2} I_2 - \frac{2}{3(b-a)} c_2. \tag{17}$$

Substituting (16) and (17) in (14) yields

$$c_2 = \frac{\gamma_2}{\gamma_1} I_2 - \frac{A_1}{\gamma_1} I_1 + \frac{1}{\gamma_1} I_3, \; \gamma_1 \neq 0, \tag{18}$$

where γ_1 and γ_2 are defined by (3) and (7) respectively. Using (18) in (16) and (17), we find that

$$c_0 = \lambda_1 I_1 + \lambda_2 I_2 + \lambda_3 I_3,$$

$$c_3 = \lambda_4 I_1 + \lambda_5 I_2 + \lambda_6 I_3.$$

Inserting the values of c_0, c_1, c_2 and c_3 in (10) together with notations (6), we obtain the solution (5). The converse of the lemma can be proved by direct computation. □

3. Existence Results

Let $\mathcal{E} = C([a,b], \mathbb{R})$ denote the Banach space of all continuous functions from $[a,b] \longrightarrow \mathbb{R}$ equipped with the sup-norm $\|x\| = \sup\{|x(t)|, t \in [a,b]\}$. For computational convenience, we introduce

$$\Lambda = \left\{ \frac{(b-a)^q}{\Gamma(q+1)} + \bar{g_1} \frac{(b-a)^q}{\Gamma(q+1)} + \bar{g_2} \frac{(b-a)^{q-1}}{\Gamma(q)} + \bar{g_3} \int_a^b \frac{(s-a)^q}{\Gamma(q+1)} dA(s) \right\}, \tag{19}$$

where $\bar{g_1} = \sup_{t\in[a,b]} |g_1(t)|, \bar{g_2} = \sup_{t\in[a,b]} |g_2(t)|, \bar{g_3} = \sup_{t\in[a,b]} |g_3(t)|$. By Lemma 2, we transform the problems (1) and (2) into an equivalent fixed-point problem as

$$x = \mathcal{J}x, \tag{20}$$

where $\mathcal{J} : \mathcal{E} \longrightarrow \mathcal{E}$ is defined by

$$\begin{aligned} (\mathcal{J}x)(t) &= \int_a^t \frac{(t-s)^{q-1}}{\Gamma(q)} f(s,x(s))ds - g_1(t) \int_a^b \frac{(b-s)^{q-1}}{\Gamma(q)} f(s,x(s))ds \\ &- g_2(t) \int_a^b \frac{(b-s)^{q-2}}{\Gamma(q-1)} f(s,x(s))ds + g_3(t) \int_a^b \Big(\int_a^s \frac{(s-u)^{q-1}}{\Gamma(q)} f(u,x(u))du \Big) dA(s), \end{aligned} \tag{21}$$

where $g_1(t), g_2(t)$ and $g_3(t)$ are given by (6).

Evidently, the existence of fixed points of the operator $\mathcal{J}$ will imply the existence of solutions for the problems (1) and (2).

Now, the platform is set to present our main results. The following known fixed-point theorem [30] will be used in the proof of our first result.

Theorem 1. *Let X be a Banach space. Assume that $\mathcal{G} : X \longrightarrow X$ is a completely continuous operator and the set $\mathcal{P} = \{x \in X | x = \beta\mathcal{G}x,\ 0 < \beta < 1\}$ is bounded. Then $\mathcal{G}$ has a fixed point in X.*

Theorem 2. *Suppose that there exists $\varrho \in C([a,b], \mathbb{R}^+)$ such that $|f(t,x(t))| \leq \varrho(t)$, $\forall t \in [a,b]$, $x \in \mathcal{E}$, with $\sup_{t\in[a,b]} |\varrho(t)| = \|\varrho\|$. Then the problems (1) and (2) has at least one solution on $[a,b]$.*

Proof. Observe that continuity of the operator $\mathcal{J}$ follows from that of f. Let $\Phi \subset \mathcal{E}$ be bounded. Then, $\forall x \in \Phi$ together with the given assumption $|f(t,x(t))| \leq \varrho(t)$, we get

$$\begin{aligned}
|(\mathcal{J}x)| &\leq \sup_{t\in[a,b]} \Bigg\{ \int_a^t \frac{(t-s)^{q-1}}{\Gamma(q)}|f(s,x(s))|ds + |g_1(t)| \int_a^b \frac{(b-s)^{q-1}}{\Gamma(q)}|f(s,x(s))|ds \\
&\quad +|g_2(t)| \int_a^b \frac{(b-s)^{q-2}}{\Gamma(q-1)}|f(s,x(s))|ds \\
&\quad +|g_3(t)| \int_a^b \Big(\int_a^s \frac{(s-u)^{q-1}}{\Gamma(q)}|f(u,x(u))|du \Big) dA(s) \Bigg\} \\
&\leq \|\varrho\| \left[\frac{(b-a)^q}{\Gamma(q+1)} + \bar{g_1}\frac{(b-a)^q}{\Gamma(q+1)} + \bar{g_2}\frac{(b-a)^{q-1}}{\Gamma(q)} + \bar{g_3}\int_a^b \frac{(s-a)^q}{\Gamma(q+1)}dA(s) \right] \\
&= \|\varrho\|\Lambda = M_1,
\end{aligned}$$

which shows that $\mathcal{J}$ is bounded. Next, for $a < t_1 < t_2 < b$, we have

$$\begin{aligned}
&|(\mathcal{J}x)(t_2) - (\mathcal{J}x)(t_1)| \\
\leq\ & \int_a^{t_1} \frac{|(t_2-s)^{q-1} - (t_1-s)^{q-1}|}{\Gamma(q)}|f(s,x(s))|ds \\
&+ \int_{t_1}^{t_2} \frac{|(t_2-s)^{q-1}|}{\Gamma(q)}|f(s,x(s))|ds + |g_1(t_2) - g_1(t_1)| \int_a^b \frac{|(b-s)^{q-1}|}{\Gamma(q)}|f(s,x(s))|ds \\
&+|g_2(t_2) - g_2(t_1)| \int_a^b \frac{|(b-s)^{q-2}|}{\Gamma(q-1)}|f(s,x(s))|ds \\
&+|g_3(t_2) - g_3(t_1)| \int_a^b \Big(\int_a^s \frac{|(s-u)^{q-1}|}{\Gamma(q)}|f(u,x(u))|du \Big) dA(s) \\
\leq\ & \|\varrho\| \Bigg[\frac{|(t_2-a)^q - (t_1-a)^q| + 2(t_2-t_1)^q}{\Gamma(q+1)} + \frac{|g_1(t_2) - g_1(t_1)|\,|(b-a)^q|}{\Gamma(q+1)} \\
&+ \frac{|g_2(t_2) - g_2(t_1)|\,|(b-a)^{q-1}|}{\Gamma(q)} + \frac{|g_3(t_2) - g_3(t_1)|}{\Gamma(q+1)} \int_a^b \Big(\int_a^s (s-a)^q) dA(s) \Big) \Bigg],
\end{aligned}$$

which tends to zero as $t_2 \longrightarrow t_1$ independent of x. Thus, $\mathcal{J}$ is equicontinuous on Φ. Hence, by Arzelá-Ascoli theorem, $\mathcal{J}$ is relatively compact on Φ. Therefore, $\mathcal{J}(\Phi)$ is a relatively compact subset of $\mathcal{E}$.

Now we consider a set $\mathcal{P} = \{x \in \mathcal{E} | x = \beta\mathcal{J}x,\ 0 < \beta < 1\}$, and show that the set $\mathcal{P}$ is bounded. Let $x \in \mathcal{P}$, then $x = \beta\mathcal{J}x$, $0 < \beta < 1$. For any $t \in [a,b]$, we have

$$|x(t)| = \beta|(\mathcal{J}x)(t)|$$

$$
\begin{aligned}
&\leq \sup_{t\in[a,b]}\Big\{\int_a^t \frac{(t-s)^{q-1}}{\Gamma(q)}|f(s,x(s))|ds + |g_1(t)|\int_a^b \frac{(b-s)^{q-1}}{\Gamma(q)}|f(s,x(s))|ds \\
&\quad +|g_2(t)|\int_a^b \frac{(b-s)^{q-2}}{\Gamma(q-1)}|f(s,x(s))|ds \\
&\quad +|g_3(t)|\int_a^b \Big(\int_a^s \frac{(s-u)^{q-1}}{\Gamma(q)}|f(u,x(u))|du\Big)dA(s)\Big\} \\
&\leq \|\varrho\|\Big[\frac{(b-a)^q}{\Gamma(q+1)} + \bar{g_1}\frac{(b-a)^q}{\Gamma(q+1)} + \bar{g_2}\frac{(b-a)^{q-1}}{\Gamma(q)} + \bar{g_3}\int_a^b \frac{(s-a)^q}{\Gamma(q+1)}dA(s)\Big] \\
&= \|\varrho\|\Lambda,
\end{aligned}
$$

where Λ given by (19). Thus, $\|x\| \leq \|\varrho\|\Lambda$ for any $t \in [a,b]$. Therefore, the set $\mathcal{P}$ is bounded. In consequence, the conclusion of Theorem 1 applies and that the operator $\mathcal{J}$ has at least one fixed point. Thus, there exists at least one solution for the problems (1) and (2) on $[a,b]$. □

Example 1. *Consider the fractional boundary value problem*

$$
\begin{cases}
{}^cD^{\frac{11}{3}}x(t) = \dfrac{6\,e^{-x^2}}{\sqrt{t^4+24}} + \dfrac{\cos t}{t^2+1}\Big(\dfrac{|x|^3}{1+|x|^3}\Big) + t^3 + 6, \; t \in [1,2], \\
x(1) = \displaystyle\sum_{i=1}^{4}\alpha_i x(\eta_i) + \int_1^2 x(s)dA(s), \; x'(1) = 0, \; x(2) = 0, \; x'(2) = 0,
\end{cases}
\tag{22}
$$

where $q = 11/3$, $a = 1$, $b = 2$, $\alpha_1 = -1/2$, $\alpha_2 = -1/6$, $\alpha_3 = 1/6$, $\alpha_4 = 2$, $\eta_1 = 6/5$, $\eta_2 = 7/5$, $\eta_3 = 8/5$, $\eta_4 = 9/5$ *and* $f(t,x) = \dfrac{6\,e^{-x^2}}{\sqrt{t^4+24}} + \dfrac{\cos t}{t^2+1}\Big(\dfrac{|x|^3}{1+|x|^3}\Big) + t^3 + 6$.

Clearly, $|f(t,x)| \leq \frac{6}{\sqrt{t^4+24}} + \frac{\cos t}{t^2+1} + t^3 + 6 = \varrho(t) > 0$. *Therefore, there exists at least one solution for the problem (22) on* $[1,2]$ *by the conclusion of Theorem 2.*

Our next existence result is based on the following fixed-point theorem [30].

Theorem 3. *Let* Ω *be an open bounded subset of a Banach space* X *with* $0 \in \Omega$ *and the operator* $\mathcal{F} : \overline{\Omega} \longrightarrow X$ *is completely continuous satisfying* $\|\mathcal{F}x\| \leq \|x\|$, $\forall\, x \in \partial\Omega$. *Then the operator* $\mathcal{F}$ *has a fixed point in* $\overline{\Omega}$.

Theorem 4. *Let* $|f(t,x)| \leq \xi|x|$ *for* $0 < |x| < \tau$, *where* τ *and* ξ *are positive constants. Then the problems (1) and (2) has at least one solution for small values of* ξ.

Proof. Let us choose ξ such that

$$\Lambda\xi < 1, \tag{23}$$

where Λ given by (19). Define $B_{r_2} = \{x \in \mathcal{E}; \|x\| \leq r_2\}$ and take $x \in \mathcal{E}$ such that $\|x\| = r_2$, that is, $x \in \partial B_{r_2}$. As we argued in Theorem 2, it can be shown that $\mathcal{J}$ is completely continuous and

$$
\begin{aligned}
|(\mathcal{J}x)(t)| &\leq \sup_{t\in[a,b]}\Big\{\int_a^t \frac{(t-s)^{q-1}}{\Gamma(q)}|f(s,x(s))|ds + |g_1(t)|\int_a^b \frac{(b-s)^{q-1}}{\Gamma(q)}|f(s,x(s))|ds \\
&\quad +|g_2(t)|\int_a^b \frac{(b-s)^{q-2}}{\Gamma(q-1)}|f(s,x(s))|ds \\
&\quad +|g_3(t)|\int_a^b \Big(\int_a^s \frac{(s-u)^{q-1}}{\Gamma(q)}|f(u,x(u))|du\Big)dA(s)\Big\}
\end{aligned}
$$

$$\begin{aligned} &\leq \xi\|x\|\left[\frac{(b-a)^q}{\Gamma(q+1)}+\bar{g_1}\frac{(b-a)^q}{\Gamma(q+1)}+\bar{g_2}\frac{(b-a)^{q-1}}{\Gamma(q)}+\bar{g_3}\int_a^b\frac{(s-a)^q}{\Gamma(q+1)}dA(s)\right] \\ &= \xi\Lambda\|x\|. \end{aligned}$$

Using (23) and the norm $\|x\| = \sup_{t\in[a,b]}|x(t)|$, we get $\|\mathcal{J}x\| \leq \|x\|$, $x \in \partial B_{r_2}$. Therefore, conclusion of Theorem 3 applies and hence the problems (1) and (2) has a solution on $[a,b]$. □

Example 2. *Consider the fractional boundary value problem.*

$$\begin{cases} {}^cD^{\frac{27}{7}}x(t) = \dfrac{2|x|}{\sqrt{100+t^2}}\Big(1+\dfrac{|x|}{1+|x|}\Big), \ t\in[0,1], \\ x(0) = \displaystyle\sum_{i=1}^{4}\alpha_i x(\eta_i)+\int_0^1 x(s)dA(s), \ x'(0)=0, \ x(1)=0, \ x'(1)=0, \end{cases} \tag{24}$$

where $q = 27/7$, $a = 0$, $b = 1$, $\alpha_1 = -2$, $\alpha_2 = -1/6$, $\alpha_3 = 1/6$, $\alpha_4 = 15/4$, $\eta_1 = 1/7$, $\eta_2 = 2/7$, $\eta_3 = 3/7$, $\eta_4 = 4/7$, *and* $f(t,x(t)) = \dfrac{2|x|}{\sqrt{100+t^2}}\Big(1+\dfrac{|x|}{1+|x|}\Big)$. *Let us take* $A(s) = \frac{s^2}{2}$. *Using the given data, we have that* $A_1 \approx -1.25$, $A_2 \approx 1.45068$, $A_3 \approx 0.903110$, $\gamma_1 \approx -0.43194$, $\gamma_2 \approx -0.115630$, $\lambda_1 \approx 1.96464$, $\lambda_2 \approx -0.422566$, $\lambda_3 \approx 0.771712$, $\lambda_4 \approx 1.92928$, $\lambda_5 \approx 0.154867$, $\lambda_6 \approx 1.54342$, $\bar{g_1} \approx 1.96464$, $\bar{g_2} \approx 0.422566$, $\bar{g_3} \approx 0.771712$, $\Lambda \approx 0.243646$, *where* Λ *is given by (19). Clearly the hypothesis of Theorem 4 is satisfied with* $\xi = \frac{2}{5}$. *Also* $\xi\Lambda \approx 0.097458 < 1$. *Therefore, the problem (24) has at least one solution on* $[0,1]$.

In the next result, we apply a fixed-point theorem due to Krasnoselskii [31] to establish the existence of solutions for the problems (1) and (2).

Theorem 5. *(Krasnoselskii [31]) Let M be a closed, convex, bounded and nonempty subset of a Banach space X and let* $\mathcal{F}_1, \mathcal{F}_2$ *be the operators defined from M to X such that: (i)* $\mathcal{F}_1x+\mathcal{F}_2y \in M$ *wherever* $x, y \in M$*; (ii)* $\mathcal{F}_1$ *is compact and continuous; (iii)* $\mathcal{F}_2$ *is a contraction. Then there exists* $z \in M$ *such that* $z = \mathcal{F}_1z+\mathcal{F}_2z$.

Theorem 6. *Assume that* $f : [a,b]\times\mathbb{R} \longrightarrow \mathbb{R}$ *is a continuous function such that the following conditions hold:*

(H_1) $|f(t,x)-f(t,y)| \leq L|x-y|$, $L>0$, $\forall\, t\in[a,b]$, $x,y\in\mathbb{R}$;
(H_2) $|f(t,x)| \leq \mu(t)$, $\forall\,(t,x)\in[a,b]\times\mathbb{R}$, $\mu\in C([a,b],\mathbb{R}^+)$.

Then the problems (1) and (2) has at least one solution on $[a,b]$ *if*

$$\left(\Lambda-\frac{(b-a)^q}{\Gamma(q+1)}\right)L<1, \tag{25}$$

where Λ *is defined by (19).*

Proof. Consider a closed ball $B_r = \{x\in\mathcal{E} : \|x\|\leq r\}$ with $r \geq \Lambda\|\mu\|$, $\sup_{t\in[a,b]}|\mu(t)| = \|\mu\|$, and Λ is given by (19). Define operators $\mathcal{J}_1$ and $\mathcal{J}_2$ on B_r as

$$\begin{aligned} (\mathcal{J}_1x)(t) &= \int_a^t\frac{(t-s)^{q-1}}{\Gamma(q)}f(s,x(s))ds, \\ (\mathcal{J}_2x)(t) &= -g_1(t)\int_a^b\frac{(t-s)^{q-1}}{\Gamma(q)}f(s,x(s))ds-g_2(t)\int_a^b\frac{(t-s)^{q-2}}{\Gamma(q-1)}f(s,x(s))ds \\ &\quad +g_3(t)\int_a^b\Big(\int_a^s\frac{(s-u)^{q-1}}{\Gamma(q)}f(u,x(u))du\Big)dA(s). \end{aligned}$$

Please note that $\mathcal{J} = \mathcal{J}_1 + \mathcal{J}_2$. For $x, y \in B_r$, we have

$$\begin{aligned}
\|\mathcal{J}_1 x + \mathcal{J}_2 y\| &\leq \sup_{t\in[a,b]} \Big\{ \int_a^t \frac{(t-s)^{q-1}}{\Gamma(q)} |f(s,x(s))| ds + |g_1(t)| \int_a^b \frac{(t-s)^{q-1}}{\Gamma(q)} |f(s,y(s))| ds \\
&\quad + |(g_2(t)| \int_a^b \frac{(t-s)^{q-2}}{\Gamma(q-1)} |f(s,y(s))| ds \\
&\quad + |g_3(t)| \int_a^b \Big(\int_a^s \frac{(s-u)^{q-1}}{\Gamma(q)} |f(u,y(u)|) du \Big) dA(s) \Big\} \\
&\leq \|\mu\| \sup_{t\in[a,b]} \Big\{ \int_a^t \frac{(t-s)^{q-1}}{\Gamma(q)} ds + |g_1(t)| \int_a^b \frac{(t-s)^{q-1}}{\Gamma(q)} ds \\
&\quad + |g_2(t)| \int_a^b \frac{(t-s)^{q-2}}{\Gamma(q-1)} ds + |g_3(t)| \int_a^b \Big(\int_a^s \frac{(s-u)^{q-1}}{\Gamma(q)} du \Big) dA(s) \Big\} \\
&\leq \|\mu\| \Lambda \leq r,
\end{aligned}$$

where we have used (19). Thus, $\mathcal{J}_1 x + \mathcal{J}_2 y \in B_r$. Next we show that $\mathcal{J}_2$ is a contraction. For $x, y \in B_r$, we have

$$\begin{aligned}
\|\mathcal{J}_2 x - \mathcal{J}_2 y\| &= \sup_{t\in[a,b]} \Big| (\mathcal{J}_2 x)(t) - (\mathcal{J}_2 y)(t) \Big| \\
&\leq \sup_{t\in[a,b]} \Big\{ |g_1(t)| \int_a^b \frac{(b-s)^{q-1}}{\Gamma(q)} |f(s,x(s)) - f(s,y(s))| ds \\
&\quad + |g_2(t)| \int_a^b \frac{(b-s)^{q-2}}{\Gamma(q-1)} |f(s,x(s)) - f(s,y(s))| ds \\
&\quad + |g_3(t)| \int_a^b \Big(\int_a^s \frac{(s-u)^{q-1}}{\Gamma(q)} |f(u,x(u)) - f(u,y(u))| du \Big) dA(s) \Big\} \\
&\leq L\|x-y\| \Big[\bar{g}_1 \frac{(b-a)^q}{\Gamma(q+1)} + \bar{g}_2 \frac{(b-a)^{q-1}}{\Gamma(q)} + \bar{g}_3 \int_a^b \frac{(s-a)^q}{\Gamma(q+1)} dA(s) \Big] \\
&= L\Big(\Lambda - \frac{(b-a)^q}{\Gamma(q+1)} \Big) \|x-y\|,
\end{aligned}$$

which shows that $\mathcal{J}_2$ is a contraction by the condition (25). Continuity of f implies that the operator $\mathcal{J}_1$ is continuous. Also, $\mathcal{J}_1$ is uniformly bounded on B_r as

$$\|\mathcal{J}_1 x\| \leq \frac{(b-a)^q}{\Gamma(q+1)} \|\mu\|.$$

Next, we establish that the operator $\mathcal{J}_1$ is compact. Setting $\mathcal{S} = [a,b] \times B_r$, we define $\sup_{(t,x)\in\mathcal{S}} |f(t,x)| = M_r$. For $a < t_2 < t_1 < b$, we get

$$\begin{aligned}
|(\mathcal{J}_1 x)(t_1) - (\mathcal{J}_1 x)(t_2)| &= \Big| \int_a^{t_2} \frac{(t_1-s)^{q-1} - (t_2-s)^{q-1}}{\Gamma(q)} f(s,x(s)) ds \\
&\quad + \int_{t_2}^{t_1} \frac{(t_1-s)^{q-1}}{\Gamma(q)} f(s,x(s)) ds \Big| \\
&\leq \frac{M_r}{\Gamma(q+1)} \Big[|(t_1-a)^q - (t_2-a)^q| + 2(t_1-t_2)^q \Big] \\
&\longrightarrow 0 \text{ when } t_1 - t_2 \longrightarrow 0,
\end{aligned}$$

independent of x. Thus, $\mathcal{J}_1$ is equicontinuous on B_r. Hence, by Arzelá-Ascoli theorem, $\mathcal{J}_1$ is compact on B_r. Therefore, the conclusion of Theorem 5 applies to the problems (1) and (2). $\square$

Remark 1. *By interchanging the role of the operators $\mathcal{J}_1$ and $\mathcal{J}_2$ in Theorem 6, the condition (25) becomes:*

$$\frac{L(b-a)^q}{\Gamma(q+1)} < 1.$$

Example 3. *Consider the fractional differential equation:*

$$^cD^{\frac{11}{3}}x(t) = \frac{\delta}{\sqrt{t^2+8}}\, tan^{-1}x + e^{2t},\ t \in [1,2]. \tag{26}$$

subject to the boundary conditions of Example 1.

Let us take $A(s) = \frac{(s-1)^2}{2} + \frac{1}{2}$. Using the given data (from Example 1), it is found that $A_1 = -1$, $A_2 \approx 1.543333$, $A_3 \approx 1.245333$, $\gamma_1 \approx -0.379778$, $\gamma_2 \approx 0.081778$, $\lambda_1 \approx 1.877706$, $\lambda_2 \approx -0.261556$, $\lambda_3 \approx 0.877706$, $\lambda_4 \approx 1.755413$, $\lambda_5 \approx 0.476887$, $\lambda_6 \approx 1.755413$, $\bar{g_1} \approx 1.877706$, $\bar{g_2} \approx 0.264808$, $\bar{g_3} \approx 0.877706$, $\Lambda \approx 0.272140$ (Λ is given by (19) and $\Lambda - \frac{(b-a)^2}{\Gamma(q+1)} \approx 0.204166$.

Clearly the hypotheses of Theorem 6 are satisfied with $L = \delta/3$ and $\mu(t) = \frac{\delta\pi}{2\sqrt{t^2+8}} + e^{2t}$. Also $L\Big(\Lambda - \frac{(b-a)^2}{\Gamma(q+1)}\Big) < 1$ for $\delta < 14.693960$. Therefore, there exists at least one solution for the problem (26) on $[1,2]$.

4. Uniqueness of Solution

Here, we prove the uniqueness of solutions for the problems (1) and (2).

Theorem 7. *Assume that $f : [a,b] \times \mathbb{R} \longrightarrow \mathbb{R}$ is a continuous function satisfying the condition (H_1). Then the problems (1) and (2) has a unique solution on $[a,b]$ if*

$$L\Lambda < 1, \tag{27}$$

where Λ is given by (19).

Proof. Setting $\sup_{t\in[a,b]} |f(t,0)| = N < \infty$, and selecting

$$r_1 \geq N\Lambda(1 - L\Lambda)^{-1},$$

we define $B_{r_1} = \{x \in \mathcal{E} : \|x\| \leq r_1\}$, and show that $\mathcal{J}B_{r_1} \subset B_{r_1}$, where the operator $\mathcal{J}$ is defined by (21). For $x \in B_{r_1}$,

$$\begin{aligned} |f(t,x(t))| &= |f(t,x(t)) - f(t,0) + f(t,0)| \leq |f(t,x(t)) - f(t,0)| + |f(t,0)| \\ &\leq L|x(t)| + N \leq L\|x\| + N \leq Lr_1 + N. \end{aligned}$$

Then,

$$\begin{aligned} \|\mathcal{J}x\| \leq & \sup_{t\in[a,b]} \Big\{ \int_a^t \frac{(t-s)^{q-1}}{\Gamma(q)} |f(s,x(s))| ds + |g_1(t)| \int_a^b \frac{(b-s)^{q-1}}{\Gamma(q)} |f(s,x(s))| ds \\ & + |g_2(t)| \int_a^b \frac{(b-s)^{q-2}}{\Gamma(q-1)} |f(s,x(s))| ds \\ & + |g_3(t)| \int_a^b \Big(\int_a^s \frac{(s-u)^{q-1}}{\Gamma(q)} |f(u,x(u))| du \Big) dA(s) \Big\} \end{aligned}$$

$$
\begin{aligned}
&\leq (L\|x\|+N)\left[\frac{(b-a)^q}{\Gamma(q+1)}+\bar{g_1}\frac{(b-a)^q}{\Gamma(q+1)}+\bar{g_2}\frac{(b-a)^{q-1}}{\Gamma(q)}+\bar{g_3}\int_a^b\frac{(s-a)^q}{\Gamma(q+1)}dA(s)\right]\\
&\leq (Lr_1+N)\Lambda\leq r_1.
\end{aligned}
$$

This shows that $\mathcal{J}x\in B_{r_1}$ for any $x\in B_{r_1}$. Therefore, $\mathcal{J}B_{r_1}\subset B_{r_1}$. Now, we show that $\mathcal{J}$ is a contraction. For $x,y\in\mathcal{E}$ and $t\in[a,b]$, we obtain

$$
\begin{aligned}
\|(\mathcal{J}x)-(\mathcal{J}y)\| &= \sup_{t\in[a,b]}\left|(\mathcal{J}x)(t)-(\mathcal{J}y)(t)\right|\\
&\leq \sup_{t\in[a,b]}\left\{\int_a^t\frac{(t-s)^{q-1}}{\Gamma(q)}\left|f(s,x(s))-f(s,y(s))\right|ds\right.\\
&\quad +|g_1(t)|\int_a^b\frac{(b-s)^{q-1}}{\Gamma(q)}\left|f(s,x(s))-f(s,y(s))\right|ds\\
&\quad +|g_2(t)|\int_a^b\frac{(b-s)^{q-2}}{\Gamma(q-1)}\left|f(s,x(s))-f(s,y(s))\right|ds\\
&\quad \left.+|g_3(t)|\int_a^b\Big(\int_a^s\frac{(s-u)^{q-1}}{\Gamma(q)}\left|f(u,x(u))-f(u,y(u))\right|du\Big)dA(s)\right\}\\
&\leq L\|x-y\|\left[\frac{(b-a)^q}{\Gamma(q+1)}+\bar{g_1}\frac{(b-a)^q}{\Gamma(q+1)}+\bar{g_2}\frac{(b-a)^{q-1}}{\Gamma(q)}+\bar{g_3}\int_a^b\frac{(s-a)^q}{\Gamma(q+1)}dA(s)\right]\\
&= L\Lambda\|x-y\|.
\end{aligned}
$$

By the condition (27), we deduce from the above inequality that $\mathcal{J}$ is a contraction. Thus, by the conclusion of Banach fixed-point theorem, the problems (1) and (2) has a unique solution on $[a,b]$. □

Example 4. *Let us take the problem considered in Example 1, and note that $L\Lambda<1$ for $\delta<11.023738$. Clearly the hypothesis of Theorem 7 is satisfied. Hence it follows by the conclusion of Theorem 7 that the problem (22) has a unique solution on $[1,2]$.*

5. Ulam Stability

In this section, we discuss the Ulam stability for the problems (1) and (2) by means of integral representation of its solution given by

$$
\begin{aligned}
y(t) &= \int_a^t\frac{(t-s)^{q-1}}{\Gamma(q)}f(s,y(s))ds-g_1(t)\int_a^b\frac{(b-s)^{q-1}}{\Gamma(q)}f(s,y(s))ds\\
&\quad -g_2(t)\int_a^b\frac{(b-s)^{q-2}}{\Gamma(q-1)}f(s,y(s))ds\\
&\quad +g_3(t)\int_a^b\Big(\int_a^s\frac{(s-u)^{q-1}}{\Gamma(q)}f(u,y(u))du\Big)dA(s).
\end{aligned}
\tag{28}
$$

Here $y\in C([a,b],\mathbb{R})$ possesses a fractional derivative of order $3<q\leq 4$ and $f:[a,b]\times\mathbb{R}\longrightarrow\mathbb{R}$ is a continuous function. Then the nonlinear operator $\mathcal{Q}:C([a,b],\mathbb{R})\longrightarrow C([a,b],\mathbb{R})$ defined by

$$
\mathcal{Q}y(t)={}^cD^qy(t)-f(t,y(t))
$$

is continuous.

Definition 3. *For each $\epsilon>0$ and for each solution y of (1) and (2) such that*

$$
\|\mathcal{Q}y\|\leq\epsilon, \tag{29}
$$

the problems (1) and (2) is said to be Ulam-Hyers stable if we can find a positive real number ν *and a solution* $x \in C([a,b],\mathbb{R})$ *of (1) and (2) satisfying the inequality:*

$$\|x - y\| \le \nu\, \epsilon_*,$$

where ϵ_* *is a positive real number depending on* ϵ.

Definition 4. *Let there exists* $\kappa \in C(\mathbb{R}^+,\mathbb{R}^+)$ *such that for each solution* y *of (1) and (2), we can find a solution* $x \in C([a,b],\mathbb{R})$ *of (1) and (2) such that*

$$|x(t) - y(t)| \le \kappa(\epsilon),\ t \in [a,b].$$

Then the problems (1) and (2) is said to be generalized Ulam-Hyers stable

Definition 5. *For each* $\epsilon > 0$ *and for each solution* y *of (1) and (2), the problems (1) and (2) is called Ulam-Hyers-Rassias stable with respect to* $\sigma \in C([a,b],\mathbb{R}^+)$ *if*

$$|\mathcal{Q}y(t)| \le \epsilon\sigma(t),\ t \in [a,b], \tag{30}$$

and there exist a real number $\nu > 0$ *and a solution* $x \in C([a,b],\mathbb{R})$ *of (1) and (2) such that*

$$|x(t) - y(t)| \le \nu\epsilon_*\sigma(t),\ t \in [a,b],$$

where ϵ_* *is a positive real number depending on* ϵ.

Theorem 8. *Let the assumption* (H_1) *hold with* $L\Lambda < 1$*, where* Λ *is defined by (19). Then the problems (1) and (2) is both Ulam-Hyers and generalized Ulam-Hyers stable.*

Proof. Let $x \in C([a,b],\mathbb{R})$ be a solution of (1) and (2) satisfying (21) by Theorem 7. Let y be any solution satisfying (29). Then by Lemma 2, y satisfies the integral equation (28). Furthermore, the equivalence in Lemma 2 implies the equivalence between the operators $\mathcal{Q}$ and $\mathcal{J} - I$ (where I is identity operator) for every solution $y \in C([a,b],\mathbb{R})$ of (1) and (2) satisfying (27). Therefore, we deduce by the fixed-point property of the operator $\mathcal{J}$ (given by (21)) and (29) that

$$\begin{aligned} |y(t) - x(t)| &= |y(t) - \mathcal{J}y(t) + \mathcal{J}y(t) - \mathcal{J}x(t)| \\ &\le |\mathcal{J}x(t) - \mathcal{J}y(t)| + |\mathcal{J}y(t) - y(t)| \\ &\le L\Lambda\|x - y\| + \epsilon, \end{aligned}$$

where $\epsilon > 0$ and $L\Lambda < 1$. In consequence, we get

$$\|x - y\| \le \frac{\epsilon}{1 - L\Lambda}.$$

Fixing $\epsilon_* = \frac{\epsilon}{1-L\Lambda}$ and $\nu = 1$, we obtain the Ulam-Hyers stability condition. In addition, the generalized Ulam-Hyers stability follows by taking $\kappa(\epsilon) = \frac{\epsilon}{1-L\Lambda}$. □

Theorem 9. *Assume that* (H_1) *holds with* $L < \Lambda^{-1}$ *(where* Λ *is defined by (19)), and there exists a function* $\sigma \in C([a,b],\mathbb{R}^+)$ *satisfying the condition (30). Then the problems (1) and (2) is Ulam-Hyers-Rassias stable with respect to* σ.

Proof. Following the arguments employed in the proof of Theorem 8, we have

$$\|x - y\| \le \epsilon_*\sigma(t),$$

where $\epsilon_* = \frac{\epsilon}{1-L\Lambda}$. This completes the proof. □

Example 5. *Consider the following fractional differential equation*

$$^cD^{\frac{27}{7}}x(t) = \frac{3}{t^2+16}\Big(\cos x + \frac{x}{1+x}\Big),\ t \in [0,1], \tag{31}$$

subject to the same data and the boundary conditions given in Example 1 with $f(t,x(t)) = \frac{3}{t^2+16}\Big(\cos x + \frac{x}{1+x}\Big)$. *Obviously* $|f(t,x) - f(t,y)| \le \frac{3}{8}\|x-y\|$, *so,* $L = 3/8$ *and* $L\Lambda \approx 0.102053 < 1$. *Then the problem (31) is Ulam-Hyers stable, and generalized Ulam-Hyers stable. In addition, if there exists a continuous and positive function* $\sigma = e^{1+t^2} + 5$ *satisfying the condition (30), then the problem (31) is Ulam-Hyers-Rassias stable with the given value of* $f(t,x)$.

6. Conclusions

We have obtained several existence results for a new class of Caputo-type fractional differential equations of order $q \in (3,4]$, supplemented with non-conjugate Riemann-Stieltjes integro-multipoint boundary conditions on an arbitrary domain by imposing different kinds of conditions on the nonlinear function involved in the problem. The existence results, relying on different fixed-point theorems, are presented in Section 3. The uniqueness of solution for the given problem is studied in Section 4 with the aid of Banach fixed-point theorem. Section 5 is concerned with different kinds of Ulam stability for the problem at hand. Some new results follow as special cases of the ones presented in this paper. For example, taking $A(s) = s$, our results correspond to the ones for non-conjugate integro-multipoint boundary conditions of the form: $x(a) = \sum_{i=1}^{n-2}\alpha_i x(\eta_i) + \int_a^b x(s)ds,\ x'(a) = 0,\ x(b) = 0,\ x'(b) = 0$ and the value of Λ given by (19) takes the following form in this situation:

$$\Lambda = \Big\{\frac{(b-a)^q}{\Gamma(q+1)} + \bar{g}_1\frac{(b-a)^q}{\Gamma(q+1)} + \bar{g}_2\frac{(b-a)^{q-1}}{\Gamma(q)} + \bar{g}_3\frac{(b-a)^{q+1}}{\Gamma(q+2)}\Big\}.$$

Letting $A(s) = 0$ in our results, we get the ones for non-conjugate multipoint boundary conditions of the form: $x(a) = \sum_{i=1}^{n-2}\alpha_i x(\eta_i),\ x'(a) = 0,\ x(b) = 0,\ x'(b) = 0$. In case we take $\alpha_i = 0$ for all $i = 1,\dots,n-2$, our results reduce to the ones with non-conjugate Riemann-Stieltjes boundary conditions. By fixing $A(s) = 0$ and $\alpha_i = 0$ for all $i = 1,\dots,n-2$, our results correspond to a boundary value problem of fractional order $q \in (3,4]$ with conjugate boundary conditions.

Author Contributions: Formal Analysis, B.A., Y.A., S.K.N. and A.A.

Acknowledgments: The authors acknowledge with thanks DSR technical and financial support. The authors also thank the reviewers for their constructive remarks on our work.

References

1. Zaslavsky, G.M. *Hamiltonian Chaos and Fractional Dynamics*; Oxford University Press: Oxford, UK, 2005.
2. Zhang, D.; Zhang, Y.; Zhang, Z.; Ahmed, N.; Zhang, Y.; Li, F.; Belic, M.R.; Xiao, M. Unveiling the link between fractional Schrödinger equation and light propagation in honeycomb lattice. *Ann. Phys.* **2017**, *529*, 1700149. [CrossRef]
3. Fallahgoul, H.A.; Focardi, S.M.; Fabozzi, F.J. *Fractional Calculus and Fractional Processes with Applications to Financial Economics. Theory and Application*; Elsevier/Academic Press: London, UK, 2017.
4. Javidi, M.; Ahmad, B. Dynamic analysis of time fractional order phytoplankton-toxic phytoplankton-zooplankton system. *Ecol. Model.* **2015**, *318*, 8–18. [CrossRef]
5. Magin, R.L. *Fractional Calculus in Bioengineering*; Begell House Publishers: Danbury, CT, USA, 2006.

6. Kilbas, A.A.; Srivastava, H.M.; Trujillo, J.J. *Theory and Applications of Fractional Differential Equations*; North-Holland Mathematics Studies, 204; Elsevier Science B.V.: Amsterdam, The Netherlands, 2006.
7. Lakshimikantham, V.; Leela, S.; Devi, J.V. *Theory of Fractional Dynamic Systems*; Cambridge Academic Publishers: Cambridge, UK, 2009.
8. Diethelm, K. *The Analysis of Fractional Differential Equations. An Application—Oriented Exposition Using Differential Operators of Caputo Type*; Lecture Notes in Mathematics 2004; Springer: Berlin, Germany, 2010.
9. Ahmad, B.; Alsaedi, A.; Ntouyas, S.K.; Tariboon, J. *Hadamard-Type Fractional Differential Equations, Inclusions and Inequalities*; Springer: Cham, Switzerland, 2017.
10. Ahmad, B.; Nieto, J.J. Riemann-Liouville fractional integro-differential equations with fractional nonlocal integral boundary conditions. *Bound. Value Probl.* **2011**, *2011*, 36. [CrossRef]
11. Liang, S.; Zhang, J. Existence of multiple positive solutions for m-point fractional boundary value problems on an infinite interval. *Math. Comput. Model.* **2011**, *54*, 1334–1346. [CrossRef]
12. Bai, Z.B.; Sun, W. Existence and multiplicity of positive solutions for singular fractional boundary value problems. *Comput. Math. Appl.* **2012**, *63*, 1369–1381. [CrossRef]
13. Agarwal, R.P.; O'Regan, D.; Stanek, S. Positive solutions for mixed problems of singular fractional differential equations. *Math. Nachr.* **2012**, *285*, 27–41. [CrossRef]
14. Ahmad, B.; Ntouyas, S.K.; Alsaedi, A. A study of nonlinear fractional differential equations of arbitrary order with Riemann-Liouville type multistrip boundary conditions. *Math. Probl. Eng.* **2013**, *2013*, 320415. [CrossRef]
15. Ahmad, B.; Ntouyas, S.K. Existence results for higher order fractional differential inclusions with multi-strip fractional integral boundary conditions. *Electron. J. Qual. Theory Differ. Equ.* **2013**, *2013*, 20. [CrossRef]
16. O'Regan, D.; Stanek, S. Fractional boundary value problems with singularities in space variables. *Nonlinear Dyn.* **2013**, *71*, 641–652. [CrossRef]
17. Zhai, C.; Xu, L. Properties of positive solutions to a class of four-point boundary value problem of Caputo fractional differential equations with a parameter. *Commun. Nonlinear Sci. Numer. Simul.* **2014**, *19*, 2820–2827. [CrossRef]
18. Graef, J.R.; Kong, L.; Wang, M. Existence and uniqueness of solutions for a fractional boundary value problem on a graph. *Fract. Calc. Appl. Anal.* **2014**, *17*, 499–510. [CrossRef]
19. Wang, G.; Liu, S.; Zhang, L. Eigenvalue problem for nonlinear fractional differential equations with integral boundary conditions. *Abstr. Appl. Anal.* **2014**, *2014*, 916260. [CrossRef]
20. Henderson, J.; Kosmatov, N. Eigenvalue comparison for fractional boundary value problems with the Caputo derivative. *Fract. Calc. Appl. Anal.* **2014**, *17*, 872–880. [CrossRef]
21. Zhang, L.; Ahmad, B.; Wang, G. Successive iterations for positive extremal solutions of nonlinear fractional differential equations on a half line. *Bull. Aust. Math. Soc.* **2015**, *91*, 116–128. [CrossRef]
22. Henderson, J.; Luca, R. Nonexistence of positive solutions for a system of coupled fractional boundary value problems. *Bound. Value Probl.* **2015**, *2015*, 138. [CrossRef]
23. Ntouyas, S.K.; Etemad, S. On the existence of solutions for fractional differential inclusions with sum and integral boundary conditions. *Appl. Math. Comput.* **2015**, *266*, 235–243. [CrossRef]
24. Mei, Z.D.; Peng, J.G.; Gao, J.H. Existence and uniqueness of solutions for nonlinear general fractional differential equations in Banach spaces. *Indag. Math.* **2015**, *26*, 669–678. [CrossRef]
25. Ahmad, B.; Ntouyas, S.K. Existence results for fractional differential inclusions with Erdelyi-Kober fractional integral conditions. *Analele Universitatii Ovidius Constanta-Seria Matematica* **2017**, *25*, 5–24. [CrossRef]
26. Srivastava, H.M. Remarks on some families of fractional-order differential equations. *Integr. Transf. Spec. Funct.* **2017**, *28*, 560–564. [CrossRef]
27. Wang, G.; Pei, K.; Agarwal, R.P.; Zhang, L.; Ahmad, B. Nonlocal Hadamard fractional boundary value problem with Hadamard integral and discrete boundary conditions on a half-line. *J. Comput. Appl. Math.* **2018**, *343*, 230–239. [CrossRef]
28. Ahmad, B.; Luca, R. Existence of solutions for sequential fractional integro-differential equations and inclusions with nonlocal boundary conditions. *Appl. Math. Comput.* **2018**, *339*, 516–534. [CrossRef]

29. Zhang, Y.; Wang, R.; Zhong, H.; Zhang, J.; Belic, M.R.; Zhang, Y. Resonant mode conversions and Rabi oscillations in a fractional Schrödinger equation. *Opt. Express* **2017**, *25*, 32401. [CrossRef]
30. Sun, J.X. *Nonlinear Functional Analysis and Its Application*; Science Press: Bejing, China, 2018.
31. Krasnoselskii, M.A. Two remarks on the method of successive approximations. *Uspekhi Matematicheskikh Nauk* **1955**, *10*, 123–127.

2

Continuous Dependence of Solutions of Integer and Fractional Order Non-Instantaneous Impulsive Equations with Random Impulsive and Junction Points

Yu Chen [1] and Jin Rong Wang [1,2,*]

1 Department of Mathematics, Guizhou University, Guiyang 550025, China; cyrainie@126.com
2 School of Mathematical Sciences, Qufu Normal University, Qufu 273165, China
* Correspondence: jrwang@gzu.edu.cn

Abstract: This paper gives continuous dependence results for solutions of integer and fractional order, non-instantaneous impulsive differential equations with random impulse and junction points. The notion of the continuous dependence of solutions of these equations on the initial point is introduced. We prove some sufficient conditions that ensure the solutions to perturbed problems have a continuous dependence. Finally, we use numerical examples to demonstrate the obtained theoretical results.

Keywords: non-instantaneous impulsive equations; random impulsive and junction points; continuous dependence

---

1. Introduction

Impulsive differential equations (IDEs) are applied in many fields, such as mechanical engineering, biology, and medical science. Generally speaking, there are two classes of impulsive equations. One is composed of instantaneous IDEs, for which the duration of the impulsive perturbation is very short compared to the entire evolution process, see for example, References [1,2]. The other class is composed of non-instantaneous IDEs, for which the impulsive action starts at a fixed point, and remains active over a period of time that may be related to the previous state.

Non-instantaneous IDEs were introduced in Reference [3] and address the shortcomings of instantaneous IDEs, which do not seem to describe some of the dynamics of evolution in pharmacotherapy. Wang and Fečkan [4] corrected non-instantaneous impulsive equations in Reference [3] and proposed new and generalized non-instantaneous IDEs by considering the impact from the previous system state. Wang [5] used the notion of a non-instantaneous impulsive operator to represent the solutions of linear problems, which are simplified from the model in Reference [4]. The existence and stability of solutions and control problems for these non-instantaneous IDEs, as well as inclusions have been studied in References [6–25]. Meanwhile, fractional differential equations provide an alternative model and are gaining much importance and attention. The qualitative theory of fractional differential equations was studied extensively in the literature; see References [26–33] and the references therein.

Recently, Dishlieva [34] studied a class of instantaneous IDEs with random impulsive effects and established sufficient conditions to ensure continuous dependence of the solutions. Motivated by

Reference [34], we investigate the continuous dependence of the solutions of the following first-order nonlinear differential equations with random non-instantaneous impulsive effects:

$$\begin{cases} x'(t) = \mathbf{f}(t, x(t)), \ \ t \in (s_i, t_{i+1}], \ i \in \mathbf{N} := \{0, 1, 2, \cdots\}, \\ x(t_i^+) = \mathbf{h_i}(t_i, x(t_i^-)), \ \ i \in \mathbf{N}^+ := \{1, 2, \cdots\}, \\ x(t) = \mathbf{h_i}(t, x(t_i^-)), \ \ t \in (t_i, s_i], \ i \in \mathbf{N}^+, \\ x(0) = x_0, \end{cases} \tag{1}$$

and also of the fractional-order random non-instantaneous IDEs:

$$\begin{cases} {}^C\mathbf{D}^{\alpha}_{s_i,t} x(t) = \mathbf{f}(t, x(t)), \ \ t \in (s_i, t_{i+1}], \ i \in \mathbf{N}, \ \alpha \in (0,1), \\ x(t_i^+) = \mathbf{h_i}(t_i, x(t_i^-)), \ \ i \in \mathbf{N}^+, \\ x(t) = \mathbf{h_i}(t, x(t_i^-)), \ \ t \in (t_i, s_i], \ i \in \mathbf{N}^+, \\ x(0) = x_0, \end{cases} \tag{2}$$

where ${}^C\mathbf{D}^{\alpha}_{s_i,t}$ denotes the classical Caputo fractional derivative of order α, by changing the lower limit s_i, as in Reference [35]. The random impulse and junction points, t_i and s_i, respectively, satisfy $t_0 = s_0 = 0 < t_1 < s_1 < t_2 < \cdots < s_i < t_{i+1} < \cdots, t_i \to \infty$. The symbol $x(t_i^+)$ and $x(t_i^-)$ represent the right and left limits of $x(t)$ at $t = t_i$, respectively. In addition, we set $x(t_i^-) = x(t_i)$. The function $\mathbf{f} : [0, \infty) \times \mathbf{R} \to \mathbf{R}$ is continuous, and the function $\mathbf{h_i} : [t_i, s_i] \times \mathbf{R} \to \mathbf{R}$ is continuous for all $i \in \mathbf{N}^+$. The piecewise continuous solutions of Equations (1) and (2) have been represented in Reference [14] [Equations (5) and (7), therein].

We also introduce the following related original and perturbed equations without impulses:

$$\begin{cases} X'(t) = \mathbf{f}(t, X(t)), \ \ t \in [s_i, t_{i+1}], \ i \in \mathbf{N}, \\ X(s_i) = x_{s_i}, \end{cases} \tag{3}$$

$$\begin{cases} X'(t) = \mathbf{f}(t, X(t)), \ \ t \in [s_i, t_{i+1}], \ i \in \mathbf{N}, \\ X(s_i) = \widetilde{x}_{s_i}, \end{cases} \tag{4}$$

$$\begin{cases} {}^C\mathbf{D}^{\alpha}_{s_i,t} X(t) = \mathbf{f}(t, X(t)), \ \ t \in [s_i, t_{i+1}], \ i \in \mathbf{N}, \ \alpha \in (0,1), \\ X(s_i) = x_{s_i}, \end{cases} \tag{5}$$

and:

$$\begin{cases} {}^C\mathbf{D}^{\alpha}_{s_i,t} X(t) = \mathbf{f}(t, X(t)), \ \ t \in [s_i, t_{i+1}], \ i \in \mathbf{N}, \ \alpha \in (0,1), \\ X(s_i) = \widetilde{x}_{s_i}. \end{cases} \tag{6}$$

Denote any solution of Equations (1) or (2) by $x(\cdot; 0, x_0) \in PC(\mathbf{J}, \mathbf{R}) := \{x : \mathbf{J} \to \mathbf{R} : x \in \mathbf{C}((t_k, t_{k+1}], \mathbf{R}), k = 0, 1, \cdots$ and there exists $x(t_k^+)$ and $x(t_k^-), k = 1, 2, \cdots$ with $x(t_k^-) = x(t_k)\}$, where $\mathbf{J} = [0, \infty)$ and $\mathbf{C}((t_k, t_{k+1}], \mathbf{R})$ denotes the space of all continuous functions from $(t_k, t_{k+1}]$ into $\mathbf{R}$. Additionally, denote any solution of Equations (3) or (5) by $X(\cdot; s_i, x_{s_i}) \in \mathbf{C}([s_i, t_{i+1}], \mathbf{R})$. For the interval $(t_i, s_i]$, we denote its solutions by $X(\cdot; t_i, x(t_i^+))$. Then, the following relationship is valid:

$$x(t; 0, x_0) = \begin{cases} X(t; 0, x_0), \ t \in [0, t_1]; \\ X(t; t_1, x(t_1^+)), \ t \in (t_1, s_1]; \\ X(t; s_1, x_{s_1}), \ t \in (s_1, t_2]; \\ \cdots\cdots \end{cases} \tag{7}$$

The main objective of this article is to present the continuous dependence of solutions with respect to the initial condition when random impulse and junction points are incorporated in Equations (1) and (2). We will take notice of the fact that the location and the number of the impulse points and junction points are not determined in a finite time interval, and so, we can assume that the impulse and junction points are random.

The main contributions of this paper are two folds. We extend the concept and results in Reference [34] to random non-instantaneous impulsive cases by imposing different conditions on the nonlinear term. We also extend the continuous dependence of solutions of first-order non-instantaneous impulsive equations to fractional-order non-instantaneous impulsive equations.

The rest of this paper is organized as follows. Section 2 gives the relevant definitions and notions for the continuous dependence of solutions and contains the main results. Section 3 gives two examples to demonstrate the application of our results. In Section 4, conclusions are drawn.

2. Main Results

Based on Reference [12] (Definition 2.1, therein) and Reference [34] (Definition 2, therein), we give the following definitions for the continuous dependence of solutions.

Definition 1. *The solution $x(\cdot;0,x_0) \in PC(\mathbf{J},\mathbf{R})$ of Equations* (1) *or* (2) *depends continuously on the initial point $(0,x_0)$, if for any $\varepsilon > 0$, $\mathbf{T} > 0$, there exists a $\delta = \delta(\varepsilon,\mathbf{T}) > 0$, such that for any $(0,\widetilde{x}_0) \in [0,\mathbf{T}] \times \mathbf{R}$, and $|\widetilde{x}_0 - x_0| < \delta$, then:*

$$|x(t;0,\widetilde{x}_0) - x(t;0,x_0)| < \varepsilon,\ t \in [0,t_1] \bigcup (t_i,s_i] \bigcup (s_i,t_{i+1}],\ i \in \mathbf{N}^+.$$

Definition 2. *The solution $X(\cdot;s_i,x_{s_i}) \in \mathbf{C}([s_i,t_{i+1}],\mathbf{R})$ of Equations* (3) *and* (5) *depends uniformly and continuously on the initial point (s_i,x_{s_i}), if for any $\varepsilon > 0$, $\mathbf{T} > 0$, there exists a $\delta = \delta(\varepsilon,\mathbf{T}) > 0$, such that for any $(s_i,\widetilde{x}_{s_i}) \in [0,\mathbf{T}] \times \mathbf{R}$, and $|\widetilde{x}_{s_i} - x_{s_i}| < \delta$, then:*

$$|X(t;s_i,\widetilde{x}_{s_i}) - X(t;s_i,x_{s_i})| < \varepsilon,\ t \in [s_i,t_{i+1}],\ i \in \mathbf{N}^+.$$

We introduce the following assumptions for further discussion:

$[H_1]$ The function $\mathbf{f} : \mathbf{J} \times \mathbf{R} \to \mathbf{R}$ is continuous and $\mathbf{h_i} \in \mathbf{C}([t_i,s_i] \times \mathbf{R},\mathbf{R}), i \in \mathbf{N}^+$.

$[H_2]$ There exists an $\mathbf{L_f} > 0$, such that $|\mathbf{f}(t,x) - \mathbf{f}(t,y)| \leq \mathbf{L_f}|x-y|$, for each $t \in [s_i,t_{i+1}]$, $i \in \mathbf{N}$, and for all $x,y \in \mathbf{R}$.

$[H_3]$ There exists a positive constant $\mathbf{L_{h_i}}$, $i \in \mathbf{N}^+$, such that $|\mathbf{h_i}(t,x) - \mathbf{h_i}(t,y)| \leq \mathbf{L_{h_i}}|x - y|$, for each $t \in [t_i,s_i]$, $i \in \mathbf{N}^+$, and for all $x,y \in \mathbf{R}$.

$[H_4]$ The solutions of Equations (3) or (5) depend uniformly and continuously on the initial point.

$[H_5]$ The functions $\mathbf{h_i}(t,x), i \in \mathbf{N}^+$, are uniformly bounded, i.e., for any $i \in \mathbf{N}^+$, there exists an $\mathbf{M} > 0$, for any $x \in \mathbf{R}$, $t \in \mathbf{J}$, such that $|\mathbf{h_i}(t,x)| \leq \mathbf{M}$.

Theorem 1. *Assume that $[H_1] - [H_3]$ are satisfied. Then the solution of Equation (1) depends continuously on the initial point $(0,x_0)$ at the random impulse and junction points.*

Proof. Let ε and $\mathbf{T}$ be two arbitrary positive constants and $\mathbf{\Omega} = \{t_1,s_1,t_2,s_2,\cdots\}$ be an arbitrary set of impulse and junction points. Note that if $t_i \to \infty$ as $i \to \infty$, then for any selection of the set of impulse points and junction points, there exists $k \in \mathbf{N}$ such that $s_k < \mathbf{T} \leq t_{k+1}$, i.e., there exists at most k impulse points and at most k junction points belonging to the interval $[0,\mathbf{T}]$. Without loss of generality, we assume that $\mathbf{T} = t_{k+1}$.

We divide the proof into several cases.

Case 1. For the interval $(s_k,t_{k+1}]$, the solutions of Equations (3) and (4) are given in Reference [14]:

$$X(t;s_k,x_{s_k}) = x_{s_k} + \int_{s_k}^{t} \mathbf{f}(s,X(s;s_k,x_{s_k}))ds, \tag{8}$$

and:

$$X(t;s_k,\widetilde{x}_{s_k}) = \widetilde{x}_{s_k} + \int_{s_k}^{t} \mathbf{f}(s,\widetilde{X}(s;s_k,x_{s_k}))ds. \tag{9}$$

Assume that there exists a $\delta_{k,k+1} = \delta_{k,k+1}(\varepsilon, \mathbf{T})$, where $0 < \delta_{k,k+1} < \varepsilon$ and $|\widetilde{x}_{s_k} - x_{s_k}| < \delta_{k,k+1}$, linking Equations (8) and (9), we get:

$$\begin{aligned} & |X(t; s_k, \widetilde{x}_{s_k}) - X(t; s_k, x_{s_k})| \\ \leq \; & |\widetilde{x}_{s_k} - x_{s_k}| + \mathbf{L_f} \int_{s_k}^{t} |X(s; s_k, \widetilde{x}_{s_k}) - X(s; s_k, x_{s_k})| ds \\ \leq \; & \delta_{k,k+1} + \mathbf{L_f} \int_{s_k}^{t} |X(s; s_k, \widetilde{x}_{s_k}) - X(s; s_k, x_{s_k})| ds. \end{aligned} \tag{10}$$

From Reference [36] (Theorem 1.1, therein), we get:

$$|X(t; s_k, \widetilde{x}_{s_k}) - X(t; s_k, x_{s_k})| \leq \delta_{k,k+1} e^{\mathbf{L_f}(t_{k+1} - s_k)}.$$

Since $0 < \delta_{k,k+1} < \varepsilon$, we have:

$$|X(t; s_k, \widetilde{x}_{s_k}) - X(t; s_k, x_{s_k})| < \varepsilon, \; t \in (s_k, t_{k+1}].$$

Case 2. For the interval $(t_k, s_k]$, we have the following expression of the solutions, respectively:

$$X(t; t_k, x(t_k^+)) = \mathbf{h_k}(t, x(t_k^-)) \quad \text{and} \quad X(t; t_k, \widetilde{x}(t_k)) = \mathbf{h_k}(t, \widetilde{x}(t_k^-)).$$

Assume there exists $\delta_{k,k} < \delta_{k,k+1}$, where $|\widetilde{x}(t_k^+) - x(t_k^+)| < \delta_{kk}$, then we have:

$$|X(t; t_k, \widetilde{x}(t_k^+)) - X(t; t_k, x(t_k^+))| \leq \mathbf{L_{h_k}} |\widetilde{x}(t_k^-) - x(t_k^-)|. \tag{11}$$

For $(s_{k-1}, t_k]$, similar to Equation (10), we assume there exists a $\delta_{k-1,k} < \frac{\delta_{k,k}}{\mathbf{L_{h_k}} e^{\mathbf{L_f}(t_k - s_{k-1})}}$. Then we obtain:

$$|X(t; s_{k-1}, \widetilde{x}_{s_{k-1}}) - X(t; s_{k-1}, x_{s_{k-1}})| \leq \delta_{k-1,k-1} e^{\mathbf{L_f}(t_k - s_{k-1})} < \frac{\delta_{k,k}}{\mathbf{L_{h_k}}} < \varepsilon, \; t \in (s_{k-1}, t_k]. \tag{12}$$

For $t = t_k$, we get:

$$|X(t_k; s_{k-1}, \widetilde{x}_{s_{k-2}}) - X(t_k; s_{k-1}, x_{s_{k-1}})| < \frac{\delta_{k,k}}{\mathbf{L_{h_k}}},$$

and:

$$|\mathbf{h_k}(t_k, \widetilde{x}(t_k^-)) - \mathbf{h_k}(t_k, x(t_k^-))| < \mathbf{L_{h_k}} \frac{\delta_{k,k}}{\mathbf{L_{h_k}}} < \delta_{k,k}.$$

Therefore, Equation (11) becomes:

$$|X(t; t_k, \widetilde{x}(t_k^+)) - X(t; t_k, x(t_k^+))| < \delta_{k,k} < \delta_{k,k+1} < \varepsilon, \; t \in (t_k, s_k]. \tag{13}$$

From the procedure of Equations (12) and (13), we finally establish the following facts:

For the interval $(t_1, s_1]$, if there is a $\delta_{11} < \delta_{12}$, where $|\widetilde{x}(t_1^+) - x(t_1^+)| < \delta_{11}$, then:

$$|X(t; t_1, \widetilde{x}(t_1^+)) - X(t; t_1, x(t_1^+))| \leq \mathbf{L_{h_1}} |\widetilde{x}(t_1^-) - x(t_1^-)|. \tag{14}$$

For the interval $[0, t_1]$, we assume there is a $\delta_{01} < \frac{\delta_{11}}{\mathbf{L_{h_1}} e^{\mathbf{L_f} t_1}}$, where $|\widetilde{x}_0 - x_0| < \delta_{01}$. Thus, we obtain:

$$|X(t; 0, \widetilde{x}_0) - X(t; 0, x_0)| \leq \delta_{01} e^{\mathbf{L_f} t_1} < \frac{\delta_{11}}{\mathbf{L_{h_1}}} < \varepsilon, \quad t \in [0, t_1]. \tag{15}$$

For $t = t_1$, we get:

$$|X(t_1; 0, \widetilde{x}_0) - X(t_1; 0, x_0)| < \frac{\delta_{11}}{\mathbf{L_{h_1}}},$$

and hence, we have:

$$|h_1(t_1, x(t_1^-)) - h_1(t_1, x(t_1^-))| < \mathbf{L}_{\mathbf{h_1}} \frac{\delta_{11}}{\mathbf{L}_{\mathbf{h_1}}} < \delta_{11}. \tag{16}$$

Thus, Equation (14) becomes:

$$|X(t; t_1, \tilde{x}(t_1^+)) - X(t; t_1, x(t_1^+))| < \delta_{11} < \delta_{12} < \varepsilon, \quad t \in (t_1, s_1]. \tag{17}$$

Next, we notice that:

$$\begin{aligned} \delta_{01} &= \delta_{01}(\varepsilon, \mathbf{T}, \delta_{11}), \\ &\vdots \\ \delta_{k-1,k} &= \delta_{k-1,k}(\varepsilon, \mathbf{T}, \delta_{kk}), \\ \delta_{kk} &= \delta_{kk}(\varepsilon, \mathbf{T}, \delta_{k,k+1}), \\ \delta_{k,k+1} &= \delta_{k,k+1}(\varepsilon, \mathbf{T}). \end{aligned}$$

Therefore, we have $\delta_{01} = \delta_{01}(\varepsilon, \mathbf{T})$. Considering Equations (7) and (15), we obtain, for any $\varepsilon > 0$, there exists a $\delta_{01} > 0$, where $|\tilde{x}_0 - x_0| < \delta_{01}$, thus:

$$|x(t; 0, \tilde{x}_0) - x(t; 0, x_0)| < \varepsilon, \ t \in [0, t_1], \tag{18}$$

and from Equation (16) we get:

$$|\mathbf{h_1}(t_1, \tilde{x}(t_1^-)) - \mathbf{h_1}(t_1, x(t_1^-))| = |\tilde{x}(t_1^+) - x(t_1^+)| < \delta_{11}.$$

Taking into account Equations (17) and (7), we find that:

$$|x(t; 0, \tilde{x}_0) - x(t; 0, x_0)| < \varepsilon, \ t \in (t_1, s_1], \tag{19}$$

and:

$$|\mathbf{h_1}(s_1, \tilde{x}(t_1^-)) - \mathbf{h_1}(s_1, x(t_1^-))| = |\tilde{x}_{s_1} - x_{s_1}| < \delta_{12}.$$

According to Equation (7) again, we obtain:

$$|x(t; 0, \tilde{x}_0) - x(t; 0, x_0)| < \varepsilon, \ t \in (s_1, t_2].$$

Similarly, we achieve the following conclusion:

$$|x(t; 0, \tilde{x}_0) - x(t; 0, x_0)| < \varepsilon, \ t \in (t_k, s_k], \tag{20}$$

and:

$$|\mathbf{h_k}(s_k, \tilde{x}(t_k^-)) - \mathbf{h_k}(s_k, x(t_k^-))| = |\tilde{x}_{s_k} - x_{s_k}| < \delta_{k,k+1}.$$

Then, it is further determined:

$$|x(t; 0, \tilde{x}_0) - x(t; 0, x_0)| < \varepsilon, \ t \in (s_k, t_{k+1}]. \tag{21}$$

By Equations (18)–(21), we get that for any $\varepsilon > 0$ and $\tilde{x}_0, x_0 \in \mathbf{R}$, there exists a $\delta > 0$, where $|\tilde{x}_0 - x_0| < \delta$, such that $|x(t; 0, \tilde{x}_0) - x(t; 0, x_0)| < \varepsilon$ for $t \in [0, t_1] \bigcup (t_i, s_i] \bigcup (s_i, t_{i+1}], i = 1, 2, \cdots, k$, where $\delta = \delta_{01}(\varepsilon, \mathbf{T})$. By Definition 1, the proof is completed. □

Theorem 2. *Suppose that $[H_1] - [H_3]$ are satisfied. Then the solution of Equation (2) depends continuously on the initial point $(0, x_0)$ for random impulse and junction points.*

Proof. Let ε and $\mathbf{T}$ be two arbitrary positive constants like in Theorem 1. We divide the proof into two cases.

Case 1. For the interval $(s_k, t_{k+1}]$, the representation of the solutions of Equations (5) and (6) are given by [14]:

$$X(t; s_k, x_{s_k}) = x_{s_k} + \frac{1}{\Gamma(\alpha)} \int_{s_k}^{t} (t-s)^{\alpha-1} \mathbf{f}(s, X(s; s_k, x_{s_k})) ds, \tag{22}$$

and:

$$X(t; s_k, \widetilde{x}_{s_k}) = \widetilde{x}_{s_k} + \frac{1}{\Gamma(\alpha)} \int_{s_k}^{t} (t-s)^{\alpha-1} \mathbf{f}(s, X(s; s_k, \widetilde{x}_{s_k})) ds. \tag{23}$$

Assume that there exists a $\delta_{k,k+1} = \delta_{k,k+1}(\varepsilon, \mathbf{T})$, $0 < \delta_{k,k+1} < \varepsilon$, where $|\widetilde{x}_{s_k} - x_{s_k}| < \delta_{k,k+1}$, then from Equations (22) and (23), we have:

$$\begin{aligned} & |X(t; s_k, \widetilde{x}_{s_k}) - X(t; s_k, x_{s_k})| \\ \leq \; & |\widetilde{x}_{s_k} - x_{s_k}| + \frac{\mathbf{L_f}}{\Gamma(\alpha)} \int_{s_k}^{t} (t-s)^{\alpha-1} |X(s; s_k, \widetilde{x}_{s_k}) - X(s; s_k, x_{s_k})| ds \\ \leq \; & \delta_{k,k+1} + \frac{\mathbf{L_f}}{\Gamma(\alpha)} \int_{s_k}^{t} (t-s)^{\alpha-1} |X(s; s_k, \widetilde{x}_{s_k}) - X(s; s_k, x_{s_k})| ds. \end{aligned}$$

Using Reference [37] (Corollary 2, therein), we get:

$$|X(t; s_k, \widetilde{x}_{s_k}) - X(t; s_k, x_{s_k})| \leq \delta_{k,k+1} \mathbf{E}_\alpha (\mathbf{L_f} (t_{k+1} - s_k)^\alpha),$$

where $\mathbf{E}_\alpha$ is the standard Mittag–Leffler function [35] defined as:

$$\mathbf{E}_\alpha(z) = \sum_{k=0}^{\infty} \frac{z^k}{\Gamma(k\alpha + 1)}, \quad z \in \mathbb{C}.$$

Owing to $0 < \delta_{k,k+1} < \varepsilon$, one can get:

$$|X(t; s_k, \widetilde{x}_{s_k}) - X(t; s_k, x_{s_k})| < \varepsilon, \; t \in (s_k, t_{k+1}].$$

Case 2. From the interval $(t_k, s_k]$, the expression of the solutions are given by:

$$X(t; t_k, x(t_k^+)) = \mathbf{h_k}(t, x(t_k^-)) \quad \text{and} \quad X(t; t_k, \widetilde{x}(t_k)) = \mathbf{h_k}(t, \widetilde{x}(t_k^-)).$$

Assume that there exists a $\delta_{kk} < \delta_{k,k+1}$, where $|\widetilde{x}(t_k^+) - x(t_k^+)| < \delta_{kk}$, then we have:

$$|X(t; t_k, \widetilde{x}(t_k^+)) - X(t; t_k, x(t_k^+))| \leq \mathbf{L_{h_k}} |\widetilde{x}(t_k^-) - x(t_k^-)|. \tag{24}$$

Similar to the above procedure, for the interval $(s_{k-1}, t_k]$, one has:

$$|X(t; s_{k-1}, \widetilde{x}_{s_{k-1}}) - X(t; s_{k-1}, x_{s_{k-1}})| \leq \delta_{k-1,k} \mathbf{E}_\alpha (L_f (t_k - s_{k-1})^\alpha).$$

Since $\delta_{k-1,k} < \frac{\delta_{kk}}{\mathbf{L_{h_k}} \mathbf{E}_\alpha (\mathbf{L_f} (t_k - s_{k-1})^\alpha)}$, one can obtain:

$$|X(t; s_{k-1}, \widetilde{x}_{s_{k-1}}) - X(t; s_{k-1}, x_{s_{k-1}})| < \frac{\delta_{kk}}{\mathbf{L_{h_k}}} < \varepsilon, \; t \in (s_{k-1}, t_k].$$

For $t = t_k$, we get:

$$|X(t_k; s_{k-1}, \widetilde{x}_{s_{k-1}}) - X(t_k; s_{k-1}, x_{s_{k-1}})| < \frac{\delta_{kk}}{\mathbf{L_{h_k}}},$$

and then:

$$|g_k(t_k, x(t_k^-)) - g_k(t_k, x(t_k^-))| < \mathbf{L_{h_k}} \frac{\delta_{kk}}{\mathbf{L_{h_k}}} < \delta_{kk}.$$

Therefore, (24) becomes:

$$|X(t; t_k, \widetilde{x}(t_k^+)) - X(t; t_k, x(t_k^+))| < \delta_{kk} < \varepsilon.$$

From the above, one can deduce that for the interval $(t_1, s_1]$, there exists a $\delta_{11} < \delta_{12}$, and if $|\widetilde{x}(t_1^-) - x(t_1^-)| < \delta_{11}$, then:

$$|X(t; t_1, x(t_1^+)) - X(t; t_1, x(t_1^+))| \leq \mathbf{L_{h_1}} |\widetilde{x}(t_1^-) - x(t_1^-)|.$$

For $[0, t_1]$, there is $\delta_{01} < \frac{\delta_{11}}{\mathbf{L_{h_1}} \mathbf{E}_\alpha(\mathbf{L_f} t_1^\alpha)}$, if $|\widetilde{x}_0 - x_0| < \delta_{01}$, then:

$$|X(t; 0, \widetilde{x}_0) - X(t; 0, x_0)| \leq \delta_{01} \mathbf{E}_\alpha(\mathbf{L_f} t_1^\alpha) < \frac{\delta_{11}}{\mathbf{L_{h_1}}} < \varepsilon, \ t \in [0, t_1].$$

Thus, for $t = t_1$, we have:

$$|X(t_1; 0, \widetilde{x}_0) - X(t_1; 0, x_0)| < \frac{\delta_{11}}{\mathbf{L_{h_1}}},$$

and then:

$$|\mathbf{h_1}(t_1, \widetilde{x}(t_1)) - \mathbf{h_1}(t_1, x(t_1))| < \delta_{11}.$$

and:

$$|X(t; t_1, x(t_1^+)) - X(t; t_1, x(t_1^+))| < \delta_{11} < \delta_{12} < \varepsilon, \ t \in (t_1, s_1].$$

Similar to Theorem 1, we reach the conclusion. □

Theorem 3. *Assume that $[H_1]$, $[H_4]$, and $[H_5]$ are satisfied. The solution of Equation (1) depends continuously on the initial point $(0, x_0)$ at the random impulse and junction points provided that $\mathbf{2M} \leq |\widetilde{x}_{s_i} - x_{s_i}|$.*

Proof. We divide proof into several steps.

Step 1. According to $[H_4]$, assume there is a $\delta_{k,k+1} = \delta_{k,k+1}(\varepsilon, \mathbf{T})$, where $0 < \delta_{k,k+1} < \varepsilon$, if $|\widetilde{x}_{s_k} - x_{s_k}| < \delta_{k,k+1}$, then:

$$|X(t; s_k, \widetilde{x}_{s_k}) - X(t; s_k, x_{s_k})| < \varepsilon, \ t \in (s_k, t_{k+1}].$$

Step 2. For the interval $(t_k, s_k]$, according to $[H_5]$, we obtain:

$$\begin{aligned} |X(t; t_k, \widetilde{x}(t_k^+)) - X(t; t_k, x(t_k^+))| &= |\mathbf{h_k}(t, \widetilde{x}(t_k^-)) - \mathbf{h_k}(t, x(t_k^-))| \\ &\leq \mathbf{2M} \leq |\widetilde{x}_{s_k} - x_{s_k}| < \delta_{k,k+1} < \varepsilon, \ t \in (t_k, s_k]. \end{aligned}$$

For $t = s_k$, we get:

$$|\mathbf{h_k}(s_k, \widetilde{x}(t_k^-)) - \mathbf{h_k}(s_k, x(t_k^-))| < \delta_{k,k+1}. \tag{25}$$

Step 3. In this step, we check the continuity of the solutions in the interval $(s_{k-1}, t_k]$. From Equation (25), we assume there exists a $\delta_{k-1,k}$ depending on $\varepsilon, \mathbf{T}, \delta_{k,k+1}$. For brevity, we denote

$\delta_{k-1,k} = \delta_{k-1,k}(\varepsilon, \mathbf{T}, \delta_{k,k+1})$, where $0 < \delta_{k-1,k} < \varepsilon$. If $|\widetilde{x}_{s_{k-1}} - x_{s_{k-1}}| < \delta_{k-1,k}$, then by $[H_4]$ we deduce that:

$$|X(t; s_{k-1}, \widetilde{x}_{s_{k-1}}) - X(t; s_{k-1}, x_{s_{k-1}})| < \varepsilon,\ t \in (s_{k-1}, t_k].$$

Step 4. For $(t_{k-1}, s_{k-1}]$, we get:

$$\begin{aligned}|X(t; t_{k-1}, \widetilde{x}(t_{k-1}^+)) - X(t; t_{k-1}, x(t_{k-1}^+))| &= |\mathbf{h_{k-1}}(t, \widetilde{x}(t_{k-1}^-)) - \mathbf{h_{k-1}}(t, x(t_{k-1}^-))| \\ &\leq 2\mathbf{M} \leq |\widetilde{x}_{s_{k-1}} - x_{s_{k-1}}| < \delta_{k-1,k} < \varepsilon,\ t \in (t_{k-1}, s_{k-1}].\end{aligned}$$

For $t = s_{k-1}$, we have:

$$|\mathbf{h_{k-1}}(s_{k-1}, \widetilde{x}(t_{k-1}^-)) - \mathbf{h_{k-1}}(s_{k-1}, x(t_{k-1}^-))| < \delta_{k-1,k}.$$

Similar to the above steps, we have the following general results.

Step $2k$. For the interval $(t_1, s_1]$, it follows that:

$$|X(t; t_1, \widetilde{x}(t_1^+)) - X(t; t_1, x(t_1^+))| = |\mathbf{h_1}(t, \widetilde{x}(t_1^-)) - \mathbf{h_1}(t, x(t_1^-))| < \delta_{12} < \varepsilon,\ t \in (t_1, s_1].$$

For $t = s_1$, we have:

$$|\mathbf{h_1}(s_1, \widetilde{x}(t_1^-)) - \mathbf{h_1}(s_1, x(t_1^-))| < \delta_{12}. \tag{26}$$

Step $2k+1$. For $[0, t_1]$, assume there exists a $\delta_{01} = \delta_{01}(\varepsilon, \mathbf{T}, \delta_{12})$, where $0 < \delta_{01} < \varepsilon$, if $|\widetilde{x}_0 - x_0| < \varepsilon$, then:

$$|X(t; 0, \widetilde{x}_0) - X(t; 0, x_0)| < \varepsilon,\ t \in [0, t_1].$$

Finally, we notice that:

$$\begin{aligned}\delta_{01} &= \delta_{01}(\varepsilon, \mathbf{T}, \delta_{12}), \\ &\vdots \\ \delta_{k-1,k} &= \delta_{k-1,k}(\varepsilon, \mathbf{T}, \delta_{k,k+1}), \\ \delta_{k,k+1} &= \delta_{k,k+1}(\varepsilon, \mathbf{T}).\end{aligned}$$

Therefore, we have $\delta_{01} = \delta_{01}(\varepsilon, \mathbf{T})$.

Now we apply the above results to the impulse case.

Step 1′. Based on Equation (7) and Step 2k + 1, we obtain for any $\varepsilon > 0$, there exists a $\delta_{01} > 0$, such that if $|\widetilde{x}_0 - x_0| < \delta_{01}$, then:

$$|x(t; 0, \widetilde{x}_0) - x(t; 0, x_0)| < \varepsilon,\ t \in [0, t_1].$$

Step 2′. Using Equation (7) and taking Step 2k into account, we find that:

$$|x(t; 0, \widetilde{x}_0) - x(t; 0, x_0)| < \varepsilon,\ t \in (t_1, s_1].$$

From Equation (26), we have:

$$|\mathbf{h_1}(s_1, \widetilde{x}(t_1^-)) - \mathbf{h_1}(s_1, x(t_1^-))| = |\widetilde{x}_{s_1} - x_{s_1}| < \delta_{12}.$$

Step 3′. According to Equation (7) and Step 2k − 1, it follows that:

$$|x(t; 0, \widetilde{x}_0) - x(t; 0, x_0)| < \varepsilon,\ t \in (s_1, t_2].$$

Step 4′. Again by Equation (7) and considering Step 2k − 2, we have:

$$|x(t;0,\widetilde{x}_0) - x(t;0,x_0)| < \varepsilon,\ t \in (t_2, s_2],$$

and:

$$|\mathbf{h_2}(s_2, \widetilde{x}(t_2^-)) - \mathbf{h_2}(s_2, x(t_2^-))| = |\widetilde{x}_{s_2} - x_{s_2}| < \delta_{23}.$$

Similarly, we arrive at:

Step $(2k)'$. $|x(t;0,\widetilde{x}_0) - x(t;0,x_0)| < \varepsilon,\ t \in (t_k, s_k]$.

Thus, from Equation (25), we get:

$$|\mathbf{h_k}(s_k, \widetilde{x}(t_k^-)) - \mathbf{h_k}(s_k, x(t_k^-))| = |\widetilde{x}_{s_k} - x_{s_k}| < \delta_{k,k+1}.$$

Step $(2k+1)'$. $|x(t;0,\widetilde{x}_0) - x(t;0,x_0)| < \varepsilon,\ t \in (s_k, t_{k+1}]$.

From Step $1'$ to Step $(2k+1)'$, we get that for any $\varepsilon > 0$, $\widetilde{x}_0, x_0 \in \mathbf{R}$, there exists a $\delta > 0$, such that if $|\widetilde{x}_0 - x_0| < \delta$, then $|x(t;0,\widetilde{x}_0) - x(t;0,x_0)| < \varepsilon$, for $t \in [0,t_1] \bigcup (t_i, s_i] \bigcup (s_i, t_{i+1}], i = 1, 2, \cdots, k$, and $\delta = \delta_{01}(\varepsilon, \mathbf{T})$. □

By repeating the same proof procedure in Theorem 3, we have the result:

Remark 1. *Assume that $[H_1]$, $[H_4]$, and $[H_5]$ are satisfied. Then the solution of Equation (2) depends continuously on the initial point $(0, x_0)$ at the random impulse and junction points, provided that $2\mathbf{M} \le |\widetilde{x}_{s_i} - x_{s_i}|$.*

3. Numerical Examples

Let ε and $\mathbf{T}$ be two arbitrary positive constants.

Example 1. *Consider:*

$$\begin{cases} x'(t) = ax(t),\ t \in (s_i, t_{i+1}],\ i \in \mathbf{N},\ a > 0, \\ x(t_i^+) = \frac{\varepsilon}{2e^{at}} \cos \sqrt{t_i |x(t_i^-)|},\ i \in \mathbf{N}^+, \\ x(t) = \frac{\varepsilon}{2e^{at}} \cos \sqrt{t |x(t_i^-)|},\ t \in (t_i, s_i],\ i \in \mathbf{N}^+, \\ x(0) = x_0. \end{cases} \tag{27}$$

Then the solution of Equation (27) *can be analytically determined, namely:*

$$x(t) = \begin{cases} e^{at} x_0,\ \text{for } t \in (s_0, t_1], \\ \frac{\varepsilon}{2e^{at}} \cos \sqrt{t|x(t_1^-)|},\ \text{for } t \in (t_1, s_1], \\ \frac{\varepsilon}{2e^{as_1}} \cos \sqrt{s_1 |x(t_1^-)|} e^{a(t-s_1)},\ \text{for } t \in (s_1, t_2], \\ \vdots \\ \frac{\varepsilon}{2e^{at}} \cos \sqrt{t|x(t_{m-1}^-)|},\ \text{for } t \in (t_{m-1}, s_m], \\ \frac{\varepsilon}{2e^{as_m}} (\cos \sqrt{s_m |x(t_{m-1}^-)|} e^{a(t-s_m)},\ \text{for } t \in (s_m, t_{m+1}], \\ \vdots \end{cases} \tag{28}$$

Let $\mathbf{f}(t,x) = ax$ and $\mathbf{h_i}(t,x) = \cos\sqrt{t|x|}$. Note that $\mathbf{h_i} \in \mathbf{C}([t_i, s_i] \times \mathbf{R}, \mathbf{R}), i \in \mathbf{N}^+$. For any $x, y \in \mathbf{R}$, $|\mathbf{f}(t,x) - \mathbf{f}(t,y)| \le a|x-y|$ and $|\mathbf{h_i}(t,x) - \mathbf{h_i}(t,y)| \le \frac{\varepsilon}{2ea}|x-y|$. Set $\mathbf{L_f} = a$ and $\mathbf{L_{h_i}} = \frac{\varepsilon}{2ea}$. So, $[H_1]$–$[H_3]$ all hold. Therefore, all the assumptions in Theorem 1 are satisfied.

From Definition 2, choosing $\delta = \frac{\varepsilon}{e^{a\mathbf{T}}}$, we find that the solution of Equation (27) *without impulses satisfies uniform and continuous dependence on the initial point. Choosing $2\mathbf{M} \le \frac{\varepsilon}{e^{a\mathbf{T}}}$, then $\mathbf{h}_i(t,x)$ are uniformly bounded. Thus, the conditions $[H_1]$, $[H_4]$, and $[H_5]$ hold, and all the assumptions of Theorem 3 are satisfied.*

The solutions of Equation (27) *and the corresponding perturbation problem (with* $x_0 = 1, \widetilde{x_0} = 1.3, \mathbf{T} = 1, a = \frac{1}{2}, \varepsilon = \frac{1}{2}$*) are shown in Figure 1.*

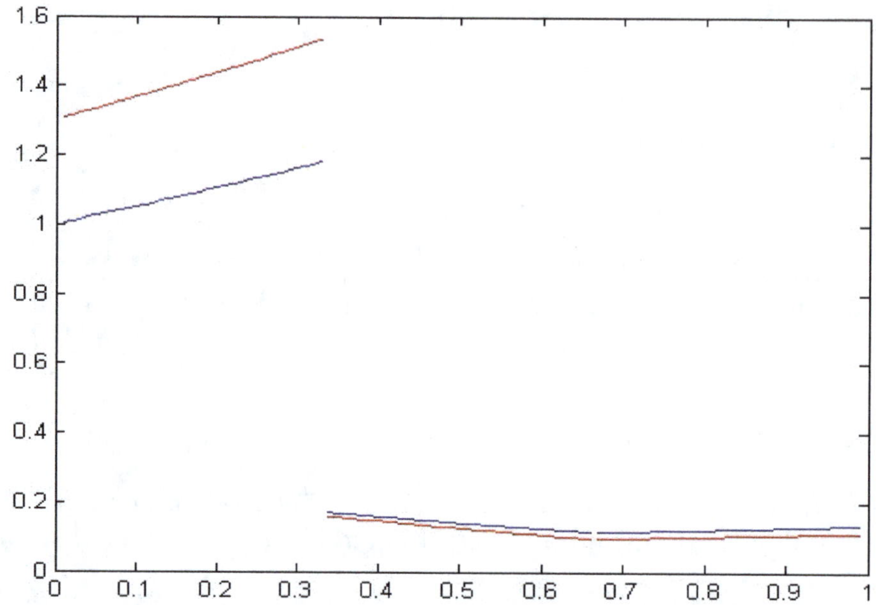

Figure 1. The blue line denotes the solution of Equation (27) and the red line denotes the corresponding perturbation problem.

Example 2. *Consider:*

$$\begin{cases} {}^{C}\mathbf{D}^{\frac{1}{2}}_{s_i,t}x(t) = ax(t),\ t \in (s_i, t_{i+1}],\ i \in \mathbf{N},\ a > 0, \\ x(t_i^+) = \frac{\varepsilon}{2\mathbf{E}_{\frac{1}{2}}(at^{\frac{1}{2}})}(\cos\sqrt{t_i|x(t_i^-)|}),\ i \in \mathbf{N}^+, \\ x(t) = \frac{\varepsilon}{2\mathbf{E}_{\frac{1}{2}}(at^{\frac{1}{2}})}(\cos\sqrt{t|x(t_i)^-|}),\ t \in (t_i, s_i],\ i \in \mathbf{N}^+, \\ x(0) = x_0. \end{cases} \tag{29}$$

Like Equation (27)*, let* $\mathbf{f}(t,x) = ax$*, and* $\mathbf{h_i}(t,x) = \frac{\varepsilon}{2\mathbf{E}_{\frac{1}{2}}(at^{\frac{1}{2}})}(\cos\sqrt{t|x|})$*. Choosing* $\mathbf{L_f} = a$*,* $\mathbf{L_{h_i}} = \frac{\varepsilon}{2a^2}$*, therefore,* $[H_1] - [H_3]$ *hold. Then, one can obtain the solution of Equation* (29)*, namely:*

$$x(t) = \begin{cases} x_0\mathbf{E}_{\frac{1}{2}}(at^{\frac{1}{2}}),\ \textit{for } t \in (s_0, t_1], \\ \frac{\varepsilon}{2\mathbf{E}_{\frac{1}{2}}(at^{\frac{1}{2}})}\cos\sqrt{t|x(t_1^-)|},\ \textit{for } t \in (t_1, s_1], \\ \frac{\varepsilon}{2\mathbf{E}_{\frac{1}{2}}(as_1^{\frac{1}{2}})}\cos\sqrt{s_1|x(t_1^-)|}E_{\frac{1}{2}}(a(t-s_1)^{\frac{1}{2}}),\ \textit{for } t \in (s_1, t_2], \\ \vdots \\ \frac{\varepsilon}{2\mathbf{E}_{\frac{1}{2}}(at^{\frac{1}{2}})}\cos\sqrt{t|x(t_{m-1}^-)|},\ \textit{for } t \in (t_{m-1}, s_m], \\ \frac{\varepsilon}{2\mathbf{E}_{\frac{1}{2}}(as_m^{\frac{1}{2}})}\cos\sqrt{s_m|x(t_{m-1}^-)|}E_{\frac{1}{2}}(a(t-s_m)^{\frac{1}{2}}),\ \textit{for } t \in (s_m, t_{m+1}], \\ \vdots \end{cases} \tag{30}$$

From Definition 2, choosing $\delta = \frac{\varepsilon}{\mathbf{E}_{\frac{1}{2}}(a\mathbf{T}^{\frac{1}{2}})}$*, we get that the solution of Equation* (29) *without impulses satisfies uniform and continuous dependence on the initial point. Choosing* $2\mathbf{M} \leq \frac{\varepsilon}{\mathbf{E}_{\frac{1}{2}}(a\mathbf{T}^{\frac{1}{2}})}$*, then the* $\mathbf{h}_i(t,x)$ *are uniformly bounded. Thus, the conditions* $[H_1], [H_4]$*, and* $[H_5]$ *hold, and all the assumptions in Remark 1 are satisfied.*

The solutions of Equation (29) *and the corresponding perturbation problem (with* $x_0 = 1, \widetilde{x_0} = 1.3, \mathbf{T} = \frac{3}{2}, a = \frac{1}{5}, \varepsilon = \frac{1}{2}$*) are shown in Figure 2.*

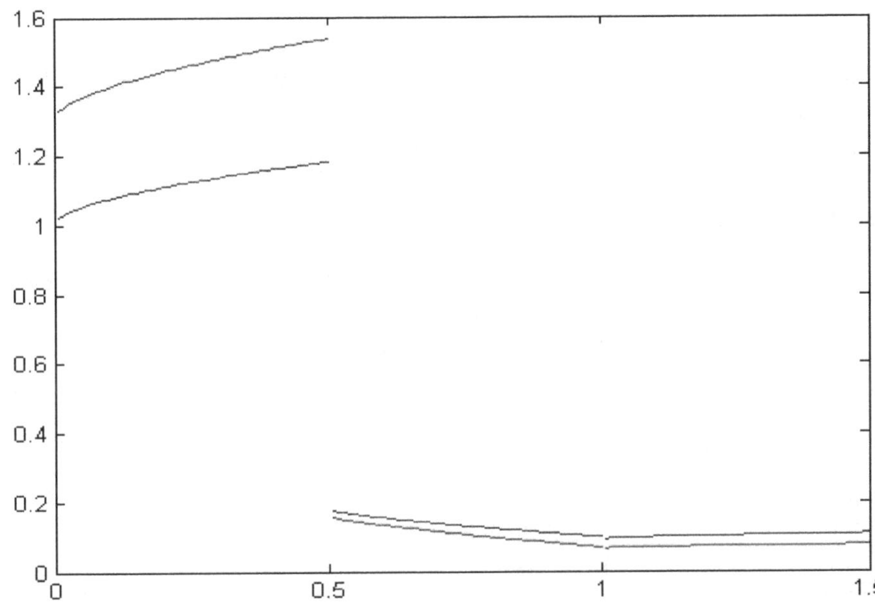

Figure 2. The blue line denotes the solution of Equation (29) and the red line denotes the corresponding perturbation problem.

Remark 2. *Equations* (27) *and* (29) *are called non-instantaneous impulsive logistic models, which are motivated from the instantaneous impulsive logistic equations. For more details of the models, one can refer to Reference [12] (Section 4, therein).*

4. Conclusions

In this paper, we presented the continuous dependence of the solutions to first order non-instantaneous IDEs with random impulse and junction points. Then, we extended the results to study the same problem for fractional order cases. The backward checking approach [34] (from the last subinterval to the first subinterval) is extended to differential and algebra equations and is used to prove the main results. The approach is different from Reference [13].

Author Contributions: The contributions of both authors (Y.C. and J.W.) are equal. All the main results and numerical examples were developed together.

Acknowledgments: The authors are grateful to the referees for their careful reading of the manuscript and their valuable comments. We also thank the editor.

References

1. Chalishajar, D.N.; Malar, K.; Karthikeyan, K. Approximate controllability of abstract impulsive fractional neutral evolution equations with infinite delay in Banach spaces. *Electron. J. Differ. Equ.* **2013**, *2013*, 1–21.
2. Chalishajar, D.N.; Karthikeyan, K.; Anguraj, A. Existence results for impulsive perturbed partial neutral functional differential equations in Frechet spaces. *Dyn. Contin. Discret. Impuls. Syst. Ser. Math. Anal.* **2015**, *22*, 25–45.
3. Hernández, E.; O'Regan, D. On a new class of abstract impulsive differential equations. *Proc. Am. Math. Soc.* **2013**, *141*, 1641–1649. [CrossRef]
4. Wang, J.; Fečkan, M. A general class of impulsive evolution equations. *Topol. Meth. Nonlinear Anal.* **2015**, *46*, 915–934. [CrossRef]
5. Wang, J. Stability of noninstantaneous impulsive evolution equations. *Appl. Math. Lett.* **2017**, *73*, 157–162. [CrossRef]
6. Chen, P.; Li, Y.; Yang, H. Perturbation method for nonlocal impulsive evolution equations. *Nonlinear Anal. Hybrid Syst.* **2013**, *8*, 22–30. [CrossRef]
7. Bai, L.; Nieto, J.J. Variational approach to differential equations with not instantaneous impulses. *Appl. Math. Lett.* **2017**, *73*, 44–48. [CrossRef]
8. Pierri, M.; O'Regan, D.; Rolnik, V. Existence of solutions for semi-linear abstract differential equations with not instantaneous impulses. *Appl. Math. Comput.* **2013**, *219*, 6743–6749. [CrossRef]
9. Pierri, M.; Henríquez, H.R.; Prokczyk, A. Global solutions for abstract differential equations with non-instantaneous impulses. *Mediterr. J. Math.* **2016**, *34*, 1685–1708. [CrossRef]
10. Hernández, E.; Pierri, M.; O'Regan, D. On abstract differential equations with non instantaneous impulses. *Topol. Methods Nonlinear Anal.* **2015**, *46*, 1067–1085.
11. Yang, D.; Wang, J. Non-instantaneous impulsive fractional-order implicit differential equations with random effects. *Stoch. Anal. Appl.* **2017**, *35*, 719–741. [CrossRef]
12. Yang, D.; Wang, J.; O'Regan, D. A class of nonlinear non-instantaneous impulsive differential equations involving parameters and fractional order. *Appl. Math. Comput.* **2018**, *321*, 654–671. [CrossRef]
13. Yang, D.; Wang, J.; O'Regan, D. Asymptotic properties of the solutions of nonlinear non-instantaneous impulsive differential equations. *J. Frankl. Inst.* **2017**, *354*, 6978–7011. [CrossRef]
14. Yang, D.; Wang, J.; O'Regan, D. On the orbital Hausdorff dependence of differential equations with non-instantaneous impulses. *C. R. Acad. Sci. Paris Ser. I* **2018**, *356*, 150–171. [CrossRef]
15. Abbas, S.; Benchohra, M. Uniqueness and Ulam stabilities results for partial fractional differential equations with not instantaneous impulses. *Appl. Math. Comput.* **2015**, *257*, 190–198. [CrossRef]
16. Muslim, M.; Kumar, A.; Fečkan, M. Existence, uniqueness and stability of solutions to second order nonlinear differential equations with non-instantaneous impulses. *J. King Saud Univ.* **2018**, *30*, 204–213. [CrossRef]
17. Colao, V.; Muglia, L.; Xu, H.K. An existence result for a new class of impulsive functional differential equations with delay. *J. Math. Anal. Appl.* **2016**, *441*, 668–683. [CrossRef]
18. Wang, J.; Fečkan, M. *Non-Instantaneous Impulsive Differential Equations Basic Theory and Computation*; IOP Publishing: Bristol, UK, 2018.
19. Wang, J.; Zhou, Y.; Lin, Z. On a new class of impulsive fractional differential equations. *Appl. Math. Comput.* **2014**, *242*, 649–657. [CrossRef]
20. Wang, J.; Fečkan, M.; Tian, Y. Stability analysis for a general class of non-instantaneous impulsive differential equations. *Mediterr. J. Math.* **2017**, *14*, 1–21. [CrossRef]
21. Wang, J.; Ibrahim, A.G.; O'Regan, D. Topological structure of the solution set for fractional non-instantaneous impulsive evolution inclusions. *J. Fixed Point Theory Appl.* **2018**, *20*, 1–25. [CrossRef]
22. Wang, J.; Ibrahim, A.G.; O'Regan, D.; Zhou, Y. Controllability for noninstantaneous impulsive semilinear functional differential inclusions without compactness. *Indag. Math.* **2018**, *29*, 1362–1392. [CrossRef]
23. Wang, J.; Ibrahim, A.G.; O'Regan, D. Hilfer type fractional differential switched inclusions with noninstantaneous impulsive and nonlocal conditions. *Nonlinear Anal. Model. Contr.* **2018**, *23*, 921–941. [CrossRef]
24. Wang, J.; Ibrahim, A.G.; O'Regan, D. Nonempty and compactness of solution set for fractional semilinear evolution inclusions with non-instantaneous impulses. *Electr. J. Differ. Equ.* **2019**, *2019*, 1–17.

25. Liu, S.; Wang, J.; Shen, D.; O'Regan, D. Iterative learning control for parabolic partial differential inclusions with noninstantaneous impulses. *Appl. Math. Comput.* **2019**, *350*, 48–59. [CrossRef]
26. Luo, D.; Wang, J.; Shen, D. Learning formation control for fractional-order multi-agent systems. *Math. Meth. Appl. Sci.* **2018**, *41*, 5003–5014. [CrossRef]
27. Zhang, J.; Wang, J. Numerical analysis for a class of Navier–Stokes equations with time fractional derivatives. *Appl. Math. Comput.* **2018**, *336*, 481–489.
28. Zhu, B.; Liu, L.; Wu, Y. Local and global existence of mild solutions for a class of nonlinear fractional reaction-diffusion equation with delay. *Appl. Math. Lett.* **2016**, *61*, 73–79. [CrossRef]
29. Ren, L.; Wang, J.; Fečkan, M. Asymptotically periodic solutions for Caputo type fractional evolution equations. *Fract. Calc. Appl. Anal.* **2018**, *21*, 1294–1312. [CrossRef]
30. Wang, X.; Wang, J.; Shen, D.; Zhou, Y. Convergence analysis for iterative learning control of conformable fractional differential equations. *Math. Meth. Appl. Sci.* **2018**, *41*, 8315–8328. [CrossRef]
31. Wang, Y.; Liu, L.; Zhang, X.; Wu, Y. Positive solutions of a fractional semipositone differential system arising from the study of HIV infection models. *Appl. Math. Comput.* **2015**, *258*, 312–324.
32. Zhang, X.; Mao, C.; Liu, L.; Wu, Y. Exact iterative solution for an abstract fractional dynamic system model for bioprocess. *Qual. Theory Dyn. Syst.* **2017**, *16*, 205–222. [CrossRef]
33. Zhang, X.; Liu, L.; Wu, B.Y. Wiwatanapataphee, Nontrivial solutions for a fractional advection dispersion equation in anomalous diffusion. *Appl. Math. Lett.* **2017**, *66*, 1–8. [CrossRef]
34. Dishlieva, K. On the qualitative theory of differential equaitons with random impulsive moments. *Int. J. Sci. Tech. Manag.* **2015**, *4*, 172–180.
35. Kilbas, A.A.; Srivastava, H.M.; Trujillo, J.J. *Theory and Applications of Fractional Differential Equations*; Elsevier: Amsterdam, The Netherlands, 2006.
36. Bainov, D.; Simeonov, P. *Integral Inequalities and Applications*; Kluwer Academic Publishers: Dordrecht, The Netherlands,1992.
37. Ye, H.; Gao, J.; Ding, Y. A generalized Gronwall inequality and its application to a fractional differential equation. *J. Math. Anal. Appl.* **2007**, *328*, 1075–1081. [CrossRef]

3

Stability, Existence and Uniqueness of Boundary Value Problems for a Coupled System of Fractional Differential Equations

Nazim I Mahmudov * and Areen Al-Khateeb

Eastern Mediterranean University, Gazimagusa 99628, T.R. North Cyprus, Mersin 10, Turkey; khteb1987@live.com

* Correspondence: Nazim.mahmudov@emu.edu.tr;

Abstract: The current article studies a coupled system of fractional differential equations with boundary conditions and proves the existence and uniqueness of solutions by applying Leray-Schauder's alternative and contraction mapping principle. Furthermore, the Hyers-Ulam stability of solutions is discussed and sufficient conditions for the stability are developed. Obtained results are supported by examples and illustrated in the last section.

Keywords: fractional derivative; fixed point theorem; fractional differential equation

---

1. Introduction

Fractional calculus is undoubtedly one of the very fast-growing fields of modern mathematics, due to its broad range of applications in various fields of science and its unique efficiency in modeling complex phenomena [1,2]. In particular, fractional differential equations with boundary conditions are widely employed to build complex mathematical models for numerous real-life problems such as blood flow problem, underground water flow, population dynamics, and bioengineering. As an example, consider the following equation that describes a thermostat model

$$-x'' = g(t)f(t,x), x(0) = 0, \beta x\prime(1) = x(\eta),$$

where $t \in (0,1), \eta \in (0,1]$ and β is a positive constant. Note that solutions of the above equation with the specified integral boundary conditions are in fact solutions of the one-dimensional heat equation describing a heated bar with a controller at point 1, which increases or reduces heat based on the temperature picked by a sensor at η. A few of the relevant studies on coupled systems of fractional differential equations with integral boundary conditions are briefly reviewed below and for further information on this topic, refer to References [3,4].

In Reference [5], Ntouyas and Obaid used Leray-Schauder's alternative and Banach's fixed-point theorem to prove the existence and uniqueness of solutions for the following coupled fractional differential equations with Riemann-Liouville integral boundary conditions:

$$\begin{cases} {}^cD^{\alpha}_{0+}u(t) = g(t,u(t),v(t)), t \in [0,1], \\ {}^cD^{\beta}_{0+}v(t) = g(t,u(t),v(t)), t \in [0,1], \\ u(0) = \gamma I^p u(\eta) = \gamma \int_0^{\eta} \frac{(\eta-s)^{p-1}}{\Gamma(p)} u(s)ds, 0 < \eta < 1, \\ v(0) = \delta I^q v(\zeta) = \delta \int_0^{\zeta} \frac{(\zeta-s)^{q-1}}{\Gamma(q)} v(s)ds, 0 < \zeta < 1. \end{cases}$$

Here, ${}^cD^{\alpha}_{0+}$ and ${}^cD^{\beta}_{0+}$ are Caputo fractional derivatives, $0 < \alpha, \beta \leq 1$, $f, g \in C\big([0,1] \times \mathbb{R}^2, \mathbb{R}\big)$ and $p, q, \gamma, \delta \in \mathbb{R}$.

Similarly, Ahmed and Ntouyas [6] employed Banach fixed-point theorem and Leray-Schauder's alternative to prove the existence and uniqueness of solutions for the following coupled fractional differential system:

$$\begin{cases} {}^cD^q x(t) = f(t,x(t),y(t)), & t \in [0,1], \quad 1 < q \leq 2, \\ {}^cD^p y(t) = g(t,x(t),y(t)), & t \in [0,1], \quad 1 < q \leq 2, \end{cases}$$

supplemented with coupled and uncoupled slit-strips-type integral boundary conditions, respectively, given by

$$\begin{cases} x(0) = 0, \quad x(\zeta) = a\int_0^\eta y(s)ds + b\int_\xi^1 y(s)ds, & 0 < \eta < \zeta < \xi < 1, \\ y(0) = 0, \quad y(\zeta) = a\int_0^\eta x(s)ds + b\int_\xi^1 x(s)ds, & 0 < \eta < \zeta < \xi < 1, \end{cases}$$

and

$$\begin{cases} x(0) = 0, \quad x(\zeta) = a\int_0^\eta x(s)ds + b\int_\xi^1 x(s)ds, & 0 < \eta < \zeta < \xi < 1, \\ y(0) = 0, \quad y(\zeta) = a\int_0^\eta y(s)ds + b\int_\xi^1 y(s)ds, & 0 < \eta < \zeta < \xi < 1. \end{cases}$$

Furthermore, Alsulami et al. [7] investigated the following coupled system of fractional differential equations:

$$\begin{cases} {}^cD^\alpha x(t) = f(t,x(t),y(t)), t \in [0,T], 1 < \alpha \leq 2, \\ {}^cD^\beta y(t) = g(t,x(t),y(t)), t \in [0,T], 1 < \beta \leq 2, \end{cases}$$

subject to the following non-separated coupled boundary conditions:

$$\begin{cases} x(0) = \lambda_1 y(T), x'(0) = \lambda_2 y'(T), \\ y(0) = \mu_1 x(T), y'(0) = \mu_2 x'(T). \end{cases}$$

Note that ${}^cD^\alpha$ and ${}^cD^\beta$ denote Caputo fractional derivatives of order α and β. Moreover, λ_i, μ_i, $i = 1, 2$, are real constants with $\lambda_i\mu_i \neq 1$ and $f, g : [0,T] \times \mathbb{R} \times \mathbb{R} \to \mathbb{R}$ are appropriately chosen functions. For further details on this topic, refer to References [8–21].

The current paper studies the following coupled system of nonlinear fractional differential equations:

$$\begin{cases} {}^cD^\alpha x(t) = f(t,x(t),y(t)), & t \in [0,T], \quad 1 < \alpha \leq 2, \\ \\ {}^cD^\beta y(t) = g(t,x(t),y(t)), & t \in [0,T], \quad 1 < \beta \leq 2, \end{cases} \tag{1}$$

supplemented with boundary conditions of the form:

$$x(T) = \eta y'(\rho), \quad y(T) = \zeta x\prime(\mu), \quad x(0) = 0, \quad y(0) = 0, \rho, \mu \in [0,T] \tag{2}$$

Here, ${}^cD^k$ denotes Caputo fractional derivative of order k ($k = \alpha, \beta$); and $f, g \in C\big([0,T] \times \mathbb{R}^2, \mathbb{R}\big)$ are given continuous functions. Note that η, ζ are real constants such that $T^2 - \eta\zeta \neq 0$.

The rest of this paper is organized in the following manner: In Section 2, we briefly review some of the relevant definitions from fractional calculus and prove an auxiliary lemma that will be used later. Section 3 deals with proving the existence and uniqueness of solutions for the given problem, and Section 4 discusses the Hyers-Ulam stability of solutions and presents sufficient conditions for the stability. The paper concludes with supporting examples and obtained results.

2. Preliminaries

We begin this section by reviewing the definitions of fractional derivative and integral [1,2].

Definition 1. *The Riemann-Liouville fractional integral of order τ for a continuous function h is given by*

$$I^\tau h(s) = \frac{1}{\Gamma(\tau)} \int_0^s \frac{h(t)}{(s-t)^{1-\tau}} dt, \quad \tau > 0,$$

provided that the right-hand side is point-wise defined on $[0, \infty)$.

Definition 2. *The Caputo fractional derivatives of order* τ *for* $(h-1)$*—times absolutely continuous function* $g : [0, \infty) \to \mathbb{R}$ *is defined as*

$$^{c}D^{\tau} g(s) = \frac{1}{\Gamma(h-\tau)} \int_0^s (s-t)^{h-\tau-1} g^{(h)}(t)dt, \quad h-1 < \tau < h, \quad h = [\tau] + 1,$$

where $[\tau]$ *is the integer part of real number* τ.

Here we prove the following auxiliary lemma that will be used in the next section.

Lemma 1. *Let* $u, v \in C([0,T], \mathbb{R})$ *then the unique solution for the problem*

$$\begin{cases} ^{c}D^{\alpha} x(t) = u(t), & t \in [0,T], \quad 1 < \alpha \leq 2, \\ ^{c}D^{\beta} y(t) = v(t), & t \in [0,T], \quad 1 < \beta \leq 2, \\ x(T) = \eta y'(\rho), & y(T) = \zeta x'(\mu), \quad x(0) = 0, \quad y(0) = 0, \rho, \mu \in [0,T] \end{cases} \tag{3}$$

is

$$\begin{aligned} x(t) = \; & \tfrac{t}{\Delta}\Big(\eta T \int_0^{\rho} \tfrac{(\rho-s)^{\beta-2}}{\Gamma(\beta-1)} v(s)ds - T \int_0^{T} \tfrac{(T-s)^{\alpha-1}}{\Gamma(\alpha)} u(s)ds + \eta\zeta \int_0^{\mu} \tfrac{(\mu-s)^{\alpha-2}}{\Gamma(\alpha-1)} u(s)ds - \eta \int_0^{T} \tfrac{(T-s)^{\beta-1}}{\Gamma(\beta)} v(s)ds\Big) \\ & + \int_0^t \tfrac{(t-s)^{\alpha-1}}{\Gamma(\alpha)} u(s)ds, \end{aligned} \tag{4}$$

and

$$\begin{aligned} y(t) = \; & \tfrac{t}{\Delta}\Big(\eta\zeta \int_0^{\rho} \tfrac{(\rho-s)^{\beta-2}}{\Gamma(\beta-1)} v(s)ds - \zeta \int_0^{T} \tfrac{(T-s)^{\alpha-1}}{\Gamma(\alpha)} u(s)ds + T\zeta \int_0^{\mu} \tfrac{(\mu-s)^{\alpha-2}}{\Gamma(\alpha-1)} u(s) - T \int_0^{T} \tfrac{(T-s)^{\beta-1}}{\Gamma(\beta)} v(s)ds\Big) \\ & + \int_0^t \tfrac{(t-s)^{\beta-1}}{\Gamma(\beta)} v(s)ds \end{aligned} \tag{5}$$

where $\Delta = T^2 - \eta\zeta \neq 0$.

Proof. General solutions of the fractional differential equations in (3) are known [6] as

$$\begin{aligned} x(t) &= at + b + \tfrac{1}{\Gamma(\alpha)} \int_0^t (t-s)^{\alpha-1} u(s)ds, \\ y(t) &= ct + d + \tfrac{1}{\Gamma(\beta)} \int_0^t (t-s)^{\beta-1} v(s)ds, \end{aligned} \tag{6}$$

where $a, b, c,$ and d are arbitrary constants.

Apply conditions $x(0) = 0$ and $y(0) = 0$, and we obtain $b = d = 0$.

Here

$$x\prime(t) = a + \frac{1}{\Gamma(\alpha-1)} \int_0^t (t-s)^{\alpha-2} u(s)ds,$$

$$y'(t) = c + \frac{1}{\Gamma(\beta-1)} \int_0^t (t-s)^{\beta-2} v(s)ds.$$

Considering boundary conditions

$$x(T) = \eta y'(\rho), \quad y(T) = \zeta x\prime(\mu)$$

we get

$$aT + \int_0^T \frac{(T-s)^{\alpha-1}}{\Gamma(\alpha)} u(s)ds = \eta c + \eta \int_0^{\rho} \frac{(\rho-s)^{\beta-2}}{\Gamma(\beta-1)} v(s)ds,$$

and

$$cT + \int_0^T \frac{(T-s)^{\beta-1}}{\Gamma(\beta)} v(s)ds = a\zeta + \zeta \int_0^\mu \frac{(\mu-s)^{\alpha-2}}{\Gamma(\alpha-1)} u(s)ds,$$

so

$$a = \frac{1}{T}\left(\eta c + \eta \int_0^\rho \frac{(\rho-s)^{\beta-2}}{\Gamma(\beta-1)} v(s)ds - \int_0^T \frac{(T-s)^{\alpha-1}}{\Gamma(\alpha)} u(s)ds\right),$$

$$c = \frac{1}{T}\left(a\zeta + \zeta \int_0^\mu \frac{(\mu-s)^{\alpha-2}}{\Gamma(\alpha-1)} u(s)ds - \int_0^T \frac{(T-s)^{\beta-1}}{\Gamma(\beta)} v(s)ds\right).$$

Hence, by substituting the value of a into c, we obtain the final result for these constants as

$$c = \tfrac{1}{T}\Big(\tfrac{\zeta}{T}\Big[\eta c + \eta \int_0^\rho \tfrac{(\rho-s)^{\beta-2}}{\Gamma(\beta-1)} v(s)ds - \int_0^T \tfrac{(T-s)^{\alpha-1}}{\Gamma(\alpha)} u(s)ds\Big] + \zeta \int_0^\mu \tfrac{(\mu-s)^{\alpha-2}}{\Gamma(\alpha-1)} u(s)ds - \int_0^T \tfrac{(T-s)^{\beta-1}}{\Gamma(\beta)} v(s)ds\Big),$$

$$c - \tfrac{\zeta\eta c}{T^2} = \tfrac{1}{T}\Big(\tfrac{\zeta}{T}\Big[\eta \int_0^\rho \tfrac{(\rho-s)^{\beta-2}}{\Gamma(\beta-1)} v(s)ds - \int_0^T \tfrac{(T-s)^{\alpha-1}}{\Gamma(\alpha)} u(s)ds\Big] + \zeta \int_0^\mu \tfrac{(\mu-s)^{\alpha-2}}{\Gamma(\alpha-1)} u(s)ds - \int_0^T \tfrac{(T-s)^{\beta-1}}{\Gamma(\beta)} v(s)ds\Big),$$

$$c\Big(\tfrac{T^2-\zeta\eta}{T^2}\Big) = \tfrac{1}{T}\Big(\tfrac{\zeta}{T}\Big[\eta \int_0^\rho \tfrac{(\rho-s)^{\beta-2}}{\Gamma(\beta-1)} v(s)ds - \int_0^T \tfrac{(T-s)^{\alpha-1}}{\Gamma(\alpha)} u(s)ds\Big] + \zeta \int_0^\mu \tfrac{(\mu-s)^{\alpha-2}}{\Gamma(\alpha-1)} u(s)ds - \int_0^T \tfrac{(T-s)^{\beta-1}}{\Gamma(\beta)} v(s)ds\Big),$$

$$c = \tfrac{T}{T^2-\zeta\eta}\Big(\tfrac{\zeta}{T}\Big[\eta \int_0^\rho \tfrac{(\rho-s)^{\beta-2}}{\Gamma(\beta-1)} v(s)ds - \int_0^T \tfrac{(T-s)^{\alpha-1}}{\Gamma(\alpha)} u(s)ds\Big] + \zeta \int_0^\mu \tfrac{(\mu-s)^{\alpha-2}}{\Gamma(\alpha-1)} u(s)ds - \int_0^T \tfrac{(T-s)^{\beta-1}}{\Gamma(\beta)} v(s)ds\Big),$$

$$c = \tfrac{1}{T^2-\zeta\eta}\Big(\eta\zeta \int_0^\rho \tfrac{(\rho-s)^{\beta-2}}{\Gamma(\beta-1)} v(s)ds - \zeta \int_0^T \tfrac{(T-s)^{\alpha-1}}{\Gamma(\alpha)} u(s)ds + T\zeta \int_0^\mu \tfrac{(\mu-s)^{\alpha-2}}{\Gamma(\alpha-1)} u(s)ds - T \int_0^T \tfrac{(T-s)^{\beta-1}}{\Gamma(\beta)} v(s)ds\Big)$$

$$c = \tfrac{1}{\Delta}\Big(\eta\zeta \int_0^\rho \tfrac{(\rho-s)^{\beta-2}}{\Gamma(\beta-1)} v(s)ds - \zeta \int_0^T \tfrac{(T-s)^{\alpha-1}}{\Gamma(\alpha)} u(s)ds + T\zeta \int_0^\mu \tfrac{(\mu-s)^{\alpha-2}}{\Gamma(\alpha-1)} u(s)ds - T \int_0^T \tfrac{(T-s)^{\beta-1}}{\Gamma(\beta)} v(s)ds\Big),$$

and

$$a = \tfrac{1}{\Delta}\Big(\eta T \int_0^\rho \tfrac{(\rho-s)^{\beta-2}}{\Gamma(\beta-1)} v(s)ds - T \int_0^T \tfrac{(T-s)^{\alpha-1}}{\Gamma(\alpha)} u(s)ds + \eta\zeta \int_0^\mu \tfrac{(\mu-s)^{\alpha-2}}{\Gamma(\alpha-1)} u(s)ds - \eta \int_0^T \tfrac{(T-s)^{\beta-1}}{\Gamma(\beta)} v(s)ds\Big),$$

Substituting the values of $a, b, c,$ and d in (6) and (7) we get (4) and (5). The converse follows by direct computation. This completes the proof. □

3. Existence and Uniqueness of Solutions

Consider the space $C([0,T],\mathbb{R})$ endowed with norm $\|x\| = \sup_{0\le t\le T} |x(t)|$. Consequently, the product space $C([0,T],\mathbb{R}) \times C([0,T],\mathbb{R})$ is a Banach Space (endowed with $\|(x,y)\| = \|x\| + \|y\|$).

In view of Lemma 1, we define the operator $G : C([0,T],\mathbb{R}) \times C([0,T],\mathbb{R}) \to C([0,T],\mathbb{R}) \times C([0,T],\mathbb{R})$ as:

$$G(x,y)(t) = (G_1(x,y)(t), G_2(x,y)(t)),$$

where

$$\begin{aligned} G_1(x,y)(t) = \tfrac{t}{\Delta} \ &\Big(\eta T \int_0^\rho \tfrac{(\rho-s)^{\beta-2}}{\Gamma(\beta-1)} g(s,x(s),y(s))ds - T \int_0^T \tfrac{(T-s)^{\alpha-1}}{\Gamma(\alpha)} f(s,x(s),y(s))ds \\ &+\eta\zeta \int_0^\mu \tfrac{(\mu-s)^{\alpha-2}}{\Gamma(\alpha-1)} f(s,x(s),y(s))ds - \eta \int_0^T \tfrac{(T-s)^{\beta-1}}{\Gamma(\beta)} g(s,x(s),y(s))ds\Big) \\ &+ \int_0^t \tfrac{(t-s)^{\alpha-1}}{\Gamma(\alpha)} f(s,x(s),y(s))ds, \end{aligned} \tag{7}$$

and

$$
\begin{aligned}
G_2(x,y)(t) = \tfrac{t}{\Delta} & \Big(\eta\zeta \int_0^{\rho} \tfrac{(\rho-s)^{\beta-2}}{\Gamma(\beta-1)} g(s,x(s),y(s))ds - \zeta \int_0^{T} \tfrac{(T-s)^{\alpha-1}}{\Gamma(\alpha)} f(s,x(s),y(s))ds \\
& + T\zeta \int_0^{\mu} \tfrac{(\mu-s)^{\alpha-2}}{\Gamma(\alpha-1)} f(s,x(s),y(s))ds - T\int_0^{T} \tfrac{(T-s)^{\beta-1}}{\Gamma(\beta)} g(s,x(s),y(s))ds\Big) \\
& + \int_0^{t} \tfrac{(t-s)^{\beta-1}}{\Gamma(\beta)} f(s,x(s),y(s))ds,
\end{aligned} \tag{8}
$$

Here we establish the existence of the solutions for the boundary value problem (1) and (2) by using Banach's contraction mapping principle.

Theorem 1. *Assume* $f,g : C([0,T] \times \mathbb{R}^2 \to \mathbb{R}$ *are jointly continuous functions and there exist constants* $\phi, \psi \in \mathbb{R}$, *such that* $\forall\, x_1, x_2, y_1, y_2 \in \mathbb{R}, \forall t \in [0,T]$, *we have*

$$
\big|f(t,x_1,x_2) - f(t,y_1,y_2)\big| \le \phi\big(|x_2 - x_1| + |y_2 - y_1|\big),
$$

$$
\big|g(t,x_1,x_2) - f(t,y_1,y_2)\big| \le \psi\big(|x_2 - x_1| + |y_2 - y_1|\big),
$$

where

$$
\phi(Q_1 + Q_3) + \psi(Q_2 + Q_4) < 1,
$$

then the BVP (1) and (2) has a unique solution on $[0,T]$. *Here*

$$
\begin{aligned}
Q_1 &= \tfrac{T}{|\Delta|}\Big(\tfrac{T^{\alpha+1}}{\Gamma(\alpha+1)} + \tfrac{|\eta\zeta|\mu^{\alpha-1}}{\Gamma(\alpha)}\Big) + \tfrac{T^{\alpha}}{\Gamma(\alpha+1)}, \\
Q_2 &= \tfrac{T}{|\Delta|}\Big(\tfrac{|\eta|T\rho^{\beta-1}}{\Gamma(\beta)} + \tfrac{|\eta|T^{\beta}}{\Gamma(\beta+1)}\Big), \\
Q_3 &= \tfrac{T}{|\Delta|}\Big(\tfrac{|\zeta|T^{\alpha}}{\Gamma(\alpha+1)} + \tfrac{T|\zeta|\mu^{\alpha-1}}{\Gamma(\alpha)}\Big), \\
Q_4 &= \tfrac{T}{|\Delta|}\Big(\tfrac{|\eta\zeta|\rho^{\beta-1}}{\Gamma(\beta)} + \tfrac{T^{\beta+1}}{\Gamma(\beta+1)}\Big) + \tfrac{T^{\beta}}{\Gamma(\beta+1)}.
\end{aligned} \tag{9}
$$

Proof. Define $\underset{0 \le t \le T}{sup} |f(t,0,0)| = f_0 < \infty$, $\underset{0 \le t \le T}{sup} |g(t,0,0)| = g_0 < \infty$ and $\Omega_\varepsilon = \{(x,y) \in C([0,T],\mathbb{R}) \times C([0,T],\mathbb{R}) : \|(x,y)\| \le \varepsilon\}$, and $\varepsilon > 0$, such that

$$
\varepsilon \ge \frac{(Q_1 + Q_3)f_0 + (Q_2 + Q_4)g_0}{1 - [\phi(Q_1 + Q_3) + \psi(Q_2 + Q_4)]}.
$$

Firstly, we show that $G\Omega_\varepsilon \subseteq \Omega_\varepsilon$.
By our assumption, for $(x,y) \in \Omega_\varepsilon, t \in [0,T]$, we have

$$
\begin{aligned}
|f(t,x(t),y(t))| &\le |f(t,x(t),y(t)) - f(t,0,0)| + |f(t,0,0)|, \\
&\le \phi\big(|x(t)| + |y(t)|\big) + f_0 \le \phi(\|x\| + \|y\|) + f_0, \\
&\le \phi\varepsilon + f_0,
\end{aligned}
$$

and

$$
\begin{aligned}
|g(t,x(t),y(t))| &\le \psi\big(|x(t)| + |y(t)|\big) + g_0 \le \psi(\|x\| + \|y\|) + g_0, \\
&\le \psi\varepsilon + g_0,
\end{aligned}
$$

which lead to

$$\begin{aligned}
\left|G_1(x,y)(t)\right| \leq\; & \tfrac{T}{|\Delta|}\Big(|\eta| T \int_0^{\rho} \tfrac{(\rho-s)^{\beta-2}}{\Gamma(\beta-1)} ds(\psi(\|x\|+\|y\|)+g_0) \\
& +T \int_0^{T} \tfrac{(T-s)^{\alpha-1}}{\Gamma(\alpha)} ds(\phi(\|x\|+\|y\|)+f_0) \\
& +|\eta\zeta| \int_0^{\mu} \tfrac{(\mu-s)^{\alpha-2}}{\Gamma(\alpha-1)} ds(\phi(\|x\|+\|y\|)+f_0) \\
& +|\eta| \int_0^{T} \tfrac{(T-s)^{\beta-1}}{\Gamma(\beta)} ds(\psi(\|x\|+\|y\|)+g_0)\Big) \\
& + \sup_{0 \leq t \leq T} \int_0^{t} \tfrac{(t-s)^{\alpha-1}}{\Gamma(\alpha)} ds(\phi(\|x\|+\|y\|)+f_0) \\
\leq\; & (\phi(\|x\|+\|y\|)+f_0)\left[\tfrac{T}{|\Delta|}\left(\tfrac{T^{\alpha+1}}{\Gamma(\alpha+1)}+\tfrac{|\eta\zeta|\mu^{\alpha-1}}{\Gamma(\alpha)}\right)+\tfrac{T^{\alpha}}{\Gamma(\alpha+1)}\right] \\
& +(\psi(\|x\|+\|y\|)+g_0)\left[\tfrac{T}{|\Delta|}\left(\tfrac{|\eta| T\rho^{\beta-1}}{\Gamma(\beta)}+\tfrac{|\eta| T^{\beta}}{\Gamma(\beta+1)}\right)\right] \\
\leq\; & (\phi(\|x\|+\|y\|)+f_0)Q_1+(\psi(\|x\|+\|y\|)+g_0)Q_2 \\
\leq\; & (\phi\varepsilon+f_0)Q_1+(\psi\varepsilon+g_0)Q_2.
\end{aligned}$$

In a similar manner:

$$\left|G_2(x,y)(t)\right| \leq (\phi(\|x\|+\|y\|)+f_0)Q_3+(\psi(\|x\|+\|y\|)+g_0)Q_4 \leq (\phi\varepsilon+f_0)Q_3+(\psi\varepsilon+g_0)Q_4.$$

Hence,

$$\|G_1(x,y)\| \leq (\phi\varepsilon+f_0)Q_1+(\psi\varepsilon+g_0)Q_2,$$

and

$$\|G_2(x,y)\| \leq (\phi\varepsilon+f_0)Q_3+(\psi\varepsilon+g_0)Q_4.$$

Consequently,

$$\|G(x,y)\| \leq (\phi\varepsilon+f_0)(Q_1+Q_3)+(\psi\varepsilon+g_0)(Q_2+Q_4) \leq \varepsilon\,.$$

and we get $\|G(x,y)\| \leq \varepsilon$ that is $G\Omega_{\varepsilon} \subseteq \Omega_{\varepsilon}$.

Now let $(x_1,y_1),(x_2,y_2) \in C([0,T],\mathbb{R}) \times C([0,T],\mathbb{R}), \forall t \in [0,T]$.

Then we have

$$\begin{aligned}
\left|G_1(x_1,y_1)(t)-\right.\; & \left.G_1(x_2,y_2)(t)\right| \\
& \leq \tfrac{T}{|\Delta|}\Big(|\eta| T \int_0^{\rho} \tfrac{(\rho-s)^{\beta-2}}{\Gamma(\beta-1)} ds\psi(\|x_2-x_1\|+\|y_2-y_1\|) \\
& +T \int_0^{T} \tfrac{(T-s)^{\alpha-1}}{\Gamma(\alpha)} ds\phi(\|x_2-x_1\|+\|y_2-y_1\|) \\
& +|\eta\zeta| \int_0^{\mu} \tfrac{(\mu-s)^{\alpha-2}}{\Gamma(\alpha-1)} ds\phi(\|x_2-x_1\|+\|y_2-y_1\|) \\
& +|\eta| \int_0^{T} \tfrac{(T-s)^{\beta-1}}{\Gamma(\beta)} ds\psi(\|x_2-x_1\|+\|y_2-y_1\|)\Big) \\
& + \sup_{0 \leq t \leq T} \int_0^{t} \tfrac{(t-s)^{\alpha-1}}{\Gamma(\alpha)} ds\phi(\|x_2-x_1\|+\|y_2-y_1\|),
\end{aligned}$$

$$\|G_1(x_1,y_1)-G_1(x_2,y_2)\| \leq Q_1\phi(\|x_2-x_1\|+\|y_2-y_1\|)+Q_2\psi(\|x_2-x_1\|+\|y_2-y_1\|). \tag{10}$$

and likewise

$$\|G_2(x_1,y_1)-G_2(x_2,y_2)\| \leq Q_3\phi(\|x_2-x_1\|+\|y_2-y_1\|)+Q_4\psi(\|x_2-x_1\|+\|y_2-y_1\|). \tag{11}$$

From (11) and (12) we have

$$\|G(x_1,y_1)-G(x_2,y_2)\| \leq (\phi(Q_1+Q_3)+\psi(Q_2+Q_4))(\|x_2-x_1\|+\|y_2-y_1\|).$$

Since $\phi(Q_1+Q_3)+\psi(Q_2+Q_4)<1$, therefore, the operator G is a contraction operator. Hence, by Banach's fixed-point theorem, the operator G has a unique fixed point, which is the unique solution of the BVP (1) and (2). This completes the proof. □

Next we will prove the existence of solutions by applying the Leray-Schauder alternative.

Lemma 2. *"(Leray-Schauder alternative [7], p. 4) Let $F : E \to E$ be a completely continuous operator (i.e., a map restricted to any bounded set in E is compact). Let $E(F)=\{x\in E : x=\lambda F(x)\ for\ some\ 0<\lambda<1\}$. Then either the set $E(F)$ is unbounded or F has at least one fixed point)".*

Theorem 2. *Assume $f,g : C([0,T]\times\mathbb{R}^2\to\mathbb{R}$ are continuous functions and there exist $\theta_1,\theta_2,\lambda_1,\lambda_2\geq 0$ where $\theta_1,\theta_2,\lambda_1,\lambda_2$ are real constants and $\theta_0,\lambda_0>0$ such that $\forall x_i,y_i\in\mathbb{R}$, $(i=1,2)$, we have*

$$\left|f(t,x_1,x_2)\right|\leq\theta_0+\theta_1|x_1|+\theta_2|x_2|,$$

$$\left|g(t,x_1,x_2)\right|\leq\lambda_0+\lambda_1|x_1|+\lambda_2|x_2|,$$

If

$$(Q_1+Q_3)\theta_1+(Q_2+Q_4)\lambda_1<1,$$

and

$$(Q_1+Q_3)\theta_2+(Q_2+Q_4)\lambda_2<1,$$

where $Q_i, i=1,2,3,4$ are defined in (10), then the problem (1) and (2) has at least one solution.

Proof. This proof will be presented in two steps.

Step 1: We will show that $G : C([0,T],\mathbb{R})\times C([0,T],\mathbb{R})\to C([0,T],\mathbb{R})\times C([0,T],\mathbb{R})$ is completely continuous. The continuity of the operator G holds by the continuity of the functions f,g.

Let $B\subseteq C([0,T],\mathbb{R})\times C([0,T],\mathbb{R})$ be bounded. Then there exists positive constants k_1,k_2 such that

$$\left|f(t,x(t),y(t))\right|\leq k_1,\quad \left|g(t,x(t),y(t))\right|\leq k_2,\quad \forall t\in[0,T].$$

Then $\forall(x,y)\in B$, and we have

$$\left|G_1(x,y)(t)\right|\leq Q_1k_1+Q_2k_2,$$

which implies

$$||G_1(x,y)||\leq Q_1k_1+Q_2k_2,$$

and similarly

$$||G_2(x,y)||\leq Q_3k_1+Q_4k_2.$$

Thus, from the above inequalities, it follows that the operator G is uniformly bounded, since

$$||G(x,y)||\leq(Q_1+Q_3)k_1+(Q_2+Q_4)k_2.$$

Next, we will show that operator G is equicontinuous. Let $\omega_1, \omega_2 \in [0,T]$ with $\omega_1 < \omega_2$. This yields

$$\begin{aligned}
\left|G_1(x,y)(\omega_2)-\right. & \left. G_1(x,y)(\omega_1)\right| \\
& \le \frac{\omega_2-\omega_1}{|\Delta|}\Big(|\eta|T\int_0^{\rho}\frac{(\rho-s)^{\beta-2}}{\Gamma(\beta-1)}\big|g(s,x(s),y(s))\big|ds \\
& +T\int_0^{T}\frac{(T-s)^{\alpha-1}}{\Gamma(\alpha)}\big|f(s,x(s),y(s))\big|ds+|\eta\zeta|\int_0^{\mu}\frac{(\mu-s)^{\alpha-2}}{\Gamma(\alpha-1)}\big|f(s,x(s),y(s))\big|ds \\
& +|\eta|\int_0^{T}\frac{(T-s)^{\beta-1}}{\Gamma(\beta)}\big|g(s,x(s),y(s))\big|ds\Big) \\
& +\Big|\int_0^{\omega_2}\frac{(\omega_2-s)^{\alpha-1}}{\Gamma(\alpha)}f(s,x(s),y(s))ds \\
& -\int_0^{\omega_1}\frac{(\omega_1-s)^{\alpha-1}}{\Gamma(\alpha)}f(s,x(s),y(s))ds\Big| \\
& \le \frac{\omega_2-\omega_1}{|\Delta|}\Big(|\eta|Tk_2\int_0^{\rho}\frac{(\rho-s)^{\beta-2}}{\Gamma(\beta-1)}ds+Tk_1\int_0^{T}\frac{(T-s)^{\alpha-1}}{\Gamma(\alpha)}ds+|\eta\zeta|k_1\int_0^{\mu}\frac{(\mu-s)^{\alpha-2}}{\Gamma(\alpha-1)}ds \\
& +|\eta|k_2\int_0^{T}\frac{(T-s)^{\beta-1}}{\Gamma(\beta)}ds\Big)+\Big|\int_0^{\omega_1}\Big(\frac{(\omega_2-s)^{\alpha-1}}{\Gamma(\alpha)}-\frac{(\omega_1-s)^{\alpha-1}}{\Gamma(\alpha)}\Big)f(s,x(s),y(s))ds\Big| \\
& +\Big|\int_{\omega_1}^{\omega_2}\frac{(\omega_2-s)^{\alpha-1}}{\Gamma(\alpha)}f(s,x(s),y(s))ds\Big|, \\
& \le \frac{\omega_2-\omega_1}{|\Delta|}\Big(\frac{k_2|\eta|T\rho^{\beta-1}}{\Gamma(\beta)}+\frac{k_1T^{\alpha+1}}{\Gamma(\alpha+1)}+\frac{k_1|\eta\zeta|\mu^{\alpha-1}}{\Gamma(\alpha)}+\frac{k_2|\eta|T^{\beta}}{\Gamma(\beta+1)}\Big) \\
& +\frac{k_1}{\Gamma(\alpha)}\Big(\int_0^{\omega_1}\big((\omega_2-s)^{\alpha-1}-(\omega_1-s)^{\alpha-1}\big)ds+\int_{\omega_1}^{\omega_2}(\omega_2-s)^{\alpha-1}ds\Big).
\end{aligned}$$

And we obtain

$$\begin{aligned}
\left|G_1(x,y)(\omega_2)-G_1(x,y)(\omega_1)\right| & \le \frac{\omega_2-\omega_1}{|\Delta|}\Big(\frac{k_2|\eta|T\rho^{\beta-1}}{\Gamma(\beta)}+\frac{k_1T^{\alpha+1}}{\Gamma(\alpha+1)}+\frac{k_1|\eta\zeta|\mu^{\alpha-1}}{\Gamma(\alpha)}+\frac{k_2|\eta|T^{\beta}}{\Gamma(\beta+1)}\Big) \\
& +\frac{k_1}{\Gamma(\alpha+1)}[\omega_2{}^{\alpha}-\omega_1{}^{\alpha}].
\end{aligned}$$

Hence, we have $||G_1(x,y)(\omega_2)-G_1(x,y)(\omega_1)|| \to 0$ independent of x and y as $\omega_2 \to \omega_1$. Furthermore, we obtain

$$\begin{aligned}
\left|G_2(x,y)(\omega_2)-G_2(x,y)(\omega_1)\right| & \le \frac{\omega_2-\omega_1}{|\Delta|}\Big(\frac{k_2|\eta\zeta|\rho^{\beta-1}}{\Gamma(\beta)}+\frac{k_1|\zeta|T^{\alpha}}{\Gamma(\alpha+1)}+\frac{k_1T|\zeta|\mu^{\alpha-1}}{\Gamma(\alpha)}+\frac{k_2T^{\beta+1}}{\Gamma(\beta+1)}\Big) \\
& +\frac{k_2}{\Gamma(\beta+1)}\big[\omega_2{}^{\beta}-\omega_1{}^{\beta}\big],
\end{aligned}$$

which implies that $||G_2(x,y)(\omega_2)-G_2(x,y)(\omega_1)|| \to 0$ independent of x and y as $\omega_2 \to \omega_1$.

Therefore, operator $G(x,y)$ is equicontinuous, and thus $G(x,y)$ is completely continuous.

Step 2: (Boundedness of operator)

Finally, we will show that $Z = \{(x,y) \in C([0,T],\mathbb{R}) \times C([0,T],\mathbb{R}) : (x,y) = hG(x,y), h \in [0,1]\}$ is bounded. Let $(x,y) \in \mathbb{R}$, with $(x,y) = hG(x,y)$ for any $t \in [0,T]$, we have

$$x(t) = hG_1(x,y)(t), \ \ y(t) = hG_2(x,y)(t).$$

Then

$$|x(t)| \le Q_1\big(\theta_0+\theta_1|x(t)|+\theta_2|y(t)|\big)+Q_2\big(\lambda_0+\lambda_1|x(t)|+\lambda_2|y(t)|\big),$$

and

$$|y(t)| \le Q_3\big(\theta_0+\theta_1|x(t)|+\theta_2|y(t)|\big)+Q_4\big(\lambda_0+\lambda_1|x(t)|+\lambda_2|y(t)|\big).$$

Hence,

$$||x|| \le Q_1(\theta_0+\theta_1||x||+\theta_2||y||)+Q_2(\lambda_0+\lambda_1||x||+\lambda_2||y||),$$

and

$$||y|| \le Q_3(\theta_0+\theta_1||x||+\theta_2||y||)+Q_4(\lambda_0+\lambda_1||x||+\lambda_2||y||),$$

which implies

$$\begin{aligned}\|x\|+\|y\| &\leq (Q_1+Q_3)\theta_0+(Q_2+Q_4)\lambda_0+((Q_1+Q_3)\theta_1+(Q_2+Q_4)\lambda_1)\|x\| \\ &+((Q_1+Q_3)\theta_2+(Q_2+Q_4)\lambda_2)\|y\|.\end{aligned}$$

Therefore,

$$\|(x,y)\| \leq \frac{(Q_1+Q_3)\theta_0+(Q_2+Q_4)\lambda_0}{Q_0},$$

where $Q_0 = min\{1-(Q_1+Q_3)\theta_1-(Q_2+Q_4)\lambda_1, 1-(Q_1+Q_3)\theta_2-(Q_2+Q_4)\lambda_2\}$. This proves that Z is bounded and hence by Leray-Schauder alternative theorem, operator G has at least one fixed point. Therefore, the BVP (1) and (2) has at least one solution on $[0,T]$. This completes the proof. □

4. Hyers-Ulam Stability

In this section, we will discuss the Hyers-Ulam stability of the solutions for the BVP (1) and (2) by means of integral representation of its solution given by

$$x(t) = G_1(x,y)(t), y(t) = G_2(x,y)(t),$$

where G_1 and G_2 are defined by (8) and (9).

Define the following nonlinear operators $N_1, N_2 \in C([0,T],\mathbb{R}) \times C([0,T],\mathbb{R}) \to C([0,T],\mathbb{R})$;

$$\begin{aligned}{}^cD^\alpha x(t) - f(t,x(t),y(t)) &= N_1(x,y)(t), \quad t \in [0,T], \\ {}^cD^\beta y(t) - g(t,x(t),y(t)) &= N_2(x,y)(t), \quad t \in [0,T].\end{aligned}$$

For some $\varepsilon_1, \varepsilon_2 > 0$, we consider the following inequality:

$$N_1(x,y) \leq \varepsilon_1, \quad N_2(x,y) \leq \varepsilon_2. \tag{12}$$

Definition 3. **([8,9]).** *The coupled system (1) and (2) is said to be Hyers-Ulam stable, if there exist $M_1, M_2 > 0$, such that for every solution $(x^*,y^*) \in C([0,T],\mathbb{R}) \times C([0,T],\mathbb{R})$ of the inequality (13), there exists a unique solution $(x,y) \in C([0,T],\mathbb{R}) \times C([0,T],\mathbb{R})$ of problems (1) and (2) with*

$$\|(x,y)-(x^*,y^*)\| \leq M_1\varepsilon_1 + M_2\varepsilon_2.$$

Theorem 3. *Let the assumptions of Theorem 1 hold. Then the BVP (1) and (2) is Hyers-Ulam-stable.*

Proof. Let $(x,y) \in C([0,T],\mathbb{R}) \times C([0,T],\mathbb{R})$ be the solution of the problems (1) and (2) satisfying (8) and (9). Let (x^*,y^*) be any solution satisfying (13):

$$\begin{aligned}{}^cD^\alpha x^*(t) &= f(t,x^*(t),y^*(t)) + N_1(x^*,y^*)(t), \quad t \in [0,T], \\ {}^cD^\beta y^*(t) &= g(t,x^*(t),y^*(t)) + N_2(x^*,y^*)(t), \quad t \in [0,T].\end{aligned}$$

So

$$\begin{aligned}x^*(t) = \; & G_1(x^*,y^*)(t) \\ & + \frac{t}{\Delta}\Big(\eta T \int_0^\rho \frac{(\rho-s)^{\beta-2}}{\Gamma(\beta-1)} N_2(x^*,y^*)(s)ds - T\int_0^T \frac{(T-s)^{\alpha-1}}{\Gamma(\alpha)} N_1(x^*,y^*)(s)ds \\ & + \eta\zeta \int_0^\mu \frac{(\mu-s)^{\alpha-2}}{\Gamma(\alpha-1)} N_1(x^*,y^*)(s)ds - \eta \int_0^T \frac{(T-s)^{\beta-1}}{\Gamma(\beta)} N_2(x^*,y^*)(s)ds\Big) \\ & + \int_0^t \frac{(t-s)^{\alpha-1}}{\Gamma(\alpha)} N_1(x^*,y^*)(s)ds,\end{aligned}$$

It follows that

$$\begin{aligned}|G_1(x^*,y^*)(t)- \quad & x^*(t)| \\ & \leq \frac{T}{|\Delta|}\Big(|\eta|T\int_0^{\rho}\frac{(\rho-s)^{\beta-2}}{\Gamma(\beta-1)}ds\varepsilon_2 + T\int_0^{T}\frac{(T-s)^{\alpha-1}}{\Gamma(\alpha)}ds\varepsilon_1 + |\eta\zeta|\int_0^{\mu}\frac{(\mu-s)^{\alpha-2}}{\Gamma(\alpha-1)}ds\varepsilon_1 \\ & +|\eta|\int_0^{T}\frac{(T-s)^{\beta-1}}{\Gamma(\beta)}ds\varepsilon_2\Big) + \int_0^{T}\frac{(T-s)^{\alpha-1}}{\Gamma(\alpha)}ds\varepsilon_1, \\ & \leq \left[\frac{T}{|\Delta|}\Big(\frac{T^{\alpha+1}}{\Gamma(\alpha+1)} + \frac{|\eta\zeta|\mu^{\alpha-1}}{\Gamma(\alpha)}\Big) + \frac{T^{\alpha}}{\Gamma(\alpha+1)}\right]\varepsilon_1 + \frac{T}{|\Delta|}\Big(\frac{|\eta|T\rho^{\beta-1}}{\Gamma(\beta)} + \frac{|\eta|T^{\beta}}{\Gamma(\beta+1)}\Big)\varepsilon_2, \\ & \leq Q_1\varepsilon_1 + Q_2\varepsilon_2.\end{aligned}$$

Similarly,

$$\begin{aligned}|G_1(x^*,y^*)(t)-x^*(t)| & \leq \frac{T}{|\Delta|}\Big(\frac{|\zeta|T^{\alpha}}{\Gamma(\alpha+1)} + \frac{T|\zeta|\mu^{\alpha-1}}{\Gamma(\alpha)}\Big)\varepsilon_1 + \left[\frac{T}{|\Delta|}\Big(\frac{|\eta\zeta|\rho^{\beta-1}}{\Gamma(\beta)} + \frac{T^{\beta+1}}{\Gamma(\beta+1)}\Big) + \frac{T^{\beta}}{\Gamma(\beta+1)}\right], \\ & \leq Q_3\varepsilon_1 + Q_4\varepsilon_2,\end{aligned}$$

where $Q_i, i = 1,\ 2,\ 3,\ 4$ are defined in (10).

Therefore, we deduce by the fixed-point property of operator G, that is given by (8) and (9), which

$$\begin{aligned}|x(t)-x^*(t)| & = |x(t)-G_1(x^*,y^*)(t)+G_1(x^*,y^*)(t)-x^*(t)| \\ & \leq |G_1(x,y)(t)-G_1(x^*,y^*)(t)| + |G_1(x^*,y^*)(t)-x^*(t)| \\ & \leq (Q_1\phi+Q_2\psi)(x,y)-(x^*,y^*)+Q_1\varepsilon_1+Q_2\varepsilon_2,\end{aligned} \tag{13}$$

and similarly

$$\begin{aligned}|y(t)-y^*(t)| & = |y(t)-G_2(x^*,y^*)(t)+G_2(x^*,y^*)(t)-y^*(t)| \\ & \leq |G_2(x,y)(t)-G_2(x^*,y^*)(t)| + |G_2(x^*,y^*)(t)-y^*(t)| \\ & \leq (Q_3\phi+Q_4\psi)(x,y)-(x^*,y^*)+Q_3\varepsilon_1+Q_4\varepsilon_2,\end{aligned} \tag{14}$$

From (14) and (15) it follows that

$$\|(x,y)-(x^*,y^*)\| \leq (Q_1\phi+Q_2\psi+Q_3\phi+Q_4\psi)\|(x,y)-(x^*,y^*)\| + (Q_1+Q_3)\varepsilon_1 + (Q_2+Q_4)\varepsilon_2,$$

$$\begin{aligned}\|(x,y)-(x^*,y^*)\| & \leq \frac{(Q_1+Q_3)\varepsilon_1+(Q_2+Q_4)\varepsilon_2}{1-((Q_1+Q_3)\phi+(Q_2+Q_4)\psi)}, \\ & \leq M_1\varepsilon_1+M_2\varepsilon_2.\end{aligned}$$

with

$$M_1 = \frac{(Q_1+Q_3)}{1-((Q_1+Q_3)\phi+(Q_2+Q_4)\psi)},$$

$$M_2 = \frac{(Q_2+Q_4)}{1-((Q_1+Q_3)\phi+(Q_2+Q_4)\psi)}.$$

Thus, sufficient conditions for the Hyers-Ulam stability of the solutions are obtained. □

5. Examples

Example 1. *Consider the following coupled system of fractional differential equations*

$$\begin{cases} {}^cD^{\frac{3}{2}}x(t) = \frac{1}{6\pi\sqrt{81+t^2}}\Big(\frac{|x(t)|}{3+|x(t)|} + \frac{|y(t)|}{5+|x(t)|}\Big), \\ {}^cD^{\frac{7}{4}}y(t) = \frac{1}{12\pi\sqrt{64+t^2}}(sin(x(t))+sin(y(t))), \\ x(1) = 2y'(1),\ \ y(1) = -x'(1/2),\ \ x(0) = 0,\ \ y(0) = 0, \end{cases} \tag{15}$$

$$\alpha = \frac{3}{2}, \beta = \frac{7}{4}, T = 1, \eta = 2, \zeta = -1, \mu = \frac{1}{2}, \rho = 1.$$

Using the given data, we find that $\Delta = 3, Q_1 = 1.269, Q_2 = 1.1398, Q_3 = 0.5167, Q_4 = 1.554, \phi = \frac{1}{54\pi}, \psi = \frac{1}{48\pi}$.

It is clear that

$$f(t,x(t),y(t)) = \frac{1}{6\pi\sqrt{81+t^2}}\left(\frac{|x(t)|}{3+|x(t)|} + \frac{|y(t)|}{5+|x(t)|}\right),$$

and

$$g(t,x(t),y(t)) = \frac{1}{12\pi\sqrt{64+t^2}}(sin(x(t)) + sin(y(t))),$$

are jointly continuous functions and Lipschitz function with $\phi = \frac{1}{54\pi}, \psi = \frac{1}{48\pi}$. *Moreover,*

$$\frac{1}{54\pi}(1.269 + 0.5167) + \frac{1}{48\pi}(1.1398 + 1.554) = 0.0283 < 1.$$

Thus, all the conditions of Theorem 1 are satisfied, then problem (16) has a unique solution on $[0,1]$, *which is Hyers-Ulam-stable.*

Example 2. *Consider the following system of fractional differential equation*

$$\begin{cases} {}^cD^{5/3}x(t) = \frac{1}{80+t^4} + \frac{|x(t)|}{120(1+y^2(t))} + \frac{1}{4\sqrt{2500+t^2}}e^{-3t}\cos(y(t)),\ t \in [0,1] \\ {}^cD^{6/5}y(t) = \frac{1}{\sqrt{16+t^2}}\cos t + \frac{1}{150}e^{-3t}sin(y(t)) + \frac{1}{180}x(t),\ t \in [0,1] \\ x(1) = -3y'(1/3),\quad y(1) = x'(1),\quad x(0) = 0,\quad y(0) = 0, \end{cases} \tag{16}$$

$$\alpha = \frac{5}{3}, \beta = \frac{6}{5}, T = 1, \eta = -3, \zeta = 1, \mu = 1, \rho = 1/3.$$

Using the given data, we find that $\Delta = 3, Q_1 = 1.269, Q_2 = 1.1398, Q_3 = 0.5167, Q_4 = 1.554, \phi = \frac{1}{54\pi}, \psi = \frac{1}{48\pi}$.

It is clear that

$$|f(t,x,y)| \leq \frac{1}{80} + \frac{1}{120}|x| + \frac{1}{200}|y|,$$

$$|g(t,x,y)| \leq \frac{1}{4} + \frac{1}{180}|x| + \frac{1}{150}|y|.$$

Thus, $\theta_0 = \frac{1}{80}, \theta_1 = \frac{1}{120}, \theta_2 = \frac{1}{200}, \lambda_0 = \frac{1}{4}, \lambda_1 = \frac{1}{180}, \lambda_2 = \frac{1}{150}$.

Note that $(Q_1+Q_3)\theta_1 + (Q_2+Q_4)\lambda_1 = 0.0298 < 1$ *and* $(Q_1+Q_3)\theta_2 + (Q_2+Q_4)\lambda_2 = 0.0269 < 1$, *and hence by Theorem 2, problem (17) has at least one solution on* $[0,1]$.

6. Conclusions

In this paper, the existence, uniqueness and the Hyers-Ulam stability of solutions for a coupled system of nonlinear fractional differential equations with boundary conditions were established and discussed.

Future studies may focus on different concepts of stability and existence results to a neutral time-delay system/inclusion, time-delay system/inclusion with finite delay.

Author Contributions: The authors have made the same contribution. All authors read and approved the final manuscript.

Acknowledgments: The authors wish to thank the anonymous reviewers for their valuable comments and suggestions.

References

1. Podlubny, I. *Fractional Differential Equations*; Academic Press: San Diego, CA, USA, 1999.
2. Kilbas, A.A.; Srivastava, H.M.; Trujillo, J.J. *Theory and Applications of Fractional Differential Equations. North-Holland Mathematics Studies*; Elsevier: Amsterdam, The Netherlands, 2006; Volume 204.
3. Chalishajar, D.; Raja, D.S.; Karthikeyan, K.; Sundararajan, P. Existence results for nonautonomous impulsive fractional evolution equations. *Res. Nonlinear Anal.* **2018**, *1*, 133–147.
4. Chalishajar, D.; Kumar, A. Existence, uniqueness and Ulam's stability of solutions for a coupled system of fractional differential equations with integral boundary conditions. *Mathematics* **2018**, *6*, 96. [CrossRef]
5. Ntouyas, S.K.; Obaid, M. A coupled system of fractional differential equations with nonlocal integral boundary conditions. *Adv. Differ. Equ.* **2012**, *2012*, 130–139. [CrossRef]
6. Ahmad, B.; Ntouyas, S.K. A Coupled system of nonlocal fractional differential equations with coupled and uncoupled slit-strips-type integral boundary conditions. *J. Math. Sci.* **2017**, *226*, 175–196. [CrossRef]
7. Alsulami, H.H.; Ntouyas, S.K.; Agarwal, R.P.; Ahmad, B.; Alsaedi, A. A study of fractional-order coupled systems with a new concept of coupled non-separated boundary conditions. *Bound. Value Probl.* **2017**, *2017*, 68–74. [CrossRef]
8. Zhang, Y.; Bai, Z.; Feng, T. Existence results for a coupled system of nonlinear fractional three-point boundary value problems at resonance. *Comput. Math. Appl.* **2011**, *61*, 1032–1047. [CrossRef]
9. Granas, A.; Dugundji, J. *Fixed Point Theory*; Springer: New York, NY, USA, 2005.
10. Hyers, D.H. On the stability of the linear functional equation. *Proc. Nat. Acad. Sci. USA* **1941**, *27*, 222. [CrossRef] [PubMed]
11. Rus, I.A. Ulam stabilities of ordinary differential equations in a Banach space. *Carpathian J. Math.* **2010**, 103–107.
12. Cabada, A.; Wang, G. Positive solutions of nonlinear fractional differential equations with integral boundary value conditions. *J. Math. Anal. Appl.* **2012**, *389*, 403–411. [CrossRef]
13. Graef, J.R.; Kong, L.; Wang, M. Existence and uniqueness of solutions for a fractional boundary value problem on a graph. *Fract. Calc. Appl. Anal.* **2014**, *17*, 499–510. [CrossRef]
14. Ahmad, B.; Nieto, J.J. Existence results for a coupled system of nonlinear fractional differential equations with three-point boundary conditions. *Comput. Math. Appl.* **2009**, *58*, 1838–1843. [CrossRef]
15. Su, X. Boundary value problem for a coupled system of nonlinear fractional differential equations. *Appl. Math. Lett.* **2009**, *22*, 64–69. [CrossRef]
16. Wang, J.; Xiang, H.; Liu, Z. Positive solution to nonzero boundary values problem for a coupled system of nonlinear fractional differential equations. *Int. J. Differ. Equ.* **2010**, *10*, 12. [CrossRef]
17. Ahmad, B.; Ntouyas, S.K.; Alsaedi, A. On a coupled system of fractional differential equations with coupled nonlocal and integral boundary conditions. *Chaos Solitons Fractals* **2016**, *83*, 234–241. [CrossRef]
18. Zhai, C.; Xu, L. Properties of positive solutions to a class of four-point boundary value problem of Caputo fractional differential equations with a parameter. *Commun. Nonlinear Sci. Numer. Simul.* **2014**, *19*, 2820–2827. [CrossRef]
19. Ahmad, B.; Ntouyas, S.K. Existence results for a coupled system of Caputo type sequential fractional differential equations with nonlocal integral boundary conditions. *Appl. Math. Comput.* **2015**, *266*, 615–622. [CrossRef]
20. Tariboon, J.; Ntouyas, S.K.; Sudsutad, W. Coupled systems of Riemann-Liouville fractional differential equations with Hadamard fractional integral boundary conditions. *J. Nonlinear Sci. Appl.* **2016**, *9*, 295–308. [CrossRef]
21. Mahmudov, N.I.; Bawaneh, S.; Al-Khateeb, A. On a coupled system of fractional differential equations with four point integral boundary conditions. *Mathematics* **2019**, *7*, 279. [CrossRef]

4

Existence Result and Uniqueness for some Fractional Problem

Guotao Wang [1,5,6,*], Abdeljabbar Ghanmi [2], Samah Horrigue [3] and Samar Madian [4]

1 School of Mathematics and Computer Science, Shanxi Normal University, Linfen, Shanxi 041004, China
2 Faculté des Sciences de Tunis, Université de Tunis El Manar, Tunis 1060, Tunisia; abdeljabbar.ghanmi@lamsin.rnu.tn
3 Department of Mathematics, Higher Institute of Applied Science and Technology, University of Monastir, Monastir 5000, Tunisia; samah.horrigue@fst.rnu.tn
4 Basic Sciences Department, Higher Institute for Engineering and Technology, New Damietta 34517, Egypt; samar_math@yahoo.com
5 College of Mathematics and System Science, Shandong University of Science and Technology, Qingdao, Shandong 266590, China
6 Nonlinear Analysis and Applied Mathematics (NAAM) Research Group, Department of Mathematics, Faculty of Science, King Abdulaziz University, Jeddah 21589, Saudi Arabia
* Correspondence: wgt2512@163.com

Abstract: In this article, by the use of the lower and upper solutions method, we prove the existence of a positive solution for a Riemann–Liouville fractional boundary value problem. Furthermore, the uniqueness of the positive solution is given. To demonstrate the serviceability of the main results, some examples are presented.

Keywords: positive solution; green function; fractional differential equation; the method of lower and upper solutions

---

1. Introduction

The aim of this work is to study the existence and uniqueness of the positive solution for the following problem:

$$\begin{cases} D^{\alpha}_{0+}k(t) = j(t,k(t)), \quad 1<\alpha\leq 2, \quad 0<t<1 \quad , \\ k(0)=0, \quad \beta k(1)-\gamma k(\eta)=0, \quad \eta\in[0,1] \quad , \end{cases} \tag{1}$$

where β, γ, and η are positive real numbers such that $\beta - 2\gamma\eta^{\alpha-1} > 0$, j is a nonnegative continuous function on $[0,1]\times[0,\infty)$, and D^{α}_{0+} is the fractional derivative in the sense of Riemann–Liouville. This type of equation is important in many disciplines such as chemistry, aerodynamics, polymer rheology, etc.

Different techniques are used in such problems to obtain the existence of solutions, for example the variational method, the Adomian decomposition method, etc.; we refer the reader to [1–8] and references therein.

Existence results of nonlinear fractional problems are given by the use of fixed point theorems; see [9–15]. More precisely, the authors in [12] gave the existence of positive solutions for the following equation:

$$D^{\alpha}_{0+}k(t) + j(t,k(t)) = 0, \quad 1<\alpha\leq 2, \quad 0<t<1,$$

according to some boundary conditions.

Using the theory of the fixed point index, the author in [11], presented the existence of the positive solution for the following system:

$$\begin{cases} D^{\alpha}_{0+}k(t) + j(t, k(t)) = 0, \quad 0 < t < 1, \quad 1 < \alpha \leq 2, \\ k(0) = 0, \quad \beta k(\eta) = k(1). \end{cases}$$

Recently, the upper solution method and lower solution method have been the aim of many papers; see for example the book [16] and the recent papers [17–30]. The main idea of this method is to study some modified problem and, then, give the existence results for the principal problem.

Motivated by the above works, in this article, we will present a new method to study the given problem, that is we combine the lower and upper solution method with the fixed point theorem method in order to prove the existence and uniqueness of the positive solution. Let us assume the following:

Hypothesis (H1). *The function j is nonnegative and continuous on $[0,1] \times [0,+\infty)$.*

Hypothesis (H2). *For each $t \in [0,1]$, the function $j(t,.)$ is bounded and increasing on $[0,+\infty)$.*

Hypothesis (H3). *There exists a function $a : [0,1] \to [0,\infty)$ such that the function j satisfies:*

$$|j(s,x) - j(s,y)| \leq a(s)|x-y|, \; \forall\, s \in [0,1], \; \forall\, x, y \geq 0. \tag{2}$$

The main theorems of this paper are summarized as follows.

Theorem 1. *Under Hypotheses (H_1)–(H_2). If $\beta - 2\gamma\eta^{\alpha-1} > 0$, then Equation (1) admits a positive solution.*

Theorem 2. *Under hypothesis (H_3), if $\beta - 2\gamma\eta^{\alpha-1} > 0$ and if:*

$$\int_0^1 [1 + \frac{\beta}{(\beta - \gamma\eta^{\alpha-1})}(1-\delta)^{\alpha-1}]\delta^{\alpha-1}a(\delta)d\delta < \Gamma(\alpha), \tag{3}$$

then Equation (1) admits a unique positive solution.

2. Preliminaries

In this section, we collect some basic results and notations that will be used in the forthcoming sections.

We denote by $L(0,1)$ the set of all integrable functions on $(0,1)$ and by $C(0,1)$ the set of functions that are continuous on $(0,1)$.

Lemma 1 ([31]). *Let $\alpha > 0$, $N = [\alpha] + 1$. Assume that the function k is in $C(0,1) \cap L(0,1)$. Then, the following equation:*

$$D^{\alpha}_{0+}k(t) = 0,$$

admits a unique solution. Moreover, this solution is given by:

$$k(t) = C_1 t^{\alpha-1} + C_2 t^{\alpha-2} + \ldots + C_N t^{\alpha-N},$$

for some $C_i \in \mathbb{R}$, where $i = 1, 2, ..., N$.

Lemma 2 ([31]). *Let $\alpha > 0$ and $N = [\alpha] + 1$. Assume that either k and $D^{\alpha}_{0+}k$ are in $C(0,1) \cap L(0,1)$. Then, there exists $C_i \in \mathbb{R}$, for $i = 1, 2, ..., N$, such that:*

$$I^{\alpha}_{0+} D^{\alpha}_{0+} k(t) = k(t) - C_1 t^{\alpha-1} - C_2 t^{\alpha-2} - \ldots - C_N t^{\alpha-N}.$$

Now, we give the Green function associated with the problem (1).

Theorem 3. *If $1 < \alpha < 2$ and h is continuous in $[0,1]$, then equation:*

$$D^{\alpha}_{0+}k(t) + h(t) = 0, \quad 0 < t < 1, \tag{4}$$

with the following conditions:

$$k(0) = 0, \quad \beta k(1) - \gamma k(\eta) = 0, \quad \eta \in [0,1], \tag{5}$$

admits a unique solution, which is given by:

$$k(t) = \int_0^1 G(t,\delta)h(\delta)d\delta,$$

with $G(t,\delta)$ being the Green function defined by:

$$\begin{aligned} &\Gamma(\alpha)G(t,\delta) \\ &= (t-\delta)^{\alpha-1}\chi_{[0,t]}(\delta) + \frac{t^{\alpha-1}}{\beta - \gamma\eta^{\alpha-1}}\left[\beta(1-\delta)^{\alpha-1} - \gamma(\eta-\delta)^{\alpha-1}\chi_{[0,\eta]}(\delta)\right], \end{aligned} \tag{6}$$

where χ_A is the function defined by:

$$\chi_A(x) = \begin{cases} 1 & \text{if } x \in A, \\ 0 & \text{if } x \notin A. \end{cases}$$

Proof. From Equation (4) and using Lemma 2, there exist two real numbers C_1 and C_2 such that:

$$k(t) = -I^{\alpha}_{0+}h(t) + C_1 t^{\alpha-1} + C_2 t^{\alpha-2}.$$

It follows that:

$$k(t) = \int_0^t \frac{(t-\delta)^{\alpha-1}}{\Gamma(\alpha)}h(\delta)d\delta + C_1 t^{\alpha-1} + C_2 t^{\alpha-2}. \tag{7}$$

Since $k(0) = 0$, then $C_2 = 0$.

On the other hand:

$$\begin{cases} k(1) = \int_0^1 \frac{(1-\delta)^{\alpha-1}}{\Gamma(\alpha)}h(\delta)d\delta + C_1, \\ k(\eta) = \int_0^{\eta} \frac{(\eta-\delta)^{\alpha-1}}{\Gamma(\alpha)}h(\delta)d\delta + C_1\eta^{\alpha-1}. \end{cases}$$

As $\beta k(1) - \gamma k(\eta) = 0$, then we have:

$$C_1 = \frac{\beta \int_0^1 \frac{(1-\delta)^{\alpha-1}}{\Gamma(\alpha)}h(\delta) - \gamma \int_0^{\eta} \frac{(\eta-\delta)^{\alpha-1}}{\Gamma(\alpha)}h(\delta)d\delta}{\beta - \gamma\eta^{\alpha-1}}. \tag{8}$$

By substituting the values of C_1 and C_2 into Equation (7), we get:

$$\begin{aligned} &k(t) \\ &= \int_0^t \frac{(t-\delta)^{\alpha-1}}{\Gamma(\alpha)}h(\delta)d\delta + \frac{1}{\beta - \gamma\eta^{\alpha-1}}\left[\beta\int_0^1 \frac{(1-\delta)^{\alpha-1}}{\Gamma(\alpha)}h(\delta)d\delta - \gamma\int_0^{\eta} \frac{(\eta-\delta)^{\alpha-1}}{\Gamma(\alpha)}h(\delta)d\delta\right]t^{\alpha-1} \\ &= \frac{1}{\Gamma(\alpha)}\int_0^1 \left[(t-\delta)^{\alpha-1}\chi_{[0,t]}(\delta) + \frac{1}{\beta - \gamma\eta^{\alpha-1}}\left\{\beta(1-\delta)^{\alpha-1} - \gamma(\eta-\delta)^{\alpha-1}\chi_{[0,\eta]}(\delta)\right\}t^{\alpha-1}\right]h(\delta)d\delta. \end{aligned}$$

That is:

$$k(t) = \int_0^t G(t,\delta)h(\delta)d\delta.$$

It follows that for all real numbers $t, \delta \in [0,1]$, we have:

$$G(t,\delta) = \frac{(t-\delta)^{\alpha-1}}{\Gamma(\alpha)}\chi_{[0,t]}(\delta) + \frac{1}{\Gamma(\alpha)(\beta-\gamma\eta^{\alpha-1})}\left[\beta(1-\delta)^{\alpha-1} - \gamma(\eta-\delta)^{\alpha-1}\chi_{[0,\eta]}(\delta)\right]t^{\alpha-1}.$$

□

Proposition 1. *Let G be the function given by Equation (6), then we have the following properties:*

(i) *Put $q(\delta) = \frac{2\beta-\gamma\eta^{\alpha-1}}{\Gamma(\alpha)(\beta-\gamma\eta^{\alpha-1})}(1-\delta)^{\alpha-1}$, then we have:*

$$G(t,\delta) \leq q(\delta), \forall\ t,\delta \in [0,1].$$

(ii) *Put $p(t) = \frac{\beta}{\beta-\gamma\eta^{\alpha-1}}t^{\alpha-1}$, then we obtain:*

$$G(t,\delta) \geq q(\delta)p(t),\ \forall\ t,\delta \in [0,1].$$

Proof. **(i)** Firstly, we remark that $G(1,\delta)$ is given by:

$$G(1,\delta) = \frac{1}{\Gamma(\alpha)}\begin{cases} (1-\delta)^{\alpha-1} + \frac{1}{\beta-\gamma\eta^{\alpha-1}}\left\{\beta(1-\delta)^{\alpha-1} - \gamma(\eta-\delta)^{\alpha-1}\right\} & \text{, if } 0 \leq \delta \leq \eta \\ (1-\delta)^{\alpha-1} + \frac{1}{\beta-\gamma\eta^{\alpha-1}}\beta(1-\delta)^{\alpha-1} & \text{, if } \eta \leq \delta \leq 1. \end{cases}$$

On the other hand, for all $\delta \in [0,1]$, the function $t \longmapsto G(t,\delta)$ is increasing, so, for any $t,\delta \in [0,1]$, we have:

$$G(t,\delta) \leq G(1,\delta).$$

As $\gamma > 0$, it is easy to see that:

$$\begin{aligned}\Gamma(\alpha)G(t,\delta) &\leq (1-\delta)^{\alpha-1} + \frac{1}{\beta-\gamma\eta^{\alpha-1}}\beta(1-\delta)^{\alpha-1} \\ &= \frac{2\beta-\gamma\eta^{\alpha-1}}{\beta-\gamma\eta^{\alpha-1}}(1-\delta)^{\alpha-1}, \quad \forall t,\delta \in [0,1].\end{aligned}$$

That is, if $t,\delta \in [0,1]$, then we have:

$$G(t,\delta) \leq q(\delta).$$

(ii) If $0 \le \eta \le t < 1$, then using (6) and the fact that $\beta - 2\gamma\eta^{\alpha-1} > 0$, we obtain:

$$\begin{aligned}
&G(t,\delta)\\
&= q(\delta)\begin{cases} \frac{\beta-\gamma\eta^{\alpha-1}}{2\beta-\gamma\eta^{\alpha-1}}\left(\frac{t-\delta}{1-\delta}\right)^{\alpha-1} + \frac{1}{2\beta-\gamma\eta^{\alpha-1}}\left\{\beta-\gamma(\frac{\eta-\delta}{1-\delta})^{\alpha-1}\right\}t^{\alpha-1}, & \text{if } 0\le\delta\le\eta\\ \frac{\beta-\gamma\eta^{\alpha-1}}{2\beta-\gamma\eta^{\alpha-1}}\left(\frac{t-\delta}{1-\delta}\right)^{\alpha-1} + \frac{\beta}{2\beta-\gamma\eta^{\alpha-1}}t^{\alpha-1}, & \text{if } \eta\le\delta\le t\\ \frac{\beta}{2\beta-\gamma\eta^{\alpha-1}}t^{\alpha-1}, & \text{if } t\le\delta<1 \end{cases}\\
&\ge q(\delta)\begin{cases} p(t) + \frac{\beta-\gamma\eta^{\alpha-1}}{2\beta-\gamma\eta^{\alpha-1}}\left(\frac{t-\delta}{1-\delta}\right)^{\alpha-1} - \frac{\gamma}{2\beta-\gamma\eta^{\alpha-1}}(\frac{\eta-\delta}{1-\delta})^{\alpha-1}t^{\alpha-1}, & \text{if } 0\le\delta\le\eta\\ p(t) + \frac{\beta-\gamma\eta^{\alpha-1}}{2\beta-\gamma\eta^{\alpha-1}}\left(\frac{t-\delta}{1-\delta}\right)^{\alpha-1}, & \text{if } \eta\le\delta\le t\\ p(t), & \text{if } t\le\delta<1, \end{cases}\\
&\ge q(\delta)\begin{cases} p(t) + \dfrac{(\beta-\gamma\eta^{\alpha-1})(t-\delta)^{\alpha-1} - \gamma(\eta-\delta)^{\alpha-1}t^{\alpha-1}}{2\beta-\gamma\eta^{\alpha-1}}, & \text{if } 0\le\delta\le\eta\\ p(t), & \text{if } \eta\le\delta<1, \end{cases}\\
&\ge q(\delta)\begin{cases} p(t) + \frac{1}{2\beta-\gamma\eta^{\alpha-1}}\left[(\beta-\gamma\eta^{\alpha-1})t^{\alpha-1} - \gamma\eta^{\alpha-1}t^{\alpha-1}\right], & \text{if } 0\le\delta\le\eta\\ p(t), & \text{if } \eta\le\delta<1, \end{cases}\\
&\ge q(\delta)\begin{cases} p(t) + \frac{1}{2\beta-\gamma\eta^{\alpha-1}}(\beta-2\gamma\eta^{\alpha-1})t^{\alpha-1}, & \text{if } 0\le\delta\le\eta\\ p(t), & \text{if } \eta\le\delta<1, \end{cases}\\
&\ge p(t)q(\delta),
\end{aligned}$$

where $p(t) = \frac{\beta}{2\beta-\gamma\eta^{\alpha-1}}t^{\alpha-1}$. The proof of Proposition 1 is now completed. □

3. Proof of the Main Results

This section is devoted to proving our main results. To this aim, we will apply the following lemma.

Lemma 3 (See [32]). *Let E be a semi-order Banach space and P be a cone in E. Let $D \subset P$ and a nondecreasing operator $T : D \to E$. Assume that the equation $x - T(x) = 0$ admits a lower solution $x_0 \in D$ and an upper solution $y_0 \in D$, with $x_0 \le y_0$. Assume that if $x_0 \le x \le y_0$, then $x \in D$. If one of the following statements holds:*

(i) *P is normal, and T is compact continuous.*
(ii) *P is regular, and T is continuous.*
(ii) *E is reflexive, P normal, and T continuous or weak continuous.*

Then, the equation

$$T(x) = x,$$

admits a maximum solution x^* and admits a minimum solution y^* such that $x_0 \le x^* \le y^* \le y_0$.

Note that a function v (resp. A function w) is called the lower solution (resp. upper solution) of operator T if:

$$v(t) \le Tv(t), (\text{ resp. } w(t) \ge Tw(t)).$$

Let $E = C[0,1]$, equipped with the supremum norm. Put $P = \{k \in E\, k(t) \ge 0; 0 \le t \le 1\}$, which is a cone in E. Define T on P by:

$$T(k)(t) = \int_0^1 G(t,\delta)j(\delta,k(\delta))d\delta,$$

so it is not difficult to see that k is a solution for Equation (1) if and only if $T(k) = k$.

Proof of Theorem 1. We divide the proof into four steps.

Step 1: We will prove that T maps P into itself and that it is completely continuous.

First, since G and j are nonnegative and continuous, it is easy to see that T maps P into itself and that it is continuous. Let Ω be a bounded subset of P, which is to say the existence of $M > 0$ with:

$$\|k\| \leq M, \forall \ k \in \Omega.$$

Put:

$$L = \max_{0 \leq t \leq 1, k \in \Omega} |h(t,k)|.$$

Then, for all k in Ω, we get:

$$|Tk(t)| \leq \int_0^1 G(t,\delta)|h(\delta,k(\delta))|d\delta \leq L \int_0^1 G(t,\delta)d\delta.$$

That is, $T(\Omega)$ is a bounded subset of P.

Now, for any $k \in \Omega$ and $0 \leq t_1 < t_2 \leq 1$, we have:

$$\begin{aligned}
&|Tk(t_2) - Tk(t_1)| = |\int_0^1 G(t_2,\delta)h(\delta,k(\delta))d\delta - \int_0^1 G(t_1,\delta)h(\delta,k(\delta))d\delta| \\
&= |\int_0^1 (G(t_2,\delta) - G(t_1,\delta))h(\delta,k(\delta))d\delta| \\
&\leq \int_0^1 |G(t_2,\delta) - G(t_1,\delta)||h(\delta,k(\delta))|d\delta \\
&\leq L \int_0^1 |G(t_2,\delta) - G(t_1,\delta)|d\delta \\
&\leq L\left(\int_0^{t_1} |G(t_2,\delta) - G(t_1,\delta)|d\delta + \int_{t_1}^{t_2} |G(t_2,\delta) - G(t_1,\delta)|d\delta + \int_{t_2}^1 |G(t_2,\delta) - G(t_1,\delta)|d\delta \right) \\
&\leq L\left(I_1 + I_2 + I_3 \right)
\end{aligned}$$

where:

$$\begin{aligned}
I_1 &= \int_0^{t_1} |G(t_2,\delta) - G(t_1,\delta)|d\delta \\
&\leq \int_0^{t_1} |(t_2-\delta)^{\alpha-1}\chi_{[0,t_2]}(\delta) - (t_1-\delta)^{\alpha-1}\chi_{[0,t_1]}(\delta)|d\delta \\
&+ \frac{(t_2^{\alpha-1} - t_1^{\alpha-1})}{\beta - \gamma\eta^{\alpha-1}} \int_0^{t_1} |\beta(1-\delta)^{\alpha-1} - \gamma(\eta-\delta)^{\alpha-1}\chi_{[0,\eta]}(\delta)|d\delta \\
&= \begin{cases} \frac{1}{\alpha}\left(t_2^\alpha - (t_2-t_1)^\alpha - t_1^\alpha\right) + \frac{(t_2^{\alpha-1}-t_1^{\alpha-1})}{\beta-\gamma\eta^{\alpha-1}}\left(\frac{\beta}{\alpha}(1-(1-t_1)^\alpha) + \frac{\gamma\eta^\alpha}{\alpha}\right), & \text{if } \eta < t_1 \\ \frac{t_2^\alpha - t_1^\alpha}{\alpha} + \frac{(t_2^{\alpha-1}-t_1^{\alpha-1})}{\beta-\gamma\eta^{\alpha-1}}\left(\frac{\beta}{\alpha}(1-(1-t_1)^\alpha) + \gamma\frac{\eta^\alpha - (\eta-t_1)^\alpha}{\alpha}\right), & \text{if } \eta > t_1 \end{cases}
\end{aligned}$$

$$I_2 = \int_{t_1}^{t_2} |G(t_2,\delta) - G(t_1,\delta)|d\delta$$
$$\leq \int_{t_1}^{t_2} |(t_2-\delta)^{\alpha-1}\chi_{[0,t_2]}(\delta) - (t_1-\delta)^{\alpha-1}\chi_{[0,t_1]}(\delta)|d\delta$$
$$+ \frac{(t_2^\alpha - (t_2-t_1)^\alpha)}{\beta-\gamma\eta^{\alpha-1}} \int_{t_1}^{t_2} |\beta(1-\delta)^{\alpha-1} - \gamma(\eta-\delta)^{\alpha-1}\chi_{[0,\eta]}(\delta)|d\delta$$
$$= \begin{cases} \frac{1}{\alpha}(t_2-t_1)^\alpha + \frac{(t_2^{\alpha-1}-t_1^{\alpha-1})}{\beta-\gamma\eta^{\alpha-1}}\frac{\beta}{\alpha}((1-t_1)^\alpha - (1-t_2)^\alpha), & \text{if } 0 \leq \eta < t_1 \\ \frac{t_2^\alpha - t_1^\alpha}{\alpha} + + \frac{(t_2^{\alpha-1}-t_1^{\alpha-1})}{\beta-\gamma\eta^{\alpha-1}}\left\{\frac{\beta}{\alpha}((1-t_1)^\alpha - (1-t_2)^\alpha) - \frac{\gamma}{\alpha}(t_1-\eta)^\alpha\right\}, & \text{if } t_1 \leq \eta \leq t_2 \\ \frac{t_2^\alpha - t_1^\alpha}{\alpha} + + \frac{(t_2^{\alpha-1}-t_1^{\alpha-1})}{\beta-\gamma\eta^{\alpha-1}}\left\{\frac{\beta}{\alpha}((1-t_1)^\alpha - (1-t_2)^\alpha) + \frac{\gamma}{\alpha}[(t_2-\eta)^\alpha - (t_1-\eta)^\alpha]\right\}, & \text{if } t_2 \leq \eta \leq 1 \end{cases},$$

$$I_3 = \int_{t_2}^{1} |G(t_2,\delta) - G(t_1,\delta)|d\delta$$
$$\leq \int_{t_1}^{t_2} |(t_2-\delta)^{\alpha-1}\chi_{[0,t_2]}(\delta) - (t_1-\delta)^{\alpha-1}\chi_{[0,t_1]}(\delta)|d\delta$$
$$\leq \int_{t_1}^{t_2} |(t_2-\delta)^{\alpha-1}\chi_{[0,t_2]}(\delta) - (t_1-\delta)^{\alpha-1}\chi_{[0,t_1]}(\delta)|d\delta$$
$$+ \frac{(t_2^\alpha - (t_2-t_1)^\alpha)}{\beta-\gamma\eta^{\alpha-1}} \int_{t_1}^{t_2} |\beta(1-\delta)^{\alpha-1} - \gamma(\eta-\delta)^{\alpha-1}\chi_{[0,\eta]}(\delta)|d\delta$$
$$= \begin{cases} \frac{t_2^{\alpha-1}-t_1^{\alpha-1}}{\beta-\gamma\eta^{\alpha-1}}\frac{\beta}{\alpha}(1-t_2)^\alpha, & \text{if } \eta < t_2 \\ \frac{t_2^{\alpha-1}-t_1^{\alpha-1}}{\beta-\gamma\eta^{\alpha-1}}\left\{\frac{\beta}{\alpha}(1-t_2)^\alpha - \frac{\gamma}{\alpha}(\eta-t_2)^\alpha\right\}, & \text{if } t_2 < 1. \end{cases}$$

Then, we obtain that:

$$|Tk(t_2) - Tk(t_1)| \leq L(I_1 + I_2 + I_3)$$
$$\leq L\left(\frac{t_2^\alpha - t_1^\alpha}{\alpha} + \frac{t_2^{\alpha-1} - t_1^{\alpha-1}}{\beta-\gamma\eta^{\alpha-1}}\left(\frac{\beta-\gamma\eta^\alpha}{\alpha}\right)\right).$$

Since t^α and $t^{\alpha-1}$ are uniformly continuous when $t \in [0,1]$ and $1 < \alpha < 2$, it is easy to prove that $T(\Omega)$ is equicontinuous. From the Arzela–Ascoli theorem (see [33], we deduce that $\overline{T(\Omega)}$ is a compact subset. That is, $T : P \to P$ is a completely continuous operator.

Step 2: T is an increasing operator.

Let $0 \leq t \leq 1$. Since the function $\delta \longmapsto j(t,\delta)$ is nondecreasing, then there exists $a > 0$, such that the function $[0,a] \ni \delta \longmapsto j(t,\delta)$ is strictly increasing. It follows that for $k_1 \leq k_2$, we have:

$$Tk_1(t) = \int_0^1 G(t,\delta)h(\delta,k_1(\delta))d\delta \leq \int_0^1 G(t,\delta)h(\delta,k_2(\delta))d\delta = Tk_2(t).$$

Step 3: For each $t \in [0,1]$ and from (H_2), there exists $M > 0$ with $0 < j(t,k(t)) < M$. It follows by applying Theorem 3 that equation:

$$\begin{cases} D_{0+}^\alpha w(t) + M = 0, & 0 < t < 1,\ 1 < \alpha \leq 2, \\ w(0) = 0,\ \beta w(1) - \gamma w(\eta) = 0 & ,\ \eta \in [0,1]. \end{cases}$$

has a solution w. Moreover, this solution satisfies:

$$w(t) = \int_0^1 G(t,\delta)Md\delta \geq \int_0^1 G(t,\delta)j(t,w(\delta))d\delta = Tw(t).$$

That is, the operator T admits w as an upper solution.
On the other hand, the operator T admits the zero function as a lower solution; moreover:

$$0 \leq w(t) \ \forall\, 0 \leq t \leq 1.$$

Step 4: Since P is a normal cone, Lemma 3 implies that T admits a fixed point $k \in \langle 0, w(t) \rangle$. Therefore, Equation (1) admits a positive solution. □

Proof of Theorem 2. To prove Theorem 2, we begin to prove that T has a fixed point. We remark that if for n large enough, T^n is a contraction operator, then T has a unique fixed point. Indeed, assume that for n large enough, T^n is a contraction operator, and fix $x \in E$. Since T is an increasing operator, which is uniformly bounded, then the sequence $\{T^m x\}_{m\in\mathbb{N}}$ is convergent, that is there is $p \in E$ such that $\lim_{m\to\infty} T^m x = p$. Since T is continuous, we get:

$$Tp = T \lim_{m\to\infty} T^m p = \lim_{m\to\infty} T^{m+1} = p,$$

On the other hand, if p is a fixed point for the operator T, then it is also a fixed point for the operator T^n, so we obtain the uniqueness of p.

Now, let us prove that for n large enough, the operator T^n is a contraction. Let $k, v \in P$, then we have:

$$\begin{aligned}
|Tk(t) - Tv(t)| &= \int_0^1 G(t,\delta)|j(\delta,k(\delta)) - j(\delta,v(\delta))|d\delta \\
&\leq \int_0^1 G(t,\delta)a(\delta)|k(\delta) - v(\delta)|d\delta \\
&\leq \frac{\|k-v\|}{\Gamma(\alpha)} \int_0^1 [(t-\delta)^{\alpha-1} + \frac{t^{\alpha-1}\beta}{(\beta - \gamma\eta^{\alpha-1})}(1-\delta)^{\alpha-1}]a(\delta)d\delta \\
&\leq \frac{\|k-v\|t^{\alpha-1}}{\Gamma(\alpha)} \int_0^1 [1 + \frac{\beta}{(\beta - \gamma\eta^{\alpha-1})}(1-\delta)^{\alpha-1}]a(\delta)d\delta \\
&\leq \frac{\|k-v\|t^{\alpha-1}}{\Gamma(\alpha)} K,
\end{aligned}$$

where $K = \int_0^1 [1 + \frac{\beta}{(\beta-\gamma\eta^{\alpha-1})}(1-\delta)^{\alpha-1}]a(\delta)d\delta$.

Similarly, we have:

$$\begin{aligned}
|T^2k(t) - T^2v(t)| &= \int_0^1 G(t,\delta)|j(\delta,Tk(\delta)) - j(\delta,Tv(\delta))|d\delta \\
&\leq \int_0^1 G(t,\delta)a(\delta)|Tk(\delta) - Tv(\delta)|d\delta \\
&\leq \int_0^1 G(t,\delta)a(\delta)\frac{\|k-v\|\delta^{\alpha-1}}{\Gamma(\alpha)}Kd\delta \\
&\leq \frac{\|k-v\|t^{\alpha-1}}{\Gamma(\alpha)^2} \int_0^1 [1 + \frac{\beta}{(\beta - \gamma\eta^{\alpha-1})}(1-\delta)^{\alpha-1}]\delta^{\alpha-1}a(\delta)d\delta \\
&\leq \frac{\|k-v\|t^{\alpha-1}}{\Gamma(\alpha)^2} KH,
\end{aligned}$$

where $H = \int_0^1 [1 + \frac{\beta}{(\beta - \gamma\eta^{\alpha-1})}(1-\delta)^{\alpha-1}]\delta^{\alpha-1}a(\delta)d\delta$.

By mathematical induction, it follows that:

$$|T^n k(t) - T^n v(t)| \leq \frac{\|k - v\|t^{\alpha-1}}{\Gamma(\alpha)^n} KH^{n-1}.$$

By using (3), we get:

$$\frac{H}{\Gamma(\alpha)} < 1.$$

Then, for n large enough, it follows that:

$$\frac{KH^{n-1}}{\Gamma(\alpha)^n} = \frac{K}{\Gamma(\alpha)}\left(\frac{H}{\Gamma(\alpha)}\right)^{n-1} < 1.$$

Hence, it holds that:

$$|T^n k(t) - T^n v(t)| < \|k - v\|t^{\alpha-1} \leq \|k - v\|,$$

and this completes the proof. □

4. Examples

In this section, some examples are presented in order to illustrate the usefulness of our main results.

Example 1. *Consider the system:*

$$\begin{cases} D^{\alpha}_{0+}k(t) = \sqrt{t}e^{-k(t)}, & 0 < t < 1, \quad 1 < \alpha \leq 2, \\ k(0) = 0, \quad \beta k(1) - \gamma k(\eta) = 0, & \eta \in [0,1] \end{cases}, \tag{9}$$

where $\beta, \gamma > 0$, $\beta - 2\gamma\eta^{\alpha-1} > 0$.

Note that since for any $t \in [0,1]$, *we have* $0 < \sqrt{t}e^{-k(t)} < \sqrt{t}$, *which implies that Conditions* (H_1) *and* (H_2) *hold. On the other hand, there is an equivalence between the solution of Problem (9) and the fixed point of the operator T given by:*

$$Tk(t) = \int_0^1 G(t,\delta)\sqrt{\delta}e^{-k(\delta)}d\delta.$$

Take $w(t) = \int_0^1 G(t,\delta)\sqrt{\delta}d\delta$ *and* $v(t) \equiv 0$, *then:*

$$w(t) \geq \int_0^1 G(t,\delta)\sqrt{\delta}e^{-w(\delta)}d\delta = Tw(t),$$

which implies that the operator T admits the function w as an upper solution. Moreover, it is obvious that the zero function is a lower solution for T. Hence, from Theorem 1, we conclude that Problem (9) admits a positive solution.

Example 2. *In the second example, we study the following problem:*

$$\begin{cases} D^{3/2}_{0+}k(t) = \frac{\sin t}{(1+t^2)}\arctan(1 + k(t)), & 0 < t < 1, \\ k(0) = 0, \quad \beta k(1) - \gamma k(\eta) = 0, \quad \eta \in [0,1] \end{cases}, \tag{10}$$

where $\beta, \gamma > 0$, $\beta - 2\gamma\sqrt{\eta} > 0$.

Note that $\arctan(1 + k(t)) < \frac{\pi}{2}$ *for each* $t \in [0,1]$, *then* $j(t, k(t)) = \frac{\sin t}{(1+t^2)}\arctan(1 + k(t))$ *is an increasing and bounded function on k. Therefore, we can easily prove that Conditions* (H_1) *and* (H_2) *are*

satisfied.
Take $w(t) = \int_0^1 G(t,\delta)\frac{\sin\delta}{(1+\delta^2)}d\delta$ *and* $v(t) \equiv 0$*, then:*

$$w(t) \geq \int_0^1 G(t,\delta)\sqrt{\delta}\frac{\sin\delta}{(1+\delta^2)}\arctan(1+k(\delta))d\delta = Tw(t),$$

which implies that T admits the function w as an upper solution and the zero function as a lower solution. Thus, from Theorem 1, we obtain a positive solution to Problem (9).

Example 3. *In this example, we take* $\beta = 1$*,* $\gamma = 0$*, and* $\eta \in [0,1]$*, and we consider the following problem:*

$$\begin{cases} D^{\frac{3}{2}}y(t) = \lambda y(t) + f(t), \\ y(0) = 0, \quad y(1) = 0, \end{cases} \tag{11}$$

where f is a nonnegative function. It is clear that we have:

$$j(t,x) = \lambda x + f(t),$$

and $\alpha = \frac{3}{2}$*. Moreover, for all* $t \in [0,1]$*, one has:*

$$|j(t,x_1) - j(t,x_2)| \leq \lambda|x_1 - x_2|, \text{ that is } a(t) = \lambda,$$

and:

$$\begin{aligned} \int_0^1 [1 + \frac{\beta}{(\beta - \gamma\eta^{\alpha-1})}(1-\delta)^{\alpha-1}]\delta^{\alpha-1}a(\delta)d\delta &= \lambda\int_0^1 \left(1 + (1-\delta)^{\alpha-1}\right)\delta^{\alpha-1}d\delta \\ &= \lambda(B(1,\alpha) + B(\alpha,\alpha)) \\ &= \lambda\Gamma(\alpha)\left(\frac{1}{\Gamma(\alpha+1)} + \frac{\Gamma(\alpha)}{\Gamma(2\alpha)}\right) \\ &= \lambda\Gamma(\frac{3}{2})\left(\frac{1}{\Gamma(\frac{5}{2})} + \frac{\Gamma(\frac{3}{2})}{\Gamma(3)}\right) < \Gamma(\frac{3}{2}), \end{aligned}$$

for all $0 < \lambda < \frac{12\sqrt{\pi}}{16+3\pi}$*, where* $B(.,.)$ *is the beta function.*

Finally, all conditions of Theorem 2 are satisfied. Therefore, for $0 < \lambda < \frac{12\sqrt{\pi}}{16+3\pi}$*, Problem (11) admits a unique solution.*

Author Contributions: All authors contributed equally to this work.

References

1. Ben Ali, K.; Ghanmi, A.; Kefi, K. Existence of solutions for fractional differential equations with dirichlet boundary conditions. *Electr. J. Differ. Equ.* **2016**, *2016*, 1–11.
2. Cabada, A.; Wang, G. Positive solutions of nonlinear fractional differential equations with integral boundary value conditions. *J. Math. Anal. Appl.* **2012**, *389*, 403–411. [CrossRef]
3. Jafari, H.; Gejji, V.D. Positive solutions of nonlinear fractional boundary value problems using adomian decomposition method. *Appl. Math. Comput.* **2006**, *180*, 700–706. [CrossRef]
4. Liang, S.H.; Zhang, J.H. Positive solutions for boundary value problems of nonlinear fractional differential equation. *Nonlinear Anal.* **2009**′ *71*, 5545–5550. [CrossRef]

5. Miller, K.; Ross, B. *An Introduction to the Fractional Calculus and Fractional Differential Equations*; Wiley and Sons: New York, NY, USA, 1993.
6. Nieto, J.J. Maximum principles for fractional differential equations derived from Mittag-Leffler functions. *Appl. Math. Lett.* **2010**, *23*, 1248–1251. [CrossRef]
7. Podlubny, I. Fractional Differential Equations. In *Mathematics in Science and Engineering*; Academic Press: New York, NY, USA, 1999.
8. Wei, Z. Dong, W.; Che, J. Periodic boundary value problems for fractional differential equations involving a Riemann–Liouville fractional derivative. *Nonlinear Anal.* **2010**, *73*, 3232–3238. [CrossRef]
9. Ahmad, B.; Nieto, J.J. Existence results for nonlinear boundary value problems of fractional integro-differential equations with integral boundary conditions. *Bound Value Probl.* **2009**, *2009*, 708576.
10. Bǎleanu, D.; Octavian, G.M.; O'Regan, D. On a fractional differential equation with infinitely many solutions. *Adv. Diff. Equ.* **2012**, *2012*, 145. [CrossRef]
11. Bai, Z. On positive solutions of a nonlocal fractional boundary value problem. *Nonlinear Anal.* **2010**, *72*, 916–924. [CrossRef]
12. Bai, Z.; Lü, H.S. Positive solutions of boundary value problems of nonlinear fractional differential equation. *J. Math. Anal. Appl.* **2005**, *311*, 495–505. [CrossRef]
13. Ghanmi, A.; Kratou, M.; Saoudi, K. A Multiplicity Results for a Singular Problem Involving a Riemann–Liouville Fractional Derivative. *FILOMAT* **2018**, *32*, 653–669. [CrossRef]
14. Saoudi, K.; Agarwal, P.; Kumam, P.; Ghanmi, A.; Thounthong, P. The Nehari manifold for a boundary value problem involving Riemann–Liouville fractional derivative. *Adv. Diff. Equ.* **2018**, *2018*, 263. [CrossRef]
15. Ghanmi, A.; Horrigue, S. Existence Results for Nonlinear Boundary Value Problems. *FILOMAT* **2018**, *32*, 609–618. [CrossRef]
16. Lakshmikantham, V.; Leela, S.; Vasundhara, J. *Theory of Fractional Dynamic Systems*; Cambridge Academic Publishers: Cambridge, UK, 2009.
17. Wang, G.; Pei, K.; Agarwal, R.P.; Zhang, L.; Ahmad, B. Nonlocal Hadamard fractional boundary value problem with Hadamard integral and discrete boundary conditions on a half-line. *J. Comput. Appl. Math.* **2018**, *343*, 230–239. [CrossRef]
18. Wang, G.; Ren, X.; Bai, Z.; Hou, W. Radial symmetry of standing waves for nonlinear fractional Hardy-Schrödinger equation. *Appl. Math. Lett.* **2019**, *96*, 131–137. [CrossRef]
19. Pei, K.; Wang, G.; Sun, Y. Successive iterations and positive extremal solutions for a Hadamard type fractional integro-differential equations on infinite domain. *Appl. Math. Comput.* **2017**, *312*, 158–168. [CrossRef]
20. Bai, Z.; Zhang, S.; Sun, S.; Yin, C. Monotone iterative method for a class of fractional differential equations. *Electron. J. Differ. Equ.* **2016**, *2016*, 1–8.
21. Ding, Y.; Wei, Z.; Xu, J.; O'Regan, D. Extremal solutions for nonlinear fractional boundary value problems with p-Laplacian. *J. Comput. Appl. Math.* **2015**, *288*, 151–158. [CrossRef]
22. Wang, G. Twin iterative positive solutions of fractional q-difference Schrödinger equations. *Appl. Math. Lett.* **2018**, *76*, 103–109. [CrossRef]
23. Wang, G.; Sudsutad, W.; Zhang, L.; Tariboon, J. Monotone iterative technique for a nonlinear fractional q-difference equation of Caputo type. *Adv. Differ. Equ.* **2016** *211*. [CrossRef]
24. Hu, C.; Liu, B.; Xie, S. Monotone iterative solutions for nonlinear boundary value problems of fractional differential equation with deviating arguments. *Appl. Math. Comput.* **2013**, *222*, 72–81. [CrossRef]
25. Liu, Z.; Sun, J.; Szanto, I. Monotone iterative technique for Riemann–Liouville fractional integro- differential equations with advanced arguments. *Results Math.* **2013**, *63*, 1277–1287. [CrossRef]
26. Zhang, X.; Liu, L.; Wu, Y.; Lu, Y. The iterative solutions of nonlinear fractional differential equations. *Appl. Math. Comput.* **2013**, *219*, 4680–4691. [CrossRef]
27. Zhang, X.; Liu, L.; Wu, Y. The uniqueness of positive solution for a fractional order model of turbulent flow in a porous medium. *Appl. Math. Lett.* **2014**, *37*, 26–33. [CrossRef]
28. Zhang, L.; Ahmad, B.; Wang, G.; Agarwal, R.P. Nonlinear fractional integro-differential equations on unbounded domains in a Banach space. *J. Comput. Appl. Math.* **2013**, *249*, 51–56. [CrossRef]
29. Zhang, L.; Ahmad, B.; Wang, G. Successive iterations for positive extremal solutions of nonlinear fractional differential equations on a half-line. *Bull. Aust. Math. Soc.* **2015**, *91*, 116–128. [CrossRef]
30. Wang, G. Explicit iteration and unbounded solutions for fractional integral boundary value problem on an infinite interval. *Appl. Math. Lett.* **2015**, *47*, 1–7. [CrossRef]

31. Kilbas, A.A.; Srivastava, H.M.; Trujillo, J.J. Theory and Applications of Fractional Differential Equations. In *North-Holland Mathematics Studies*; Elsevier Science BV: Amsterdam, The Netherlands, 2006; Volume 204.
32. Zhong, C.; Fan, X.; Chen, W. *Nonlinear Functional Analysis and Its Application*; Lan Zhou University Press: Lanzhou, China, 1998.
33. Rudin, W. *Principles of Mathematical Analysis*; McGraw-Hill New York: 1976; ISBN 978-0-07-054235-8.

5

The Langevin Equation in terms of Generalized Liouville–Caputo Derivatives with Nonlocal Boundary Conditions Involving a Generalized Fractional Integral

Bashir Ahmad [1], Madeaha Alghanmi [1], Ahmed Alsaedi [1], Hari M. Srivastava[2,3,*] and Sotiris K. Ntouyas [1,4]

1 Nonlinear Analysis and Applied Mathematics (NAAM)-Research Group, Department of Mathematics, Faculty of Science, King Abdulaziz University, P.O. Box 80203, Jeddah 21589, Saudi Arabia; bashirahmad_qau@yahoo.com (B.A.); madeaha@hotmail.com (M.A.); aalsaedi@hotmail.com (A.A.); sntouyas@uoi.gr (S.K.N.)

2 Department of Mathematics and Statistics, University of Victoria, Victoria, BC V8W 3R4, Canada

3 Department of Medical Research, China Medical University Hospital, China Medical University, Taichung 40402, Taiwan

4 Department of Mathematics, University of Ioannina, 451 10 Ioannina, Greece

* Correspondence: harimsri@math.uvic.ca

Abstract: In this paper, we establish sufficient conditions for the existence of solutions for a nonlinear Langevin equation based on Liouville-Caputo-type generalized fractional differential operators of different orders, supplemented with nonlocal boundary conditions involving a generalized integral operator. The modern techniques of functional analysis are employed to obtain the desired results. The paper concludes with illustrative examples.

Keywords: Langevin equation; generalized fractional integral; generalized Liouville–Caputo derivative; nonlocal boundary conditions; existence; fixed point

---

1. Introduction

The topic of fractional calculus has emerged as an interesting area of investigation in view of its widespread applications in social sciences, engineering and technical sciences. Mathematical models based on fractional order differential and integral operators are considered to be more realistic and practical than their integer-order counterparts as such models can reveal the history of the ongoing phenomena in systems and processes. This branch of mathematical analysis is now very developed and covers a wide range of interesting results, for instance [1–7].

The Langevin equation is an effective tool of mathematical physics, which can describe processes like anomalous diffusion in a descent manner. Examples of such processes include price index fluctuations [8], harmonic oscillators [9], etc. A generic Langevin equation for noise sources with correlations also plays a central role in the theory of critical dynamics [10]. The nature of the quantum noise can be understood better by means of a generalized Langevin equation [11]. The role of the Langevin equation in fractional systems, such as fractional reaction-diffusion systems [12,13], is very rich and beautiful. The fractional analogue (also known as the stochastic differential equation) of the usual Langevin equation is suggested for systems in which the separation between microscopic and macroscopic time scales is not observed; for example, see [8]. In [14], the author investigated moments, variances, position and velocity correlation for a Riemann-Liouville-type fractional Langevin equation in time and compared the results obtained with the ones derived for the same generalized Langevin equation involving the Liouville-Caputo fractional derivative. Some recent results on the Langevin equation with different boundary conditions can be found in the papers [15–20] and the references cited therein.

Motivated by the aforementioned work on the Langevin equation and its variants, in this paper, we introduce and study a new form of Langevin equation involving generalized Liouville-Caputo derivatives of different orders and solve it with nonlocal generalized fractional integral boundary conditions. In precise terms, we investigate the problem:

$$\begin{cases} {}^{\rho}_{c}D^{\alpha}_{a+}({}^{\rho}_{c}D^{\beta}_{a+} + \lambda)x(t) = f(t, x(t)), \ \ t \in J := [a, T], \ \ \lambda \in \mathbb{R}, \\ x(a) = 0, \ \ x(\eta) = 0, \ \ x(T) = \mu \, {}^{\rho}I^{\gamma}_{a+}x(\xi), \ \ a < \eta < \xi < T, \end{cases} \tag{1}$$

where ${}^{\rho}_{c}D^{\alpha}_{a+}$, ${}^{\rho}_{c}D^{\beta}_{a+}$ denote the Liouville–Caputo-type generalized fractional differential operators of order $1 < \alpha \le 2$, $0 < \beta < 1$, $\rho > 0$, respectively, ${}^{\rho}I^{\gamma}_{a+}$ is the generalized fractional integral operator of order $\gamma > 0$ and $\rho > 0$, and $f : [a, T] \times \mathbb{R} \to \mathbb{R}$ is a given continuous function.

Here, we emphasize that the present work may have useful applications in fractional quantum mechanics and fractional statistical mechanics, in relation to further generalization of the Feynman and Weiner path integrals [21].

We compose the rest of the article as follows. Section 2 contains the basic concepts of generalized fractional calculus and an auxiliary lemma dealing with the linear variant of the given problem. In Section 3, we present the main results and illustrative examples.

2. Preliminaries

Definition 1 ([22]). *The generalized left-sided fractional integral of order $\beta > 0$ and $\rho > 0$ of $g \in X^p_c(a, b)$ for $-\infty < a < t < b < \infty$, is defined by:*

$$({}^{\rho}I^{\beta}_{a+}g)(t) = \frac{\rho^{1-\beta}}{\Gamma(\beta)} \int_a^t \frac{s^{\rho-1}}{(t^{\rho} - s^{\rho})^{1-\beta}} g(s)ds, \tag{2}$$

where $X^p_c(a, b)$ denotes the space of all complex-valued Lebesgue measurable functions ϕ on (a, b) equipped with the norm:

$$\|\phi\|_{X^p_c} = \Big(\int_a^b |x^c \phi(x)|^p \frac{dx}{x} \Big)^{1/p} < \infty, \ \ c \in \mathbb{R}, 1 \le p \le \infty.$$

Similarly, the right-sided fractional integral ${}^{\rho}I^{\beta}_{b-}g$ is defined by:

$$({}^{\rho}I^{\beta}_{b-}g)(t) = \frac{\rho^{1-\alpha}}{\Gamma(\beta)} \int_t^b \frac{s^{\rho-1}}{(s^{\rho} - t^{\rho})^{1-\beta}} g(s)ds. \tag{3}$$

Definition 2 ([23]). *For $\beta > 0$, $n = [\beta] + 1$, $\rho > 0$ and $0 \le a < x < b < \infty$, we define the generalized fractional derivatives in terms of the generalized fractional integrals (2) and (3) as:*

$$\begin{aligned} ({}^{\rho}D^{\beta}_{a+}g)(t) &= \Big(t^{1-\rho}\frac{d}{dt}\Big)^n ({}^{\rho}I^{n-\beta}_{a+}g)(t) \\ &= \frac{\rho^{\beta-n+1}}{\Gamma(n-\beta)} \Big(t^{1-\rho}\frac{d}{dt}\Big)^n \int_a^t \frac{s^{\rho-1}}{(t^{\rho} - s^{\rho})^{\beta-n+1}} g(s)ds, \end{aligned} \tag{4}$$

and:

$$\begin{aligned} ({}^{\rho}D^{\beta}_{b-}g)(t) &= \Big(-t^{1-\rho}\frac{d}{dt}\Big)^n ({}^{\rho}I^{n-\beta}_{b-}g)(t) \\ &= \frac{\rho^{\beta-n+1}}{\Gamma(n-\beta)} \Big(-t^{1-\rho}\frac{d}{dt}\Big)^n \int_t^b \frac{s^{\rho-1}}{(s^{\rho} - t^{\rho})^{\beta-n+1}} g(s)ds, \end{aligned} \tag{5}$$

if the integrals in the above expressions exist.

Definition 3 ([24]). *For $\beta > 0, n = [\beta] + 1$ and $g \in AC^n_\delta[a,b]$, the Liouville–Caputo-type generalized fractional derivatives ${}^{\rho}_{c}D^{\beta}_{a+}g$ and ${}^{\rho}_{c}D^{\beta}_{b-}g$ are respectively defined via* (4) *and* (5) *as follows:*

$$ {}^{\rho}_{c}D^{\beta}_{a+}g(x) = {}^{\rho}D^{\beta}_{a+}\Big[g(t) - \sum_{k=0}^{n-1} \frac{\delta^k g(a)}{k!}\Big(\frac{t^\rho - a^\rho}{\rho}\Big)^k\Big](x), \ \ \delta = x^{1-\rho}\frac{d}{dx}, \tag{6} $$

$$ {}^{\rho}_{c}D^{\beta}_{b-}g(x) = {}^{\rho}D^{\beta}_{b-}\Big[g(t) - \sum_{k=0}^{n-1} \frac{(-1)^k\delta^k g(b)}{k!}\Big(\frac{b^\rho - t^\rho}{\rho}\Big)^k\Big](x), \ \ \delta = x^{1-\rho}\frac{d}{dx}, \tag{7} $$

where $AC^n_\delta[a,b]$ denotes the class of all absolutely-continuous functions g possessing δ^{n-1}-derivative ($\delta^{n-1}g \in AC([a,b],\mathbb{R})$), equipped with the norm $\|g\|_{AC^n_\delta} = \sum_{k=0}^{n-1}\|\delta^k g\|_C$.

Remark 1 ([24]). *For $\alpha \geq 0$ and $g \in AC^n_\delta[a,b]$, the left and right generalized Liouville–Caputo derivatives of g are respectively defined by the expressions:*

$$ {}^{\rho}_{c}D^{\beta}_{a+}g(t) = \frac{1}{\Gamma(n-\beta)}\int_a^t \Big(\frac{t^\rho - s^\rho}{\rho}\Big)^{n-\beta-1}\frac{(\delta^n g)(s)ds}{s^{1-\rho}}, \tag{8} $$

$$ {}^{\rho}_{c}D^{\beta}_{b-}g(t) = \frac{1}{\Gamma(n-\beta)}\int_t^b \Big(\frac{s^\rho - t^\rho}{\rho}\Big)^{n-\alpha-1}\frac{(-1)^n(\delta^n g)(s)ds}{s^{1-\rho}}. \tag{9} $$

Lemma 1 ([24]). *Let $g \in AC^n_\delta[a,b]$ or $C^n_\delta[a,b]$ and $\beta \in \mathbb{R}$. Then:*

$$ {}^{\rho}I^{\beta}_{a+}\,{}^{\rho}_{c}D^{\beta}_{a+}g(x) = g(x) - \sum_{k=0}^{n-1}\frac{(\delta^k g)(a)}{k!}\Big(\frac{x^\rho - a^\rho}{\rho}\Big)^k, $$

$$ {}^{\rho}I^{\beta}_{b-}\,{}^{\rho}_{c}D^{\beta}_{b-}g(x) = g(x) - \sum_{k=0}^{n-1}\frac{(-1)^k(\delta^k g)(a)}{k!}\Big(\frac{b^\rho - x^\rho}{\rho}\Big)^k. $$

In particular, for $0 < \beta \leq 1$, we have:

$$ {}^{\rho}I^{\beta}_{a+}\,{}^{\rho}_{c}D^{\beta}_{a+}g(x) = g(x) - g(a), \ {}^{\rho}I^{\beta}_{b-}\,{}^{\rho}_{c}D^{\beta}_{b-}g(x) = g(x) - g(b). $$

Definition 4. *A function $x \in C([a,T],\mathbb{R})$ is called a solution of* (1) *if x satisfies the equation ${}^{\rho}_{c}D^{\alpha}_{a+}({}^{\rho}_{c}D^{\beta}_{a+} + \lambda)x(t) = f(t,x(t))$ on $[a,T]$, and the conditions $x(a) = 0, x(\eta) = 0, x(T) = \mu^\rho I^\gamma_{a+}x(\xi)$.*

In the next lemma, we solve the linear variant of Problem (1).

Lemma 2. *Let $h \in C([a,T],\mathbb{R})$, $x \in AC^3_\delta(J)$ and:*

$$ \Omega = \Big[\frac{(T^\rho - a^\rho)^\beta(T^\rho - \eta^\rho)}{\rho^{\beta+1}\Gamma(\beta+2)} - \frac{\mu(\xi^\rho - a^\rho)^{\beta+\gamma}[(\beta+1)(\xi^\rho - \eta^\rho) - \gamma(\eta^\rho - a^\rho)]}{\rho^{\beta+\gamma+1}\Gamma(\beta+\gamma+2)(\beta+1)}\Big] \neq 0. \tag{10} $$

Then, the unique solution of linear problem:

$$ \begin{cases} {}^{\rho}_{c}D^{\alpha}_{a+}({}^{\rho}_{c}D^{\beta}_{a+} + \lambda)x(t) = h(t), \ \ t \in J := [a,T], \\ x(a) = 0, \ \ x(\eta) = 0, \ \ x(T) = \mu^\rho I^\gamma_{a+}x(\xi), \ \ a < \eta < \xi < T, \end{cases} \tag{11} $$

is given by:

$$\begin{aligned} x(t) &= {}^{\rho}I_{a+}^{\alpha+\beta}h(t) - \lambda^{\rho}I_{a+}^{\beta}x(t) + \frac{(t^{\rho}-a^{\rho})^{\beta}(\eta^{\rho}-t^{\rho})}{\rho^{\beta+1}\Gamma(\beta+2)\Omega}\Big\{{}^{\rho}I_{a+}^{\alpha+\beta}h(T) - \lambda^{\rho}I_{a+}^{\beta}x(T) \\ &\quad -\mu^{\rho}I_{a+}^{\alpha+\beta+\gamma}h(\xi) + \mu\lambda^{\rho}I_{a+}^{\beta+\gamma}x(\xi)\Big\} - \frac{(t^{\rho}-a^{\rho})^{\beta}}{\Omega(\eta^{\rho}-a^{\rho})^{\beta}}\Big(\frac{(T^{\rho}-a^{\rho})^{\beta}(T^{\rho}-t^{\rho})}{\rho^{\beta+1}\Gamma(\beta+2)} \\ &\quad -\frac{\mu(\xi^{\rho}-a^{\rho})^{\beta+\gamma}[(\beta+1)(\xi^{\rho}-t^{\rho})-\gamma(t^{\rho}-a^{\rho})]}{\rho^{\beta+\gamma+1}\Gamma(\beta+\gamma+2)(\beta+1)}\Big)\Big\{{}^{\rho}I_{a+}^{\alpha+\beta}h(\eta) - \lambda^{\rho}I_{a+}^{\beta}x(\eta)\Big\}. \end{aligned} \tag{12}$$

Proof. Applying ${}^{\rho}I_{a+}^{\alpha}$ on the fractional differential equation in (11) and using Lemma 1 yield:

$$({}_{c}^{\rho}D_{a+}^{\beta}+\lambda)x(t) = {}^{\rho}I_{a+}^{\alpha}h(t) + c_1 + c_2\frac{(t^{\rho}-a^{\rho})}{\rho}, \tag{13}$$

for some $c_1, c_2 \in \mathbb{R}$.

Applying ${}^{\rho}I_{a+}^{\beta}$ to both sides of Equation (13), the general solution of the Langevin equation in (11) is found to be:

$$x(t) = {}^{\rho}I_{a+}^{\alpha+\beta}h(t) - \lambda^{\rho}I_{a+}^{\beta}x(t) + c_1\frac{(t^{\rho}-a^{\rho})^{\beta}}{\rho^{\beta}\Gamma(\beta+1)} + c_2\frac{(t^{\rho}-a^{\rho})^{\beta+1}}{\rho^{\beta+1}\Gamma(\beta+2)} + c_3, \tag{14}$$

where $c_3 \in \mathbb{R}$.

Using the condition $x(a) = 0$ in (14), we find that $c_3 = 0$. Inserting the value of c_3 in (14) and then applying the operator ${}^{\rho}I_{a+}^{\gamma}$ on the resulting equation, we get:

$${}^{\rho}I_{a}^{\gamma}x(t) = {}^{\rho}I_{a+}^{\alpha+\beta+\gamma}h(t) - \lambda^{\rho}I_{a+}^{\beta+\gamma}x(t) + c_1\frac{(t^{\rho}-a^{\rho})^{\beta+\gamma}}{\rho^{\beta+\gamma}\Gamma(\beta+\gamma+1)} + c_2\frac{(t^{\rho}-a^{\rho})^{\beta+\gamma+1}}{\rho^{\beta+\gamma+1}\Gamma(\beta+\gamma+2)}. \tag{15}$$

Using the boundary conditions $x(\eta) = 0$ and $x(T) = \mu^{\rho}I_{a+}^{\gamma}x(\xi)$ together with (14) and (15) leads to a system of algebraic equations in c_1 and c_2, which, upon solving, yields:

$$\begin{aligned} c_1 &= \frac{\rho^{\beta}\Gamma(\beta+1)}{\Omega(\eta^{\rho}-a^{\rho})^{\beta}}\Big\{\frac{(\eta^{\rho}-a^{\rho})^{\beta+1}}{\rho^{\beta+1}\Gamma(\beta+2)}\Big({}^{\rho}I_{a+}^{\alpha+\beta}h(T) - \lambda^{\rho}I_{a+}^{\beta}x(T) - \mu^{\rho}I^{\alpha+\beta+\gamma}h(\xi) + \mu\lambda^{\rho}I^{\beta+\gamma}x(\xi)\Big) \\ &\quad -\Big(\frac{(T^{\rho}-a^{\rho})^{\beta+1}}{\rho^{\beta+1}\Gamma(\beta+2)} - \frac{\mu(\xi^{\rho}-a^{\rho})^{\beta+\gamma+1}}{\rho^{\beta+\gamma+1}\Gamma(\beta+\gamma+2)}\Big)\Big({}^{\rho}I_{a+}^{\alpha+\beta}h(\eta) - \lambda^{\rho}I_{a+}^{\beta}x(\eta)\Big)\Big\}, \\ c_2 &= -\frac{\rho^{\beta}\Gamma(\beta+1)}{\Omega(\eta^{\rho}-a^{\rho})^{\beta}}\Big\{\frac{(\eta^{\rho}-a^{\rho})^{\beta}}{\rho^{\beta}\Gamma(\beta+1)}\Big({}^{\rho}I_{a+}^{\alpha+\beta}h(T) - \lambda^{\rho}I_{a+}^{\beta}x(T) - \mu^{\rho}I^{\alpha+\beta+\gamma}h(\xi) + \mu\lambda^{\rho}I^{\beta+\gamma}x(\xi)\Big) \\ &\quad -\Big(\frac{(T^{\rho}-a^{\rho})^{\beta}}{\rho^{\beta}\Gamma(\beta+1)} - \frac{\mu(\xi^{\rho}-a^{\rho})^{\beta+\gamma}}{\rho^{\beta+\gamma}\Gamma(\beta+\gamma+1)}\Big)\Big({}^{\rho}I_{a+}^{\alpha+\beta}h(\eta) - \lambda^{\rho}I_{a+}^{\beta}x(\eta)\Big)\Big\}. \end{aligned}$$

Inserting the values of c_1, c_2 and c_3 in (13) yields the solution (12). The converse of the Lemma 2, can be obtained by direct computation. This finishes the proof. □

3. Existence and Uniqueness Results

In view of Lemma 2, we introduce an operator $\mathcal{F} : \mathcal{C} \to \mathcal{C}$ by:

$$\begin{aligned} \mathcal{F}(x)(t) &= {}^{\rho}I_{a+}^{\alpha+\beta}f(t,x(t)) - \lambda^{\rho}I_{a+}^{\beta}x(t) + \frac{(t^{\rho}-a^{\rho})^{\beta}(\eta^{\rho}-t^{\rho})}{\rho^{\beta+1}\Gamma(\beta+2)\Omega}\Big\{{}^{\rho}I_{a+}^{\alpha+\beta}f(T,x(T)) - \lambda^{\rho}I_{a+}^{\beta}x(T) \\ &\quad - \mu^{\rho}I_{a+}^{\alpha+\beta+\gamma}f(\xi,x(\xi)) + \mu\lambda^{\rho}I_{a+}^{\beta+\gamma}x(\xi)\Big\} - \frac{(t^{\rho}-a^{\rho})^{\beta}}{\Omega(\eta^{\rho}-a^{\rho})^{\beta}}\Big[\frac{(T^{\rho}-a^{\rho})^{\beta}(T^{\rho}-t^{\rho})}{\rho^{\beta+1}\Gamma(\beta+2)} \\ &\quad - \frac{\mu(\xi^{\rho}-a^{\rho})^{\beta+\gamma}[(\beta+1)(\xi^{\rho}-t^{\rho})-\gamma(t^{\rho}-a^{\rho})]}{\rho^{\beta+\gamma+1}\Gamma(\beta+\gamma+2)(\beta+1)}\Big]\Big\{{}^{\rho}I_{a+}^{\alpha+\beta}f(\eta,x(\eta)) - \lambda^{\rho}I_{a+}^{\beta}x(\eta)\Big\}. \end{aligned} \tag{16}$$

Here, $\mathcal{C}$ denotes the Banach space of all continuous functions from $[a,T]$ to $\mathbb{R}$ equipped with the norm $\|x\| = \sup_{t\in[a,T]} |x(t)|$.

For the sake of computational convenience, we set:

$$\begin{aligned}\Lambda_1 &= \frac{(T^\rho - a^\rho)^{\alpha+\beta}}{\rho^{\alpha+\beta}\Gamma(\alpha+\beta+1)}\left[1+\frac{\zeta_1}{\rho^{\beta+1}\Gamma(\beta+2)|\Omega|}\right]+\frac{|\mu|(\xi^\rho - a^\rho)^{\alpha+\beta+\gamma}\zeta_1}{\rho^{\alpha+2\beta+\gamma+1}\Gamma(\alpha+\beta+\gamma+1)\Gamma(\beta+2)|\Omega|}\\ &\quad+\frac{(\eta^\rho - a^\rho)^{\alpha}\zeta_2}{\rho^{\alpha+\beta}\Gamma(\alpha+\beta+1)|\Omega|},\end{aligned} \tag{17}$$

$$\begin{aligned}\Lambda_2 &= \frac{|\lambda|(T^\rho - a^\rho)^{\beta}}{\rho^{\beta}\Gamma(\beta+1)}\left[1+\frac{\zeta_1}{\rho^{\beta+1}\Gamma(\beta+2)|\Omega|}\right]+\frac{|\mu||\lambda|(\xi^\rho - a^\rho)^{\beta+\gamma}\zeta_1}{\rho^{2\beta+\gamma+1}\Gamma(\beta+\gamma+1)\Gamma(\beta+2)|\Omega|}\\ &\quad+\frac{|\lambda|\zeta_2}{\rho^{\beta}\Gamma(\beta+1)|\Omega|},\end{aligned} \tag{18}$$

where:

$$\zeta_1 := \max_{t\in[a,T]}\left|(t^\rho - a^\rho)^\beta(\eta^\rho - t^\rho)\right|, \tag{19}$$

$$\zeta_2 := \max_{t\in[a,T]}\left|(t^\rho - a^\rho)^\beta\left[\frac{(T^\rho - a^\rho)^\beta(T^\rho - t^\rho)}{\rho^{\beta+1}\Gamma(\beta+2)} - \frac{\mu(\xi^\rho - a^\rho)^{\beta+\gamma}[(\beta+1)(\xi^\rho - t^\rho)-\gamma(t^\rho - a^\rho)]}{\rho^{\beta+\gamma+1}\Gamma(\beta+\gamma+2)(\beta+1)}\right]\right|. \tag{20}$$

Now, we are in a position to present our main results. Our first existence result for the problem (1) is based on Krasnoselskii's fixed point theorem [25], which is stated below.

Lemma 3. *(Krasnoselskii's fixed point theorem) Let $\mathcal{S}$ be a closed convex and non-empty subset of a Banach space E. Let $\mathcal{G}_1, \mathcal{G}_2$ be the operators from $\mathcal{S}$ to E such that (a) $\mathcal{G}_1 x + \mathcal{G}_2 y \in \mathcal{S}$ whenever $u, v \in \mathcal{S}$; (b) $\mathcal{G}_1$ is compact and continuous; and (c) $\mathcal{G}_2$ is a contraction mapping. Then, there exists a fixed point $\omega \in \mathcal{S}$ such that $\omega = \mathcal{G}_1\omega + \mathcal{G}_2\omega$.*

Theorem 1. *Let $f : J\times\mathbb{R}\to\mathbb{R}$ be a continuous function such that the following condition holds:*

(A_1) There exists a continuous function $\phi \in C([a,T],\mathbb{R}^+)$ such that:

$$|f(t,u)| \le \phi(t), \quad \forall (t,u) \in J\times\mathbb{R}.$$

Then, the problem (1) has at least one solution on J, provided that:

$$\Lambda_2 < 1. \tag{21}$$

Proof. Introduce a closed ball $B_r = \{x \in \mathcal{C} : \|x\| \le r\}$, with $r > \frac{\|\phi\|\Lambda_1}{1-\Lambda_2}$, $\|\phi\| = \sup_{t\in[a,T]}|\phi(t)|$, where Λ_2 is given by (18). Then, we define operators $\mathcal{F}_1$ and $\mathcal{F}_2$ from B_r to $\mathcal{C}$ by:

$$\begin{aligned}\mathcal{F}_1(x)(t) &= {}^\rho I_{a+}^{\alpha+\beta} f(t,x(t)) + \frac{(t^\rho - a^\rho)^\beta(\eta^\rho - t^\rho)}{\rho^{\beta+1}\Gamma(\beta+2)\Omega}\left\{{}^\rho I_{a+}^{\alpha+\beta} f(T,x(T)) - \mu^\rho I_{a+}^{\alpha+\beta+\gamma} f(\xi,x(\xi))\right\}\\ &\quad-\frac{(t^\rho - a^\rho)^\beta}{\Omega(\eta^\rho - a^\rho)^\beta}\left[\frac{(T^\rho - a^\rho)^\beta(T^\rho - t^\rho)}{\rho^{\beta+1}\Gamma(\beta+2)} - \frac{\mu(\xi^\rho - a^\rho)^{\beta+\gamma}[(\beta+1)(\xi^\rho - t^\rho)-\gamma(t^\rho - a^\rho)]}{\rho^{\beta+\gamma+1}\Gamma(\beta+\gamma+2)(\beta+1)}\right]\\ &\quad\times{}^\rho I_{a+}^{\alpha+\beta} f(\eta,x(\eta)),\\ \mathcal{F}_2(x)(t) &= -\lambda^\rho I_{a+}^{\beta} x(t) - \frac{(t^\rho - a^\rho)^\beta(\eta^\rho - t^\rho)}{\rho^{\beta+1}\Gamma(\beta+2)\Omega}\left\{\lambda^\rho I_{a+}^{\beta} x(T) - \mu\lambda^\rho I_{a+}^{\beta+\gamma} x(\xi)\right\} + \frac{\lambda(t^\rho - a^\rho)^\beta}{\Omega(\eta^\rho - a^\rho)^\beta}\times\\ &\quad\times\left[\frac{(T^\rho - a^\rho)^\beta(T^\rho - t^\rho)}{\rho^{\beta+1}\Gamma(\beta+2)} - \frac{\mu(\xi^\rho - a^\rho)^{\beta+\gamma}[(\beta+1)(\xi^\rho - t^\rho)-\gamma(t^\rho - a^\rho)]}{\rho^{\beta+\gamma+1}\Gamma(\beta+\gamma+2)(\beta+1)}\right]{}^\rho I_{a+}^{\beta} x(\eta).\end{aligned}$$

Note that $\mathcal{F} = \mathcal{F}_1 + \mathcal{F}_2$ on B_r. For $x, y \in B_r$, we find that:

$$\begin{aligned}
&\|\mathcal{F}_1 x + \mathcal{F}_2 y\| \\
\leq\ & \sup_{t\in J} \Big\{ {}^\rho I_{a+}^{\alpha+\beta} |f(t,x(t))| + |\lambda| {}^\rho I_{a+}^{\beta} |y(t)| \\
&+ \frac{(t^\rho - a^\rho)^\beta |(\eta^\rho - t^\rho)|}{\rho^{\beta+1}\Gamma(\beta+2)|\Omega|} \Big\{ {}^\rho I_{a+}^{\alpha+\beta} |f(T,x(T))| + |\lambda| {}^\rho I_{a+}^{\beta} |y(T)| \\
&+ |\mu| {}^\rho I_{a+}^{\alpha+\beta+\gamma} |f(\xi,x(\xi))| + |\mu||\lambda| {}^\rho I_{a+}^{\beta+\gamma} |x(\xi)| \Big\} + \frac{(t^\rho - a^\rho)^\beta}{|\Omega|(\eta^\rho - a^\rho)^\beta} \Big| \frac{(T^\rho - a^\rho)^\beta (T^\rho - t^\rho)}{\rho^{\beta+1}\Gamma(\beta+2)} \\
&- \frac{\mu(\xi^\rho - a^\rho)^{\beta+\gamma}[(\beta+1)(\xi^\rho - t^\rho) - \gamma(t^\rho - a^\rho)]}{\rho^{\beta+\gamma+1}\Gamma(\beta+\gamma+2)(\beta+1)} \Big| \Big\{ {}^\rho I_{a+}^{\alpha+\beta} |f(\eta,x(\eta))| + |\lambda| {}^\rho I_{a+}^{\beta} |y(\eta)| \Big\} \\
\leq\ & \|\phi\| \Big\{ \frac{(T^\rho - a^\rho)^{\alpha+\beta}}{\rho^{\alpha+\beta}\Gamma(\alpha+\beta+1)} \Big[1 + \frac{\zeta_1}{\rho^{\beta+1}\Gamma(\beta+2)|\Omega|} \Big] + \frac{|\mu|(\xi^\rho - a^\rho)^{\alpha+\beta+\gamma}\zeta_1}{\rho^{\alpha+2\beta+\gamma+1}\Gamma(\alpha+\beta+\gamma+1)\Gamma(\beta+2)|\Omega|} \\
&+ \frac{(\eta^\rho - a^\rho)^\alpha \zeta_2}{\rho^{\alpha+\beta}\Gamma(\alpha+\beta+1)|\Omega|} \Big\} + \|x\| \Big\{ \frac{|\lambda|(T^\rho - a^\rho)^\beta}{\rho^\beta\Gamma(\beta+1)} \Big[1 + \frac{\zeta_1}{\rho^{\beta+1}\Gamma(\beta+2)|\Omega|} \Big] \\
&+ \frac{|\mu||\lambda|(\xi^\rho - a^\rho)^{\beta+\gamma}\zeta_1}{\rho^{2\beta+\gamma+1}\Gamma(\beta+\gamma+1)\Gamma(\beta+2)|\Omega|} + \frac{|\lambda|\zeta_2}{\rho^\beta\Gamma(\beta+1)|\Omega|} \Big\} \\
\leq\ & \|\phi\|\Lambda_1 + r\Lambda_2 < r.
\end{aligned}$$

Thus, $\mathcal{F}_1 x + \mathcal{F}_2 y \in B_r$.
Next, it will be shown that $\mathcal{F}_2$ is a contraction. For that, let $x, y \in C$. Then:

$$\begin{aligned}
&\|\mathcal{F}_2 x - \mathcal{F}_2 y\| \\
\leq\ & \sup_{t\in J} \Big\{ |\lambda| {}^\rho I_{a+}^{\beta} |x(t) - y(t)| + \frac{|(t^\rho - a^\rho)^\beta (\eta^\rho - t^\rho)|}{\rho^{\beta+1}\Gamma(\beta+2)|\Omega|} \times \\
&\times \Big\{ |\lambda| {}^\rho I_{a+}^{\beta} |x(T) - y(T)| + |\mu||\lambda| {}^\rho I_{a+}^{\beta+\gamma} |x(\xi) - y(\xi)| \Big\} \\
&+ \frac{|\lambda||(t^\rho - a^\rho)^\beta|}{|\Omega|(\eta^\rho - a^\rho)^\beta} \Big| \frac{(T^\rho - a^\rho)^\beta (T^\rho - t^\rho)}{\rho^{\beta+1}\Gamma(\beta+2)} - \frac{\mu(\xi^\rho - a^\rho)^{\beta+\gamma}[(\beta+1)(\xi^\rho - t^\rho) - \gamma(t^\rho - a^\rho)]}{\rho^{\beta+\gamma+1}\Gamma(\beta+\gamma+2)(\beta+1)} \Big| \times \\
&\times {}^\rho I_{a+}^{\beta} |x(\eta) - y(\eta)| \Big\} \\
\leq\ & \Big\{ \frac{|\lambda|(T^\rho - a^\rho)^\beta}{\rho^\beta\Gamma(\beta+1)} \Big[1 + \frac{\zeta_1}{\rho^{\beta+1}\Gamma(\beta+2)|\Omega|} \Big] + \frac{|\mu||\lambda|(\xi^\rho - a^\rho)^{\beta+\gamma}\zeta_1}{\rho^{2\beta+\gamma+1}\Gamma(\beta+\gamma+1)\Gamma(\beta+2)|\Omega|} \\
&+ \frac{|\lambda|\zeta_2}{\rho^\beta\Gamma(\beta+1)|\Omega|} \Big\} \|x - y\| \\
=\ & \Lambda_2 \|x - y\|,
\end{aligned}$$

which, by the condition (21), implies that $\mathcal{F}_2$ is a contraction. The continuity of the operator $\mathcal{F}_1$ follows from that of f. Furthermore, $\mathcal{F}_1$ is uniformly bounded on B_r as:

$$\|\mathcal{F}_1 x\| \leq \|\phi\|\Lambda_1.$$

Finally, we establish the compactness of the operator $\mathcal{F}_1$. Let us set $\sup_{(t,x)\in J\times B_r} |f(t,x)| = \bar{f} < \infty$. Then, for $t_1, t_2 \in J$, $t_1 < t_2$, we have:

$$|(\mathcal{F}_1 x)(t_2) - (\mathcal{F}_1 x)(t_1)|$$

$$
\begin{aligned}
&= \Big|\frac{\rho^{1-(\alpha+\beta)}}{\Gamma(\alpha+\beta)}\Big[\int_0^{t_1} s^{\rho-1}[(t_2^\rho-s^\rho)^{\alpha+\beta-1}-(t_1^\rho-s^\rho)^{\alpha+\beta-1}]f(s,x(s))ds \\
&\quad+\int_{t_1}^{t_2} s^{\rho-1}(t_2^\rho-s^\rho)^{\alpha+\beta-1}f(s,x(s))ds\Big] \\
&\quad+\Big[\frac{(t_2^\rho-a^\rho)^\beta(\eta^\rho-t_2^\rho)}{\rho^{\beta+1}\Gamma(\beta+2)\Omega}-\frac{(t_1^\rho-a^\rho)^\beta(\eta^\rho-t_1^\rho)}{\rho^{\beta+1}\Gamma(\beta+2)\Omega}\Big]\Big\{{}^\rho I_{a+}^{\alpha+\beta}f(T,x(T))-\mu^\rho I_{a+}^{\alpha+\beta+\gamma}f(\xi,x(\xi))\Big\} \\
&\quad-\Big[\frac{(t_2^\rho-a^\rho)^\beta}{\Omega(\eta^\rho-a^\rho)^\beta}\Big[\frac{(T^\rho-a^\rho)^\beta(T^\rho-t_2^\rho)}{\rho^{\beta+1}\Gamma(\beta+2)}-\frac{\mu(\xi^\rho-a^\rho)^{\beta+\gamma}[(\beta+1)(\xi^\rho-t_2^\rho)-\gamma(t^\rho-a^\rho)]}{\rho^{\beta+\gamma+1}\Gamma(\beta+\gamma+2)(\beta+1)}\Big] \\
&\quad-\frac{(t_1^\rho-a^\rho)^\beta}{\Omega(\eta^\rho-a^\rho)^\beta}\Big[\frac{(T^\rho-a^\rho)^\beta(T^\rho-t_1^\rho)}{\rho^{\beta+1}\Gamma(\beta+2)}-\frac{\mu(\xi^\rho-a^\rho)^{\beta+\gamma}[(\beta+1)(\xi^\rho-t_1^\rho)-\gamma(t_1^\rho-a^\rho)]}{\rho^{\beta+\gamma+1}\Gamma(\beta+\gamma+2)(\beta+1)}\Big]\Big]\times \\
&\quad\times{}^\rho I_{a+}^{\alpha+\beta}f(\eta,x(\eta))\Big| \\
&\le \frac{\bar f}{\rho^{\alpha+\beta}\Gamma(\alpha+\beta+1)}\Big\{|t_2^{\rho(\alpha+\beta)}-t_1^{\rho(\alpha+\beta)}|+2(t_2^\rho-t_1^\rho)^{\alpha+\beta}\Big\} \\
&\quad+\Big|\frac{(t_2^\rho-a^\rho)^\beta(\eta^\rho-t_2^\rho)}{\rho^{\beta+1}\Gamma(\beta+2)\Omega}-\frac{(t_1^\rho-a^\rho)^\beta(\eta^\rho-t_1^\rho)}{\rho^{\beta+1}\Gamma(\beta+2)\Omega}\Big|\Big\{{}^\rho I_{a+}^{\alpha+\beta}|f(T,x(T))|+\mu^\rho I_{a+}^{\alpha+\beta+\gamma}|f(\xi,x(\xi))|\Big\} \\
&\quad+\Big|\frac{(t_2^\rho-a^\rho)^\beta}{\Omega(\eta^\rho-a^\rho)^\beta}\Big[\frac{(T^\rho-a^\rho)^\beta(T^\rho-t_2^\rho)}{\rho^{\beta+1}\Gamma(\beta+2)}-\frac{\mu(\xi^\rho-a^\rho)^{\beta+\gamma}[(\beta+1)(\xi^\rho-t_2^\rho)-\gamma(t_2^\rho-a^\rho)]}{\rho^{\beta+\gamma+1}\Gamma(\beta+\gamma+2)(\beta+1)}\Big] \\
&\quad-\frac{(t_1^\rho-a^\rho)^\beta}{\Omega(\eta^\rho-a^\rho)^\beta}\Big[\frac{(T^\rho-a^\rho)^\beta(T^\rho-t_1^\rho)}{\rho^{\beta+1}\Gamma(\beta+2)}-\frac{\mu(\xi^\rho-a^\rho)^{\beta+\gamma}[(\beta+1)(\xi^\rho-t_1^\rho)-\gamma(t_1^\rho-a^\rho)]}{\rho^{\beta+\gamma+1}\Gamma(\beta+\gamma+2)(\beta+1)}\Big]\Big|\times \\
&\quad\times{}^\rho I_{a+}^{\alpha+\beta}|f(\eta,x(\eta))|,
\end{aligned}
$$

which tends to zero as $t_2 \to t_1$, independently of $x \in B_r$. Thus, $\mathcal{F}_1$ is equicontinuous. Therefore, $\mathcal{F}_1$ is relatively compact on B_r. As a consequence, we deduce by the the Arzelá–Ascoli theorem that $\mathcal{F}_1$ is compact on B_r. Thus, the hypothesis of Lemma 3 is satisfied. Therefore, the conclusion of Lemma 3 applies, and hence, there exists at least one solution for the problem (1) on J. □

In the next result, the uniqueness of solutions for the problem (1) is shown by means of the Banach contraction mapping principle.

Theorem 2. *Let $f : J \times \mathbb{R} \to \mathbb{R}$ be a continuous function satisfying the Lipschitz condition:*

(A_2)

$$|f(t,u)-f(t,v)| \le L|u-v|, L>0, /, \quad \text{for } t\in J \text{ and every } u,v\in\mathbb{R}.$$

Then, there exists a unique solution for the problem (1) on $[a,T]$, provided that:

$$L\Lambda_1+\Lambda_2<1, \tag{22}$$

where Λ_1 and Λ_2 are respectively given by (17) and (18).

Proof. In the first step, we show that $\mathcal{F}B_{\bar r} \subset B_{\bar r}$, where $B_{\bar r} = \{x \in C([a,T],\mathbb{R}) : \|x\| \le \bar r\}$, $M = \sup_{t\in[a,T]}|f(t,0)|$, $\bar r \ge \frac{\Lambda_1 M}{1-L\Lambda_1-\Lambda_2}$, and the operator $\mathcal{F} : C \to C$ is given by (16). For $x \in B_{\bar r}$, using (A_2), we get:

$$
\begin{aligned}
&|\mathcal{F}(x)(t)| \\
&\le {}^\rho I_{a+}^{\alpha+\beta}[|f(t,x(t))-f(t,0)|+|f(t,0)|]+|\lambda|^\rho I_{a+}^{\beta}|x(t)| \\
&\quad+\frac{(t^\rho-a^\rho)^\beta|(\eta^\rho-t^\rho)|}{\rho^{\beta+1}\Gamma(\beta+2)|\Omega|}\Big\{{}^\rho I_{a+}^{\alpha+\beta}[|f(T,x(T))-f(T,0)|+|f(T,0)|]+|\lambda|^\rho I_{a+}^{\beta}|x(T)| \\
&\quad+|\mu|^\rho I_{a+}^{\alpha+\beta+\gamma}[|f(\xi,x(\xi))-f(\xi,0)|+|f(\xi,0)|]+|\mu||\lambda|^\rho I_{a+}^{\beta+\gamma}|x(\xi)|\Big\}
\end{aligned}
$$

$$
\begin{aligned}
&+\frac{(t^\rho-a^\rho)^\beta}{|\Omega|(\eta^\rho-a^\rho)^\beta}\Big|\frac{(T^\rho-a^\rho)^\beta(T^\rho-t^\rho)}{\rho^{\beta+1}\Gamma(\beta+2)}-\frac{\mu(\xi^\rho-a^\rho)^{\beta+\gamma}[(\beta+1)(\xi^\rho-t^\rho)-\gamma(t^\rho-a^\rho)]}{\rho^{\beta+\gamma+1}\Gamma(\beta+\gamma+2)(\beta+1)}\Big| \\
&\times\Big\{{}^\rho I_{a+}^{\alpha+\beta}[|f(\eta,x(\eta))-f(\eta,0)|+|f(\eta,0)|]+|\lambda|{}^\rho I_{a+}^{\beta}|x(\eta)|\Big\} \\
\leq\ & (L\bar{r}+M)\Big(\frac{(T^\rho-a^\rho)^{\alpha+\beta}}{\rho^{\alpha+\beta}\Gamma(\alpha+\beta+1)}\Big[1+\frac{\zeta_1}{\rho^{\beta+1}\Gamma(\beta+2)|\Omega|}\Big]+\frac{|\mu|(\xi^\rho-a^\rho)^{\alpha+\beta+\gamma}\zeta_1}{\rho^{\alpha+2\beta+\gamma+1}\Gamma(\alpha+\beta+\gamma+1)\Gamma(\beta+2)|\Omega|} \\
&+\frac{(\eta^\rho-a^\rho)^\alpha\zeta_2}{\rho^{\alpha+\beta}\Gamma(\alpha+\beta+1)|\Omega|}\Big) \\
&+\bar{r}\Big(\frac{|\lambda|(T^\rho-a^\rho)^\beta}{\rho^\beta\Gamma(\beta+1)}\Big[1+\frac{\zeta_1}{\rho^{\beta+1}\Gamma(\beta+2)|\Omega|}\Big]+\frac{|\mu||\lambda|(\xi^\rho-a^\rho)^{\beta+\gamma}\zeta_1}{\rho^{2\beta+\gamma+1}\Gamma(\beta+\gamma+1)\Gamma(\beta+2)|\Omega|}+\frac{|\lambda|\zeta_2}{\rho^\beta\Gamma(\beta+1)|\Omega|}\Big) \\
=\ & (L\bar{r}+M)\Lambda_1+\Lambda_2\bar{r}\leq\bar{r},
\end{aligned}
$$

which, on taking the norm for $t\in[a,T]$, implies that $\|\mathcal{F}(x)\|\leq\bar{r}$. Thus, the operator $\mathcal{F}$ maps $B_{\bar{r}}$ into itself. Now, we proceed to prove that the operator $\mathcal{F}$ is a contraction. For $x,y\in C([a,T],\mathbb{R})$ and $t\in[a,T]$, we have:

$$
\begin{aligned}
&|\mathcal{F}(x)(t)-\mathcal{F}(y)(t)| \\
\leq\ & {}^\rho I_{a+}^{\alpha+\beta}|f(t,x(t))-f(t,y(t))|+|\lambda|{}^\rho I_{a+}^{\beta}|x(t)-y(t)| \\
&+\frac{(t^\rho-a^\rho)^\beta|(\eta^\rho-t^\rho)|}{\rho^{\beta+1}\Gamma(\beta+2)|\Omega|}\Big\{{}^\rho I_{a+}^{\alpha+\beta}|f(T,x(T))-f(T,y(T))|+|\lambda|{}^\rho I_{a+}^{\beta}|x(T)-y(T)| \\
&+|\mu|{}^\rho I_{a+}^{\alpha+\beta+\gamma}|f(\xi,x(\xi))-f(\xi,y(\xi))|+|\mu||\lambda|{}^\rho I_{a+}^{\beta+\gamma}|x(\xi)-y(\xi)|\Big\} \\
&+\frac{(t^\rho-a^\rho)^\beta}{|\Omega|(\eta^\rho-a^\rho)^\beta}\Big|\frac{(T^\rho-a^\rho)^\beta(T^\rho-t^\rho)}{\rho^{\beta+1}\Gamma(\beta+2)}-\frac{\mu(\xi^\rho-a^\rho)^{\beta+\gamma}[(\beta+1)(\xi^\rho-t^\rho)-\gamma(t^\rho-a^\rho)]}{\rho^{\beta+\gamma+1}\Gamma(\beta+\gamma+2)(\beta+1)}\Big|\times \\
&\times\Big\{{}^\rho I_{a+}^{\alpha+\beta}|f(\eta,x(\eta))-f(\eta,y(\eta))|+|\lambda|{}^\rho I_{a+}^{\beta}|x(\eta)-y(\eta)|\Big\} \\
\leq\ & L\|x-y\|\Big(\frac{(T^\rho-a^\rho)^{\alpha+\beta}}{\rho^{\alpha+\beta}\Gamma(\alpha+\beta+1)}\Big[1+\frac{\zeta_1}{\rho^{\beta+1}\Gamma(\beta+2)|\Omega|}\Big] \\
&+\frac{|\mu|(\xi^\rho-a^\rho)^{\alpha+\beta+\gamma}\zeta_1}{\rho^{\alpha+2\beta+\gamma+1}\Gamma(\alpha+\beta+\gamma+1)\Gamma(\beta+2)|\Omega|} \\
&+\frac{(\eta^\rho-a^\rho)^\alpha\zeta_2}{\rho^{\alpha+\beta}\Gamma(\alpha+\beta+1)|\Omega|}\Big)+\|x-y\|\Big(\frac{|\lambda|(T^\rho-a^\rho)^\beta}{\rho^\beta\Gamma(\beta+1)}\Big[1+\frac{\zeta_1}{\rho^{\beta+1}\Gamma(\beta+2)|\Omega|}\Big] \\
&+\frac{|\mu||\lambda|(\xi^\rho-a^\rho)^{\beta+\gamma}\zeta_1}{\rho^{2\beta+\gamma+1}\Gamma(\beta+\gamma+1)\Gamma(\beta+2)|\Omega|}+\frac{|\lambda|\zeta_2}{\rho^\beta\Gamma(\beta+1)|\Omega|}\Big) \\
=\ & (L\Lambda_1+\Lambda_2)\|x-y\|.
\end{aligned}
$$

Taking the norm of the above inequality for $t\in[a,T]$, we get:

$$\|\mathcal{F}(x)-\mathcal{F}(y)\|\leq(L\Lambda_1+\Lambda_2)\|x-y\|,$$

which implies that the operator $\mathcal{F}$ is a contraction on account of the condition (22). Thus, we deduce by the Banach contraction mapping principle that the operator $\mathcal{F}$ has a unique fixed point. Hence, there exists a unique solution for the problem (1). The proof is complete. □

Example 1. *Let us consider the following boundary value problem:*

$$
\begin{cases}
{}^{1/3}_{\ c}D^{5/4}\big({}^{1/3}_{\ c}D^{1/4}+1/5\big)x(t)=\dfrac{1}{\sqrt{400+t}}\Big(\dfrac{|x(t)|+2}{|x(t)|+1}+e^{-t}\Big), & t\in J:=[1,2], \\
x(1)=0,\quad x(3/2)=0,\quad x(2)=2/7\,{}^{1/3}I^{3/4}x(7/4). &
\end{cases}
\tag{23}
$$

Here, $\rho = 1/3, \alpha = 5/4$, $\beta = 1/4$, $\gamma = 3/4$, $\lambda = 1/5$, $\mu = 2/7$, $a = 1$, $\eta = 3/2$, $\xi = 7/4$, $T = 2$ and $f(t,x) = \frac{1}{\sqrt{400+t}}\left(\frac{|x|+2}{|x|+1} + e^{-t}\right)$. Using the given data, we find that $|\Omega| \approx 0.293634, \Lambda_1 \approx 1.336009, \Lambda_2 \approx 0.673563, \zeta_1 \approx 0.082260$, $\zeta_2 \approx 0.232036$, where $\Omega, \Lambda_1, \Lambda_2, \zeta_1$, and ζ_2 are given by (10), (17), (18), (19) and (20) respectively.

For illustrating Theorem 1, we show that all the conditions of Theorem 1 are satisfied. Clearly, $f(t,x)$ is continuous and satisfies the condition (A_1) with $\phi(t) = \frac{2+e^{-t}}{\sqrt{400+t}}$. Furthermore, $\Lambda_2 \approx 0.673563 < 1$. Thus, all the conditions of Theorem 1 are satisfied, and consequently, the problem (23) has at least one solution on $[1,2]$.

Furthermore, Theorem 2 is applicable to the problem (23) with $L = 1/20$ as $L\Lambda_1 + \Lambda_2 \approx 0.740363 < 1$. Thus, all the assumptions of Theorem 2 are satisfied. Therefore, the conclusion of Theorem 2 applies to the problem (23) on $[1,2]$.

4. Conclusions

We have introduced a new type of nonlinear Langevin equation in terms of Liouville-Caputo-type generalized fractional differential operators of different orders and solved it with nonlocal generalized integral boundary conditions. The existence result was obtained by applying the Krasnoselskii fixed point theorem without requiring the nonlinear function to be of the Lipschitz type, while the uniqueness of solutions for the given problem was based on a celebrated fixed point theorem due to Banach. Here, we remark that many known existence results, obtained by means of the Krasnoselskii fixed point theorem, demand the associated nonlinear function to satisfy the Lipschitz condition. Moreover, by fixing the parameters involved in the given problem, we can obtain some new results as special cases of the ones presented in this paper. For example, letting $\rho = 1$, $\mu = 0$, $a = 0$ and $T = 1$ in the results of Section 3, we get the ones derived in [15].

Author Contributions: Formal analysis, B.A., M.A., A.A., H.M.S. and S.K.N.

References

1. Zaslavsky, G.M. *Hamiltonian Chaos and Fractional Dynamics*; Oxford University Press: Oxford, UK, 2005.
2. Magin, R.L. *Fractional Calculus in Bioengineering*; Begell House Publishers: Danbury, CT, USA, 2006.
3. Kilbas, A.A.; Srivastava, H.M.; Trujillo, J.J. *Theory and Applications of Fractional Differential Equations*; North-Holland Mathematics Studies, 204; Elsevier Science B.V.: Amsterdam, The Netherlands, 2006.
4. Diethelm, K. *The Analysis of Fractional Differential Equations. An Application-Oriented Exposition Using Differential Operators of Liouville-Caputo Type. Lecture Notes in Mathematics 2004*; Springer: Berlin, Germany, 2010.
5. Javidi, M.; Ahmad, B. Dynamic analysis of time fractional order phytoplankton-toxic phytoplankton-zooplankton system. *Ecol. Model.* **2015**, *318*, 8–18. [CrossRef]
6. Fallahgoul, H.A.; Focardi, S.M.; Fabozzi, F.J. *Fractional Calculus and Fractional Processes with Applications to Financial Economics. Theory and Application*; Elsevier/Academic Press: London, UK, 2017.
7. Ahmad, B.; Alsaedi, A.; Ntouyas, S.K.; Tariboon, J. *Hadamard-Type Fractional Differential Equations, Inclusions and Inequalities*; Springer: Cham, Switzerland, 2017.
8. West, B.J.; Picozzi, S. Fractional Langevin model of memory in financial time series. *Phys. Rev. E* **2002**, *65*, 037106. [CrossRef] [PubMed]
9. Vinales, A.D.; Desposito, M.A. Anomalous diffusion: Exact solution of the generalized Langevin equation for harmonically bounded particle. *Phys. Rev. E* **2006**, *73*, 016111. [CrossRef] [PubMed]
10. Hohenberg, P.C.; Halperin, B.I. Theory of dynamic critical phenomena. *Rev. Mod. Phys.* **1977**, *49*, 435–479. [CrossRef]
11. Metiu, H.; Schon, G. Description of Quantum noise by a Langevin equation. *Phys. Rev. Lett.* **1984**, *53*, 13. [CrossRef]
12. Datsko, B.; Gafiychuk, V. Complex nonlinear dynamics in subdiffusive activator–inhibitor systems. *Commun. Nonlinear Sci. Numer. Simul.* **2012**, *17*, 1673–1680. [CrossRef]

13. Datsko, B.; Gafiychuk, V. Complex spatio-temporal solutions in fractional reaction-diffusion systems near a bifurcation point. *Fract. Calc. Appl. Anal.* **2018**, *21*, 237–253. [CrossRef]
14. Fa, K.S. Fractional Langevin equation and Riemann–Liouville fractional derivative. *Eur. Phys. J. E* **2007**, *24*, 139–143.
15. Ahmad, B.; Nieto, J.J.; Alsaedi, A.; El-Shahed, M. A study of nonlinear Langevin equation involving two fractional orders in different intervals. *Nonlinear Anal. Real World Appl.* **2012**, *13*, 599–606. [CrossRef]
16. Wang, G.; Zhang, L.; Song, G. Boundary value problem of a nonlinear Langevin equation with two different fractional orders and impulses. *Fixed Point Theory Appl.* **2012**, *2012*, 200. [CrossRef]
17. Ahmad, B.; Ntouyas, S.K. New existence results for differential inclusions involving Langevin equation with two indices. *J. Nonlinear Convex Anal.* **2013**, *14*, 437–450.
18. Muensawat, T.; Ntouyas, S.K.; Tariboon, J. Systems of generalized Sturm-Liouville and Langevin fractional differential equations. *Adv. Differ. Equ.* **2017**, *2017*, 63. [CrossRef]
19. Fazli, H.; Nieto, J.J. Fractional Langevin equation with anti-periodic boundary conditions. *Chaos Solitons Fractals* **2018**, *114*, 332–337. [CrossRef]
20. Ahmad, B.; Alsaedi, A.; Salem, S. On a nonlocal integral boundary value problem of nonlinear Langevin equation with different fractional orders. *Adv. Differ. Equ.* **2019**, *2019*, 57. [CrossRef]
21. Laskin, N. Fractional quantum mechanics and Levy path integrals. *Phys. Lett. A* **2000**, *268*, 298–305. [CrossRef]
22. Katugampola, U.N. New Approach to a generalized fractional integral. *Appl. Math. Comput.* **2015**, *218*, 860–865. [CrossRef]
23. Katugampola, U.N. A new approach to generalized fractional derivatives. *Bull. Math. Anal. Appl.* **2014**, *6*, 1–15.
24. Jarad, F.; Abdeljawad, T.; Baleanu, D. On the generalized fractional derivatives and their Caputo modification. *J. Nonlinear Sci. Appl.* **2017**, *10*, 2607–2619. [CrossRef]
25. Krasnoselskii, M.A. Two remarks on the method of successive approximations. *Uspekhi Matematicheskikh Nauk* **1955**, *10*, 123–127.

6

Nonlocal Fractional Evolution Inclusions of Order $\alpha \in (1, 2)$

Jia Wei He [1], Yong Liang [2], Bashir Ahmad [3] and Yong Zhou [1,3,*]

1 Faculty of Mathematics and Computational Science, Xiangtan University, Xiangtan 411105, China; hjw.haoye@163.com
2 Faculty of Information Technology, Macau University of Science and Technology, Macau 999078, China; yliang@must.edu.mo
3 Nonlinear Analysis and Applied Mathematics (NAAM) Research Group, Faculty of Science, King Abdulaziz University, Jeddah 21589, Saudi Arabia; bashirahmad_qau@yahoo.com
* Correspondence: yzhou@xtu.edu.cn

Abstract: This paper studies the existence of mild solutions and the compactness of a set of mild solutions to a nonlocal problem of fractional evolution inclusions of order $\alpha \in (1,2)$. The main tools of our study include the concepts of fractional calculus, multivalued analysis, the cosine family, method of measure of noncompactness, and fixed-point theorem. As an application, we apply the obtained results to a control problem.

Keywords: fractional evolution inclusions; mild solutions; condensing multivalued map

---

1. Introduction

In the past several decades, there has been a significant development in the theory and applications for fractional evolution equations and inclusions; for example, see the monographs by Miller and Ross [1], Podlubny [2], Kilbas et al. [3], Zhou [4], and the recent papers [5,6]. More recently, time-fractional diffusion and wave equations have been attracting the widespread attention of many fields of science and engineering [7,8]. The interest in the study of these topics arises from the fact that fractional diffusion equations $\alpha \in (0,1)$ or fractional wave equations $\alpha \in (1,2)$ can capture some nonlocal aspects of phenomena or systems. Examples of these phenomena include porous media, memory effects, anomalous diffusion, viscoelastic media, and so on. The papers [9–11] cover many of these applications.

By virtue of semigroup theory and the operator theoretical method, some fractional diffusion and wave equations can be abstracted as fractional evolution equations. Bajlekova [12] exploited the concept of the fractional resolvent solution operator to investigate the associated fractional abstract Cauchy problem. A number of papers [13–17] and the references therein were inspired by this concept, and the topic of the existence of mild solutions to fractional abstract equations of order $\alpha \in (1,2)$ was also studied. For further discussion in [18], the authors considered the controllability results for fractional evolution equations of order $\alpha \in (1,2)$ by applying the concepts of Mainardi's Wright function (a probability density function) and strongly continuous cosine families.

The study of fractional evolution inclusions of order $\alpha \in (0,1)$ also gained significant importance (see, e.g., [19,20]). However, the study of fractional evolution inclusions of order $\alpha \in (1,2)$ supplemented with nonlocal conditions is yet to be initiated. We need to point out that the work spaces are of finite dimension if the strongly continuous cosine families are compact (see, e.g., [21,22]). Motivated by this fact and the above-mentioned works and relying on the known material, we aim to develop a suitable definition for mild solutions of fractional evolution equations in terms of Mainardi's Wright function. For this purpose, we consider the following nonlocal problem of fractional evolution inclusions without further assumptions regarding the compactness of the cosine families or the associated sine families.

$$\begin{cases} {}^C D_t^\alpha x(t) \in Ax(t) + F(t, x(t)), & t \in J = [0, a],\ a > 0, \\ x(0) + g(x) = x_0,\ x'(0) = x_1, \end{cases} \tag{1}$$

where ${}^C D_t^\alpha$ is a Caputo fractional derivative of order $1 < \alpha < 2$; A is the infinitesimal generator of a strongly continuous cosine family $\{C(t)\}_{t \geq 0}$ of uniformly bounded linear operators in a Banach space X; $F : [0, a] \times X \to X$ is a multivalued map; g is a given appropriate function; and x_0, x_1 are elements of space X.

Here, we emphasize that the present work is also motivated by an inclusion of the following partial differential model:

$$\begin{cases} \partial_t^\alpha u(t,z) \in \partial_z^2 u(t,z) + F(t, z, u(t,z)), & z \in [0, \pi],\ t \in [0, a], \\ u(t, 0) = u(t, \pi) = 0, & t \in [0, a], \\ u(0, z) + g(u) = u_0(z),\ u'(0, z) = u_1(z), & z \in [0, \pi], \end{cases}$$

where ∂_t^α is a Caputo fractional partial derivative. This model includes a class of fractional wave equations that have a memory effect and are not observed in integer-order differential equations; further, this class of equations indicates the coexistence of finite wave speed and absence of a wavefront (see, e.g., [9]). It is interesting that for the case of $\alpha = 2$, the above fractional partial differential inclusion reduces to a second-order differential inclusion involving one-dimensional wave equations with nonlocal initial-boundary conditions. For the case of $\alpha = 1$ or $\alpha \in (0, 1)$ with the initial value $u_1(z)$ vanished, the model contains the classical diffusion equations or fractional diffusion equations. In addition, these types of equations can be handled by the method of semigroup theory (see, e.g., [20]) but not cosine families.

The rest of this paper is organized as follows. In Section 2, we recall some preliminary concepts related to our study. In Section 3, we establish an existence result for mild solutions of Equation (1) and discuss the compactness of the set of mild solutions. In Section 4, we show the utility of the obtained work by applying it to a control problem.

2. Preliminaries

Let X be a Banach space with the norm $\|\cdot\|$. Denote by $\mathcal{L}(X)$ the space of all bounded linear operators from X to X equipped with the norm $\|\cdot\|_{\mathcal{L}(X)}$. Let $C(J, X)$ denote the space of all continuous functions from J into X equipped with the usual sup-norm $\|x\|_C = \sup_{t \in J} \|x(t)\|$, where $J = [0, a], a > 0$. A measurable function $f : J \to X$ is Bochner integrable if $\|f\|$ is Lebesgue integrable. Let $L^p(J, X)$ $(p \geq 1)$ be the Banach space of measurable functions (defined in the sense of Bochner integral) endowed with the norm

$$\|f\|_p = \left(\int_J \|f(t)\|^p dt \right)^{\frac{1}{p}}.$$

Definition 1. *The fractional integral with the lower limit zero for a function $u : [0, \infty) \to X$ is given by*

$$I_{0+}^\alpha u(t) = \frac{1}{\Gamma(\alpha)} \int_0^t (t - s)^{\alpha - 1} u(s) ds, \quad t > 0,\ \alpha \in \mathbb{R}_+,$$

provided the right side is point-wise defined on $[0, \infty)$, where $\Gamma(\cdot)$ is the gamma function.

Definition 2. *The Riemann–Liouville derivative with the lower limit zero for a function* $u : [0, \infty) \to X$ *is defined by*

$$^{L}D^{\alpha}_{0+}u(t) = \frac{1}{\Gamma(n-\alpha)} \frac{d^n}{dt^n} \int_0^t (t-s)^{n-\alpha-1} u(s) ds, \quad t > 0,\ n-1 < \alpha < n,\ \alpha \in \mathbb{R}_+.$$

Definition 3. *The Caputo derivative with the lower limit zero for a function u is defined by*

$$^{C}D^{\alpha}_{0+}u(t) = {}^{L}D^{\alpha}_{0+}\Big(u(t) - \sum_{k=0}^{n-1} \frac{u^{(k)}(0)}{k!} t^k\Big), \quad t > 0,\ n-1 < \alpha < n, \alpha \in \mathbb{R}_+.$$

Definition 4. [23] *A family of bounded linear operators* $\{C(t)\}_{t\in\mathbb{R}}$ *mapping the Banach space X into itself is called a strongly continuous cosine family if and only if* $C(0) = I$, $C(s+t) + C(s-t) = 2C(s)C(t)$ *for all* $s, t \in \mathbb{R}$, *and the map* $t \mapsto C(t)x$ *is strongly continuous for each* $x \in X$.

Let $\{S(t)\}_{t\in\mathbb{R}}$ denote the strongly continuous sine families associated with the strongly continuous cosine families $\{C(t)\}_{t\in\mathbb{R}}$, where

$$S(t)x = \int_0^t C(s)x ds, \quad x \in X,\ t \in \mathbb{R}. \tag{2}$$

In addition, an operator A is said to be an infinitesimal generator of cosine families $\{C(t)\}_{t\in\mathbb{R}}$ if

$$Ax = \frac{d^2}{dt^2} C(t)x\Big|_{t=0}, \quad \text{for all } x \in \mathcal{D}(A),$$

where the domain of A is given by $\mathcal{D}(A) = \{x \in X : C(t)x \in C^2(\mathbb{R}, X)\}$.

A multivalued map G is called upper semicontinuous (u.s.c.) on X if, for each $x_* \in X$, the set $G(x_*)$ is a nonempty subset of X, and for every open set $B \subseteq X$ such that $G(x_*) \subset B$, there exists a neighborhood V of x_* with the property that $G(V(x_*)) \subset B$. G is convex-valued if $G(x)$ is convex for all $x \in X$. G is closed if its graph $\Gamma_G = \{(x,y) \in X \times X : y \in G(x)\}$ is a closed subset of the space $X \times X$. The map G is bounded if $G(B)$ is bounded in X for every bounded set $B \subseteq X$. We say that G is completely continuous if $G(B)$ is relatively compact for every bounded subset B of X. Furthermore, if G is completely continuous with nonempty values, then G is u.s.c. if and only if G has a closed graph. If there exists an element $x \in X$ such that $x \in G(x)$, then G has a fixed point.

Let B be a subset of X. Then, we define

$$\begin{aligned}
&\mathcal{P}(X) = \{B \subseteq X : B \text{ is nonempty}\}, \quad \mathcal{P}_{cv}(X) = \{B \in \mathcal{P}(X) : B \text{ is convex}\},\\
&\mathcal{P}_{cl}(X) = \{B \in \mathcal{P}(X) : B \text{ is closed}\}, \quad \mathcal{P}_{bd}(X) = \{B \in \mathcal{P}(X) : B \text{ is bounded}\},\\
&\mathcal{P}_{cp}(X) = \{B \in \mathcal{P}(X) : B \text{ is compact}\}, \quad \mathcal{P}_{cl,cv}(X) = \mathcal{P}_{cl}(X) \cap \mathcal{P}_{cv}(X).
\end{aligned}$$

In addition, let $co(B)$ be the convex hull of a subset B, and let $\overline{co}(B)$ be the closed convex hull in X. A multivalued map $G : J \to \mathcal{P}_{cl}(X)$ is said to be measurable if, for each $x \in X$, the function $Z : J \to X$ defined by $Z(t) = d(x, G(t)) = \inf\{\|x - z\| : z \in G(t)\}$ is Lebesgue measurable. Let $G : J \to \mathcal{P}(X)$. A single-valued map $f : J \to X$ is called a selection of G if $f(t) \in G(t)$ for every $t \in J$.

Definition 5. *A multivalued map* $F : J \times X \to \mathcal{P}(X)$ *is called* L^1*-Carathéodory if*

(i) *the map* $t \mapsto F(t,x)$ *is measurable for each* $x \in X$;
(ii) *the map* $u \mapsto F(t,x)$ *is upper semicontinuous on X for almost all* $t \in J$;
(iii) *for each positive real number r, there exists* $h_r \in L^1(J, \mathbb{R}_+)$ *such that*

$$\|F(t,x)\|_{\mathcal{P}(X)} = \sup\{\|v\| : v(t) \in F(t,x)\} \le h_r(t), \ for\ \|x\| \le r,\ for\ a.e.\ t \in J.$$

For every $\Omega \in \mathcal{P}(X)$, the Hausdorff measure of noncompactness (MNC) is defined by

$$\chi(\Omega) = \inf\{\varepsilon > 0 : \ \Omega \text{ has a finite } \varepsilon\text{-net}\},$$

and the Kuratowski MNC is defined by

$$\tau(\Omega) = \inf\{d > 0 : \ \Omega \subset \bigcup_{j=1}^{n} M_j \text{ and } \operatorname{diam}(M_j) \leq d\},$$

where the diameter of M_j is given by $\operatorname{diam}(M_j) = \sup\{\|x - y\| : x, y \in M_j\}$, $j = 1, \ldots, n$. The Hausdorff and Kuratowski MNCs are connected by the relations:

$$\chi(\Omega) \leq \tau(\Omega) \leq 2\chi(\Omega).$$

A measure of noncompactness χ (or τ) is called: monotone if $\Omega_1, \Omega_2 \in \mathcal{P}(X)$ with $\Omega_1 \subseteq \Omega_2$ implies $\chi(\Omega_1) \leq \chi(\Omega_2)$; nonsingular if $\chi(\{c\} \cup \Omega) = \chi(\Omega)$ for every $c \in X$, $\Omega \in \mathcal{P}(X)$; regular if $\chi(\Omega) = 0$ is equivalent to the relative compactness of Ω.

We now introduce the MNC ν as follows: for a bounded set $D \subset C(J, X)$, we define

$$\nu(D) = \max_{D \in \Theta(D)} \left(\sup_{t \in J} \chi(D(t)), \ \operatorname{mod}_C(D) \right),$$

where $\Theta(D)$ is the collection of all denumerable subsets of D and $\operatorname{mod}_C(D)$ is the modulus of equicontinuity of the set of functions D that have the following form

$$\operatorname{mod}_C(D) = \lim_{\delta \to 0} \sup_{x \in D} \max_{|t_2 - t_1| < \delta} \|x(t_2) - x(t_1)\|.$$

It is known that the MNC ν is monotone, nonsingular, and regular. For more details on the MNC, we refer to [24,25].

Lemma 1. ([24]). *Let $W \subset X$ be bounded. Then, for each $\varepsilon > 0$, there exists a sequence $\{x_n\}_{n=1}^{\infty} \subset W$ such that*

$$\chi(W) \leq 2\chi\left(\{x_n\}_{n=1}^{\infty}\right) + \varepsilon.$$

Lemma 2. ([26]). *Let χ_C be the Hausdorff MNC on $C(J, X)$, and let $W(t) = \{x(t) : x \in W\}$. If $W \subset C(J, X)$ is bounded, then for every $t \in J$,*

$$\chi(W(t)) \leq \chi_C(W).$$

Furthermore, if W is equicontinuous, then the map $t \mapsto \chi(W(t))$ is continuous on J and

$$\chi_C(W) = \sup_{t \in J} \chi(W(t)).$$

Lemma 3. ([26]). *Let $\{x_n\}_{n=1}^{\infty}$ be a sequence of Bochner integrable functions from J into X. If there exists a function $\rho(\cdot) \in L^1(J, \mathbb{R}_+)$ satisfying $\|x_n(t)\| \leq \rho(t)$ for almost all $t \in J$ and for every $n \geq 1$, then the function $\psi(t) = \chi(\{x_n(t)\}_{n=1}^{\infty}) \in L^1(J, \mathbb{R}_+)$ satisfies*

$$\chi\left(\left\{\int_0^t x_n(s)ds : \ n \geq 1\right\}\right) \leq 2\int_0^t \psi(s)ds.$$

Lemma 4. ([27, Lemma 4]). *Let $\{f_n\}_{n=1}^{\infty} \subset L^p(J, X)$ $(p \geq 1)$ be an integrable bounded sequence satisfying*

$$\chi(\{f_n\}_{n=1}^{\infty}) \leq \gamma(t), \quad a.e., t \in J,$$

where $\gamma(\cdot) \in L^1(J, \mathbb{R}_+)$. *Then, for each* $\epsilon > 0$, *there exists a compact* $K_\epsilon \subseteq X$, *a measurable set* $J_\epsilon \subset J$ *with measure less than* ϵ, *and a sequence of functions* $\{g_n^\epsilon\}_{n=1}^\infty \subset L^p(J, X)$ *such that* $\{g_n^\epsilon(t)\}_{n=1}^\infty \subseteq K_\epsilon$, *for* $t \in J$, *and*

$$\|f_n(t) - g_n^\epsilon(t)\| < 2\gamma(t) + \epsilon, \quad for\ each\ n \geq 1\ and\ for\ every\ t \in J - J_\epsilon.$$

Lemma 5. ([28]) *Let* χ *be the Hausdorff MNC on* X. *If* $\{W_n\}_{n=1}^\infty \subset X$ *is a nonempty decreasing closed sequence and* $\lim_{n\to\infty} \chi(W_n) = 0$, *then* $\bigcap_{n=1}^\infty W_n$ *is nonempty and compact.*

Definition 6. *Let* D *be a subset of a Banach space* X. *A multivalued function* $F : D \to \mathcal{P}(X)$ *is said to be* ν*-condensing if* $\nu(F(\Omega)) \ngeq \nu(\Omega)$ *for every bounded and not relatively compact set* $\Omega \subseteq D$.

Lemma 6. ([25, Corollary 3.3.1]). *Let* Ω *be a convex closed subset of a Banach space* X *and* ν *be a nonsingular MNC defined on subsets of* Ω. *If* $F : \Omega \to P_{cv,cp}(\Omega)$ *is a closed* ν*-condensing multivalued map, then* F *has a fixed point.*

Lemma 7. ([25, Proposition 3.5.1]). *Let* Ω *be a closed subset of a Banach space* X *and* $F : \Omega \to P_{cp}(X)$ *be a closed multivalued function that is* ν*-condensing on every bounded subset of* Ω, *where* ν *is a monotone MNC in* X. *If the set of fixed points of* F *is bounded, then it is compact.*

Throughout this paper, we suppose that A is the infinitesimal generator of a strongly continuous cosine family of uniformly bounded linear operators $\{C(t)\}_{t\geq 0}$ in a Banach space X: that is, there exists $M \geq 1$ such that $\|C(t)\|_{\mathcal{L}(X)} \leq M$ for $t \geq 0$. In the sequel, we always set $q = \frac{\alpha}{2}$ for $\alpha \in (1, 2)$.

As argued in [18], we define a mild solution of Equation (1) as follows.

Definition 7. *A function* $x \in C(J, X)$ *is said to be a mild solution of Equation* (1) *if* $x(0) + g(x) = x_0$, $x'(0) = x_1$ *and there exists* $f \in L^1(J, X)$ *such that* $f(t) \in F(t, x(t))$ *on a.e.* $t \in J$ *and*

$$x(t) = C_q(t)(x_0 - g(x)) + K_q(t)x_1 + \int_0^t (t-s)^{q-1} P_q(t-s) f(s) ds,$$

where

$$C_q(t) = \int_0^\infty M_q(\theta) C(t^q\theta) d\theta, \quad K_q(t) = \int_0^t C_q(s) ds, \quad P_q(t) = \int_0^\infty q\theta M_q(\theta) S(t^q\theta) d\theta,$$

$$M_q(\theta) = \frac{1}{q}\theta^{-1-\frac{1}{q}} \xi_q(\theta^{-\frac{1}{q}}),\ \xi_q(\theta) = \frac{1}{\pi}\sum_{n=1}^\infty (-1)^{n-1}\theta^{-nq-1}\frac{\Gamma(nq+1)}{n!}\sin(n\pi q), \theta \in (0, \infty),$$

and $M_q(\cdot)$ *is the Mainardi's Wright-type function defined on* $(0, \infty)$ *such that*

$$M_q(\theta) \geq 0\ for\ \theta \in (0, \infty)\ and\ \int_0^\infty M_q(\theta) d\theta = 1.$$

Remark 1. *In considering the case of* $\alpha \in (0, 1)$, *we know from the references that there is a similar representation of mild solutions if the initial value* $x_1 = 0$ *for the case of* $\alpha \in (1, 2)$. *However, the biggest difference is that the operator* A *(typically the Laplacian operator) generates a* C_0*-semigroup, and one can use the method of semigroup theory to obtain some well-known results for the case of* $\alpha \in (0, 1)$ *instead of cosine families. Further, if* α *tends to 1, the method of semigroup theory can be also used to deal with first-order evolution problems; if* α *tends to 2, we can directly solve an evolution problem by using the concept of cosine families. Thus, the studied evolution problem in Equation (1) is more different from the case of* $\alpha \in (0, 1]$, *and it is valuable to consider the existence of Equation (1).*

Remark 2. *The setting $q = \alpha/2$ for $\alpha \in (1,2)$ is derived from the constraint of the Laplace transform of Mainardi's Wright-type function and the resolvent of cosine families (see [18]). This reflects the fact that the probability density function is closely related to the mild solutions of the corresponding evolution problems.*

Lemma 8. ([18]) *The operators $C_q(t)$, $K_q(t)$, and $P_q(t)$ (appearing in Definition 7) have the following properties:*

(i) *For any $t \geq 0$, the operators $C_q(t)$, $K_q(t)$, and $P_q(t)$ are linear operators;*
(ii) *For any fixed $t \geq 0$ and for any $x \in X$, the following estimates hold:*

$$\|C_q(t)x\| \leq M\|x\|, \quad \|K_q(t)x\| \leq M\|x\|t, \quad \|P_q(t)x\| \leq \frac{M}{\Gamma(2q)}\|x\|t^q;$$

(iii) $\{C_q(t), t \geq 0\}$, $\{K_q(t), t \geq 0\}$, *and* $\{t^{q-1}P_q(t), t \geq 0\}$ *are strongly continuous.*

Lemma 9. ([29]) *Let X be a separable metric space and let $G : \Omega \to \mathcal{P}_{cl}(X)$ be a multivalued map with nonempty closed images. Then, G is measurable if and only if there exist measurable single-valued maps $g_n : \Omega \to X$ such that*

$$G(\omega) = \overline{\bigcup\{g_n(\omega),\ n \geq 1\}}, \quad for\ every\ \omega \in \Omega.$$

Lemma 10. ([30, Theorem 8.2.10]) *Let $(\Omega, \mathcal{A}, \mu)$ be a complete σ-finite measurable space, and let X, Y be two complete separable metric spaces. If $F : \Omega \to \mathcal{P}(X)$ is a measurable multivalued map with nonempty closed images and $G : \Omega \times X \mapsto Y$ is a Carathéodory map (that is, for every $x \in X$, the multivalued map $\omega \mapsto G(\omega, x)$ is measurable, and for every $\omega \in \Omega$, the multivalued map $x \mapsto G(\omega, x)$ is continuous), then for every measurable map $h : \Omega \mapsto Y$ satisfying $h(\omega) \in G(\omega, F(\omega))$ for almost all $\omega \in \Omega$, there exists a measurable selection $f(\omega) \in F(\omega)$ such that $h(\omega) = G(\omega, f(\omega))$ for almost all $\omega \in \Omega$.*

3. Main Results

We need to state the following hypotheses for the forthcoming analysis.

Hypothesis 1. *The operator A is the infinitesimal generator of a uniformly bounded cosine family $\{C(t)\}_{t \geq 0}$ in X.*

Hypothesis 2. *The multivalued map $F : J \times X \to \mathcal{P}_{cl,cv}(X)$ is an L^1-Carathéodory multivalued map satisfying the following conditions:*

(i) *For every $t \in J$, the map $F(t, \cdot) : X \to \mathcal{P}_{cl,cv}(X)$ is u.s.c.;*
(ii) *For each $x \in X$, the map $F(\cdot, x) : J \to \mathcal{P}_{cl,cv}(X)$ is measurable and the set*

$$S_{F,x} = \{f \in L^1(J, X) :\ f(t) \in F(t, x(t)) \text{ for a.e. } t \in J\}$$

is nonempty.

Hypothesis 3. *There exists a function $k_f(\cdot) \in L^1(J, \mathbb{R}_+)$ such that*

$$\|F(t,x)\| = \sup\{\|f\| :\ f \in F(t,x)\} \leq k_f(t)(1 + \|x\|), \text{ for a.a. } t \in J \text{ and all } x \in X.$$

Hypothesis 4. *There exists a function $\beta(\cdot) \in L^1(J, \mathbb{R}_+)$ such that $\chi(F(t, D)) \leq \beta(t)\chi(D)$ for every bounded subset $D \subset C(J, X)$.*

Hypothesis 5. *$g : C(J, X) \to X$ is a continuous and compact function, and there exist constants N_{g1}, N_{g2} such that $\|g(x)\| \leq N_{g1}\|x\|_C + N_{g2}$* for $x \in C(J, X)$.

Remark 3. *If X is a finite dimension Banach space, then for each $x \in C(J,X)$, $S_{F,x} \neq \emptyset$ (see, e.g., Lasota and Opial [31]). If X is an infinite dimension Banach space and $x \in C(J,X)$, it follows from Hu and Papageorgiou [32] that $S_{F,x} \neq \emptyset$ if and only if the function $\varsigma : J \to \mathbb{R}_+$ given by $\varsigma(t) := \inf\{\|v\| : v \in F(t,x)\}$ belongs to $L^1(J,\mathbb{R}_+)$.*

Lemma 11. ([31]). *Let X be a Banach space, let $F : J \times X \to \mathcal{P}_{cp,cv}(X)$ be a L^1-Carathéodory multivalued map with $S_{F,x} \neq \emptyset$ (see (H2)), and let Ψ be a linear continuous operator from $L^1(J,X)$ to $C(J,X)$. Then,*

$$\Psi \circ S_F : C(J,X) \to \mathcal{P}_{cp,cv}(C(J,X)), \quad x \mapsto (\Psi \circ S_F)(x) := \Psi(S_{F,x}),$$

is a closed graph operator in $C(J,X) \times C(J,X)$.

Theorem 1. *Assume that* (H1)–(H5) *are satisfied. Then, Equation (1) has at least one mild solution provided that $\|k_f\|_1 < (1 - MN_{g1})M^{-1}a^{1-2q}\Gamma(2q)$ and $\|\beta\|_1 < (8M)^{-1}a^{1-2q}\Gamma(2q)$.*

Proof. By (H2), we can define a multivalued map $\mathscr{P} : C(J,X) \to \mathcal{P}(C(J,X))$ as follows: for $x \in C(J,X)$, $\mathscr{P}(x)$ is the set of all functions $y \in \mathscr{P}(x)$ satisfying

$$y(t) = C_q(t)(x_0 - g(x)) + K_q(t)x_1 + \int_0^t (t-s)^{q-1}P_q(t-s)f(s)ds, \quad t \in J,$$

where $f \in S_{F,x}$. It will be verified in several steps, claims and parts that the operator $\mathscr{P}$ has fixed points that correspond to mild solutions of Equation (1).

Step 1. $\mathscr{P}$ maps a bounded closed convex set into a bounded closed convex set.

By the hypothesis of function $k_f(\cdot)$ in (H3), there exists $r > 0$ such that

$$M\|x_0\| + MN_{g1}r + MN_{g2} + Ma\|x_1\| + \frac{Ma^{2q-1}}{\Gamma(2q)}\|k_f\|_1 + \frac{Ma^{2q-1}}{\Gamma(2q)}\|k_f\|_1 r \le r. \tag{3}$$

Furthermore, we introduce $W_0 = \{x \in C(J,X) : \|x\|_C \le r\}$ and observe that W_0 is a nonempty bounded closed and convex subset of $C(J,X)$. Let $x \in W_0$ and $y \in \mathscr{P}(x)$, then, there exists $f \in S_{F,x}$ such that for each $t \in J$ and for any $x \in W_0$, we have

$$y(t) = C_q(t)(x_0 - g(x)) + K_q(t)x_1 + \int_0^t (t-s)^{q-1}P_q(t-s)f(s)ds.$$

By (H3) and (H4), we have

$$\begin{aligned}
\|y(t)\| \le& \|C_q(t)\|_{\mathcal{L}(X)}\|x_0 - g(x)\| + \|K_q(t)\|_{\mathcal{L}(X)}\|x_1\| + \int_0^t (t-s)^{q-1}\|P_q(t-s)f(s)\|ds \\
\le& M\|x_0\| + M\|g(x)\| + Mt\|x_1\| + \frac{M}{\Gamma(2q)}\int_0^t (t-s)^{2q-1}k_f(s)(1 + |x(s)|)ds \\
\le& M\|x_0\| + MN_{g1}\|x\|_C + MN_{g2} + Mt\|x_1\| + \frac{Mt^{2q-1}}{\Gamma(2q)}\|k_f\|_1 + \frac{Mt^{2q-1}}{\Gamma(2q)}\|k_f\|_1\|x\|_C \\
\le& M\|x_0\| + MN_{g1}r + MN_{g2} + Ma\|x_1\| + \frac{Ma^{2q-1}}{\Gamma(2q)}\|k_f\|_1 + \frac{Ma^{2q-1}}{\Gamma(2q)}\|k_f\|_1 r \\
\le& r.
\end{aligned}$$

Therefore, $\|y\|_C \le r$, which implies that $\mathscr{P}(W_0) \subseteq W_0$.

Define $W_1 = \overline{co}\mathscr{P}(W_0)$. Clearly, $W_1 \subset C(J, X)$ is a nonempty bounded closed and convex set. Repeating the arguments employed in the previous step, for any $x \in W_1$, $y \in \mathscr{P}(x)$, it follows that there exists $f \in S_{F,x}$ such that for each $t \in J$ and for any $x \in W_1$,

$$y(t) = C_q(t)(x_0 - g(x)) + K_q(t)x_1 + \int_0^t (t-s)^{q-1}P_q(t-s)f(s)ds.$$

By (H3) and (H4), together with Lemma 8 (ii), we have

$$\begin{aligned}\|y(t)\| \leq & \|C_q(t)(x_0 - g(x))\| + \|K_q(t)x_1\| + \int_0^t (t-s)^{q-1}\|P_q(t-s)f(s)\|ds \\ \leq & M\|x_0\| + MN_{g1}r + MN_{g2} + Ma\|x_1\| + \frac{Ma^{2q-1}}{\Gamma(2q)}\|k_f\|_1 + \frac{Ma^{2q-1}}{\Gamma(2q)}\|k_f\|_1 r \leq r,\end{aligned}$$

which implies that $\mathscr{P}(W_1) \subseteq W_1$ and $W_1 \subset W_0$.

Next, for every $n \geq 1$, we define $W_{n+1} = \overline{co}\mathscr{P}(W_n)$. From the above proof, it is easy to see that W_n is a nonempty bounded closed and convex subset of $C(J, X)$. Furthermore, $W_2 = \overline{co}\mathscr{P}(W_1) \subset W_1$. By induction, we know that the sequence $\{W_n\}_{n=1}^{\infty}$ is a decreasing sequence of nonempty bounded closed and convex subsets of $C(J, X)$. Furthermore, we set $W = \bigcap_{n=1}^{\infty} W_n$ and note that W is bounded closed and convex since W_n is bounded closed and convex for every $n \geq 1$.

Now, we establish that $\mathscr{P}(W) \subseteq W$. Indeed, $\mathscr{P}(W) \subseteq \mathscr{P}(W_n) \subseteq \overline{co}\mathscr{P}(W_n) = W_{n+1}$ for every $n \geq 1$. Therefore, $\mathscr{P}(W) \subseteq \bigcap_{n=2}^{\infty} W_n$. On the other hand, $W_n \subset W_1$ for every $n \geq 1$. Hence,

$$\mathscr{P}(W) \subseteq \bigcap_{n=2}^{\infty} W_n = \bigcap_{n=1}^{\infty} W_n = W.$$

Step 2. The multivalued map $\mathscr{P}$ is ν-condensing.
Let $\mathbb{B} \subseteq W$ be such that

$$\nu(\mathbb{B}) \leq \nu(\mathscr{P}(\mathbb{B})). \tag{4}$$

We show below that $\mathbb{B}$ is a relatively compact set; that is, $\nu(\mathbb{B}) = 0$.

Let $\sigma(\mathbb{B}) = \sup_{t \in J} \chi(\mathbb{B}(t))$, and let $\nu(\mathscr{P}(\mathbb{B}))$ be achieved on a sequence $\{y_n\}_{n=1}^{\infty} \subset \mathscr{P}(\mathbb{B})$; that is,

$$\nu(\{y_n\}_{n=1}^{\infty}) = \max\Big(\sigma(\{y_n\}_{n=1}^{\infty}),\ \mathrm{mod}_C(\{y_n\}_{n=1}^{\infty})\Big).$$

Then,

$$y_n(t) = C_q(t)(x_0 - g(x_n)) + K_q(t)x_1 + \int_0^t (t-s)^{q-1}P_q(t-s)f_n(s)ds, \quad t \in J,$$

where $\{x_n\}_{n=1}^{\infty} \subset \mathbb{B}$ and $f_n \in Sel_{F,x_n}$ for every $n \geq 1$.

Since g is compact, the set $\{g(x_n) : n \geq 1\}$ is relatively compact and $C_q(t)$, $K_q(t)$ are strongly continuous for $t \geq 0$. Hence, for every $t \in J$, we have

$$\nu(\{C_q(t)(x_0 - g(x_n)) + K_q(t)x_1,\ n \geq 1\}) = 0.$$

Therefore, it is enough to estimate that

$$\nu\left(\left\{\int_0^t (t-s)^{q-1}P_q(t-s)f_n(s)ds,\ n \geq 1\right\}\right) = 0.$$

Claim I. $\sigma(\{y_n\}_{n=1}^{\infty}) = 0$.

For any $t \in J$, using (H4), Lemma 3, and Lemma 8 (ii), we have

$$\begin{aligned}
\sigma(\{y_n\}_{n=1}^{\infty}) =& \sup_{t\in J} \chi(\{y_n(t)\}_{n=1}^{\infty}) \le 2\sup_{t\in J}\int_0^t (t-s)^{q-1}\chi\left(\{P_q(t-s)f_n(s)\}_{n=1}^{\infty}\right) ds \\
\le& \sup_{t\in J}\frac{2M}{\Gamma(2q)}\int_0^t (t-s)^{2q-1}\beta(s)\chi\left(\{x_n(s)\}_{n=1}^{\infty}\right) ds \\
\le& \sup_{t\in J}\frac{2M}{\Gamma(2q)}\int_0^t (t-s)^{2q-1}\beta(s)ds\sigma(\{x_n\}_{n=1}^{\infty}) \\
\le& \frac{2Ma^{2q-1}}{\Gamma(2q)}\int_0^a \beta(s)ds\sigma(\{x_n\}_{n=1}^{\infty}) < \frac{1}{4}\sigma(\{x_n\}_{n=1}^{\infty}).
\end{aligned}$$

On the other hand, Equation (4) implies that $\sigma(\{y_n\}_{n=1}^{\infty}) \ge \sigma(\{x_n\}_{n=1}^{\infty})$. In consequence, we have $\sigma(\{y_n\}_{n=1}^{\infty}) = 0$.

Claim II. $\mathrm{mod}_C(\{y_n\}_{n=1}^{\infty}) = 0$; that is, the set $\mathbb{B}$ is equicontinuous.
Let

$$\widetilde{y}_n(\cdot) = \int_0^{\cdot} (\cdot - s)^{q-1}P_q(t-s)f_n(s)ds.$$

Therefore, it remains to be verified that $\mathrm{mod}_C(\{\widetilde{y}_n\}_{n=1}^{\infty}) = 0$. Then, for any $t_1, t_2 \in J$ with $t_1 < t_2$, we have

$$\begin{aligned}
\|\widetilde{y}_n(t_2) - \widetilde{y}_n(t_1)\| \le& \int_{t_1}^{t_2} \|(t_2-s)^{q-1}P_q(t_2-s)f_n(s)\|ds \\
&+ \int_0^{t_1} \|((t_2-s)^{q-1}P_q(t_2-s) - (t_1-s)^{q-1}P_q(t_2-s))f_n(s)\|ds \\
=& I_1 + I_2.
\end{aligned}$$

According to Lemma 8 (ii), we get

$$\begin{aligned}
I_1 \le& \frac{M}{\Gamma(2q)}\int_{t_1}^{t_2} (t_2-s)^{2q-1}k_f(s)(1+\|x_n(s)\|)ds \\
\le& \frac{M}{\Gamma(2q)}(t_2-t_1)^{2q-1}\int_{t_1}^{t_2} k_f(s)ds(1+\|x_n\|_C) \to 0, \quad \text{as } t_2 \to t_1.
\end{aligned}$$

Let $T_q(t) = t^{q-1}P_q(t)$ for $t \in J$. Then, we know from Lemma 8 (iii) that $T_q(t)$ is a strongly continuous operator. For I_2, taking $\varepsilon > 0$ to be small enough, we obtain

$$\begin{aligned}
I_2 \le& \int_0^{t_1-\varepsilon} \|(T_q(t_2-s) - T_q(t_1-s))f_n(s)\|ds + \int_{t_1-\varepsilon}^{t_1} \|(T_q(t_2-s) - T_q(t_1-s))f_n(s)\|ds \\
\le& \int_0^{t_1} k_f(s)(1+\|x_n(s)\|)ds \sup_{s\in[0,t_1-\varepsilon]} \|T_q(t_2-s) - T_q(t_1-s)\|_{\mathcal{L}(X)} \\
&+ \left(\frac{M\varepsilon^{2q-1}}{\Gamma(2q)} + \frac{M(t_2-t_1+\varepsilon)^{2q-1}}{\Gamma(2q)}\right)\int_{t_1-\varepsilon}^{t_1} k_f(s)(1+\|x_n(s)\|)ds \\
\le& \|k_f\|_1(1+\|x_n\|_C) \sup_{s\in[0,t_1-\varepsilon]} \|T_q(t_2-s) - T_q(t_1-s)\|_{\mathcal{L}(X)} \\
&+ \left(\frac{M\varepsilon^{2q-1}}{\Gamma(2q)} + \frac{M(t_2-t_1+\varepsilon)^{2q-1}}{\Gamma(2q)}\right)(1+\|x_n\|_C)\int_{t_1-\varepsilon}^{t_1} k_f(s)ds \\
\to& 0, \quad \text{as } t_2 \to t_1,\ \varepsilon \to 0.
\end{aligned}$$

Consequently, we have

$$\mathrm{mod}_C\left(\left\{\int_0^t (t-s)^{q-1}P_q(t-s)f_n(s)ds,\ n\geq 1\right\}\right)=0.$$

As a conclusion, it follows that $\mathrm{mod}_C(\{y_n\}_{n=1}^{\infty})=0$. Hence, the multivalued map $\mathscr{P}$ is ν-condensing.

Step 3. The multimap $\mathscr{P}(x)$ is convex and compact for each $x\in W$.

Part I. $\mathscr{P}(x)$ has convex values for each $x\in W$.

In fact, if y_1, y_2 belong to $\mathscr{P}(x)$ for each $x\in W$, then there exist $f_1, f_2\in S_{F,x}$ such that for each $t\in J$, we have

$$y_i(t)=C_q(t)(x_0-g(x))+K_q(t)x_1+\int_0^t (t-s)^{q-1}P_q(t-s)f_i(s)ds,\quad i=1,2.$$

Let $\theta\in[0,1]$. Then, for each $t\in J$, we get

$$\begin{aligned}(\theta y_1+(1-\theta)y_2)(t)=&C_q(t)(x_0-g(x))+K_q(t)x_1\\&+\int_0^t (t-s)^{q-1}P_q(t-s)(\theta f_1+(1-\theta)f_2)(s)ds.\end{aligned}$$

As F has convex values by the definition of $S_{F,x}$, we deduce that $\theta f_1(s)+(1-\theta)f_2(s)\in S_{F,x}$. Thus, $\theta y_1+(1-\theta)y_2\in\mathscr{P}(x)$.

Part II. $\mathscr{P}$ has compact values. In view of the foregoing facts, it is enough to show that W is nonempty and compact in $C(J,X)$: that is, by Lemma 5, we need to show that

$$\lim_{n\to\infty}\nu(W_n)=0. \tag{5}$$

As in Step 2, we can show that $\mathrm{mod}_C(W_n)=0$; that is, W_n is equicontinuous. Hence, it remains to be shown that $\sigma(W_n)=0$. By Lemma 1, for each $\varepsilon>0$, there exists a sequence $\{y_k\}_{k=1}^{\infty}$ in $\mathscr{P}(W_{n-1})$ such that

$$\sigma(W_n)=\sigma(\mathscr{P}(W_n))\leq 2\sigma(\{y_k\}_{k=1}^{\infty})+\varepsilon.$$

Therefore, by Lemma 2 and the nonsingularity of σ, it follows that

$$\sigma(W_n)\leq 2\sigma(\{y_k\}_{k=1}^{\infty})+\varepsilon=2\sup_{t\in J}\chi(\{y_k(t)\}_{k=1}^{\infty})+\varepsilon. \tag{6}$$

Since $y_k\in\mathscr{P}(W_{n-1})$ $(k\geq 1)$, there exists $x_k\in W_{n-1}$ such that $y_k\in\mathscr{P}(x_k)$. Hence, from the compactness of g and the strong continuity of $C_q(t)$ and $K_q(t)$ for $t\in J$, there exists $f_k\in S_{F,x_k}$ such that for every $t\in J$,

$$\begin{aligned}\chi(\{y_k(t)\}_{k=1}^{\infty})\leq&\chi(\{C_q(t)(x_0-g(\{x_k\}_{k=1}^{\infty}))+K_q(t)x_1\})\\&+\chi\left(\left\{\int_0^t (t-s)^{q-1}P_q(t-s)f_k(s)ds:\ k\geq 1\right\}\right)\\=&\chi\left(\left\{\int_0^t (t-s)^{q-1}P_q(t-s)f_k(s)ds:\ k\geq 1\right\}\right).\end{aligned}$$

By (H5) and Lemma 1, for a.e. $t\in J$, we have

$$\chi(\{f_k(t)\}_{k=1}^{\infty})\leq\chi(F(t,\{x_k(t)\}_{k=1}^{\infty}))\leq\beta(t)\chi(\{x_k(t)\}_{k=1}^{\infty})\leq\beta(t)\sigma(W_{n-1}):=\gamma(t).$$

On the other hand, by (H3), for almost all $t\in J$, $\|f_k(t)\|\leq k_f(t)(1+r)$ for every $k\geq 1$. Hence, $f_k\in L^1(J,X)$, $k\geq 1$. Note that $\gamma(\cdot)\in L^1(J,\mathbb{R}_+)$ from (H4). It follows from Lemma 4 that there exists a compact $K_\epsilon \subset X$, a measurable set $J_\epsilon \subset J$ with measure less than ϵ, and a sequence of functions $\{g_k^\epsilon\} \subset L^1(J, X)$ such that $\{g_k^\epsilon(s)\}_{k=1}^\infty \subseteq K_\epsilon$ for all $s \in J$, and

$$\|f_k(s) - g_k^\epsilon(s)\| < 2\gamma(s) + \epsilon, \quad \text{for every } k \geq 1 \text{ and every } s \in J'_\epsilon = J - J_\epsilon.$$

Then, using Minkowski's inequality and the property of the MNC, we obtain

$$\begin{aligned}
&\chi\left(\left\{\int_{J'_\epsilon}(t-s)^{q-1}P_q(t-s)(f_k(s)-g_k^\epsilon(s))ds : k \geq 1\right\}\right)\\
&\leq \frac{2M}{\Gamma(2q)}\int_{J'_\epsilon}(t-s)^{2q-1}\chi(\{(f_k(s)-g_k^\epsilon(s)) : k \geq 1\})ds\\
&\leq \frac{2M}{\Gamma(2q)}\int_{J'_\epsilon}(t-s)^{2q-1}\sup_{k\geq 1}\|f_k(s)-g_k^\epsilon(s)\|ds\\
&\leq \frac{2Ma^{2q-1}}{\Gamma(2q)}\int_{J'_\epsilon}(2\gamma(s)+\epsilon)ds\\
&\leq \frac{4Ma^{2q-1}}{\Gamma(2q)}\|\gamma\|_1 + \frac{2Ma^{2q-1}}{\Gamma(2q)}\epsilon\\
&\leq \frac{4Ma^{2q-1}}{\Gamma(2q)}\sigma(W_{n-1})\|\beta\|_1 + \frac{2Ma^{2q-1}}{\Gamma(2q)}\epsilon,
\end{aligned} \tag{7}$$

and

$$\begin{aligned}
\chi\left(\left\{\int_{J_\epsilon}(t-s)^{q-1}P_q(t-s)f_k(s)ds : k \geq 1\right\}\right) &\leq \frac{2M}{\Gamma(2q)}\int_{J_\epsilon}(t-s)^{2q-1}\chi\left(\{f_k(s)\}_{k=1}^\infty\right)ds\\
&\leq \frac{2M}{\Gamma(2q)}\int_{J_\epsilon}(t-s)^{2q-1}\sup_{k\geq 1}\|f_k(s)\|ds\\
&\leq \frac{Ma^{2q-1}}{\Gamma(2q)}(1+r)\int_{J_\epsilon}k_f(s)ds.
\end{aligned} \tag{8}$$

Using Equations (7) and (8), we have

$$\begin{aligned}
&\chi\left(\left\{\int_0^t(t-s)^{q-1}P_q(t-s)f_k(s)ds : k \geq 1\right\}\right) \leq \chi\left(\left\{\int_{J'_\epsilon}(t-s)^{q-1}P_q(t-s)f_k(s)ds : k \geq 1\right\}\right)\\
&\quad + \chi\left(\left\{\int_{J_\epsilon}(t-s)^{q-1}P_q(t-s)f_k(s)ds : k \geq 1\right\}\right)\\
&\leq \chi\left(\left\{\int_{J'_\epsilon}(t-s)^{q-1}P_q(t-s)(f_k(s)-g_k^\epsilon(s))ds : k \geq 1\right\}\right)\\
&\quad + \chi\left(\left\{\int_{J'_\epsilon}(t-s)^{q-1}P_q(t-s)g_k^\epsilon(s)ds : k \geq 1\right\}\right)\\
&\quad + \chi\left(\left\{\int_{J_\epsilon}(t-s)^{q-1}P_q(t-s)f_k(s)ds : k \geq 1\right\}\right)\\
&\leq \frac{4Ma^{2q-1}}{\Gamma(2q)}\sigma(W_{n-1})\|\beta\|_1 + \frac{2Ma^{2q-1}}{\Gamma(2q)}\epsilon + \frac{Ma^{2q-1}}{\Gamma(2q)}(1+r)\int_{J_\epsilon}k_f(s)ds.
\end{aligned}$$

As ϵ is arbitrary, for all $t \in J$, we get

$$\chi\left(\left\{\int_0^t(t-s)^{q-1}P_q(t-s)f_k(s)ds\right\}\right) \leq \frac{4Ma^{2q-1}}{\Gamma(2q)}\|\beta\|_1\sigma(W_{n-1}).$$

Therefore, for each $t \in J$, we have

$$\chi(\{y_k(t)\}_{k=1}^{\infty}) \leq \frac{4Ma^{2q-1}}{\Gamma(2q)}\|\beta\|_1\sigma(W_{n-1}).$$

By the above inequality, together with Equation (6) and the arbitrary nature of ε, we can deduce that

$$\sigma(W_n) \leq \frac{8Ma^{2q-1}}{\Gamma(2q)}\|\beta\|_1\sigma(W_{n-1}).$$

Then, by induction, we find that

$$0 \leq \sigma(W_n) \leq \left(\frac{8Ma^{2q-1}}{\Gamma(2q)}\|\beta\|_1\right)^n \sigma(W_0), \quad \text{for all } n \geq 1.$$

Since this inequality is true for every $n \geq 1$, passing on to the limit $n \to \infty$ and by (H4), we obtain Equation (5). Hence, $W = \bigcap_{n=1}^{\infty} W_n$ is a nonempty compact set of X, and $\mathscr{P}$ has compact values in W.

Step 4. The values of $\mathscr{P}$ are closed.

Let $x_n, x_* \in W$ with $x_n \to x_*$ as $n \to \infty$, $y_n \in \mathscr{P}(x_n)$, and $y_n \to y_*$ as $n \to \infty$. We show that $y_* \in \mathscr{P}(x_*)$. Indeed, $y_n \in \mathscr{P}(x_n)$ means that there exists $f_n \in S_{F,x_n}$ such that

$$y_n(t) = C_q(t)(x_0 - g(x)) + K_q(t)x_1 + \int_0^t (t-s)^{q-1}P_q(t-s)f_n(s)ds.$$

Next, we must show that there exists $f_* \in S_{F,x_*}$ such that

$$y_*(t) = C_q(t)(x_0 - g(x)) + K_q(t)x_1 + \int_0^t (t-s)^{q-1}P_q(t-s)f_*(s)ds.$$

Since $x_n \to x_*$ and $y_n \in \mathscr{P}(x_n)$, we deduce that

$$\|(y_n(t) - C_q(t)x_0 + C_q(t)g(x_n) - K_q(t)x_1) - (y_*(t) - C_q(t)x_0 + C_q(t)g(x_*) - K_q(t)x_1)\| \to 0,$$

as $n \to \infty$.

Now, we consider the linear continuous operator

$$\mathscr{F} : L^1(J, X) \to C(J, X), \quad f \mapsto (\mathscr{F}f)(t) = \int_0^t (t-s)^{q-1}P_q(t-s)f(s)ds.$$

From Step 3 and Lemma 11, it follows that $\mathscr{F} \circ S_F$ is a closed graph operator. Furthermore, in view of the definition of $\mathscr{F}$, we have

$$(y_n(t) - C_q(t)x_0 + C_q(t)g(x_n) - K_q(t)x_1) \in \mathscr{F}(S_{F,x_n}).$$

In view of the fact that $x_n \to x_*$ as $n \to \infty$, the repeated application of Lemma 11 yields

$$y_*(t) - C_q(t)x_0 + C_q(t)g(x_*) - K_q(t)x_1 = \int_0^t (t-s)^{q-1}P_q(t-s)f(s)ds$$

for some $f \in S_{F,x_*}$. Thus, $\mathscr{P}$ is a closed multivalued map.

Therefore, as an implication of Steps 1–5, we deduce that $\mathscr{P} : W \to P(W)$ is closed and ν-condensing with nonempty convex compact values. Thus, all the hypotheses of Lemma 6 are satisfied. Hence, there exists at least one fixed point $x \in W$ such that $x \in \mathscr{P}(x)$, which corresponds to a mild solution of Equation (1). □

Theorem 2. *Suppose that all the assumptions of Theorem 1 are satisfied. Then, the set of mild solutions of Equation (1) is compact in $C(J, X)$.*

Proof. Note that the set of mild solutions is nonempty by Theorem 1. Indeed, letting $r > 0$, defined by Equation (3), we can get a mild solution in W_0. Now, we show that an arbitrary number of mild solutions of Equation (1) belongs to W_0. Let x be a mild solution of Equation (1). Then,

$$x(t) = C_q(t)(x_0 - g(x)) + K_q(t)x_1 + \int_0^t (t-s)^{q-1} P_q(t-s) f(s) ds,$$

where $f \in S_{F,x} = \{f \in L^1(J, X) : f(t) \in F(t, x(t)), \text{ for a.e. } t \in J\}$. Using an argument similar to the one used in Step 1 of the proof of Theorem 1, we have

$$\begin{aligned} \|x\|_C =& \sup_{t \in J} \|x(t)\| \\ \leq& \sup_{t \in J} \|C_q(t)(x_0 - g(x))\| + \sup_{t \in J} \|K_q(t)x_1\| + \sup_{t \in J} \int_0^t (t-s)^{q-1} \|P_q(t-s)f(s)\| ds \\ \leq& M\|x_0\| + MN_{g1}r + MN_{g2} + Ma\|x_1\| + \frac{Ma^{2q-1}}{\Gamma(2q)} \|k_f\|_1 + \frac{Ma^{2q-1}}{\Gamma(2q)} \|k_f\|_1 r \leq r. \end{aligned}$$

This shows that the mild solutions of Equation (1) are bounded. Thus, the conclusion follows from Lemma 7. The proof is completed. □

4. An Application

Let $\Omega \subset \mathbb{R}^N$ $(N = 1, 2, 3)$ be an open bounded set and $X = U = L^2(\Omega)$. Let us consider the following fractional partial differential equations with the constrained control u and a finite multi-point discrete mean condition:

$$\begin{cases} \partial_t^\alpha y(t,z) = \Delta y(t,z) + G(t, z, y(t,z), u(t,z)), \quad t \in [0,1], \ z \in \Omega, \ u \in U, \\ y(t,z) = 0, \quad t \in [0,1], z \in \partial\Omega \\ y(0,z) - \sum_{i=0}^{n} \int_\Omega m(\xi, z) y(t_i, \xi) d\xi = 0, \ y'(0,z) = 0, \ z \in \Omega, \end{cases} \tag{9}$$

where ∂_t^α is the Caputo fractional partial derivative of order $\alpha \in (1,2)$, $0 \leq t_0 < t_1 < \cdots < t_n \leq 1$, $m(\xi, z) : \Omega \times \Omega \to X$ is an L^2-Lebesgue integrable function, and $G : [0,1] \times \Omega \times X \times U \to X$ is a single-valued continuous measurable function.

We define $x(t) = y(t, \cdot)$, that is, $x(t)(z) = y(t,z)$, $t \in J$, $z \in \Omega$, here $J = [0,1]$. The set of the constraint functions $U : J \to \mathcal{P}_{cl,cv}(X)$ is a measurable multivalued map. If $u \in U$, then it means that $u(t) \in U(t, x(t))$, for a.e. $t \in J$. The function $f : J \times X \times U$ is given by $f(t, x(t), u(t))(z) = G(t, z, y(t,z), u(t,z))$. Equation (9) is solved if we show that there exists a control function u such that Equation (9) admits a mild solution. Let the multivalued map be given by

$$F(t, x(t)) = \{f(t, x(t), u(t)), \quad u \in U\}. \tag{10}$$

Then, the set of mild solutions of the control problem in Equation (9), with the right-hand side given by Equation (10), coincides with the set of mild solutions of Equation (1).

Let A be the Laplace operator with Dirichlet boundary conditions defined by $A = \Delta$ with

$$\mathcal{D}(A) = \{v \in L^2(\Omega) : v \in H_0^1(\Omega) \cap H^2(\Omega)\}.$$

Let $\{-\lambda_k, \phi_k\}_{k=1}^{\infty}$ be the eigensystem of the operator A. Then, $0 < \lambda_1 \le \lambda_2 \le \cdots, \lambda_k \to \infty$ as $k \to \infty$, and $\{\phi_k\}_{k=1}^{\infty}$ forms an orthonormal basis of X. Furthermore,

$$Ax = -\sum_{k=1}^{\infty} \lambda_k(x, \phi_k)\phi_k, \quad x \in \mathcal{D}(A),$$

where $(\cdot,\cdot)$ is the inner product in X. It is known that the operator A generates a strongly continuous uniformly bounded cosine family (see, e.g., [9]), which, in this case, is defined by

$$C(t)x = \sum_{k=1}^{\infty} \cos(\sqrt{\lambda_k}t)(x, \phi_k)\phi_k, \quad x \in X,$$

and then $\|C(t)\|_{\mathcal{L}(X)} \le 1$ for every $t \ge 0$. Hence, (H1) holds.

Taking $\alpha = \frac{3}{2}$, we have $q = \frac{3}{4}$. Let $g : C(J, X) \to X$ be given by $g(x)(z) = \sum_{i=0}^{n} K_g x(t_i)(z)$ with $K_g v(z) = \int_\Omega m(\xi, z)v(\xi)d\xi$ for $v \in X$, $z \in \Omega$ (noting that $K_g : X \to X$ is completely continuous). Thus, the assumption in (H5) holds true. With the choice of operator A, Equation (9) can be reformulated in X as the following nonlocal control problem:

$$\begin{cases} {}^C D_t^\alpha x(t) = Ax(t) + f(t, x(t), u(t)), & t \in J,\ u \in U, \\ x(0) = g(x),\ x'(0) = 0. \end{cases} \tag{11}$$

Next, the results obtained in Section 4 apply to the following problem of fractional evolution inclusions:

$$\begin{cases} {}^C D_t^\alpha x(t) \in Ax(t) + F(t, x(t)), & t \in J, \\ x(0) = g(x),\ x'(0) = 0. \end{cases} \tag{12}$$

Theorem 3. *Assume that the following conditions hold:*

Hypothesis 6. *$U : J \to \mathcal{P}_{cl,cv}(X)$ is a measurable multivalued map.*

Hypothesis 7. *The function $f : J \times X \times X \to X$ is L^1-Carathéodory, linear in the third argument, and there exists a function $k_f(\cdot) \in L^1(J, \mathbb{R}_+)$ satisfying $\|k_f\|_1 < \sqrt{\pi}(1 - n\|m\|)/2$ such that $\|f(t, x, y)\| \le k_f(t)(1 + \|x\|)$ for almost all $t \in J$ and all $x \in X$.*

Hypothesis 8. *There exists a function $\beta(\cdot) \in L^1(J, \mathbb{R}_+)$ satisfying $\|\beta\|_1 < \sqrt{\pi}/16$ such that*

$$\chi(f(t, D, U(t, D))) \le \beta(t)\chi(D),$$

for every bounded subset $D \subset C(J, X)$.

Then, the control problem in Equation (9) has at least one mild solution. In addition, the set of mild solutions is compact.

Proof. From (H6) and (H7), the map $t \mapsto F(t, \cdot)$ is obviously a measurable multivalued map, and then $F(\cdot,\cdot) \in \mathcal{P}_{cv,cl}(X)$. Now, we show that the selection set of F is not empty. Since U is a measurable multivalued map, it follows by Lemma 9 that there exists a sequence of measurable selections $\{u_n\}_{n=1}^{\infty} \subset U$ such that

$$U(t) = \overline{\bigcup\{u_n(t),\ n \ge 1\}} \quad \text{for every } t \in J.$$

Let $v_n(t) = f(t, x(t), u_n(t))$ for $n \ge 1$ and $t \in J$. In view of the continuity of f, v_n is thus measurable. Hence, $\overline{\{v_n(t),\ n \ge 1\}} \subseteq F(t, x(t))$. Conversely, if $f(t, x(t), u(t)) \in F(t, x(t))$ for any $u \in U$, then there exists a subsequence in U which will be still defined by $\{u_n\}_{n=1}^{\infty}$ such that $u_n \to u$ as $n \to \infty$. It follows from the continuity of f that $f(t,x(t),u_n(t)) \to f(t,x(t),u(t))$ as $n \to \infty$. Hence, $f(t,x(t),u(t)) \in \overline{\{v_n(t),\ n \geq 1\}}$. This means that

$$F(t,x(t)) = \overline{\bigcup\{v_n(t),\ n \geq 1\}},$$

Consequently, from Lemma 9, $F(\cdot,x)$ is measurable.

Next, we show that the map $x \mapsto F(\cdot,x)$ is an u.s.c. multivalued map by means of contradiction. Firstly, we suppose that F is not u.s.c. at some point $x_0 \in \Omega$. Then, there exists an open neighborhood $W \subseteq X$ such that $F(t,x_0) \subset W$, and for every open neighborhood $V \subseteq \Omega$ of x_0, there exists $x_1 \in V$ such that $F(t,x_1) \not\subset W$. Let

$$V_n = \left\{x \in \Omega,\ \|x - x_0\| < \frac{1}{n},\ n = 1,2,\ldots\right\}.$$

Clearly, V_n is a open neighborhood of x_0. Then, for each $n \geq 1$, there exist $x_n \in V_n$, $v_n \in F(t,x_n)$, and $u_n \in U$ such that $v_n = f(t,x_n,u_n)$ and $v_n \notin W$. Moreover, as $\{u_n\}_{n=1}^{\infty} \subset U$, we set $u_n \to u$ as $n \to \infty$ for some $u \in U$. By the continuity of f, owing to $x_n \to x_0$ as $n \to \infty$, we have $v_n \to v$ as $n \to \infty$, where $v = f(t,x_0,u)$, which implies that $v \in F(t,x_0) \subset W$. This contradicts that $v_n \notin W$ for each $n \geq 1$. Thus, our supposition is false.

In addition, according to the condition in (H7), we find that F is an L^1-Carathéodory multivalued map. Hence, (H2) and (H3) are satisfied. On the other hand, the hypothesis (H8) corresponds to (H4). Thus, all of the hypotheses of Theorem 1 are satisfied. Hence, Equation (12) has at least one mild solution. Furthermore, the set of mild solutions of Equation (12) is compact by Theorem 2.

Finally, we show that the mild solutions of Equation (12) do coincide with the mild solutions of the control problem in Equation (11). Let x be a solution of Equation (12). Then, there exists a single-valued selection

$$\phi \in S_{F,x} = \{\phi \in L^1(J,X),\ \phi(t) \in F(t,x(t)),\ \text{a.e. } t \in J\}, \tag{13}$$

such that

$$^{C}D_t^{\alpha}x(t) = Ax(t) + \phi(t),\ \text{a.e. } t \in J, \text{ and } x(0) = g(x),\ x'(0) = 0.$$

Now, we introduce a map $\Psi(t,u) = f(t,x(t),u(t))$ and note that it is Carathéodory. Moreover, let the equality in Equation (10) be satisfied. Then, for a.a. $t \in J$ and for every $\phi(t) \in \{f(t,x(t),u(t)), u \in U\} := \Psi(t,U(t))$, we deduce by Lemma 10 that there exists a measurable selection $u(t) \in U(t)$ such that $\phi(t) = \Psi(t,u(t)) = f(t,x(t),u(t))$ for a.a. $t \in J$. Thus, the mild solution satisfies the control problem in Equation (11).

On the other hand, let x satisfy the control problem in Equation (11). Then, x is obviously a mild solution of Equation (12), and the proof is completed. □

5. Conclusions

In the current paper, we study a class of fractional evolution inclusions with nonlocal initial conditions. We obtain the sufficient conditions for ensuring the existence of mild solutions and the compactness for set of mild solutions. We can see that the probability density function is closely related to the mild solutions of the corresponding evolution inclusion problems, which enrich the knowledge of the fractional calculus. Moreover, an illustrative example is provided to demonstrate the applicability of the proposed problem.

On the other hand, many evolution inclusion problems are focused on a finite interval. This is because the solutions of some physical models may blow up, or we can gain a clearer understanding of the state of a physical system in finite time. If the time goes to infinity, it urges us to extend the concept of mild solutions such as Equation (1) in $[0,\infty)$ and, furthermore, to find the existence of global mild solutions. However, the technique for an infinite interval is more complex, and this topic may be a future work. In addition, our future works also include the topological properties of solution sets (including R_δ, acyclicity, connectedness, compactness, and contractibility) for fractional evolution inclusions of order $\alpha \in (1, 2)$.

Author Contributions: methodology, J.W.H. and Y.Z.; writing-original draft preparation; J.W.H. and Y.Z.; writing-review and editing, Y.L. and B.A.

References

1. Miller, K.S.; Ross, B. *An Introduction to the Fractional Calculus and Differential Equations*; John Wiley: New York, NY, USA, 1993.
2. Podlubny, I. *Fractional Differential Equations*; Academic Press: San Diego, CA, USA, 1999.
3. A. A. Kilbas, H. M. Srivastava, J. J. Trujillo, Theory and applications of fractional differential equations. In *North-Holland Mathematics Studies*; Elsevier Science B.V.: Amsterdam, The Netherlands, 2006; Volume 204.
4. Zhou, Y. *Basic Theory of Fractional Differential Equations*; World Scientific: Singapore, 2014.
5. Zhou, Y. Attractivity for fractional differential equations in Banach space. *Appl. Math. Lett.* **2018**, *75*, 1–6. [CrossRef]
6. Zhou, Y. Attractivity for fractional evolution equations with almost sectorial operators. *Fract. Calc. Appl. Anal.* **2018**, *21*, 786–800. [CrossRef]
7. Mainardi, F. *Fractional Calculus and Waves in Linear Viscoelasticity, An Introduction to Mathematical Models*; Imperial College Press: London, UK, 2010.
8. Tarasov, V.E. *Fractional Dynamics*; Springer: Berlin/Heidelberg, Germany, 2010.
9. Bazhlekova, E.; Bazhlekov, I. Subordination approach to multi-term time-fractional diffusion-wave equations. *J. Comput. Appl. Math.* **2018**, *339*, 179–192. [CrossRef]
10. Kian, Y.; Yamamoto, M. On existence and uniqueness of solutions for semilinear fractional wave equations. *Fract. Calc. Appl. Anal.* **2017**, *20*, 117–138. [CrossRef]
11. Zhou, Y.; Shangerganesh, L.; Manimaran, J.; Debbouche, A. A class of time-fractional reaction-diffusion equation with nonlocal boundary condition. *Math. Meth. Appl. Sci.* **2018**, *41*, 2987–2999. [CrossRef]
12. Bajlekova, E. Fractional Evolution Equations in Banach Spaces. Ph.D Thesis, Eindhoven University of Technology, Eindhoven, The Netherlands, 2001.
13. Fan, Z. Characterization of compactness for resolvents and its applications. *Appl. Math. Comput.* **2014**, *232*, 60–67. [CrossRef]
14. Li, K.; Peng, J.; Jia, J. Cauchy problems for fractional differential equations with Riemann-Liouville fractional derivatives. *J. Funct. Anal.* **2012**, *263*, 476–510. [CrossRef]
15. Li, Y.; Sun, H.; Feng, Z. Fractional abstract Cauchy problem with order $\alpha \in (1, 2)$. *Dyn. Part. Differ. Eq.* **2016**, *13*, 155–177. [CrossRef]
16. Shu, X.B.; Wang, Q.Q. The existence and uniqueness of mild solutions for fractional differential equations with nonlocal conditions of order $1 < \alpha < 2$. *Comput. Math. Appl.* **2012**, *64*, 2100–2110.
17. Yan, Z. Approximate controllability of fractional neutral integro-differential inclusions with state-dependent delay in Hilbert spaces. *IMA J. Math. Control Inform.* **2013**, *30*, 443–462. [CrossRef]
18. Zhou, Y.; He, J.W. New results on controllability of fractional evolution systems with order $\alpha \in (1, 2)$. *Evol. Eq. Control Theory* **2019**, in press.
19. Zhou, Y.; Peng, L.; Ahmad, B.; Alsaedi, A. Topological properties of solution sets of fractional stochastic evolution inclusions. *Adv. Differ. Equ.* **2017**, *2017*, 90. [CrossRef]
20. Zhou, Y.; Vijayakumar, V.; Murugesu, R. Controllability for fractional evolution inclusions without compactness. *Evol. Equ. Control Theory* **2015**, *4*, 507–524. [CrossRef]
21. Chalishajar, D.N. Controllability of second order impulsive neutral functional differential inclusions with infinite delay. *J. Optim. Theory Appl.* **2012**, *154*, 672–684. [CrossRef]
22. Hu, J.; Liu, X. Existence results of second-order impulsive neutral functional integrodifferential inclusions with unbounded delay in Banach spaces. *Math. Comput. Model.* **2009**, *49*, 516–526. [CrossRef]
23. Travis, C.C.; Webb, G.F. Cosine families and abstract nonlinear second order differential equations. *Acta Math. Hungar.* **1978**, *32*, 75–96. [CrossRef]

24. Bothe, D. Multivalued perturbations of m-accretive differential inclusions. *Israel J. Math.* **1998**, *108*, 109–138. [CrossRef]
25. Kamenskii, M.; Obukhowskii, V.; Zecca, P. *Condensing Multivalued Maps and Semilinear Differential Inclusions in Banach Spaces; De Gruyter Series in Nonlinear Analysis and Applications, 7*; Walter de Gruyter & Co.: Berlin, Germany, 2001.
26. Monch, H. Boundary value problems for nonlinear ordinary differential equations of second order in Banach spaces. *Nonlinear Anal.* **1980**, *49*, 985–999. [CrossRef]
27. Bader, R.; Kamenskii, M.; Obukhowskii, V. On some classes of operator inclusions with lower semicontinuous nonlinearities. *Topol. Methods Nonlinear Anal.* **2001**, *17*, 143–156. [CrossRef]
28. Cardinali, T.; Rubbioni, P. Impulsive mild solutions for semilinear differential inclusions with nonlocal conditions in Banach spaces. *Nonlinear Anal.* **2012**, *75*, 871–879. [CrossRef]
29. Castaing, C.; Valadier, M. *Convex Analysis and Measurable Multifunctions*; Lecture Notes in Mathematics; Springer: Berlin, Germany; New York, NY, USA, 1977.
30. Aubin, J.P.; Frankowska, H. *Set-Valued Analysis*; Birkhauser: Boston, MA, USA, 1990.
31. Lasota, A.; Opial, Z. An application of the Kakutani-Ky-Fan theorem in the theory of ordinary differential equations. *Bull. Acad. Pol. Sci. Ser. Sci. Math. Astronom. Phys.* **1965**, *13*, 781–786.
32. Hu, S.; Papageorgiou, N. *Handbook of Multivalued Analysis, Volume I: Theory*; Kluwer: Dordrecht, The Netherlands, 1997.

7

Existence Theory for a Fractional q-Integro-Difference Equation with q-Integral Boundary Conditions of Different Orders

Sina Etemad [1], Sotiris K. Ntouyas [2,3] and Bashir Ahmad [3,*]

1 Department of Mathematics, Azarbaijan Shahid Madani University, Azarshahr, Tabriz, Iran
2 Department of Mathematics, University of Ioannina, 45110 Ioannina, Greece
3 Nonlinear Analysis and Applied Mathematics (NAAM)-Research Group, Department of Mathematics, Faculty of Science, King Abdulaziz University, P.O. Box 80203, Jeddah 21589, Saudi Arabia
* Correspondence: bashirahmad_qau@yahoo.com

Abstract: In this paper, we study the existence of solutions for a new class of fractional q-integro-difference equations involving Riemann-Liouville q-derivatives and a q-integral of different orders, supplemented with boundary conditions containing q-integrals of different orders. The first existence result is obtained by means of Krasnoselskii's fixed point theorem, while the second one relies on a Leray-Schauder nonlinear alternative. The uniqueness result is derived via the Banach contraction mapping principle. Finally, illustrative examples are presented to show the validity of the obtained results. The paper concludes with some interesting observations.

Keywords: q-integro-difference equation; boundary value problem; existence; fixed point

---

1. Introduction and Preliminaries

Fractional calculus, dealing with differential and integral operators of arbitrary order, serves as a powerful modelling tool for many real-world phenomena. An interesting feature of such operators is their nonlocal nature that accounts for the history of the phenomena involved in the fractional models. Motivated by the extensive applications of fractional calculus, many researchers turned to the theoretical development of fractional-order initial and boundary value problems. Now, the literature on the topic contains many interesting and important results on the existence and uniqueness of solutions, and other properties of solutions for fractional-order problems. The available material includes different types of derivatives such as Riemann-Liouville, Caputo, Hadamard, etc. and a variety of boundary conditions. For some recent works on the topic, for instance, see [1–8] and the references therein.

Fractional q-difference equations (fractional analogue of q-difference equations) also received significant attention. One can find preliminary work on the topic in [9], while some interesting details about initial and boundary value problems of q-difference and fractional q-difference equations can be found in the book [10].

In 2012, Ahmad et al. [11] discussed the existence and uniqueness of solutions for the nonlocal boundary value problem of fractional q-difference equations:

$$\begin{cases} {}^cD_q^{\alpha}x(t) = f(t, x(t)), \quad 0 \le t \le 1, \ 1 < \alpha \le 2, \ 0 < q < 1, \\ \alpha_1 x(0) - \beta_1 D_q x(0) = \gamma_1 x(\eta_1), \qquad \alpha_2 x(1) - \beta_2 D_q x(1) = \gamma_2 x(\eta_2), \end{cases}$$

where $f \in C([0,1] \times \mathbb{R}, \mathbb{R})$, ${}^cD_q^{\alpha}$ is the fractional q-derivative of the Caputo type, and $\alpha_i, \beta_i, \gamma_i, \eta_i \in \mathbb{R}, i = 1, 2$.

In 2013, Zhou and Liu [12] applied Mönch's fixed point theorem together with the technique of measure of weak noncompactness to investigate the existence of solutions for the following fractional q-difference equation with boundary conditions:

$$\begin{cases} {}^cD_q^\alpha u(t) + f(t, u(t)) = 0, & 0 \le t \le 1,\ 0 < q < 1, \\ u(0) = (D_q^2 u)(0) = 0, & \gamma(D_q u)(1) + \beta(D_q^2 u)(1) = 0, \end{cases}$$

where $2 < \alpha \le 3, \gamma, \beta \ge 0$ and $f : [0,1] \times \mathbb{R} \to \mathbb{R}$ is a continuous function.

In 2014, Ahmad et al. [13] derived some existence results for a nonlinear fractional q-difference equation with four-point nonlocal integral boundary conditions given by

$$\begin{cases} {}^cD_q^\beta({}^cD_q^\gamma + \lambda)u(t) = f(t, u(t)), & 0 \le t \le 1,\ 0 < q < 1,\ \lambda \in \mathbb{R}, \\ u(0) = aI_q^{\alpha-1}u(\eta), \quad u(1) = bI_q^{\alpha-1}u(\sigma), & a, b \in \mathbb{R}, \end{cases}$$

where $0 < \beta, \gamma \le 1, 0 < \eta, \sigma < 1, \alpha > 2, f : [0,1] \times \mathbb{R} \to \mathbb{R}$ is a continuous function and I_q^α denotes the Riemann-Liouville fractional q-integral of order α.

Later, Niyom et al. [14] studied the following boundary value problem containing Riemann-Liouville fractional derivatives of different orders:

$$\begin{cases} (\lambda D^\alpha + (1-\lambda)D^\beta)u(t) = f(t, u(t)), & t \in [0, T],\ 1 < \alpha, \beta < 2, \\ u(0) = 0, \quad \mu D^{\gamma_1}u(T) + (1-\mu)D^{\gamma_2}u(T) = \gamma_3, & 0 < \gamma_1, \gamma_2 < \alpha - \beta, \end{cases}$$

where D^ϕ is the ordinary Riemann-Liouville fractional derivative of order $\phi \in \{\alpha, \beta, \gamma_1, \gamma_2\}$ such that $0 < \lambda \le 1, 0 \le \mu \le 1, \gamma_3 \in \mathbb{R}$ and $f \in C([0, T] \times \mathbb{R}, \mathbb{R})$ for $T > 0$.

Some recent results on fractional q-difference equations equipped with different kinds of boundary conditions can be found in the papers [15–25].

Now, we recall some important results on fractional q-integro-difference equations. In [26], the authors studied a nonlocal four-point boundary value problem of nonlinear fractional q-integro-difference equations given by

$$\begin{cases} {}^cD_q^\beta({}^cD_q^\gamma + \lambda)x(t) = pf(t, x(t)) + kI_q^\xi g(t, x(t)),\ 0 \le t \le 1,\ 0 < q < 1, \\ \alpha_1 x(0) - \beta_1 \left(t^{(1-\gamma)}D_q x(0)\right)\Big|_{t=0} = \sigma_1 x(\eta_1),\ \alpha_2 x(1) + \beta_2 D_q x(1) = \sigma_2 x(\eta_2), \end{cases}$$

where ${}^cD_q^\beta$ and ${}^cD_q^\gamma$ denote the fractional q-derivative of the Caputo type, $0 < \beta, \gamma \le 1$, $I_q^\xi(.)$ represents a Riemann-Liouville fractional integral of order $\xi \in (0,1)$, $f, g : [0,1] \times \mathbb{R} \to \mathbb{R}$ are continuous functions, $\lambda \ne 0$ and $p, k, \alpha_i, \beta_i, \sigma_i \in \mathbb{R}, \eta_i \in (0,1)$, $i = 1, 2$. For some recent works on boundary value problems of fractional q-integro-difference equations, for instance, see [27–31].

Motivated by aforementioned works, in this paper, we study the following nonlinear fractional q-integro-difference equation

$$(\lambda D_q^\alpha + (1-\lambda)D_q^\beta)u(t) = af(t, u(t)) + bI_q^\delta g(t, u(t)), \quad t \in [0,1], a, b \in \mathbb{R}^+, \tag{1}$$

supplemented with q-integral boundary conditions

$$u(0) = 0,\ \mu \int_0^1 \frac{(1-qs)^{(\gamma_1-1)}}{\Gamma_q(\gamma_1)}u(s)d_qs + (1-\mu)\int_0^1 \frac{(1-qs)^{(\gamma_2-1)}}{\Gamma_q(\gamma_2)}u(s)d_qs = 0,\ \gamma_1, \gamma_2 > 0, \tag{2}$$

where $0 < q < 1, 1 < \alpha, \beta < 2, 0 < \delta < 1, 0 < \lambda \le 1, 0 \le \mu \le 1, \alpha - \beta > 1$ and D_q^α denotes the Riemann-Liouville fractional q-derivative of order α and $f, g : [0,1] \times \mathbb{R} \to \mathbb{R}$ are continuous functions. Notice that Equation (1) contains q-derivatives of fractional orders α and β and a fractional q-integral of orders δ, while fractional q-integrals of orders γ_1 and γ_2 are involved in the boundary conditions (2).

We make use of Krasnoselskii's fixed point theorem and a Leray-Schauder nonlinear alternative to prove the existence results, while the uniqueness result is proved via Banach contraction mapping principle for the given problem.

Let us first recall some necessary concepts and definitions about q-fractional calculus and fixed point theory.

Let $0 < q < 1$ be an arbitrary real number. For every $a \in \mathbb{R}$, the $q-$number $[a]_q$ is defined by $[a]_q = \dfrac{1-q^a}{1-q}$ [9]. In addition, the q-shifted factorial of real number a is defined by $(a;q)_0 = 1$ and $(a;q)_n = \prod_{j=0}^{n-1}(1-aq^j)$ for $n \in \mathbb{N} \cup \{\infty\}$. For $a, b \in \mathbb{R}$, the q-analogue of the power function $(a-b)^n$ with $n \in \mathbb{N}_0 := \{0,1,2,\ldots\}$ is given by

$$(a-b)^{(0)} = 1, \quad (a-b)^{(n)} = \prod_{j=0}^{n-1}(a-bq^j).$$

In general, if α is real number, then $(a-b)^{(\alpha)} = a^\alpha \prod_{j=0}^{\infty} \dfrac{a-bq^j}{a-bq^{\alpha+j}}$ and $a^{(\alpha)} = a^\alpha$ when $b = 0$. If $\alpha > 0$ and $0 \le a \le b \le t$, then $(t-b)^{(\alpha)} \le (t-a)^{(\alpha)}$. The q-Gamma function $\Gamma_q(\alpha)$ is defined as

$$\Gamma_q(\alpha) = \frac{(1-q)^{(\alpha-1)}}{(1-q)^{\alpha-1}}, \quad \alpha \in \mathbb{R} \setminus \{0,-1,-2,\ldots\}$$

and satisfies the relation $\Gamma_q(\alpha+1) = [\alpha]_q \Gamma_q(\alpha)$ [9].

Let $\alpha \ge 0$ and $u : (0,\infty) \to \mathbb{R}$ be a continuous function. The Riemann-Liouville fractional q-integral for the function u of order α is defined by $(I_q^0 u)(t) = u(t)$ and

$$(I_q^\alpha u)(t) = \frac{1}{\Gamma_q(\alpha)} \int_0^t (t-qs)^{(\alpha-1)} u(s) d_q s, \quad \alpha > 0$$

for $t \in (0,\infty)$, provided that the right-hand side is pointwise defined on $(0,\infty)$ [9].

Recall that $I_q^\beta I_q^\alpha u(t) = I_q^{\beta+\alpha} u(t)$ for $\alpha, \beta \in \mathbb{R}^+$ [9] and

$$I_q^\alpha t^\beta = \frac{\Gamma_q(\beta+1)}{\Gamma_q(\alpha+\beta+1)} t^{\alpha+\beta}, \quad \beta \in (-1,\infty),\ \alpha \ge 0,\ t > 0.$$

If $f \equiv 1$, then $I_q^\alpha 1(t) = \dfrac{1}{\Gamma_q(\alpha+1)} t^\alpha$ for all $t > 0$.

The Riemann-Liouville fractional q-derivative of order $\alpha > 0$ for a function $u : (0,\infty) \to \mathbb{R}$ is defined by [9]

$$D_q^\alpha u(t) = \frac{1}{\Gamma_q(n-\alpha)} \int_0^t \frac{u(s)}{(t-qs)^{\alpha-n+1}} d_q s, \quad n-1 < \alpha < n.$$

Next, we state some fixed point theorems related to our work.

Lemma 1. *Let M be a closed, bounded, convex and nonempty subset of a Banach space E. Let A and B be operators mapping M into E, such that*

(i) $Ax + By \in M$, where $x, y \in M$;
(ii) A is compact and continuous;
(iii) B is a contraction mapping.

Then, there exists $z \in M$ such that $z = Az + Bz$ (Krasnoselskii's fixed point theorem [32]).

Lemma 2. *Let $\mathcal{X}$ be a closed and convex subset of a Banach space $\mathcal{E}$ and let $\mathcal{Y}$ be an open subset of $\mathcal{X}$ with $0 \in \mathcal{Y}$. Then, a continuous compact map $\mathcal{H} : \overline{\mathcal{Y}} \to \mathcal{X}$ has a fixed point in $\overline{Y}$ or there is a $y \in \partial Y$ and $\sigma \in (0,1)$ such that $y = \sigma\mathcal{H}(y)$, where $\partial\mathcal{Y}$ is the boundary of $\mathcal{Y}$ in $\mathcal{X}$ (Nonlinear alternative for single-valued maps [33]).*

2. Main Results

Let $E = C([0,1],\mathbb{R})$ be the set of continuous functions defined on $[0,1]$. The set E is a Banach space with the following norm

$$\|u\|_E = \sup_{t\in[0,1]} |u(t)|, \qquad u \in E.$$

Now, we prove the following lemma which characterizes the structure of solutions for boundary value problems (1) and (2).

Lemma 3. *Let $h \in C([0,1],\mathbb{R})$ and*

$$\Delta := \frac{\mu\Gamma_q(\alpha)}{\Gamma_q(\alpha+\gamma_1)} + \frac{(1-\mu)\Gamma_q(\alpha)}{\Gamma_q(\alpha+\gamma_2)} \neq 0. \tag{3}$$

The function u is a solution for the fractional q-difference boundary value problem

$$\begin{cases} (\lambda D_q^\alpha + (1-\lambda)D_q^\beta)u(t) = h(t), \qquad t \in [0,1], \\ u(0) = 0, \ \mu \displaystyle\int_0^1 \frac{(1-qs)^{(\gamma_1-1)}}{\Gamma_q(\gamma_1)} u(s)d_qs + (1-\mu) \int_0^1 \frac{(1-qs)^{(\gamma_2-1)}}{\Gamma_q(\gamma_2)} u(s)d_qs = 0, \end{cases} \tag{4}$$

if and only if u is a solution for the fractional q-integral equation

$$\begin{aligned} u(t) \ = \ & \frac{(\lambda-1)}{\lambda\Gamma_q(\alpha-\beta)} \int_0^t (t-qs)^{(\alpha-\beta-1)} u(s)d_qs + \frac{1}{\lambda\Gamma_q(\alpha)} \int_0^t (t-qs)^{(\alpha-1)} h(s)d_qs \\ & + \frac{t^{\alpha-1}}{\Delta}\Big[- \frac{\mu(\lambda-1)}{\lambda\Gamma_q(\alpha-\beta+\gamma_1)} \int_0^1 (1-qs)^{(\alpha-\beta+\gamma_1-1)} u(s)d_qs \\ & - \frac{\mu}{\lambda\Gamma_q(\alpha+\gamma_1)} \int_0^1 (1-qs)^{(\alpha+\gamma_1-1)} h(s)d_qs \\ & - \frac{(1-\mu)(\lambda-1)}{\lambda\Gamma_q(\alpha-\beta+\gamma_2)} \int_0^1 (1-qs)^{(\alpha-\beta+\gamma_2-1)} u(s)d_qs \\ & - \frac{(1-\mu)}{\lambda\Gamma_q(\alpha+\gamma_2)} \int_0^1 (1-qs)^{(\alpha+\gamma_2-1)} h(s)d_qs\Big]. \end{aligned} \tag{5}$$

Proof. Let u be a solution of the q-fractional boundary value problem (4). Then, we have

$$D_q^\alpha u(t) = \frac{\lambda-1}{\lambda} D_q^\beta u(t) + \frac{1}{\lambda} h(t).$$

Taking the Riemann-Liouville fractional q-integral of order α to both sides of the above equation, we get

$$u(t) = \frac{\lambda-1}{\lambda} I_q^\alpha D_q^\beta u(t) + \frac{1}{\lambda} I_q^\alpha h(t) + c_1 t^{\alpha-1} + c_2 t^{\alpha-2},$$

where $c_1, c_2 \in \mathbb{R}$ are arbitrary constants. Since $1 < \alpha < 2$, it follows from the first boundary condition that $c_2 = 0$. Thus,

$$u(t) = \frac{\lambda-1}{\lambda\Gamma_q(\alpha-\beta)} \int_0^t (t-qs)^{(\alpha-\beta-1)} u(s)d_qs + \frac{1}{\lambda\Gamma_q(\alpha)} \int_0^t (t-qs)^{(\alpha-1)} h(s)d_qs + c_1 t^{\alpha-1}. \tag{6}$$

On the other hand, if $\sigma \in \{\gamma_1, \gamma_2\}$, then we have

$$\begin{aligned} I_q^{\sigma} u(t) &= \frac{\lambda - 1}{\lambda \Gamma_q(\alpha - \beta + \sigma)} \int_0^t (t - qs)^{(\alpha - \beta + \sigma - 1)} u(s) d_q s \\ &+ \frac{1}{\lambda \Gamma_q(\alpha + \sigma)} \int_0^t (t - qs)^{(\alpha + \sigma - 1)} h(s) d_q s + c_1 \frac{\Gamma_q(\alpha)}{\Gamma_q(\alpha + \sigma)} t^{\alpha + \sigma - 1}. \end{aligned}$$

Now, by using the second boundary value condition and substituting the values $\sigma \in \{\gamma_1, \gamma_2\}$ into the above expression, we obtain

$$\begin{aligned} &\frac{\mu(\lambda - 1)}{\lambda \Gamma_q(\alpha - \beta + \gamma_1)} \int_0^1 (1 - qs)^{(\alpha - \beta + \gamma_1 - 1)} u(s) d_q s \\ &+ \frac{\mu}{\lambda \Gamma_q(\alpha + \gamma_1)} \int_0^1 (1 - qs)^{(\alpha + \gamma_1 - 1)} h(s) d_q s + c_1 \frac{\mu \Gamma_q(\alpha)}{\Gamma_q(\alpha + \gamma_1)} \\ &+ \frac{(1 - \mu)(\lambda - 1)}{\lambda \Gamma_q(\alpha - \beta + \gamma_2)} \int_0^1 (1 - qs)^{(\alpha - \beta + \gamma_2 - 1)} u(s) d_q s \\ &+ \frac{(1 - \mu)}{\lambda \Gamma_q(\alpha + \gamma_2)} \int_0^1 (1 - qs)^{(\alpha + \gamma_2 - 1)} h(s) d_q s + c_1 \frac{(1 - \mu) \Gamma_q(\alpha)}{\Gamma_q(\alpha + \gamma_2)} = 0. \end{aligned}$$

Solving the above equation for c_1, we find that

$$\begin{aligned} c_1 &= \frac{1}{\Delta} \Big[- \frac{\mu(\lambda - 1)}{\lambda \Gamma_q(\alpha - \beta + \gamma_1)} \int_0^1 (1 - qs)^{(\alpha - \beta + \gamma_1 - 1)} u(s) d_q s \\ &- \frac{\mu}{\lambda \Gamma_q(\alpha + \gamma_1)} \int_0^1 (1 - qs)^{(\alpha + \gamma_1 - 1)} h(s) d_q s \\ &- \frac{(1 - \mu)(\lambda - 1)}{\lambda \Gamma_q(\alpha - \beta + \gamma_2)} \int_0^1 (1 - qs)^{(\alpha - \beta + \gamma_2 - 1)} u(s) d_q s \\ &- \frac{(1 - \mu)}{\lambda \Gamma_q(\alpha + \gamma_2)} \int_0^1 (1 - qs)^{(\alpha + \gamma_2 - 1)} h(s) d_q s \Big], \end{aligned}$$

where Δ is defined in (3).

Substituting the value of c_1 in (6), we get the solution (5). Conversely, it is clear that u is a solution for the fractional q-difference Equation (4) whenever u is a solution for the fractional q-integral Equation (5). This completes the proof. □

In relation to the problems (1) and (2), we introduce an operator $\mathcal{T}: E \to E$ by

$$\begin{aligned}(\mathcal{T}u)(t) &= \frac{(\lambda-1)}{\lambda\Gamma_q(\alpha-\beta)}\int_0^t (t-qs)^{(\alpha-\beta-1)}u(s)d_qs + \frac{a}{\lambda\Gamma_q(\alpha)}\int_0^t (t-qs)^{(\alpha-1)}f(s,u(s))d_qs \\ &+ \frac{b}{\lambda\Gamma_q(\alpha+\delta)}\int_0^t (t-qs)^{(\alpha+\delta-1)}g(s,u(s))d_qs \\ &+ \frac{t^{\alpha-1}}{\Delta}\Big[-\frac{\mu(\lambda-1)}{\lambda\Gamma_q(\alpha-\beta+\gamma_1)}\int_0^1 (1-qs)^{(\alpha-\beta+\gamma_1-1)}u(s)d_qs \\ &- \frac{a\mu}{\lambda\Gamma_q(\alpha+\gamma_1)}\int_0^1 (1-qs)^{(\alpha+\gamma_1-1)}f(s,u(s))d_qs \\ &- \frac{b\mu}{\lambda\Gamma_q(\alpha+\delta+\gamma_1)}\int_0^1 (1-qs)^{(\alpha+\delta+\gamma_1-1)}g(s,u(s))d_qs \\ &- \frac{(1-\mu)(\lambda-1)}{\lambda\Gamma_q(\alpha-\beta+\gamma_2)}\int_0^1 (1-qs)^{(\alpha-\beta+\gamma_2-1)}u(s)d_qs \\ &- \frac{a(1-\mu)}{\lambda\Gamma_q(\alpha+\gamma_2)}\int_0^1 (1-qs)^{(\alpha+\gamma_2-1)}f(s,u(s))d_qs \\ &- \frac{b(1-\mu)}{\lambda\Gamma_q(\alpha+\delta+\gamma_2)}\int_0^1 (1-qs)^{(\alpha+\delta+\gamma_2-1)}g(s,u(s))d_qs\Big],\end{aligned} \tag{7}$$

where $u \in E$ and $t \in [0,1]$. In the sequel, we set

$$\begin{aligned}\Lambda_0 &:= \frac{|\lambda-1|}{\lambda\Gamma_q(\alpha-\beta+1)} + \frac{\mu|\lambda-1|}{\lambda|\Delta|\Gamma_q(\alpha-\beta+\gamma_1+1)} + \frac{(1-\mu)|\lambda-1|}{\lambda|\Delta|\Gamma_q(\alpha-\beta+\gamma_2+1)}, \\ \Lambda_1 &:= \frac{a}{\lambda\Gamma_q(\alpha+1)} + \frac{a\mu}{\lambda|\Delta|\Gamma_q(\alpha+\gamma_1+1)} + \frac{a(1-\mu)}{\lambda|\Delta|\Gamma_q(\alpha+\gamma_2+1)}, \\ \Lambda_2 &:= \frac{b}{\lambda\Gamma_q(\alpha+\delta+1)} + \frac{b\mu}{\lambda|\Delta|\Gamma_q(\alpha+\delta+\gamma_1+1)} + \frac{b(1-\mu)}{\lambda|\Delta|\Gamma_q(\alpha+\delta+\gamma_2+1)}.\end{aligned} \tag{8}$$

Now, we are ready to present our main results. The first existence result is based on Krasnoselskii's fixed point theorem.

Theorem 1. *Suppose that $f, g : [0,1] \times \mathbb{R} \to \mathbb{R}$ are continuous functions satisfying the following conditions:*

(i) there exists a positive constant L such that for each $u, v \in \mathbb{R}$,

$$|f(t,u) - f(t,v)| \le L|u-v|, \quad t \in [0,1];$$

(ii) For each $u \in \mathbb{R}$, there exists a continuous function m on $[0,1]$ such that

$$|g(t,u)| \le m(t), \quad t \in [0,1].$$

If $\Lambda_0 + L\Lambda_1 < 1$, then the fractional q-integro-difference Equation (1) with q-integral boundary conditions (2) has at least one solution on $[0,1]$, where Λ_0 and Λ_1 are defined by (8).

Proof. Let $\|m\| = \sup_{t\in[0,1]}|m(t)|$. Define $B_r := \{u \in E : \|u\| \le r\}$ with

$$r \ge \frac{\|m\|\Lambda_2 + K\Lambda_1}{1-(\Lambda_0 + L\Lambda_1)}, \tag{9}$$

where $K := \sup_{t\in[0,1]}|f(t,0)|$ and Λ_1 and Λ_2 are given by (8). Clearly, B_r is a closed, bounded, convex and nonempty subset of Banach space E. We consider the operator $\mathcal{T}: E \to E$ as (7). By Lemma 3, it is obvious that the fixed point of $\mathcal{T}$ is the solution of problems (1) and (2). Now, for each $t \in [0,1]$, we define two operators from B_r to E as follows:

$$\begin{aligned}
\mathcal{T}_1 u(t) &= \frac{(\lambda-1)}{\lambda\Gamma_q(\alpha-\beta)}\int_0^t (t-qs)^{(\alpha-\beta-1)}u(s)d_qs + \frac{a}{\lambda\Gamma_q(\alpha)}\int_0^t (t-qs)^{(\alpha-1)}f(s,u(s))d_qs \\
&\quad + \frac{t^{\alpha-1}}{\Delta}\Big[-\frac{\mu(\lambda-1)}{\lambda\Gamma_q(\alpha-\beta+\gamma_1)}\int_0^1 (1-qs)^{(\alpha-\beta+\gamma_1-1)}u(s)d_qs \\
&\quad - \frac{a\mu}{\lambda\Gamma_q(\alpha+\gamma_1)}\int_0^1 (1-qs)^{(\alpha+\gamma_1-1)}f(s,u(s))d_qs \\
&\quad - \frac{(1-\mu)(\lambda-1)}{\lambda\Gamma_q(\alpha-\beta+\gamma_2)}\int_0^1 (1-qs)^{(\alpha-\beta+\gamma_2-1)}u(s)d_qs \\
&\quad - \frac{a(1-\mu)}{\lambda\Gamma_q(\alpha+\gamma_2)}\int_0^1 (1-qs)^{(\alpha+\gamma_2-1)}f(s,u(s))d_qs\Big]
\end{aligned}$$

and

$$\begin{aligned}
\mathcal{T}_2 u(t) &= \frac{b}{\lambda\Gamma_q(\alpha+\delta)}\int_0^t (t-qs)^{(\alpha+\delta-1)}g(s,u(s))d_qs \\
&\quad + \frac{t^{\alpha-1}}{\Delta}\Big[\frac{b\mu}{\lambda\Gamma_q(\alpha+\delta+\gamma_1)}\int_0^1 (1-qs)^{(\alpha+\delta+\gamma_1-1)}g(s,u(s))d_qs \\
&\quad - \frac{b(1-\mu)}{\lambda\Gamma_q(\alpha+\delta+\gamma_2)}\int_0^1 (1-qs)^{(\alpha+\delta+\gamma_2-1)}g(s,u(s))d_qs\Big].
\end{aligned}$$

By the condition (H_1), we have that $|f(t,u(t))| \le |f(t,u(t)) - f(t,0)| + |f(t,0)| \le L\|u\| + K \le Lr + K$ for any $u \in \mathbb{R}$ and $t \in [0,1]$. Thus, for any $u, v \in B_r$ and $t \in [0,1]$, it follows by means of (8) and (9) that

$$\begin{aligned}
|\mathcal{T}_1 u(t) + \mathcal{T}_2 v(t)| &\le \frac{|\lambda-1|}{\lambda\Gamma_q(\alpha-\beta+1)}\|u\| + \frac{a}{\lambda\Gamma_q(\alpha+1)}\Big(L\|u\|+K\Big) \\
&\quad + \frac{\mu|\lambda-1|}{\lambda|\Delta|\Gamma_q(\alpha-\beta+\gamma_1+1)}\|u\| + \frac{a\mu}{\lambda|\Delta|\Gamma_q(\alpha+\gamma_1+1)}\Big(L\|u\|+K\Big) \\
&\quad + \frac{(1-\mu)|\lambda-1|}{\lambda|\Delta|\Gamma_q(\alpha-\beta+\gamma_2+1)}\|u\| + \frac{a(1-\mu)}{\lambda|\Delta|\Gamma_q(\alpha+\gamma_2+1)}\Big(L\|u\|+K\Big) \\
&\quad + \frac{b}{\lambda\Gamma_q(\alpha+\delta+1)}\|m\| + \frac{b\mu}{\lambda|\Delta|\Gamma_q(\alpha+\delta+\gamma_1+1)}\|m\| \\
&\quad + \frac{b(1-\mu)}{\lambda|\Delta|\Gamma_q(\alpha+\delta+\gamma_2+1)}\|m\| \\
&\le (\Lambda_0 + L\Lambda_1)r + \Lambda_2\|m\| + \Lambda_1 K \le r,
\end{aligned}$$

which implies that $\|\mathcal{T}_1 u + \mathcal{T}_2 v\| \le r$ and so $\mathcal{T}_1 u + \mathcal{T}_2 v \in B_r$ for all $u, v \in B_r$.

Now, we prove that $\mathcal{T}_2$ is continuous. Let $\{u_n\}_{n\ge1}$ be a sequence in B_r such that $u_n \to u$. Then, for each $t \in [0,1]$, we have

$$\begin{aligned}
|\mathcal{T}_2 u_n(t) - \mathcal{T}_2 u(t)| &\le \frac{b}{\lambda\Gamma_q(\alpha+\delta+1)}|g(s,u_n(s)) - g(s,u(s))| \\
&\quad + \frac{b\mu}{\lambda|\Delta|\Gamma_q(\alpha+\delta+\gamma_1+1)}|g(s,u_n(s)) - g(s,u(s))| \\
&\quad + \frac{b(1-\mu)}{\lambda|\Delta|\Gamma_q(\alpha+\delta+\gamma_2+1)}|g(s,u_n(s)) - g(s,u(s))|.
\end{aligned}$$

Since g is continuous, we get $\|\mathcal{T}_2 u_n - \mathcal{T}_2 u\| \to 0$ as $u_n \to u$. In consequence, it follows that the operator $\mathcal{T}_2$ is continuous on B_r.

In the next step, we show that the operator $\mathcal{T}_2$ is compact. Let us first show that $\mathcal{T}_2$ is uniformly bounded. For each $u \in B_r$ and $t \in [0,1]$, we have

$$\begin{aligned} |\mathcal{T}_2 u(t)| &\leq \frac{bt^{\alpha+\delta}}{\lambda\Gamma_q(\alpha+\delta+1)}|g(s,u(s))| + \frac{b\mu t^{\alpha-1}}{\lambda|\Delta|\Gamma_q(\alpha+\delta+\gamma_1+1)}|g(s,u(s))| \\ &\quad + \frac{b(1-\mu)t^{\alpha-1}}{\lambda|\Delta|\Gamma_q(\alpha+\delta+\gamma_2+1)}|g(s,u(s))| \\ &\leq \|m\|\Big[\frac{b}{\lambda\Gamma_q(\alpha+\delta+1)} + \frac{b\mu}{\lambda|\Delta|\Gamma_q(\alpha+\delta+\gamma_1+1)} + \frac{b(1-\mu)}{\lambda|\Delta|\Gamma_q(\alpha+\delta+\gamma_2+1)}\Big] \\ &= \Lambda_2\|m\|, \end{aligned}$$

which implies that $\|\mathcal{T}_2 u\| \leq \Lambda_2\|m\|$.

In order to establish the equicontinuity of the operator $\mathcal{T}_2$, we assume that $t_1, t_2 \in [0,1]$ such that $t_2 > t_1$. We will show that $\mathcal{T}_2$ maps bounded sets into equicontinuous sets. For each $u \in B_r$, we have

$$\begin{aligned} |\mathcal{T}_2 u(t_2) - \mathcal{T}_2 u(t_1)| &\leq \frac{b}{\lambda\Gamma_q(\alpha+\delta)}\int_0^{t_1}[(t_2-qs)^{(\alpha+\delta-1)} - (t_1-qs)^{(\alpha+\delta-1)}]|g(s,u(s))|d_qs \\ &\quad + \frac{b}{\lambda\Gamma_q(\alpha+\delta)}\int_{t_1}^{t_2}(t_2-qs)^{(\alpha+\delta-1)}|g(s,u(s))|d_qs \\ &\quad + \frac{|t_2^{\alpha-1} - t_1^{\alpha-1}|}{|\Delta|}\Big[\frac{b\mu}{\lambda\Gamma_q(\alpha+\delta+\gamma_1)}\int_0^1(1-qs)^{(\alpha+\delta+\gamma_1-1)}|g(s,u(s))|d_qs \\ &\quad + \frac{b(1-\mu)}{\lambda\Gamma_q(\alpha+\delta+\gamma_2)}\int_0^1(1-qs)^{(\alpha+\delta+\gamma_2-1)}|g(s,u(s))|d_qs\Big] \\ &\leq \|m\|\Big[\frac{2b(t_2-t_1)^{\alpha+\delta} + b|t_2^{\alpha+\delta} - t_1^{\alpha+\delta}|}{\lambda\Gamma_q(\alpha+\delta+1)} + \frac{b\mu|t_2^{\alpha-1} - t_1^{\alpha-1}|}{\lambda|\Delta|\Gamma_q(\alpha+\delta+\gamma_1+1)} \\ &\quad + \frac{b(1-\mu)|t_2^{\alpha-1} - t_1^{\alpha-1}|}{\lambda|\Delta|\Gamma_q(\alpha+\delta+\gamma_2+1)}\Big]. \end{aligned}$$

Observe that the right-hand side of the above inequality is independent of $u \in B_r$ and tends to zero as $t_1 \to t_2$. This shows that $\mathcal{T}_2$ is equicontinuous. Therefore, the operator $\mathcal{T}_2$ is relatively compact on B_r and the Arzelá-Ascoli theorem implies that $\mathcal{T}_2$ is completely continuous and so $\mathcal{T}_2$ is compact operator on B_r.

Finally, we prove that the operator $\mathcal{T}_1$ is a contraction. For any $u, v \in B_r$ and $t \in [0,1]$, we obtain

$$\begin{aligned} |\mathcal{T}_1 u(t) - \mathcal{T}_1 v(t)| &\leq \frac{|\lambda-1|}{\lambda\Gamma_q(\alpha-\beta+1)}|u(s)-v(s)| + \frac{a}{\lambda\Gamma_q(\alpha+1)}L|u(s)-v(s)| \\ &\quad + \frac{\mu|\lambda-1|}{\lambda|\Delta|\Gamma_q(\alpha-\beta+\gamma_1+1)}|u(s)-v(s)| + \frac{a\mu}{\lambda|\Delta|\Gamma_q(\alpha+\gamma_1+1)}L|u(s)-v(s)| \\ &\quad + \frac{(1-\mu)|\lambda-1|}{\lambda|\Delta|\Gamma_q(\alpha-\beta+\gamma_2+1)}|u(s)-v(s)| + \frac{a(1-\mu)}{\lambda|\Delta|\Gamma_q(\alpha+\gamma_2+1)}L|u(s)-v(s)| \\ &\leq (\Lambda_0 + L\Lambda_1)\|u-v\|. \end{aligned}$$

Since $\Lambda_0 + L\Lambda_1 < 1$, $\mathcal{T}_1$ is a contraction. Thus, all the assumptions of Lemma 1 are satisfied. Therefore, the fractional q-integro-difference Equation (1) with q-integral boundary conditions (2) has at least one solution on $[0,1]$ and the proof is completed. □

In the following result, we prove the existence of solutions for the problem (1) and (2) by means of a Leray-Schauder nonlinear alternative.

Theorem 2. *Let $f : [0,1] \times \mathbb{R} \to \mathbb{R}$ be a continuous function satisfying the conditions:*

(H_3) *there exist continuous nondecreasing functions $\psi_1, \psi_2 : [0,\infty) \to (0,\infty)$ and functions $\phi_1, \phi_2 \in C([0,1],\mathbb{R}^+)$ such that $|f(t,u)| \leq \phi_1(t)\psi_1(|u|)$ and $|g(t,u)| \leq \phi_2(t)\psi_2(|u|)$ for each $(t,u) \in [0,1] \times \mathbb{R}$;*

(H_4) *there exists a constant $\Xi > 0$ such that*

$$\frac{(1-\Lambda_0)\Xi}{\Lambda_1\|\phi_1\|\psi_1(\Xi) + \Lambda_2\|\phi_2\|\psi_2(\Xi)} > 1, \ \ \Lambda_0 < 1,$$

where $\Lambda_0, \Lambda_1, \Lambda_2$ are defined by (8).

Then, the fractional q-integro-difference Equation (1) with q-integral boundary conditions (2) has at least one solution on $[0,1]$.

Proof. We verify the hypothesis of a Leray-Schauder nonlinear alternative (Lemma 2) in several steps. Let us first show that the operator $\mathcal{T}$, defined by (7), *maps bounded sets (balls) into bounded sets in E.* For a positive number R, let $B_R = \{u \in E : \|u\| \leq R\}$ be a bounded ball in E. Then, for $t \in [0,1]$, we have

$$\begin{aligned}
|\mathcal{T}u(t)| \ \leq\ & \frac{|\lambda-1|}{\lambda\Gamma_q(\alpha-\beta)}\int_0^t (t-qs)^{(\alpha-\beta-1)}\|u\|d_qs + \frac{a}{\lambda\Gamma_q(\alpha)}\int_0^t (t-qs)^{(\alpha-1)}\|\phi_1\|\psi_1(\|u\|)d_qs \\
& + \frac{b}{\lambda\Gamma_q(\alpha+\delta)}\int_0^t (t-qs)^{(\alpha+\delta-1)}\|\phi_2\|\psi_2(\|u\|)d_qs \\
& + \frac{t^{\alpha-1}}{|\Delta|}\Big[\frac{\mu|\lambda-1|}{\lambda\Gamma_q(\alpha-\beta+\gamma_1)}\int_0^1 (1-qs)^{(\alpha-\beta+\gamma_1-1)}\|u\|d_qs \\
& + \frac{a\mu}{\lambda\Gamma_q(\alpha+\gamma_1)}\int_0^1 (1-qs)^{(\alpha+\gamma_1-1)}\|\phi_1\|\psi_1(\|u\|)d_qs \\
& + \frac{b\mu}{\lambda\Gamma_q(\alpha+\delta+\gamma_1)}\int_0^1 (1-qs)^{(\alpha+\delta+\gamma_1-1)}\|\phi_2\|\psi_2(\|u\|)d_qs \\
& + \frac{(1-\mu)|\lambda-1|}{\lambda\Gamma_q(\alpha-\beta+\gamma_2)}\int_0^1 (1-qs)^{(\alpha-\beta+\gamma_2-1)}\|u\|d_qs \\
& + \frac{a(1-\mu)}{\lambda\Gamma_q(\alpha+\gamma_2)}\int_0^1 (1-qs)^{(\alpha+\gamma_2-1)}\|\phi_1\|\psi_1(\|u\|)d_qs \\
& + \frac{b(1-\mu)}{\lambda\Gamma_q(\alpha+\delta+\gamma_2)}\int_0^1 (1-qs)^{(\alpha+\delta+\gamma_2-1)}\|\phi_2\|\psi_2(\|u\|)d_qs\Big] \\
\leq\ & \lambda_0\|u\| + \Lambda_1\|\phi_1\|\psi_1(\|u\|) + \Lambda_2\|\phi_2\|\psi_1(\|u\|).
\end{aligned}$$

Therefore,

$$\|\mathcal{T}u\| \leq \lambda_0 R + \Lambda_1\|\phi_1\|\psi_1(R) + \Lambda_2\|\phi_2\|\psi_1(R).$$

Secondly, we show that $\mathcal{T}$ *maps bounded sets into equicontinuous sets of E.* Let $t_1, t_2 \in [0,1]$ with $t_1 < t_2$ and $u \in B_R$. Then, we have

$$\begin{aligned}
& |\mathcal{T}u(t_2) - \mathcal{T}u(t_1)| \\
\leq\ & \frac{|\lambda-1|R}{\lambda\Gamma_q(\alpha-\beta)}\left[\int_0^{t_1}[(t_2-qs)^{(\alpha-\beta-1)} - (t_1-qs)^{(\alpha-\beta-1)}]d_qs + \int_{t_1}^{t_2}(t_2-qs)^{(\alpha-\beta-1)}d_qs\right] \\
& + \frac{a\|\phi_1\|\psi_1(R)}{\lambda\Gamma_q(\alpha)}\left[\int_0^{t_1}[(t_2-qs)^{(\alpha-1)} - (t_1-qs)^{(\alpha-1)}]d_qs + \int_{t_1}^{t_2}(t_2-qs)^{(\alpha-1)}d_qs\right] \\
& + \frac{|t_2^{\alpha-1} - t_1^{\alpha-1}|}{|\Delta|}\left[\frac{\mu|\lambda-1|R}{\lambda\Gamma_q(\alpha-\beta+\gamma_1+1)} + \frac{a\mu\|\phi_1\|\psi_1(R)}{\lambda\Gamma_q(\alpha+\gamma_1+1)}\right.
\end{aligned}$$

$$
\begin{aligned}
&+\frac{(1-\mu)|\lambda-1|R}{\lambda\Gamma_q(\alpha-\beta+\gamma_2+1)}+\frac{a(1-\mu)\|\phi_1\|\psi_1(R)}{\lambda\Gamma_q(\alpha+\gamma_2+1)}\Bigg]\\
&+\frac{b\|\phi_2\psi_2(R)}{\lambda\Gamma_q(\alpha+\delta)}\left[\int_0^{t_1}[(t_2-qs)^{(\alpha+\delta-1)}-(t_1-qs)^{(\alpha+\delta-1)}]d_qs+\int_{t_1}^{t_2}(t_2-qs)^{(\alpha+\delta-1)}d_qs\right]\\
&+\frac{|t_2^{\alpha-1}-t_1^{\alpha-1}|}{|\Delta|}\left[\frac{b\mu\|\phi_2\psi_2(R)}{\lambda\Gamma_q(\alpha+\delta+\gamma_1+1)}+\frac{b(1-\mu)\|\phi_2\psi_2(R)}{\lambda\Gamma_q(\alpha+\delta+\gamma_2+1)}\right]\\
\leq\;&\frac{|\lambda-1|R}{\lambda\Gamma_q(\alpha-\beta+1)}[2(t_2-t_1)^{\alpha-\beta}+|t_2^{\alpha-\beta}-t_1^{\alpha-\beta}|]+\frac{a\|\phi_1\|\psi_1(R)}{\lambda\Gamma_q(\alpha+1)}[2(t_2-t_1)^{\alpha}+|t_2^{\alpha}-t_1^{\alpha}|]\\
&+\frac{|t_2^{\alpha-1}-t_1^{\alpha-1}|}{|\Delta|}\Bigg[\frac{\mu|\lambda-1|R}{\lambda\Gamma_q(\alpha-\beta+\gamma_1+1)}+\frac{a\mu\|\phi_1\|\psi_1(R)}{\lambda\Gamma_q(\alpha+\gamma_1+1)}\\
&+\frac{(1-\mu)|\lambda-1|R}{\lambda\Gamma_q(\alpha-\beta+\gamma_2+1)}+\frac{a(1-\mu)\|\phi_1\|\psi_1(R)}{\lambda\Gamma_q(\alpha+\gamma_2+1)}\Bigg]\\
&+\frac{b\|\phi_2\|\psi(R)}{\lambda\Gamma_q(\alpha+\delta+1)}[2(t_2-t_1)^{\alpha+\delta}+|t_2^{\alpha+\delta}-t_1^{\alpha+\delta}|]\\
&+\frac{b\|\phi_2\|\psi(R)\mu|t_2^{\alpha-1}-t_1^{\alpha-1}|}{\lambda|\Delta|\Gamma_q(\alpha+\delta+\gamma_1+1)}+\frac{b(1-\mu)\|\phi_2\|\psi(R)|t_2^{\alpha-1}-t_1^{\alpha-1}|}{\lambda|\Delta|\Gamma_q(\alpha+\delta+\gamma_2+1)}\Bigg]\\
&\longrightarrow 0 \text{ as } t_2\to t_1 \text{ independent of } u\in B_R.
\end{aligned}
$$

Thus, the Arzelá-Ascoli theorem applies and hence $\mathcal{T}: E\to E$ is completely continuous.

In the last step, we show that all solutions to the equation $u=\theta\mathcal{T}u$ are bounded for $\theta\in[0,1]$. For that, let u be a solution of $u=\theta\mathcal{T}u$ for $\theta\in[0,1]$. Then, for $t\in[0,1]$, we apply the strategy used in the first step to obtain

$$\|u\|\leq\Lambda_0\|u\|+\Lambda_1\|\phi_1\|\psi_1(\|u\|)+\Lambda_2\|\phi_2\|\psi_1(\|u\|).$$

Consequently, we have

$$\frac{(1-\Lambda_0)\|u\|}{\Lambda_1\|\phi_1\|\psi_1(\|u\|)+\Lambda_2\|\phi_2\|\psi_1(\|u\|)}\leq 1.$$

By the condition (H_4), we can find a positive number Ξ such that $\|u\|\neq\Xi$. Introduce a set

$$U=\{u\in E:\|u\|<\Xi\},\tag{10}$$

and observe that the operator $\mathcal{T}:\overline{U}\to E$ is continuous and completely continuous. With this choice of U, we cannot find $u\in\partial U$ satisfying the relation $u=\theta\mathcal{T}x$ for some $\theta\in(0,1)$. Therefore, it follows by a nonlinear alternative of the Leray-Schauder type (Lemma 2) that the operator $\mathcal{T}$ has a fixed point in $\overline{U}$. Thus, there exists a solution of problems (1) and (2) on $[0,1]$. The proof is complete. □

In our final result, the uniqueness of solutions for the given problem is shown with the aid of a Banach contraction mapping principle [34].

Theorem 3. *Let $f:[0,1]\times\mathbb{R}\to\mathbb{R}$ be a function satisfying the assumption (H_1). In addition, assume that the function $g:[0,1]\times\mathbb{R}\to\mathbb{R}$ satisfies the condition*

(H_5) *there exists a positive constant M such that, for each $u,v\in\mathbb{R}$,*

$$|g(t,u)-g(t,v)|\leq M|u-v|,\qquad t\in[0,1].$$

Then, the fractional q-integro-difference Equation (1) with q-integral boundary conditions (2) has a unique solution on $[0,1]$, provided that $\Lambda_0 + L\Lambda_1 + M\Lambda_2 < 1$, where $\Lambda_0, \Lambda_1, \Lambda_2$ are defined by (8).

Proof. By a Banach contraction mapping principle, we will show that the operator $\mathcal{T} : E \to E$ defined by (7) has a unique fixed point which corresponds to the unique solution of problems (1) and (2). Setting $\sup_{t\in[0,1]} |f(t,0)| = K < \infty$ and $\sup_{t\in[0,1]} |g(t,0)| = N < \infty$ and selecting

$$r \geq \frac{N\Lambda_2 + K\Lambda_1}{1 - (\Lambda_0 + L\Lambda_1 + M\Lambda_2)},$$

we show that $\mathcal{T}B_r \subset B_r$, where $B_r = \{u \in E : \|u\| \leq r\}$. For any $u \in B_r$, following the arguments used in the proof of Theorem 1, one can obtain

$$\|\mathcal{T}u\| \leq (\Lambda_0 + L\Lambda_1 + M\Lambda_2)r + \Lambda_2 N + \Lambda_1 K < r,$$

which implies that $\mathcal{T}B_r \subset B_r$. For any $t \in [0,1]$ and any $u, v \in \mathbb{R}$, we obtain

$$\begin{aligned}
\|(\mathcal{T}u) - (\mathcal{T}v)\| \leq\ & \frac{|\lambda - 1|}{\lambda\Gamma_q(\alpha-\beta)} \int_0^t (t-qs)^{(\alpha-\beta-1)} |u(s) - v(s)| d_q s \\
& + \frac{a}{\lambda\Gamma_q(\alpha)} \int_0^t (t-qs)^{(\alpha-1)} |f(s,u(s)) - f(s,v(s))| d_q s \\
& + \frac{b}{\lambda\Gamma_q(\alpha+\delta)} \int_0^t (t-qs)^{(\alpha+\delta-1)} |g(s,u(s)) - g(s,v(s))| d_q s \\
& + \frac{t^{\alpha-1}}{|\Delta|} \Big[\frac{\mu(\lambda-1)}{\lambda\Gamma_q(\alpha-\beta+\gamma_1)} \int_0^1 (1-qs)^{(\alpha-\beta+\gamma_1-1)} |u(s) - v(s)| d_q s \\
& + \frac{a\mu}{\lambda\Gamma_q(\alpha+\gamma_1)} \int_0^1 (1-qs)^{(\alpha+\gamma_1-1)} |f(s,u(s)) - f(s,v(s))| d_q s \\
& + \frac{b\mu}{\lambda\Gamma_q(\alpha+\delta+\gamma_1)} \int_0^1 (1-qs)^{(\alpha+\delta+\gamma_1-1)} |g(s,u(s)) - g(s,v(s))| d_q s \\
& + \frac{(1-\mu)|\lambda-1|}{\lambda\Gamma_q(\alpha-\beta+\gamma_2)} \int_0^1 (1-qs)^{(\alpha-\beta+\gamma_2-1)} |u(s) - v(s)| d_q s \\
& + \frac{a(1-\mu)}{\lambda\Gamma_q(\alpha+\gamma_2)} \int_0^1 (1-qs)^{(\alpha+\gamma_2-1)} |f(s,u(s)) - f(s,v(s))| d_q s \\
& + \frac{b(1-\mu)}{\lambda\Gamma_q(\alpha+\delta+\gamma_2)} \int_0^1 (1-qs)^{(\alpha+\delta+\gamma_2-1)} |g(s,u(s)) - g(s,v(s))| d_q s \Big] \\
\leq\ & (\Lambda_0 + L\Lambda_1 + M\Lambda_2)\|u - v\|.
\end{aligned}$$

As $\Lambda_0 + L\Lambda_1 + M\Lambda_2 < 1$, therefore $\mathcal{T}$ is a contraction. Hence, we deduce by the conclusion of the Banach contraction mapping principle that the operator $\mathcal{T}$ has a unique fixed point, which is the unique solution of problems (1) and (2). The proof is completed. □

3. Examples

I. Illustration of Theorem 1

Example 1. *Consider the fractional q-integro-difference equation*

$$(0.9D_{0.5}^{1.5} + (1-0.9)D_{0.5}^{1.01})u(t) = 0.2\frac{|90t\sin(u(t))|}{100(|\sin(u(t))|+1)} + 0.3I_{0.5}^{0.35}\frac{\sin t|u^3(t)|}{(2+t)^3(1+|u^3(t)|)}, \qquad (11)$$

subject to q-integral boundary conditions

$$u(0) = 0, \ 0.1 \int_0^1 \frac{(1-qs)^{(0.1-1)}}{\Gamma_q(0.1)} u(s) d_q s + (1-0.1) \int_0^1 \frac{(1-qs)^{(0.1-1)}}{\Gamma_q(0.1)} u(s) d_q s = 0. \tag{12}$$

Here, $\alpha = 1.5$, $q = 0.5$, $\beta = 1.01$, $a = 0.2$, $b = 0.3$, $\delta = 0.35$, $\lambda = 0.9$, $\mu = 0.1$, $\gamma_1 = \gamma_2 = 0.1$, $t \in [0,1]$ *and* $f, g : [0,1] \times \mathbb{R} \to \mathbb{R}$ *are*

$$f(t,u) = \frac{|90t \sin u|}{100(|\sin u| + 1)}, \qquad g(t,u) = \frac{\sin t |u^3|}{(2+t)^3 (1+|u^3|)}.$$

For each $u, v \in \mathbb{R}$, *notice that* $|f(t,u) - f(t,v)| \le L|u - v|$ *with* $L = 0.9 > 0$. *On the other hand, there exists a continuous function* $m(t) = \dfrac{1}{(2+t)^3}$ *on* $[0,1]$ *such that* $|g(t,u)| \le m(t)$ *for all* $u \in \mathbb{R}$. *In addition, we have* $\|m\| = \sup_{t\in[0,1]} m(t) = 0.125$. *Using the given values, it is found that* $\Delta = 0.9935$ *and* $\Lambda_0 + L\Lambda_1 = 0.5567 < 1$. *Clearly, all the assumptions of Theorem 1 are satisfied. Therefore, the conclusion of Theorem 1 implies that the fractional q-integro-difference Equation (11) with q-integral boundary conditions (12) has at least one solution on* $[0,1]$.

II. Illustration of Theorem 2

Example 2. *We consider the fractional q-integro-difference equation*

$$(0.9 D_{0.5}^{1.5} + (1-0.9) D_{0.5}^{1.01}) u(t) = \tfrac{0.2}{\sqrt{256+t^2}} \left(\sin u(t) + \tfrac{|u(t)|}{1+|u(t)|} \right) + 0.3 I_{0.5}^{0.35} \tfrac{1}{4+t} \left(\tfrac{1}{2} + \tfrac{|\arcsin u(t)|}{1+|\arcsin u(t)|} \right) \tag{13}$$

supplemented with q-integral boundary conditions

$$u(0) = 0, \ 0.1 \int_0^1 \frac{(1-qs)^{(0.1-1)}}{\Gamma_q(0.1)} u(s) d_q s + (1-0.1) \int_0^1 \frac{(1-qs)^{(0.1-1)}}{\Gamma_q(0.1)} u(s) d_q s = 0, \tag{14}$$

where $\alpha = 1.5$, $q = 0.5$, $\beta = 1.01$, $a = 0.2$, $b = 0.3$, $\delta = 0.35$, $\lambda = 0.9$, $\mu = 0.1$, $\gamma_1 = \gamma_2 = 0.1$, $t \in [0,1]$ *and*

$$f(t,u(t)) = \frac{1}{\sqrt{256+t^2}} \left(\sin u(t) + \frac{|u(t)|}{1+|u(t)|} \right), \; g(t,u(t)) = \frac{1}{4+t} \left(\frac{1}{2} + \frac{|\arcsin u(t)|}{1+|\arcsin u(t)|} \right).$$

Obviously,

$$|f(t,u(t))| \le \frac{1}{\sqrt{256+t^2}} (1+\|u\|), \; |g(t,u(t))| \le \frac{1}{4+t} (1+\|u\|),$$

with $\phi_1(t) = \dfrac{1}{\sqrt{256+t^2}}$, $\phi_2(t) = \dfrac{1}{4+t}$ *and* $\psi_1(\|u\|) = \psi_2(\|u\|) = 1 + \|u\|$. *Note that* $\|\phi_1\| = \dfrac{1}{16} = 0.0625$, $\|\phi_2\| = \dfrac{1}{4} = 0.25$ *and* $\psi_1(\Xi) = \psi_2(\Xi) = 1 + \Xi$. *Using the given data, we find that* $\Delta = 0.9935$, $\Lambda_0 = 0.2414 < 1$, $\Lambda_1 = 0.3504$, *and* $\Lambda_2 = 0.4685$. *Then, by condition* (H_4), *we get* $\Xi > 0.22438$. *Thus, all the assumptions of Theorem 2 are satisfied. Therefore, by Theorem 2, problems (13) and (14) have at least one solution on* $[0,1]$.

III. Illustration of Theorem 3

Example 3. *Let us consider the fractional q-integro-difference equation*

$$(0.9 D_{0.5}^{1.5} + (1-0.9) D_{0.5}^{1.01}) u(t) = 0.2 \frac{0.9t |\sin(\frac{\pi}{2}(t))| |u(t)|}{5+|u(t)|} + 0.3 I_{0.5}^{0.35} \frac{80 |\cos(\pi t)| |u(t)|}{100(7+|u(t)|)} \tag{15}$$

with q-integral boundary conditions

$$u(0) = 0, \ \ 0.1\int_0^1 \frac{(1-qs)^{(0.1-1)}}{\Gamma_q(0.1)} u(s)d_qs + (1-0.1)\int_0^1 \frac{(1-qs)^{(0.1-1)}}{\Gamma_q(0.1)} u(s)d_qs = 0, \tag{16}$$

where $\alpha = 1.5, q = 0.5, \beta = 1.01, a = 0.2, b = 0.3, \delta = 0.35, \lambda = 0.9, \mu = 0.1, \gamma_1 = \gamma_2 = 0.1, t \in [0,1]$ *and*

$$f(t,u(t)) = \frac{0.9t|\sin(\frac{\pi}{2}(t))||u(t)|}{5+|u(t)|}, \ g(t,u(t)) = \frac{80|\cos(\pi t)||u(t)|}{100(7+|u(t)|)}.$$

Then, $L = 9/10, M = 8/10$ *as*

$$|f(t,u(t)) - f(t,v(t))| \leq \frac{9}{10}(|u(t)-v(t)|), \ |g(t,u(t)) - g(t,v(t))| \leq \frac{8}{10}(|u(t)-v(t)|).$$

With the given data, it is found that $\Delta = 0.9935$, $\Lambda_0 + L\Lambda_1 + M\Lambda_2 = 0.9315 < 1$. *Clearly, the assumptions of Theorem 3 hold. Thus, by the conclusion of Theorem 3, problems (15) and (16) have a unique solution* $[0,1]$.

4. Conclusions

We have derived some new existence and uniqueness results for a nonlinear fractional q-integro-difference equation equipped with q-integral boundary conditions. The obtained results significantly contribute to the literature on boundary value problems of fractional q-integro-difference equations and yield several new results as special cases. Some of these results are listed below.

(a) By letting $\lambda = 1/2$ in the results of this paper, we obtain the ones for a nonlinear fractional q-integro-difference equation of the form:

$$\left(D_q^\alpha + D_q^\beta\right)u(t) = 2af(t,u(t)) + 2bI_q^\delta g(t,u(t)), \quad t \in [0,1], a, b \in \mathbb{R}^+.$$

(b) For $\mu = 1/2$, our results correspond to the following boundary conditions:

$$u(0) = 0, \quad \int_0^1 \Big[\frac{(1-qs)^{(\gamma_1-1)}}{\Gamma_q(\gamma_1)} + \frac{(1-qs)^{(\gamma_2-1)}}{\Gamma_q(\gamma_2)}\Big]u(s)d_qs = 0, \ \ \gamma_1, \gamma_2 > 0.$$

(c) Our results with $a = 0$ and $b = 0$ correspond to the ones with purely integral nonlinearity and purely non-integral nonlinearity, respectively.

Author Contributions: Formal Analysis, S.E., S.K.N. and B.A.

Acknowledgments: The authors acknowledge the reviewers for their constructive remarks on our work.

References

1. Kilbas, A.A.; Srivastava, H.M.; Trujillo, J.J. *Theory and Applications of Fractional Differential Equations*; North-Holland Mathematics Studies; Elsevier: Amsterdam, The Netherlands, 2003; Volume 204.
2. Ahmad, B.; Sivasundaram, S. On four-point nonlocal boundary value problems of nonlinear integro-differential equations of fractional order. *Appl. Math. Comput.* **2010**, *217*, 480–487. [CrossRef]
3. Agarwal, R.P.; Ntouyas, S.K.; Ahmad, B.; Alhothuali, M. Existence of solutions for integro-differential equations of fractional order with nonlocal three-point fractional boundary conditions. *Adv. Differ. Equ.* **2013**, *2013*, 128. [CrossRef]
4. Baleanu, D.; Rezapour, S.; Etemad, S.; Alsaedi, A. On a time-fractional integro-differential equation via three-point boundary value conditions. *Math. Probl. Eng.* **2015**, *2015*, 785738. [CrossRef]

5. Ntouyas, S.K.; Etemad, S. On the existence of solutions for fractional differential inclusions with sum and integral boundary conditions. *Appl. Math. Comp.* **2015**, *266*, 235–243. [CrossRef]
6. Ahmad, B.; Alsaedi, A.; Ntouyas, S.K.; Tariboon, J. *Hadamard-Type Fractional Differential Equations, Inclusions and Inequalities*; Springer International Publishing AG: London, UK, 2017.
7. Ahmad, B.; Alghanmi, M.; Ntouyas, S.K.; Alsaedi, A. Fractional differential equations involving generalized derivative with Stieltjes and fractional integral boundary conditions. *Appl. Math. Lett.* **2018**, *84*, 111–117. [CrossRef]
8. Ahmad, B.; Alruwaily, Y.; Ntouyas, S.K.; Alsaedi, A. Existence and stability results for a fractional order differential equation with non-conjugate Riemann-Stieltjes integro-multipoint boundary conditions. *Mathematics* **2019**, *7*, 249. [CrossRef]
9. Annaby, M.H.; Mansour, Z.S. *q-Fractional Calculus and Equations*; Lecture Notes in Mathematics 2056; Springer: Berlin, Germany, 2012.
10. Ahmad, B.; Ntouyas, S.K.; Tariboon, J. *Quantum Calculus. New Concepts, Impulsive IVPs and BVPs, Inequalities. Trends in Abstract and Applied Analysis*; World Scientific Publishing Co. Pte. Ltd.: Hackensack, NJ, USA, 2016.
11. Ahmad, B.; Ntouyas, S.K.; Purnaras, I.K. Existence results for nonlocal boundary value problems of nonlinear fractional q-difference equations. *Adv. Differ. Equ.* **2012**, *2012*, 140. [CrossRef]
12. Zhou, W.X.; Liu, H.Z. Existence solutions for boundary value problem of nonlinear fractional q-difference equations. *Adv. Differ. Equ.* **2013**, *2013*, 113. [CrossRef]
13. Ahmad, B.; Nieto, J.J.; Alsaedi, A.; Al-Hutami, H. Boundary value problems of nonlinear fractional q-difference (integral) equations with two fractional orders and four-point nonlocal integral boundary conditions. *Filomat* **2014**, *28*, 1719–1736. [CrossRef]
14. Niyom, S.; Ntouyas, S.K.; Laoprasittichok, S.; Tariboon, J. Boundary value problems with four orders of Riemann-Liouville fractional derivatives. *Adv. Differ. Equ.* **2016**, *2016*, 165. [CrossRef]
15. Ferreira, R.A.C. Positive solutions for a class of boundary value problems with fractional q-differences. *Comput. Math. Appl.* **2011**, *61*, 367–373. [CrossRef]
16. Graef, J.R.; Kong, L. Positive solutions for a class of higher order boundary value problems with fractional q-derivatives. *Appl. Math. Comput.* **2012**, *218*, 9682–9689. [CrossRef]
17. Liang, S.; Zhang, J. Existence and uniqueness of positive solutions for three-point boundary value problem with fractional q-differences. *J. Appl. Math. Comput.* **2012**, *40*, 277–288. [CrossRef]
18. Almeida, R.; Martins, N. Existence results for fractional q-difference equations of order $\alpha \in [2,3]$ with three-point boundary conditions. *Commun. Nonlinear Sci. Numer. Simul.* **2014**, *19*, 1675–1685. [CrossRef]
19. Etemad, S.; Ettefagh, M.; Rezapour, S. On the existence of solutions for nonlinear fractional q-difference equations with q-integral boundary conditions. *J. Adv. Math. Stud.* **2015**, *8*, 265–285.
20. Alsaedi, A.; Ntouyas, S.K.; Ahmad, B. An existence theorem for fractional q-difference inclusions with nonlocal sub-strip type boundary conditions. *Sci. World J.* **2015**, *2015*, 424306. [CrossRef] [PubMed]
21. Ahmad, B.; Etemad, S.; Ettefagh, M.; Rezapour, S. On the existence of solutions for fractional q-difference inclusions with q-anti-periodic boundary conditions. *Bull. Math. Soc. Sci. Math. Roum.* **2016**, *59*, 119–134.
22. Ren, J.; Zhai, C. A fractional q-difference equation with integral boundary conditions and comparison theorem. *Int. J. Nonlinear Sci. Numer. Simul.* **2017**, *18*, 575–583. [CrossRef]
23. Zhai, C.B.; Ren, J. Positive and negative solutions of a boundary value problem for a fractional q-difference equation. *Adv. Differ. Equ.* **2017**, *2017*, 82. [CrossRef]
24. Abdeljawad, T.; Alzabut, J. On Riemann-Liouville fractional q-difference equations and their application to retarded logistic type model. *Math. Methods Appl. Sci.* **2018**, *41*, 8953–8962. [CrossRef]
25. Zhang, L.; Sun, S. Existence and uniqueness of solutions for mixed fractional q-difference boundary value problems. *Bound. Value Probl.* **2019**, *2019*, 100. [CrossRef]
26. Ahmad, B.; Nieto, J.J.; Alsaedi, A.; Al-Hutami, H. Existence of solutions for nonlinear fractional q-difference integral equations with two fractional orders and nonlocal four-point boundary conditions. *J. Frankl. Inst.* **2014**, *351*, 2890–2909. [CrossRef]
27. Ahmad, B.; Ntouyas, S.K.; Alsaedi, A.; Alsulami, H.H. Nonlinear q-fractional differential equations with nonlocal and sub-strip type boundary conditions. *Electron. J. Qual. Theory Differ. Equ.* **2014**, *2014*, 1–12. [CrossRef]

28. Ahmad, B.; Alsaedi, A.; Al-Hutami, H. A study of sequential fractional q-integro-difference equations with perturbed anti-periodic boundary conditions. In *Fractional Dynamics*; De Gruyter Open: Berlin, Germany, 2015; pp. 110–128.
29. Ahmad, B.; Ntouyas, S.K.; Tariboon, J.; Alsaedi, A.; Alsulami, H.H. Impulsive fractional q-integro-difference equations with separated boundary conditions. *Appl. Math. Comput.* **2016**, *281*, 199–213. [CrossRef]
30. Patanarapeelert, N.; Sriphanomwan, U.; Sitthiwirattham, T. On a class of sequential fractional q-integrodifference boundary value problems involving different numbers of q in derivatives and integrals. *Adv. Differ. Equ.* **2016**, *2016*, 148. [CrossRef]
31. Sitthiwirattham, T. On a fractional q-integral boundary value problems for fractional q-difference equations and fractional q-integro-difference equations involving different numbers of order q. *Bound. Value Probl.* **2016**, *2016*, 12. [CrossRef]
32. Krasnoselskii, M.A. Two remarks on the method of successive approximations. *Uspekhi Mat. Nauk* **1955,** *10*, 123–127.
33. Granas, A.; Dugundji, J. *Fixed Point Theory*; Springer: New York, NY, USA, 2003.
34. Deimling, K. *Nonlinear Functional Analysis*; Springer: New York, NY, USA, 1985.

8

Some Estimates for Generalized Riemann-Liouville Fractional Integrals of Exponentially Convex Functions and their Applications

Saima Rashid [1], Thabet Abdeljawad [2,*], Fahd Jarad [3] and Muhammad Aslam Noor [1]

1 Department of Mathematics, COMSATS University Islamabad, Islamabad 45550, Pakistan
2 Department of Mathematics and General Sciences, Prince Sultan University, P.O. Box 66833, Riyadh 11586, Saudi Arabia
3 Department of Mathematics, Çankaya University, Etimesgut, 06790 Ankara, Turkey
* Correspondence: tabdeljawad@psu.edu.sa

Abstract: In the present paper, we investigate some Hermite-Hadamard ($\mathcal{HH}$) inequalities related to generalized Riemann-Liouville fractional integral ($\mathcal{GRLFI}$) via exponentially convex functions. We also show the fundamental identity for $\mathcal{GRLFI}$ having the first order derivative of a given exponentially convex function. Monotonicity and exponentially convexity of functions are used with some traditional and forthright inequalities. In the application part, we give examples and new inequalities for the special means.

Keywords: convex function; exponentially convex function; fractional integrals; generalized Riemann-liouville fractional integrals

---

1. Introduction

Recently, several researchers have attracted the fractional calculus, see References [1–4]. The effect and motivation of this fractional calculus in both theoretical and applied science and engineering rose substantially. Numerous studies related to the discrete versions of this fractional calculus have been established, which benefit from countless applications in the theory of time scales, physics, different fields of engineering, chemistry and so forth (e.g, see References [4–32] and the references therein).

A few decades ago, a lot of new operator definitions were given and the properties and structures of these operators have been examined. Some of these operators are very close to classical operators in terms of their characteristics and definitions. It is known that the $\mathcal{GRLFI}$, which was introduced in reference [33], extends several well-known fractional integral operators (see Remark 1 below). Both the generalized Riemann-Liouville fractional derivative and the integral operator are useful in the study of transform theory, quantum theory and fractional intgerodifferential equations.

Almost every mathematician knows the importance of convexity theory in every field of mathematics, for example in nonlinear programming and optimization theory. By using the concept of convexity, several integral inequalities have been introduced such as Jensen, $\mathcal{HH}$ and Slater inequalities, and so forth. But the well-known one is the celebrated $\mathcal{HH}$ inequality.

Let $\mathcal{I} \subseteq \mathcal{R}$ be an interval and $\mathcal{U} : \mathcal{I} \to \mathcal{R}$ be a convex function. Then the double inequality

$$\mathcal{U}\Big(\frac{d_1 + d_2}{2}\Big) \leq \frac{1}{d_2 - d_1}\int_{d_1}^{d_2} \mathcal{U}(x)dx \leq \frac{\mathcal{U}(d_1) + \mathcal{U}(d_2)}{2}, \tag{1}$$

holds for all $d_1, d_2 \in \mathcal{I}$ with $d_1 < d_2$. It is easy to see that if $\mathcal{U}$ is concave on $\mathcal{I}$, one has the reverse of this inequality. This inequality provides bounds for the mean value of a convex function.

Recently, mathematicians have focused on obtaining new variants of the $\mathcal{HH}$ inequality by giving generalizations, improvements, refinements and extensions, see References [34–36].

Exponentially convex functions have emerged as a significant new class of convex functions, which have important applications in technology, data science and statistics. The main motivation of this paper depends on new inequalities that have been proved via $\mathcal{GRLFI}$ and applied for exponentially convex functions. This identity offers new upper bounds and estimations of $\mathcal{HH}$ type integral inequalities. Some particular cases have been discussed, which can be deduced from these consequences.

Recall the definition of an exponentially convex function, which is investigated by Dragomir and Gomm [34]:

Definition 1. *(See [34]) A positive real-valued function $\mathcal{U} : K \subseteq \mathcal{R} \longrightarrow (0, \infty)$ is said to be exponentially convex on K if the inequality*

$$e^{\mathcal{U}(\tau x+(1-\tau)y)} \leq \tau e^{\mathcal{U}(x)} + (1-\tau)e^{\mathcal{U}(y)}$$

holds for $x, y \in K$ and $\tau \in [0, 1]$.

Exponentially convex functions are used to manipulate for statistical learning, sequential prediction and stochastic optimization, see References [37–39].

After the class of exponentially convex functions was introduced by Dragomir and Gomm [34], Alirezai and Mathar [37] have investigated the mathematical perspectives along with their fertile applications in statistics and information theory, see References [37,39]. Due to its significance, Pecaric and Jaksetic [40,41] used another kind of exponentially convex function introduced in Reference [42] and have provided some applications in Euler-Radau expansions and stolarsky means. Our intention is to use the exponentially convexity property of the functions as well as the absolute values of their derivatives in order to establish estimates for $\mathcal{GRLFI}$.

Definition 2. *([1–3]) Let (d_1, d_2) $(-\infty \leq d_1 < d_2 \leq \infty)$ be a finite or infinite real interval and $\rho > 0$. Let $\Psi(x)$ be an increasing and positive monotone function on $(d_1, d_2]$ with a continuous derivative on (d_1, d_2). Then the left and right-sided generalized Riemann-Liouville fractional integrals of a function $\mathcal{U}$ on $[d_1, d_2]$ are defined by*

$$I^{\rho,\Psi}_{d_1^+}\mathcal{U}(\tau) = \frac{1}{\Gamma(\rho)}\int_{d_1}^{\tau} \Psi'(x)(\Psi(\tau) - \Psi(x))^{\rho-1}\mathcal{U}(x)dx, \tag{2}$$

and

$$I^{\rho,\Psi}_{d_2^-}\mathcal{U}(\tau) = \frac{1}{\Gamma(\rho)}\int_{\tau}^{d_2} \Psi'(x)(\Psi(x) - \Psi(\tau))^{\rho-1}\mathcal{U}(x)dx, \tag{3}$$

respectively; with $\Gamma(.)$, the classical gamma function.

Remark 1. *Many known defined fractional integral operators are just special cases of (2) and (3).*

1. *Setting $\mathcal{U}(\tau) = \tau$, it turns into the both sided (RLI).*
2. *Setting $\mathcal{U}(\tau) = \log \tau$, the Hadamard fractional integrals are obtained [1,3].*
3. *Setting $\mathcal{U}(\tau) = \frac{\tau^\beta}{\beta}$, $\beta > 0$ it turns into the both sided Katugampola fractional integrals given in Reference [33].*
4. *Setting $\mathcal{U}(\tau) = \frac{(\tau-a)^\beta}{\beta}$, $\beta > 0$, the operators in reference [43] are obtained.*
5. *Setting $\mathcal{U}(\tau) = \frac{\tau^{\epsilon+\varsigma}}{\epsilon+\varsigma}$, it turns into the both sided generalized conformable fractional integrals defined by Khan et al. in reference [44].*

The principal objective of this paper is to use a new convex class and a new integral operator to obtain new versions of $\mathcal{HH}$-inequality that give bounds for the mean value of convexity. Also, we will establish some more general estimates and related modulus inequalities for $\mathcal{GRLFI}$ via exponentially convex functions. In addition, the accuracy of the results was tested with applications of special means and with some examples.

2. $\mathcal{HH}$ Inequality for $\mathcal{GRLFI}$

Theorem 1. *Let $\mathcal{U} : [d_1, d_2] \to \mathcal{R}$ be a positive function, for $0 \le d_1 < d_2$, and $e^{\mathcal{U}} \in L_1([d_1, d_2])$. Let $\Psi(x)$ is an increasing and positive monotone function on $(d_1, d_2]$, with continuous derivative $\Psi'(x)$ on (d_1, d_2). Let $\mathcal{U}$ is an exponentially convex function and $\rho \in (0, 1)$. Then*

$$e^{\mathcal{U}\left(\frac{d_1+d_2}{2}\right)} \le \frac{\Gamma(\rho+1)}{(d_2-d_1)^\rho}\left[I^{\rho,\Psi}_{d_1^+}(e^{\mathcal{U}} \circ \Psi)(\Psi^{-1}(d_2)) + I^{\rho,\Psi}_{d_2^-}(e^{\mathcal{U}} \circ \Psi)(\Psi^{-1}(d_1))\right] \le \frac{e^{\mathcal{U}(d_1)} + e^{\mathcal{U}(d_2)}}{2}. \tag{4}$$

Proof. Since $\mathcal{U}$ is an exponentially convex function for $[d_1, d_2]$, we have

$$e^{\mathcal{U}\left(\frac{u+v}{2}\right)} \le \frac{e^{\mathcal{U}(u)} + e^{\mathcal{U}(v)}}{2}. \tag{5}$$

Let $u = \tau d_1 + (1-\tau)d_2$ and $v = (1-\tau)d_1 + \tau d_2$, we get

$$2e^{\mathcal{U}\left(\frac{d_1+d_2}{2}\right)} \le e^{\mathcal{U}(\tau d_1+(1-\tau)d_2)} + e^{\mathcal{U}((1-\tau)d_1+\tau d_2)}. \tag{6}$$

Multiplying by $\tau^{\rho-1}$ on both sides of inequality (6) and then integrating w.r.t τ over $[0, 1]$, implies

$$\frac{2}{\rho}e^{\mathcal{U}\left(\frac{d_1+d_2}{2}\right)} \le \int_0^1 \tau^{\rho-1}e^{\mathcal{U}(\tau d_1+(1-\tau)d_2)}d\tau + \int_0^1 \tau^{\rho-1}e^{\mathcal{U}((1-\tau)d_1+\tau d_2)}d\tau. \tag{7}$$

Now consider,

$$\begin{aligned}
&\frac{\Gamma(\rho+1)}{2(d_2-d_1)^\rho}\left[I^{\rho,\Psi}_{d_1^+}(e^{\mathcal{U}} \circ \Psi)(\Psi^{-1}(d_2)) + I^{\rho,\Psi}_{d_2^-}(e^{\mathcal{U}} \circ \Psi)(\Psi^{-1}(d_1))\right] \\
&= \frac{\Gamma(\rho+1)}{2(d_2-d_1)^\rho}\frac{1}{\Gamma(\rho)}\left[\int_{\Psi^{-1}(d_1)}^{\Psi^{-1}(d_2)} \Psi'(z)(d_2-\Psi(z))^{\rho-1}(e^{\mathcal{U}} \circ \Psi)(z)dz\right. \\
&\quad \left. + \int_{\Psi^{-1}(d_1)}^{\Psi^{-1}(d_2)} \Psi'(z)(\Psi(z)-d_1)^{\rho-1}(e^{\mathcal{U}} \circ \Psi)(z)dz\right] \\
&= \frac{\rho}{2(d_2-d_1)^\rho}\left[\int_{\Psi^{-1}(d_1)}^{\Psi^{-1}(d_2)} \Psi'(z)(d_2-\Psi(z))^{\rho-1}e^{\mathcal{U}}(\Psi(z))dz\right. \\
&\quad \left. + \int_{\Psi^{-1}(d_1)}^{\Psi^{-1}(d_2)} \Psi'(z)(\Psi(z)-d_1)^{\rho-1}e^{\mathcal{U}}(\Psi(z))dz\right] \\
&= \frac{\rho}{2}\left[\int_0^1 \tau^{\rho-1}e^{\mathcal{U}(\tau d_1+(1-\tau)d_2)}d\tau + \int_0^1 \tau^{\rho-1}e^{\mathcal{U}((1-\tau)d_1+\tau d_2)}d\tau\right] \\
&\ge e^{\mathcal{U}\left(\frac{d_1+d_2}{2}\right)},
\end{aligned}$$

by using (7). Thus the first inequality of (4) is proved.

Next, we again use the exponential convexity of $\mathcal{U}$, that is,

$$e^{\mathcal{U}(\tau d_1+(1-\tau)d_2)}+e^{\mathcal{U}((1-\tau)d_1+\tau d_2)}\le e^{\mathcal{U}(d_1)}+e^{\mathcal{U}(d_2)}. \tag{8}$$

Multiplying by $\tau^{\rho-1}$ on both sides of inequality (8), and then integrating w.r.t τ over $[0,1]$, implies

$$\int_0^1 \tau^{\rho-1}e^{\mathcal{U}(\tau d_1+(1-\tau)d_2)}d\tau+\int_0^1 \tau^{\rho-1}e^{\mathcal{U}((1-\tau)d_1+\tau d_2)}d\tau\le\frac{e^{\mathcal{U}(d_1)}+e^{\mathcal{U}(d_2)}}{\rho}.$$

That is,

$$\frac{\Gamma(\rho+1)}{(d_2-d_1)^\rho}\left[I^{\rho,\Psi}_{d_1^+}\left(e^{\mathcal{U}}\circ\Psi\right)\left(\Psi^{-1}(d_2)\right)+I^{\rho,\Psi}_{d_2^-}\left(e^{\mathcal{U}}\circ\Psi\right)\left(\Psi^{-1}(d_1)\right)\right]\le\frac{e^{\mathcal{U}(d_1)}+e^{\mathcal{U}(d_2)}}{2}.$$

Hence the proof is completed. □

Our next result is the subsequent lemma, which is useful for our coming results.

Lemma 1. *Let $\mathcal{U}:[d_1,d_2]\to\mathcal{R}$ be a differentiable mapping, for $0\le d_1<d_2$, and $e^{\mathcal{U}}\in L_1([d_1,d_2])$. Let $\Psi(x)$ is an increasing and positive monotone function on $(d_1,d_2]$, with continuous derivative $\Psi'(x)$ on (d_1,d_2) and $\rho\in(0,1)$. Then*

$$\begin{aligned}&\frac{e^{\mathcal{U}(d_1)}+e^{\mathcal{U}(d_2)}}{2}-\frac{\Gamma(\rho+1)}{2(d_2-d_1)^\rho}\left[I^{\rho,\Psi}_{d_1^+}\left(e^{\mathcal{U}}\circ\Psi\right)\left(\Psi^{-1}(d_2)\right)+I^{\rho,\Psi}_{d_2^-}\left(e^{\mathcal{U}}\circ\Psi\right)\left(\Psi^{-1}(d_1)\right)\right]\\&=\frac{1}{2}\Bigg[\int_{\Psi^{-1}(d_1)}^{\Psi^{-1}(d_2)}\Psi'(z)\Big(\frac{\Psi(z)-d_1}{d_2-d_1}\Big)^\rho\left(e^{\mathcal{U}}\circ\Psi\right)(z)\mathcal{U}'(\Psi(z))dz\\&-\int_{\Psi^{-1}(d_1)}^{\Psi^{-1}(d_2)}\Psi'(z)\Big(\frac{d_2-\Psi(z)}{d_2-d_1}\Big)^\rho\left(e^{\mathcal{U}}\circ\Psi\right)(z)\mathcal{U}'(\Psi(z))dz\Bigg].\end{aligned} \tag{9}$$

Proof. Consider

$$\begin{aligned}\mathbb{I}&=\frac{d_2-d_1}{2}\int_0^1\left[(1-\tau)^\rho-\tau^\rho\right]e^{\mathcal{U}\left((1-\tau)d_1+\tau d_2\right)}\mathcal{U}'((1-\tau)d_1+\tau d_2)d\tau\\&=\frac{d_2-d_1}{2}\Bigg[\int_0^1(1-\tau)^\rho e^{\mathcal{U}\left((1-\tau)d_1+\tau d_2\right)}\mathcal{U}'((1-\tau)d_1+\tau d_2)d\tau\\&-\int_0^1\tau^\rho e^{\mathcal{U}\left((1-\tau)d_1+\tau d_2\right)}\mathcal{U}'((1-\tau)d_1+\tau d_2)d\tau\Bigg].\end{aligned}$$

By making a change of variable in the above equation $\Psi(z)=(1-\tau)d_1+\tau d_2$, we have

$$\begin{aligned}\mathbb{I} &= \frac{1}{2}\Bigg[\int_{\Psi^{-1}(d_1)}^{\Psi^{-1}(d_2)} \Psi'(z)\Big(\frac{d_2-\Psi(z)}{d_2-d_1}\Big)^{\rho}(e^{\mathcal{U}}\circ\Psi)(z)\mathcal{U}'(\Psi(z))dz \\ &\quad - \int_{\Psi^{-1}(d_1)}^{\Psi^{-1}(d_2)} \Psi'(z)\Big(\frac{\Psi(z)-d_1}{d_2-d_1}\Big)^{\rho}(e^{\mathcal{U}}\circ\Psi)(z)\mathcal{U}'(\Psi(z))dz\Bigg] \\ &= I_1 + I_2. \end{aligned} \tag{10}$$

Now

$$\begin{aligned} &\frac{\Gamma(\rho+1)}{2(d_2-d_1)^{\rho}} I^{\rho,\Psi}_{d_1^+}(e^{\mathcal{U}}\circ\Psi)(\Psi^{-1}(d_2)) \\ &= \frac{\rho}{2(d_2-d_1)^{\rho}} \int_{\psi^{-1}(d_1)}^{\psi^{-1}(d_2)} \psi'(z)(d_2-\psi(z))^{\rho-1}(e^{\mathcal{U}}\circ\psi)(z)dz \\ &= -\frac{1}{2(d_2-d_1)^{\rho}} \int_{\psi^{-1}(d_1)}^{\psi^{-1}(d_2)} (e^{\mathcal{U}}\circ\psi)d(d_2-\psi(z))^{\rho} \\ &= \frac{1}{2(d_2-d_1)^{\rho}}\Bigg[(d_2-d_1)^{\rho}e^{\mathcal{U}(d_1)} + \int_{\psi^{-1}(d_1)}^{\psi^{-1}(d_2)} \psi'(z)(d_2-\Psi(z))^{\rho}(e^{\mathcal{U}}\circ\psi)(z)\mathcal{U}'(\Psi(z))dz\Bigg] \\ &= \frac{e^{\mathcal{U}(d_1)}}{2} + I_1 \end{aligned}$$

and

$$\begin{aligned} &\frac{\Gamma(\rho+1)}{2(d_2-d_1)^{\rho}} I^{\rho,\Psi}_{d_2^-}(e^{\mathcal{U}}\circ\Psi)(\Psi^{-1}(d_1)) \\ &= \frac{\rho}{2(d_2-d_1)^{\rho}} \int_{\Psi^{-1}(d_1)}^{\psi^{-1}(d_2)} \psi'(z)(\Psi(z)-d_1)^{\rho-1}(e^{\mathcal{U}}\circ\Psi)(z)dz \\ &= \frac{1}{2(d_2-d_1)^{\rho}} \int_{\Psi^{-1}(d_1)}^{\psi^{-1}(d_2)} (e^{\mathcal{U}}\circ\psi)d(\Psi(z)-d_1)^{\rho} \\ &= \frac{1}{2(d_2-d_1)^{\rho}}\Bigg[(d_2-d_1)^{\rho}e^{\mathcal{U}(d_2)} - \int_{\Psi^{-1}(d_1)}^{\Psi^{-1}(d_2)} \Psi'(z)(\Psi(z)-d_1)^{\rho}(e^{\mathcal{U}}\circ\psi)(z)\mathcal{U}'(\Psi(z))dz\Bigg] \\ &= \frac{e^{\mathcal{U}(d_2)}}{2} - I_2. \end{aligned}$$

It follows that

$$\frac{e^{\mathcal{U}}(d_2)+e^{\mathcal{U}}(d_1)}{2} - I_1 - I_2 = \mathbb{I}.$$

This completes the proof. □

Theorem 2. *Let $\mathcal{U}:[d_1,d_2]\to\mathcal{R}$ be a differentiable mapping, for $0\le d_1<d_2$, and $e^{\mathcal{U}}\in L_1([d_1,d_2])$. Let $\Psi(x)$ be an increasing and positive monotone function on $(d_1,d_2]$, with continuous derivative $\Psi'(x)$ on (d_1,d_2) and $\rho\in(0,1)$. If $|\mathcal{U}'|^q$ is exponentially convex and $q\ge 1$, then*

$$\begin{aligned}&\left|\frac{e^{\mathcal{U}(d_1)}+e^{\mathcal{U}(d_2)}}{2}-\frac{\Gamma(\rho+1)}{2(d_2-d_1)^\rho}\left[I^{\rho,\Psi}_{d_1^+}\left(e^{\mathcal{U}}\circ\Psi\right)\left(\Psi^{-1}(d_2)\right)+I^{\rho,\Psi}_{d_2^-}\left(e^{\mathcal{U}}\circ\Psi\right)\left(\Psi^{-1}(d_1)\right)\right]\right|\\&\frac{d_2-d_1}{2}\left(\frac{2}{\rho+1}\left(1-\frac{1}{2^\rho}\right)\right)^{1-\frac{1}{q}}\left[\frac{1}{\rho+3}\left(1-\frac{1}{2^{\rho+2}}\right)\right]^{\frac{1}{q}}\left[|e^{\mathcal{U}(d_1)}\mathcal{U}'(d_1)|^q+|e^{\mathcal{U}(d_2)}\mathcal{U}'(d_2)|^q\right]^{\frac{1}{q}}.\end{aligned}\tag{11}$$

Proof. First note that, for every $z\in\left(\Psi^{-1}(d_1)\Psi^{-1}(d_2)\right)$, we have $d_1<\Psi(z)<d_2$. Let $\tau=\frac{d_2-\Psi(z)}{d_2-d_1}$, then we have $\Psi(z)=\tau d_1+(1-\tau)d_2$. Applying Lemma 1, Hölder's inequality and exponentially convexity of $|\mathcal{U}'|$, we obtain

$$\begin{aligned}&\left|\frac{e^{\mathcal{U}(d_1)}+e^{\mathcal{U}(d_2)}}{2}-\frac{\Gamma(\rho+1)}{2(d_2-d_1)^\rho}\left[I^{\rho,\Psi}_{d_1^+}\left(e^{\mathcal{U}}\circ\Psi\right)\left(\Psi^{-1}(d_2)\right)+I^{\rho,\Psi}_{d_2^-}\left(e^{\mathcal{U}}\circ\Psi\right)\left(\Psi^{-1}(d_1)\right)\right]\right|\\
&\le\frac{1}{2}\left[\int_{\Psi^{-1}(d_1)}^{\Psi^{-1}(d_2)}\left|\left(\frac{\Psi(z)-d_1}{d_2-d_1}\right)^\rho-\left(\frac{d_2-\Psi(z)}{d_2-d_1}\right)^\rho\right|\left|(e^{\mathcal{U}}\circ\Psi)(z)\right|d(\mathcal{U}(\Psi(z)))\right]\\
&=\frac{d_2-d_1}{2}\left(\int_0^1|(1-\tau)^\rho-\tau^\rho|d\tau\right)^{1-\frac{1}{q}}\left(\int_0^1|(1-\tau)^\rho-\tau^\rho||e^{\mathcal{U}(\tau d_1+(1-\tau)d_2)}\mathcal{U}'(\tau d_1+(1-\tau)d_2)|^q d\tau\right)^{\frac{1}{q}}\\
&\le\frac{d_2-d_1}{2}\left(\int_0^1|(1-\tau)^\rho-\tau^\rho|d\tau\right)^{1-\frac{1}{q}}\\
&\times\left(\int_0^1|(1-\tau)^\rho-\tau^\rho||\{\tau|e^{\mathcal{U}(d_1)}|^q+(1-\tau)|e^{\mathcal{U}(d_2)}|^q\}\{\tau|\mathcal{U}'(d_1)|^q+(1-\tau)|\mathcal{U}'(d_2)|^q\}d\tau\right)^{\frac{1}{q}}\\
&=\frac{d_2-d_1}{2}\left(\frac{2}{\rho+1}\left(1-\frac{1}{2^\rho}\right)\right)^{1-\frac{1}{q}}\left(\int_0^1|(1-\tau)^\rho-\tau^\rho|\left[\tau^2|e^{\mathcal{U}(d_1)}\mathcal{U}'(d_1)|^q\right.\right.\\
&\left.\left.+(1-\tau)^2|e^{\mathcal{U}(d_2)}\mathcal{U}'(d_2)|^q+\tau(1-\tau)\{|e^{\mathcal{U}(d_1)}\mathcal{U}'(d_2)|^q+|e^{\mathcal{U}(d_2)}\mathcal{U}'(d_1)|^q\}\right]d\tau\right)^{\frac{1}{q}}\\
&=\frac{d_2-d_1}{2}\left(\frac{2}{\rho+1}\left(1-\frac{1}{2^\rho}\right)\right)^{1-\frac{1}{q}}\\
&\times\left(\int_0^1|(1-\tau)^\rho-\tau^\rho|\left[\tau^2|e^{\mathcal{U}(d_1)}\mathcal{U}'(d_1)|^q+(1-\tau)^2|e^{\mathcal{U}(d_2)}\mathcal{U}'(d_2)|^q+\tau(1-\tau)\Delta(d_1,d_2)\right]d\tau\right)^{\frac{1}{q}}\\
&:=\frac{d_2-d_1}{2}\left(\frac{2}{\rho+1}\left(1-\frac{1}{2^\rho}\right)\right)^{1-\frac{1}{q}}(S_1+S_2).\end{aligned}\tag{12}$$

where

$$\begin{aligned}S_1&:=\int_0^{\frac{1}{2}}[(1-\tau)^\rho-\tau^\rho]\left[\tau^2|e^{\mathcal{U}(d_1)}\mathcal{U}'(d_1)|^q+(1-\tau)^2|e^{\mathcal{U}(d_2)}\mathcal{U}'(d_2)|^q+\tau(1-\tau)\Delta(d_1,d_2)\right]d\tau,\\
S_2&:=\int_{\frac{1}{2}}^1[\tau^\rho-(1-\tau)^\rho]\left[\tau^2|e^{\mathcal{U}(d_1)}\mathcal{U}'(d_1)|^q+(1-\tau)^2|e^{\mathcal{U}(d_2)}\mathcal{U}'(d_2)|^q+\tau(1-\tau)\Delta(d_1,d_2)\right]d\tau.\end{aligned}$$

Now

$$\int_0^{\frac{1}{2}} \tau^2(1-\tau)^\rho d\tau = \int_{\frac{1}{2}}^1 \tau^\rho(1-\tau)^2 d\tau = \beta_{\frac{1}{2}}(3,\rho+1),$$

$$\int_0^{\frac{1}{2}} \tau^\rho(1-\tau)^2 d\tau = \int_{\frac{1}{2}}^1 \tau^2(1-\tau)^\rho d\tau = \beta_{\frac{1}{2}}(\rho+1,3),$$

$$\int_0^{\frac{1}{2}} \tau^{\rho+2} d\tau = \int_{\frac{1}{2}}^1 (1-\tau)^{\rho+2} d\tau = \frac{1}{(\rho+3)2^{\rho+3}},$$

$$\int_0^{\frac{1}{2}} (1-\tau)^{\rho+2} d\tau = \int_{\frac{1}{2}}^1 \tau^{\rho+2} d\tau = \frac{1}{\rho+3} - \frac{1}{(\rho+3)2^{\rho+3}},$$

$$\int_0^{\frac{1}{2}} \left[(1-\tau)^{\rho+1}\tau - \tau^{\rho+1}(1-\tau)\right] d\tau = \beta_{\frac{1}{2}}(\rho+2,2) - \beta_{\frac{1}{2}}(2,\rho+2) = 0,$$

$$\int_{\frac{1}{2}}^1 \left[\tau^{\rho+1}(1-\tau) - (1-\tau)^{\rho+1}\tau\right] d\tau = \beta_{\frac{1}{2}}(\rho+2,2) - \beta_{\frac{1}{2}}(2,\rho+2) = 0.$$

By substituting the above integral values in (12) and after some simplification, we get the required inequality (13). □

Corollary 1. *Letting $q = 1$, then under the assumption of Theorem 2, we have*

$$\begin{aligned}
&\left|\frac{e^{\mathcal{U}(d_1)} + e^{\mathcal{U}(d_2)}}{2} - \frac{\Gamma(\rho+1)}{2(d_2-d_1)^\rho}\left[I_{d_1^+}^{\rho,\Psi}\left(e^{\mathcal{U}} \circ \Psi\right)\left(\Psi^{-1}(d_2)\right) + I_{d_2^-}^{\rho,\Psi}\left(e^{\mathcal{U}} \circ \Psi\right)\left(\Psi^{-1}(d_1)\right)\right]\right| \\
&\leq \frac{d_2-d_1}{2}\left[\frac{1}{\rho+3}\left(1-\frac{1}{2^{\rho+2}}\right)\right]\left[\left|e^{\mathcal{U}(d_1)}\mathcal{U}'(d_1)\right| + \left|e^{\mathcal{U}(d_2)}\mathcal{U}'(d_2)\right|\right].
\end{aligned} \tag{13}$$

Proof. Since Ψ is differentiable and strictly increasing function, we can write $(\Psi(x) - \Psi(\tau))^{\rho-1} \leq (\Psi(x) - \Psi(d_1))^{\rho-1}$, where as $x \in [d_1, d_2]$ and $\tau \in [d_1, x]$, $\rho \geq 1$, and $\Psi'(\tau) > 0$. Then, the subsequent inequality holds true

$$\Psi'(\tau)(\Psi(x) - \Psi(\tau))^{\rho-1} \leq \Psi'(\tau)(\Psi(x) - \Psi(d_1))^{\rho-1}. \tag{14}$$

By exponentially convexity of $\mathcal{U}$, we have

$$e^{\mathcal{U}(\tau)} \leq \frac{x-\tau}{x-d_1} e^{\mathcal{U}(d_1)} + \frac{\tau-d_1}{x-d_1} e^{\mathcal{U}(x)}. \tag{15}$$

From (14) and (15), one has

$$\begin{aligned}
&\int_{d_1}^x \Psi'(\tau)(\Psi(x) - \Psi(\tau))^{\rho-1} e^{\mathcal{U}(\tau)} d\tau \\
&\leq \frac{(\Psi(x) - \Psi(d_1))^{\rho-1}}{x-d_1}\left[e^{\mathcal{U}(d_1)} \int_{d_1}^x (x-\tau)\Psi'(\tau) d\tau + e^{\mathcal{U}(x)} \int_{d_1}^x (\tau-d_1)\Psi'(\tau) d\tau\right].
\end{aligned}$$

By using (2) of Definition 2, we get

$$\Gamma(\rho) I_{d_1^+}^{\rho,\Psi} e^{\mathcal{U}(x)} \tag{16}$$
$$\leq \frac{(\Psi(x)-\Psi(d_1))^{\rho-1}}{x-d_1}\left[(x-d_1)\left[\Psi(x)e^{\mathcal{U}(x)} - e^{\mathcal{U}(d_1)}\Psi(d_1)\right] - \left(e^{\mathcal{U}(x)} - e^{\mathcal{U}(d_1)}\right)\int_{d_1}^{x}\Psi(\tau)d\tau\right].$$

Now for $x \in [d_1, d_2]$, $\tau \in [x, d_2]$ and $\delta \geq 1$, the subsequent inequality holds true

$$\Psi'(\tau)(\Psi(\tau)-\Psi(x))^{\delta-1} \leq \Psi'(\tau)(\Psi(d_2)-\Psi(x))^{\delta-1}. \tag{17}$$

By exponentially convexity of $\mathcal{U}$, we have

$$e^{\mathcal{U}(\tau)} \leq \frac{\tau-x}{d_2-x}e^{\mathcal{U}(d_2)} + \frac{d_2-\tau}{d_2-x}e^{\mathcal{U}(x)}. \tag{18}$$

Adopting the same procedure as we did for (14) and (15), one can get from (17) and (18) the coming inequality

$$\Gamma(\delta) I_{d_2^-}^{\delta,\Psi} e^{\mathcal{U}(x)} \tag{19}$$
$$\leq \frac{(\Psi(d_2)-\Psi(x))^{\delta-1}}{d_2-x}\left[(d_2-x)\left[\Psi(d_2)e^{\mathcal{U}(d_2)} - e^{\mathcal{U}(x)}\Psi(x)\right] - \left(e^{\mathcal{U}(d_2)} - e^{\mathcal{U}(x)}\right)\int_{x}^{d_2}\Psi(\tau)d\tau\right].$$

From inequalities (16) and (19), we get (13). Hence the proof is completed. □

Particular cases are stated as follows.

Corollary 2. *Choosing $\rho = \delta$ in (13), then we have a new inequality for $\mathcal{GRLFI}$;*

$$\Gamma(\rho) I_{d_1^+}^{\rho,\Psi} e^{\mathcal{U}(x)} + \Gamma(\rho) I_{d_2^-}^{\rho,\Psi} e^{\mathcal{U}(x)}$$
$$\leq \frac{(\Psi(x)-\Psi(d_1))^{\rho-1}}{x-d_1}\left[(x-d_1)\left[\Psi(x)e^{\mathcal{U}(x)} - e^{\mathcal{U}(d_1)}\Psi(d_1)\right] - \left(e^{\mathcal{U}(x)} - e^{\mathcal{U}(d_1)}\right)\int_{d_1}^{x}\Psi(\tau)d\tau\right]$$
$$+\frac{(\Psi(d_2)-\Psi(x))^{\rho-1}}{d_2-x}\left[(d_2-x)\left[\Psi(d_2)e^{\mathcal{U}(d_2)} - e^{\mathcal{U}(x)}\Psi(x)\right] - \left(e^{\mathcal{U}(d_2)} - e^{\mathcal{U}(x)}\right)\int_{x}^{d_2}\Psi(\tau)d\tau\right].$$

Corollary 3. *Choosing $x = d_1$ and $x = d_2$ in (13), adding the resulting inequalities, then the conditions of Theorem 1 are satisfied, we have*

$$\Gamma(\rho) I_{d_1^+}^{\rho,\Psi} e^{\mathcal{U}(d_2)} + \Gamma(\delta) I_{d_2^-}^{\delta,\Psi} e^{\mathcal{U}(d_1)} \leq \left(\frac{(\Psi(d_2)-\Psi(d_1))^{\rho-1} + (\Psi(d_2)-\Psi(d_1))^{\delta-1}}{d_2-d_1}\right) \tag{20}$$
$$\times\left((d_2-d_1)\left[e^{\mathcal{U}(d_2)}\Psi(d_2) - e^{\mathcal{U}(d_1)}\Psi(d_1)\right] - \left[e^{\mathcal{U}(d_2)} - e^{\mathcal{U}(d_1)}\right]\int_{d_1}^{d_2}\Psi(\tau)d\tau\right).$$

Corollary 4. *If we take $\rho = \delta$ in (20), then we get the following inequality for $\mathcal{GRLFI}$*

$$\Gamma(\rho)\left[I_{d_1^+}^{\rho+1,\Psi}\Psi(d_2) + I_{d_2^-}^{\rho+1,\Psi}\Psi(d_1)\right] \leq \left(\frac{2(\Psi(d_2)-\Psi(d_1))^{\rho-1}}{d_2-d_1}\right)$$
$$\times\left((d_2-d_1)\left[e^{\mathcal{U}(d_2)}\Psi(d_2) - e^{\mathcal{U}(d_1)}\Psi(d_1)\right] - \left[e^{\mathcal{U}(d_2)} - e^{\mathcal{U}(d_1)}\right]\int_{d_1}^{d_2}\Psi(\tau)d\tau\right).$$

Theorem 3. *Let $\mathcal{U}, \Psi : [d_1, d_2] \longrightarrow \mathcal{R}$ be the functions such that $e^{\mathcal{U}}$ be differentiable function, Ψ is also differentiable and strictly increasing with $\Psi' \in L_1([d_1, d_2])$. Then for $\rho, \delta > 0$, $x \in [d_1, d_2]$ we have*

$$\begin{aligned}&\left|\Gamma(\rho+1)I_{d_1^+}^{\rho,\Psi}e^{\mathcal{U}(x)} + \Gamma(\delta+1)I_{d_2^+}^{\delta,\Psi}e^{\mathcal{U}(x)} - (\Psi(x)-\Psi(d_1))^{\rho}e^{\mathcal{U}(d_1)} + (\Psi(d_2)-\Psi(x))^{\delta}e^{\mathcal{U}(d_2)}\right| \\ \leq\ & \frac{\left(\Psi(x)-\Psi(d_1)\right)^{\rho}(x-d_1)|e^{\mathcal{U}(d_1)}\mathcal{U}'(d_1)| + \left(\Psi(d_2)-\Psi(x)\right)^{\delta}(d_2-x)|e^{\mathcal{U}(d_2)}\mathcal{U}'(d_2)|}{3} \\ &+\frac{\left\{\left(\Psi(d_2)-\Psi(x)\right)^{\delta}(d_2-x) + \left(\Psi(x)-\Psi(d_1)\right)^{\rho}(x-d_1)\right\}|e^{\mathcal{U}(x)}\mathcal{U}'(x)|}{3} \\ &+\frac{\left(\Psi(x)-\Psi(d_1)\right)^{\rho}(x-d_1)\Delta(d_1,x) + \left(\Psi(d_2)-\Psi(x)\right)^{\delta}(d_2-x)\Delta(d_2,x)}{6},\end{aligned} \tag{21}$$

where

$$\Delta(d_1, x) = |e^{\mathcal{U}(x)}\mathcal{U}'(d_1)| + |e^{\mathcal{U}(d_1)}\mathcal{U}'(x)|, \tag{22}$$

$$\Delta(d_2, x) = |e^{\mathcal{U}(x)}\mathcal{U}'(d_2)| + |e^{\mathcal{U}(d_2)}\mathcal{U}'(x)|. \tag{23}$$

respectively.

Proof. From the convexity of $|(e^{\mathcal{U}})'|$, we obtain

$$\begin{aligned}|e^{\mathcal{U}(\tau)}\mathcal{U}'(\tau)| &\leq \left[\frac{x-\tau}{x-d_1}|e^{\mathcal{U}(d_1)}| + \frac{\tau-d_1}{x-d_1}|e^{\mathcal{U}(x}|\right]\left[\frac{x-\tau}{x-d_1}|\mathcal{U}'(d_1)| + \frac{\tau-d_1}{x-d_1}|\mathcal{U}'(x)|\right] \\ &\leq \left(\frac{x-\tau}{x-d_1}\right)^2|e^{\mathcal{U}(d_1)}\mathcal{U}'(d_1)| + \left(\frac{\tau-d_1}{x-d_1}\right)^2|e^{\mathcal{U}(x)}\mathcal{U}'(x)| \\ &\quad+\left(\frac{x-\tau}{x-d_1}\right)\left(\frac{\tau-d_1}{x-d_1}\right)\{|e^{\mathcal{U}(d_1)}\mathcal{U}'(x)| + |e^{\mathcal{U}(x)}\mathcal{U}'(d_1)|\} \\ &= \left(\frac{x-\tau}{x-d_1}\right)^2|e^{\mathcal{U}(d_1)}\mathcal{U}'(d_1)| + \left(\frac{\tau-d_1}{x-d_1}\right)^2|e^{\mathcal{U}(x)}\mathcal{U}'(x)| + \left(\frac{x-\tau}{x-d_1}\right)\left(\frac{\tau-d_1}{x-d_1}\right)\Delta(d_1,x).\end{aligned} \tag{24}$$

From (24), we have

$$e^{\mathcal{U}(\tau)}\mathcal{U}'(\tau) \leq \left(\tfrac{x-\tau}{x-d_1}\right)^2|e^{\mathcal{U}(d_1)}\mathcal{U}'(d_1)| + \left(\tfrac{\tau-d_1}{x-d_1}\right)^2|e^{\mathcal{U}(x)}\mathcal{U}'(x)| + \left(\tfrac{x-\tau}{x-d_1}\right)\left(\tfrac{\tau-d_1}{x-d_1}\right)\Delta(d_1,x). \tag{25}$$

Since Ψ is a differentiable and strictly increasing function, we have the subsequent inequality

$$(\Psi(x)-\Psi(\tau))^{\rho} \leq (\Psi((x)-\Psi(d_1))^{\rho}, \tag{26}$$

where as $x \in [d_1, d_2]$ and $\tau \in [d_1, x]$, $\rho > 0$.

From (25) and (26), one has

$$\begin{aligned} & (\Psi(x)-\Psi(\tau))^{\rho} e^{\mathcal{U}(\tau)}\mathcal{U}'(\tau) \\ \leq\ & \frac{(\Psi(x)-\Psi(d_1))^{\rho}}{(x-d_1)^2}\Big[(x-\tau)^2|e^{\mathcal{U}(d_1)}\mathcal{U}'(d_1)|+(\tau-d_1)^2|e^{\mathcal{U}(x)}\mathcal{U}'(x)|+(x-\tau)(\tau-d_1)\Delta(d_1,x)\Big]. \end{aligned}$$

Integrating over $[d_1, x]$, we have

$$\begin{aligned} & \int_{d_1}^{x}(\Psi(x)-\Psi(\tau))^{\rho} e^{\mathcal{U}(\tau)}\mathcal{U}'(\tau)d\tau \\ \leq\ & \frac{(\Psi(x)-\Psi(d_1))^{\rho}}{(x-d_1)^2}\Big[|e^{\mathcal{U}(d_1)}\mathcal{U}'(d_1)|\int_{d_1}^{x}(x-\tau)^2 d\tau+|e^{\mathcal{U}(x)}\mathcal{U}'(x)|\int_{d_1}^{x}(\tau-d_1)^2 d\tau+\Delta(d_1,x)\int_{d_1}^{x}(x-\tau)(\tau-d_1)d\tau\Big] \\ =\ & (\Psi(x)-\Psi(d_1))^{\rho}(x-d_1)\left[\frac{2\{|e^{\mathcal{U}(d_1)}\mathcal{U}'(d_1)|+|e^{\mathcal{U}(x)}\mathcal{U}'(x)|\}+\Delta(d_1,x)}{6}\right], \end{aligned} \tag{27}$$

and

$$\begin{aligned} & \int_{d_1}^{x}(\Psi(x)-\Psi(\tau))^{\rho} e^{\mathcal{U}(\tau)}\mathcal{U}'(\tau)d\tau = e^{\mathcal{U}(\tau)}(\Psi(x)-\Psi(\tau))^{\rho}\Big|_{d_1}^{x}+\rho\int_{d_1}^{x}(\Psi(x)-\Psi(\tau))^{\rho-1}e^{\mathcal{U}(\tau)}\Psi'(\tau)d\tau \\ =\ & -e^{\mathcal{U}(d_1)}(\Psi(x)-\Psi(d_1))^{\rho}+\Gamma(\rho+1)I^{\rho,\Psi}_{d_1^+}e^{\mathcal{U}(x)}. \end{aligned} \tag{28}$$

Therefore (28) takes the form

$$\begin{aligned} & \Gamma(\rho+1)I^{\rho,\Psi}_{d_1^+}e^{\mathcal{U}(x)}-e^{\mathcal{U}(d_1)}(\Psi(x)-\Psi(d_1))^{\rho} \\ \leq\ & (\Psi(x)-\Psi(d_1))^{\rho}(x-d_1)\left[\frac{2\{|e^{\mathcal{U}(d_1)}\mathcal{U}'(d_1)|+|e^{\mathcal{U}(x)}\mathcal{U}'(x)|\}+\Delta(d_1,x)}{6}\right] \end{aligned} \tag{29}$$

Also from (24), one has

$$e^{\mathcal{U}(\tau)}\mathcal{U}'(\tau)\geq-\left[\left(\tfrac{x-\tau}{x-d_1}\right)^2|e^{\mathcal{U}(d_1)}\mathcal{U}'(d_1)|+\left(\tfrac{\tau-d_1}{x-d_1}\right)^2|e^{\mathcal{U}(x)}\mathcal{U}'(x)|+\left(\tfrac{x-\tau}{x-d_1}\right)\left(\tfrac{\tau-d_1}{x-d_1}\right)\Delta(d_1,x)\right]. \tag{30}$$

Following the same procedure as we did for (25), we also have

$$\begin{aligned} & e^{\mathcal{U}(d_1)}(\Psi(x)-\Psi(d_1))^{\rho}-\Gamma(\rho+1)I^{\rho,\Psi}_{d_1^+}e^{\mathcal{U}(x)} \\ \leq\ & (\Psi(x)-\Psi(d_1))^{\rho}(x-d_1)\left[\frac{2\{|e^{\mathcal{U}(d_1)}\mathcal{U}'(d_1)|+|e^{\mathcal{U}(x)}\mathcal{U}'(x)|\}+\Delta(d_1,x)}{6}\right] \end{aligned} \tag{31}$$

From (29) and (31), we get

$$\begin{aligned} & |\Gamma(\rho+1)I^{\rho,\Psi}_{d_1^+}e^{\mathcal{U}(x)}-e^{\mathcal{U}(d_1)}(\Psi(x)-\Psi(d_1))^{\rho}| \\ \leq\ & (\Psi(x)-\Psi(d_1))^{\rho}(x-d_1)\left[\frac{2\{|e^{\mathcal{U}(d_1)}\mathcal{U}'(d_1)|+|e^{\mathcal{U}(x)}\mathcal{U}'(x)|\}+\Delta(d_1,x)}{6}\right]. \end{aligned} \tag{32}$$

By convexity of $|(e^{\mathcal{U}})'|$, we have

$$\begin{aligned} |e^{\mathcal{U}(\tau)}\mathcal{U}'(\tau)| \leq\ & \left(\frac{\tau-x}{d_2-x}\right)^2|e^{\mathcal{U}(d_2)}\mathcal{U}'(d_2)|+\left(\frac{d_2-\tau}{d_2-x}\right)^2|e^{\mathcal{U}(x)}\mathcal{U}'(x)| \\ & +\left(\frac{\tau-x}{d_2-x}\right)\left(\frac{d_2-\tau}{d_2-x}\right)\Delta(d_2,x). \end{aligned} \tag{33}$$

Now for $x\in[d_1,d_2]$ and $\tau\in[x,d_2]$ and $\delta>0$, the following inequality holds true

$$(\Psi(\tau)-\Psi(x))^{\delta} \leq (\Psi(d_2)-\Psi(x))^{\delta}. \tag{34}$$

Following the same way as we have done for (25), (26) and (30) we can get from (33) and (34) the subsequent inequality

$$\begin{aligned} & \left|\Gamma(\delta+1) I_{d_2^+}^{\delta,\Psi} e^{\mathcal{U}(x)} - e^{\mathcal{U}(d_2)}(\Psi(d_2)-\Psi(x))^{\delta}\right| \\ \leq\ & (\Psi(d_2)-\Psi(x))^{\delta}(d_2-x)\left[\frac{2\{|e^{\mathcal{U}(d_2)}\mathcal{U}'(d_2)| + |e^{\mathcal{U}(x)}\mathcal{U}'(x)|\} + \Delta(d_2,x)}{6}\right]. \end{aligned} \tag{35}$$

From inequalities (32) and (35) using triangular inequality, we get (21) which is desired. □

Particular cases are stated as follows.

Corollary 5. *Choosing $\rho = \delta$ in (21), then we have a new inequality for $\mathcal{GRLFI}$*

$$\begin{aligned} & \left|\Gamma(\rho+1)\left[I_{d_1^+}^{\rho,\Psi} e^{\mathcal{U}(x)} + I_{d_2^+}^{\rho,\Psi} e^{\mathcal{U}(x)}\right] - (\Psi(x)-\Psi(d_1))^{\rho} e^{\mathcal{U}(d_1)} + (\Psi(d_2)-\Psi(x))^{\rho} e^{\mathcal{U}(d_2)}\right| \\ \leq\ & \frac{(\Psi(x)-\Psi(d_1))^{\rho}(x-d_1)|e^{\mathcal{U}(d_1)}\mathcal{U}'(d_1)| + (\Psi(d_2)-\Psi(x))^{\rho}(d_2-x)|e^{\mathcal{U}(d_2)}\mathcal{U}'(d_2)|}{3} \\ & + \frac{\{(\Psi(d_2)-\Psi(x))^{\rho}(d_2-x) + (\Psi(x)-\Psi(d_1))^{\rho}(x-d_1)\}|e^{\mathcal{U}(x)}\mathcal{U}'(x)|}{3} \\ & + \frac{(\Psi(x)-\Psi(d_1))^{\rho}(x-d_1)\Delta(d_1,x) + (\Psi(d_2)-\Psi(x))^{\rho}(d_2-x)\Delta(d_2,x)}{6}. \end{aligned}$$

To prove our next result we need the following Lemma.

Lemma 3. *Suppose that $\mathcal{U} : [d_1, d_2] \to \mathcal{R}$ is an exponentially convex function which is symmetric about $\frac{d_1+d_2}{2}$. Then we have*

$$e^{\mathcal{U}\left(\frac{d_1+d_2}{2}\right)} \leq e^{\mathcal{U}(x)}, \quad x \in [d_1, d_2]. \tag{36}$$

Proof. Write

$$\frac{d_1+d_2}{2} = \frac{1}{2}\left(\frac{x-d_1}{d_2-d_1}d_2 + \frac{d_2-x}{d_2-d_1}d_1\right) + \frac{1}{2}\left(\frac{x-d_1}{d_2-d_1}d_1 + \frac{d_2-x}{d_2-d_1}d_2\right).$$

Since $e^{\mathcal{U}}$ is convex, therefore we have

$$\begin{aligned} e^{\mathcal{U}\left(\frac{d_1+d_2}{2}\right)} & \leq \frac{1}{2} e^{\mathcal{U}\left(\frac{x-d_1}{d_2-d_1}d_2 + \frac{d_2-x}{d_2-d_1}d_1\right)} + \frac{1}{2} e^{\mathcal{U}\left(\frac{x-d_1}{d_2-d_1}d_1 + \frac{d_2-x}{d_2-d_1}d_2\right)} \\ & = \frac{1}{2}\left[e^{\mathcal{U}(x)} + e^{\mathcal{U}(d_1+d_2-x)}\right]. \end{aligned}$$

Also, $e^{\mathcal{U}}$ is symmetric about $\frac{d_1+d_2}{2}$, therefore we have $e^{\mathcal{U}(x)} = e^{\mathcal{U}(d_1+d_2-x)}$ and the inequality in (36) holds. □

Theorem 4. *Suppose that $\mathcal{U} : [d_1, d_2] \longrightarrow \mathcal{R}$ be an exponentially convex function such that $e^{\mathcal{U}}$ is positive convex and symmetric about $\frac{d_1+d_2}{2}$, Ψ is a differentiable and strictly increasing function having $\Psi' \in L_1([d_1, d_2])$. Then $\rho, \delta > 0$, we have*

$$\begin{aligned} & e^{\mathcal{U}\left(\frac{d_1+d_2}{2}\right)}\left[\frac{(\Psi(d_2)-\Psi(d_1))^{\rho+1}}{\rho+1}+\frac{(\Psi(d_2)-\Psi(d_1))^{\delta+1}}{\delta+1}\right] \\ \leq\; & \Gamma(\rho+1) I_{d_1^+}^{\rho+1,\Psi}\Psi(d_2)+\Gamma(\delta+1) I_{d_2^-}^{\delta+1,\Psi}\Psi(d_1) \leq \frac{(\Psi(d_2)-\Psi(d_1))^{\delta}+(\Psi(d_2)-\Psi(d_1))^{\rho}}{d_2-d_1} \\ & \times\left[(d_2-d_1)\left[e^{\mathcal{U}(d_2)}\Psi(d_2)-e^{\mathcal{U}(d_1)}\Psi(d_1)\right]-\left[e^{\mathcal{U}(d_2)}-e^{\mathcal{U}(d_1)}\right]\int_{d_1}^{d_2}\Psi(\tau)d\tau\right]. \end{aligned} \tag{37}$$

Proof. Since Ψ is differentiable and strictly increasing function therefore $(\Psi(x)-\Psi(d_1))^{\delta} \leq (\Psi(d_2)-\Psi(d_1))^{\delta}$, where as $x \in [d_1, d_2]$, $\delta > 0$, and $\Psi'(x) > 0$. Hence, the following inequality holds true

$$\Psi'(x)(\Psi(x)-\Psi(d_1))^{\delta} \leq \Psi'(x)(\Psi(d_2)-\Psi(d_1))^{\delta}. \tag{38}$$

From the exponential convexity of $\mathcal{U}$, it can be obtained

$$e^{\mathcal{U}(x)} \leq \frac{x-d_1}{d_2-d_1}e^{\mathcal{U}(d_2)}+\frac{d_2-x}{d_2-d_1}e^{\mathcal{U}(d_1)}. \tag{39}$$

From (38) and (39), one can have

$$\begin{aligned} & \int_{d_1}^{d_2} e^{\mathcal{U}(x)}\Psi'(x)(\Psi(x)-\Psi(d_1))^{\delta}dx \\ \leq\; & \frac{(\Psi(d_2)-\Psi(d_1))^{\delta}}{d_2-d_1}\left[e^{\mathcal{U}(d_2)}\int_{d_1}^{d_2}(x-d_1)\Psi'(x)dx+e^{\mathcal{U}(d_1)}\int_{d_1}^{d_2}(d_2-x)\Psi'(x)dx\right]. \end{aligned} \tag{40}$$

By using (2) of Definition 2, we get

$$\begin{aligned} & \Gamma(\delta+1) I_{d_2^-}^{\delta+1,\Psi}e^{\mathcal{U}(d_1)} \\ \leq\; & \frac{(\Psi(d_2)-\Psi(d_1))^{\delta}}{d_2-d_1}\left[(d_2-d_1)\left[e^{\mathcal{U}(d_2)}\Psi(d_2)-e^{\mathcal{U}}(d_1)\Psi(d_1)\right]-\left[e^{\mathcal{U}(d_2)}-e^{\mathcal{U}(d_1)}\right]\int_{d_1}^{d_2}\Psi(x)dx\right]. \end{aligned} \tag{41}$$

Now for $x \in [d_1, d_2]$, $\tau \in [x, d_2]$ and $\rho > 0$, the following inequality holds true

$$\Psi'(x)(\Psi(d_2)-\Psi(x))^{\rho} \leq \Psi'(x)(\Psi(d_2)-\Psi(d_1))^{\rho}. \tag{42}$$

Adopting the same procedure as we did for (38) and (39) one can get from (40) and (42) the subsequent inequality

$$\begin{aligned} & \Gamma(\rho+1) I_{d_1^+}^{\rho+1,\Psi}e^{\mathcal{U}(d_2)} \\ \leq\; & \frac{(\Psi(d_2)-\Psi(d_1))^{\rho}}{d_2-d_1}\left[(d_2-d_1)\left[e^{\mathcal{U}(d_2)}\Psi(d_2)-e^{\mathcal{U}}(d_1)\Psi(d_1)\right]-\left[e^{\mathcal{U}(d_2)}-e^{\mathcal{U}(d_1)}\right]\int_{d_1}^{d_2}\Psi(x)dx\right]. \end{aligned} \tag{43}$$

From (41) and (43), we get

$$\Gamma(\delta+1)I_{d_2^-}^{\delta+1,\Psi}e^{\mathcal{U}(d_1)}+\Gamma(\rho+1)I_{d_1^+}^{\rho+1,\Psi}e^{\mathcal{U}(d_2)} \leq \frac{(\Psi(d_2)-\Psi(d_1))^{\delta}+(\Psi(d_2)-\Psi(d_1))^{\rho}}{d_2-d_1} \tag{44}$$

$$\times\left[(d_2-d_1)\left[e^{\mathcal{U}(d_2)}\Psi(b)-e^{\mathcal{U}(d_1)}\Psi(d_1)\right]-\left[e^{\mathcal{U}(d_2)}-e^{\mathcal{U}(d_1)}\right]\int_{d_1}^{d_2}\Psi(x)dx\right].$$

Using Lemma 3 and multiplying (36) with $(\Psi(x)-\Psi(d_1))^{\delta}\Psi'(x)$ and integrating over $[d_1,d_2]$, we get

$$e^{\mathcal{U}\left(\frac{d_1+d_2}{2}\right)}\int_{d_1}^{d_2}(\Psi(x)-\Psi(d_1))^{\delta}\Psi'(x)dx \leq \int_{d_1}^{d_2}e^{\mathcal{U}(x)}(\Psi(x)-\Psi(d_1))^{\delta}\Psi'(x)dx. \tag{45}$$

By using (2) of Definition 2 we get

$$e^{\mathcal{U}\left(\frac{d_1+d_2}{2}\right)}\left[\frac{(\Psi(d_2)-\Psi(d_1))^{\delta+1}}{\delta+1}\right] \leq \Gamma(\delta+1)I_{d_2^-}^{\delta,\Psi}e^{\mathcal{U}(d_1)}. \tag{46}$$

Similarly, using Lemma 3 and multiplying (36) with $(\Psi(d_2)-\Psi(x))^{\rho}\Psi'(x)$ and integrating over $[d_1,d_2]$, we get

$$e^{\mathcal{U}\left(\frac{d_1+d_2}{2}\right)}\left[\frac{(\Psi(d_2)-\Psi(d_1))^{\rho+1}}{\rho+1}\right] \leq \Gamma(\rho+1)I_{d_1^+}^{\rho+1,\Psi}e^{\mathcal{U}(d_2)}. \tag{47}$$

From (44) and (47), we get (37) which is the required result. □

Corollary 6. *Choosing $\rho=\delta$ in (37), then we have a new inequality for $\mathcal{GRLFI}$*

$$2\frac{(\Psi(d_2)-\Psi(d_1))^{\rho+1}}{\rho+1}e^{\mathcal{U}\left(\frac{d_1+d_2}{2}\right)} \leq \Gamma(\rho+1)\left[I_{d_1^+}^{\rho+1,\Psi}\Psi(d_2)+I_{d_2^-}^{\rho+1,\Psi}\Psi(d_1)\right]\left(\frac{2(\Psi(d_2)-\Psi(d_1))^{\rho-1}}{d_2-d_1}\right)$$

$$\times\left((d_2-d_1)\left[e^{\mathcal{U}(d_2)}\Psi(d_2)-e^{\mathcal{U}(d_1)}\Psi(d_1)\right]-\left[e^{\mathcal{U}(d_2)}-e^{\mathcal{U}(d_1)}\right]\int_{d_1}^{d_2}\Psi(\tau)d\tau\right).$$

3. Examples

Example 5. *Let $d_1=2, d_2=4, \rho=2, \delta=2, \mathcal{U}(x)=2\ln x, e^{\mathcal{U}(x)}=x^2, \Psi(x)=x^2$ and $x\in[2,4]$. Then all the assumptions in Theorem 1 are satisfied.*

Clearly,

$$\begin{aligned}\Gamma(\rho)I_{d_1^+}^{\rho,\Psi}e^{\mathcal{U}(x)} &= \int_{d_1}^{x}\Psi'(\tau)(\Psi(x)-\Psi(\tau))^{\rho-1}e^{\mathcal{U}(\tau)}d\tau \\ &= 2\int_{2}^{3}(9-\tau^2)\tau^3dt \approx 70.83,\end{aligned}$$

and

$$\Gamma(\delta) I_{d_2^-}^{\delta,\Psi} e^{\mathcal{U}(x)} = \int_x^{d_2} \Psi'(t)(\Psi(\tau) - \Psi(x))^{\rho-1} e^{\mathcal{U}(\tau)} d\tau$$
$$= 2\int_3^4 (\tau^2 - 9)\tau^3 d\tau \approx 334.83.$$

Adding the above equations, we have the left-hand side term of (13)

$$\Gamma(\rho) I_{d_1^+}^{\rho,\Psi} e^{\mathcal{U}(x)} + \Gamma(\delta) I_{d_2^-}^{\delta,\Psi} e^{\mathcal{U}(x)} \approx 405.66. \quad (48)$$

On the other hand,

$$\frac{(\Psi(x) - \Psi(d_1))^{\rho-1}}{x - d_1}\left[(x - d_1)\left[\Psi(x)e^{\mathcal{U}(x)} - e^{\mathcal{U}(d_1)}\Psi(d_1)\right] - \left(e^{\mathcal{U}(x)} - e^{\mathcal{U}(d_1)}\right)\int_{d_1}^{x} \Psi(\tau)d\tau\right]$$
$$+\frac{(\Psi(d_2) - \Psi(x))^{\delta-1}}{d_2 - x}\left[(d_2 - x)\left[\Psi(d_2)e^{\mathcal{U}(d_2)} - e^{\mathcal{U}(x)}\Psi(x)\right] - \left(e^{\mathcal{U}(d_2)} - e^{\mathcal{U}(x)}\right)\int_{x}^{d_2} \Psi(\tau)d\tau\right]$$
$$= \frac{(9-4)}{3-2}\left[(3-2)[9(9) - 4(4)] - (9-4)\int_2^3 \tau^2 d\tau\right]$$
$$+\frac{(16-9)}{4-3}\left[(4-3)[16(16) - 9(9)] - (16-9)\int_3^4 \tau^2 d\tau\right]$$
$$= 166.675 + 620.83 \approx 785.50. \quad (49)$$

It is nice to see that the following implications hold in (48) and (49),

$$405.66 < 785.50.$$

Example 6. *Let* $d_1 = 2, d_2 = 4, \rho = 2, \delta = 2, \mathcal{U}(x) = 2\ln x, e^{\mathcal{U}(x)} = x^2, \Psi(x) = x^2$ *and* $x \in [2,4]$. *where* $\Delta(x, d_1)$ *and* $\Delta(x, d_2)$ *are given in* (22) *and* (23), *respectively. Then all the assumptions in Theorem 3 are satisfied.*

Clearly,

$$\Gamma(\rho+1) I_{d_1^+}^{\rho,\Psi} e^{\mathcal{U}(x)} - (\Psi(x) - \Psi(d_1))^{\rho} e^{\mathcal{U}(a)} = \int_{d_1}^{x} (\Psi(x) - \Psi(\tau))^{\rho} e^{\mathcal{U}(\tau)} \mathcal{U}'(\tau) d\tau$$
$$= 2\int_2^3 \tau(9 - \tau^2)^2 d\tau \approx 41.67$$

and

$$\Gamma(\delta+1) I_{d_2^-}^{\delta,\Psi} e^{\mathcal{U}(x)} - (\Psi(d_2) - \Psi(x))^{\delta} e^{\mathcal{U}(d_2)} = \int_{x}^{d_2} (\Psi(\tau) - \Psi(x))^{\delta} e^{\mathcal{U}(\tau)} \mathcal{U}'(\tau) d\tau$$
$$= 2\int_3^4 \tau(\tau^2 - 9)^2 d\tau \approx 114.33,$$

Adding the above equations, we get the left-hand term of (21)

$$
\begin{aligned}
&|\Gamma(\rho+1)I^{\rho,\Psi}_{d_1^+}e^{\mathcal{U}(x)} - (\Psi(x)-\Psi(d_1))^{\rho}e^{\mathcal{U}(d_1)} + \Gamma(\delta+1)I^{\delta,\Psi}_{d_2^-}e^{\mathcal{U}(x)} - (\Psi(d_2)-\Psi(x))^{\delta}e^{\mathcal{U}(d_2)}| \\
&= 41.67 + 114.33 \approx 156. \qquad (50)
\end{aligned}
$$

Next,

$$
\begin{aligned}
&\frac{(\Psi(x)-\Psi(d_1))^{\rho}(x-d_1)|e^{\mathcal{U}(d_1)}\mathcal{U}'(d_1)| + (\Psi(d_2)-\Psi(x))^{\delta}(d_2-x)|e^{\mathcal{U}(d_2)}\mathcal{U}'(d_2)|}{3} \\
&= \frac{(9-4)^2(3-2)|4(\frac{2}{2})| + (16-9)^2(4-3)|16(\frac{2}{4})|}{3} = 164,
\end{aligned}
$$

$$
\begin{aligned}
&\frac{\{(\Psi(d_2)-\Psi(x))^{\delta}(d_2-x) + (\Psi(x)-\Psi(d_1))^{\rho}(x-d_1)\}|e^{\mathcal{U}(x)}\mathcal{U}'(x)|}{3} \\
&= \frac{(16-9)^2(4-3)+(9-4)^2(3-2)}{3}|9(\frac{2}{3})| = 148,
\end{aligned}
$$

$$
\begin{aligned}
&\frac{(\Psi(x)-\Psi(d_1))^{\rho}(x-d_1)\Delta(d_1,x) + (\Psi(d_2)-\Psi(x))^{\delta}(d_2-x)\Delta(d_2,x)}{6} \\
&\approx 172.47.
\end{aligned}
$$

Adding the above equations we get the right-hand side term of (21)

$$
\begin{aligned}
&\frac{(\Psi(x)-\Psi(d_1))^{\rho}(x-d_1)|e^{\mathcal{U}(d_1)}\mathcal{U}'(d_1)| + (\Psi(d_2)-\Psi(x))^{\delta}(d_2-x)|e^{\mathcal{U}(d_2)}\mathcal{U}'(d_2)|}{3} \\
&+ \frac{\{(\Psi(d_2)-\Psi(x))^{\delta}(d_2-x) + (\Psi(x)-\Psi(d_1))^{\rho}(x-d_1)\}|e^{\mathcal{U}(x)}\mathcal{U}'(x)|}{3} \\
&+ \frac{(\Psi(x)-\Psi(d_1))^{\rho}(x-d_1)\Delta(d_1,x) + (\Psi(d_2)-\Psi(x))^{\delta}(d_2-x)\Delta(d_2,x)}{6} \\
&= 164 + 148 + 172.47 \approx 484.47. \qquad (51)
\end{aligned}
$$

It is nice to see that the following implications hold (50) and (51),

$$156 < 484.47.$$

4. Applications

We consider the following special means for arbitrary real numbers $\mu, \nu, \mu \neq \nu$:

$$
\begin{aligned}
\mathcal{A}(\mu,\nu) &= \frac{\mu+\nu}{2}, \quad \mu,\nu \in \mathcal{R}, \\
\mathcal{H}(\mu,\nu) &= \frac{2\mu\nu}{\mu+\nu}, \quad \mu,\nu \in \mathcal{R}\setminus\{0\}, \\
\mathcal{L}(\mu,\nu) &= \frac{\nu-\mu}{\ln|\nu|-\ln|\mu|}, \quad |\mu| \neq |\nu|, \mu\nu \neq 0, \\
\mathcal{L}_n(\mu,\nu) &= \left[\frac{\nu^{n+1}-\mu^{n+1}}{(n+1)(\nu-\mu)}\right], \quad n \in \mathcal{Z}\setminus\{-1,0\}, \mu,\nu \in \mathcal{R}, \mu \neq \nu.
\end{aligned}
$$

Now using the results in Section 1, we have some applications to the special means of real numbers.

Proposition 6. *For $a_1, a_2 \in \mathcal{R}^+, \quad a_1 < a_2$, then*

$$|\mathcal{A}(e^{a_1}, e^{a_2}) - \mathcal{L}(e^{a_1}, e^{a_2})| \leq \frac{a_2 - a_1}{2^{2-\frac{1}{q}}} \left(\frac{7}{32}\right)^{\frac{1}{q}} \left[|e^{a_1}|^q + |e^{a_2}|^q\right]^{\frac{1}{q}}.$$

Proof. Apply Theorem 2 with $e^{\mathcal{U}(x)} = e^x, \Psi(x) = x, \rho = 1$, and we obtain the desired result. □

Proposition 6. *For $a_1, a_2 \in \mathcal{R}^+, \quad a_1 < a_2$, then*

$$|\mathcal{H}^{-1}(a_1, a_2) - \mathcal{L}^{-1}(a_1, a_2)| \leq \frac{a_2 - a_1}{2^{2-\frac{1}{q}}} \left(\frac{7}{32}\right)^{\frac{1}{q}} \left[|\frac{1}{a_1}|^{2q} + |\frac{1}{a_2}|^{2q}\right]^{\frac{1}{q}}.$$

Proof. Apply Theorem 2 with $e^{\mathcal{U}(x)} = \frac{1}{x}, \Psi(x) = x, \rho = 1$, and we obtain the desired result. □

Proposition 6. *Let $a_1, a_2 \in \mathcal{R}^+, a_1 < a_2$, then*

$$|\mathcal{A}(a_1^2, a_2^2) - \mathcal{L}_2^2(a_1, a_2)| \leq \frac{a_2 - a_1}{2^{1-\frac{1}{q}}} \left(\frac{7}{32}\right)^{\frac{1}{q}} \left[|a_1|^q + |a_2|^q\right]^{\frac{1}{q}}.$$

Proof. Apply Theorem 2 with $e^{\mathcal{U}(x)} = x^2, \Psi(x) = x, \rho = 1$, and we obtain the desired result. □

Proposition 6. *Let $a_1, a_2 \in \mathcal{R}^+, a_1 < a_2$, then*

$$|\mathcal{A}(a_1^n, a_2^n) - \mathcal{L}_n^n(a_1, a_2)| \leq \frac{n(a_2 - a_1)}{2^{2-\frac{1}{q}}} \left(\frac{7}{32}\right)^{\frac{1}{q}} \left[|a_1|^{(n-1)q} + |a_2|^{(n-1)q}\right]^{\frac{1}{q}}.$$

Proof. Apply Theorem 2 with $e^{\mathcal{U}(x)} = x^n, \Psi(x) = x, \rho = 1$, and we obtain the general result. □

5. Conclusions

In this article, we have investigated a few fractional integral inequalities for $\mathcal{GRLFI}$ via exponentially convexity. These inequalities have bounds of the sum of left-sided and right-sided fractional integrals and inequalities for the function, and their first derivative in absolute value is exponentially convex. Also, fractional inequalities of $\mathcal{HH}$ type for a symmetric and exponentially convex function are proved. These estimates, bounds and inequalities exist for all fractional operators are stated in Remark 1. The method followed to produce fractional inequalities is innovative and simple. It could be followed to broaden further consequences for other classes of functions related to exponentially convex functions, using convenient fractional integral operators.

Author Contributions: All authors contributed equally to the writing of this paper. All authors read and approved the final manuscript.

Acknowledgments: The authors are grateful to the reviewers for their valuable and constructive suggestions.

References

1. Samko, S.G.; Kilbas, A.A.; Marichev, O.I. *Fractional Integrals and Derivatives: Theory and Applications*; Gordon and Breach: Yverdon, Switzerland, 1993.
2. Podlubny, I. *Fractional Differential Equations*; Academic Press: San Diego, CA, USA, 1999.
3. Kilbas, A.; Srivastava, H.M.; Trujillo, J.J. *Theory and Application of Fractional Differential Equations*; North Holland Mathematics Studies: Berlin, Germany, 2006; Volume 204.

4. Miller, K.; Ross, B. *An Introduction to the Fractional Differential Equations*; John Wiley and Sons Inc.: New York, NY, USA, 1993.
5. Sonin, N.Y. On differentiation with arbitrary index. *Mosc. Matem. Sbornik.* **1869**, *6*, 1–38.
6. Letnikov, A.V. Theory of differentiation with an arbitray index. *Matem. Sbornik.* **1868**, *3*, 1–66. (In Russian)
7. Laurent, H. On the Calculation of Derivatives with Any Indices. 2004, Volume 3, pp. 240–252. Available online: http://www.numdam.org/item/NAM_1884_3_3__240_0/ (accessed on 25 July 2019). (In French)
8. Magin, R.L. Magin, R.L. Fractional Calculus in Bioengineering. *Crit. Rev. Biomed. Eng.* **2004**, *32*, 1–104. [CrossRef] [PubMed]
9. Hilfer, R. *Applications of Fractional Calculus in Physics*; Word Scientific: Singapore, 2000.
10. Srivastava, H.M.; Tomovski, Z. Fractional calculus with an integral operator containing generalized Mittag-Leffler function in the kernal. *Appl. Math. Comput.* **2009**, *211*, 198–210.
11. Tomovski, Z.; Hilfer, R.; Srivastava, H.M. Fractional and operational calculus with generalized fractional derivative operators and Mittag-Leffler function. *Integral Transforms Spec. Funct.* **2011**, *21*, 797–814. [CrossRef]
12. Belarbi, S.; Dahmani, Z. On some new fractional integral inequalities. *J. Inequal. Pure Appl. Math.* **2009**, *10*, 86.
13. Dahmani, Z. On Minkowski and Hermite-Hadamard integral inequalities via fractional integration. *Ann. Funct.* **2010**, *1*, 51–58. [CrossRef]
14. Dahmani, Z. New inequalities in fractional integrals. *Int. J. Nonlinear Sci.* **2010**, *9*, 493–497.
15. Dahmani, Z. The Riemann-Liouville operator to generate some new inequalities. *Int. J. Nonlinear Sci.* **2011**, *12*, 452–455.
16. Denton, Z.; Vatsala, A.S. Fractional integral inequalities and applications. *Comput. Math. Appl.* **2010**, *59*, 1087–1094. [CrossRef]
17. Anastassiou, G.A. *Fractional Differentiation Inequalities*; Springer: NewYork, NY, USA, 2009.
18. Abdeljawad, T. A Lyapunov type inequality for fractional operators with nonsingular Mittag-Leffler kernel. *J. Inequal. Appl.* **2017**, *2017*, 130. [CrossRef] [PubMed]
19. Abdeljawad, T.; Alzabut, J. The q-Fractional Analogue for Gronwall-Type Inequality. *J. Funct. Spaces Appl.* **2013**, *2013*, 543839. [CrossRef]
20. Adjabi, Y.; Jarad, F.; Abdeljawad, T. On generalized fractional operators and a Gronwall type inequality with applications. *Filomat* **2017**, *31*, 5457–5473. [CrossRef]
21. Jarad, F.; Abdeljawad, T.; Hammouch, Z. On a class of ordinary differential equations in the frame of Atangana-Baleanu fractional derivative. *Chaos Solitons Fractals* **2018**, *117*, 16–20. [CrossRef]
22. Abdeljawad, T.; Al-Mdallal, Q.M.; Hajji, M.A. Arbitrary order fractional difference operators with discrete exponential kernels and applications. *Discret. Dyn. Nat. Soc.* **2017**, *2017*, 4149320. [CrossRef]
23. Abdeljawad, T. Fractional operators with exponential kernels and a Lyapunov type inequality. *Adv. Differ. Equ.* **2017**, *2017*, 313. [CrossRef]
24. Alzabut, J.; Abdeljawad, T.; Jarad, F.; Sudsutad, W. A Gronwall inequality via the generalized proportional fractional derivative with applications. *J. Inequal. Appl.* **2019**, *2019*, 101. [CrossRef]
25. Khan, H.; Abdeljawad, T.; Tunç, C.; Alkhazzan, A.; Khan, A. Minkowski's inequality for the AB-fractional integral operator. *J. Inequal. Appl.* **2019**, *2019*, 96. [CrossRef]
26. Sarikaya, M.Z.; Set, E.; Yaldiz, H.; Başak, N. Hermite-Hadamard's inequalities for fractional integrals and related fractional inequalities, *Math. Comput. Model.* 2013, 57, 2403–2407. [CrossRef]
27. Chalishajar, D.; Ravichandran, C.; Dhanalakshmi, S.; Murugesu, R. Existence of fractional impulsive functional integro-differential equations in Banach spaces. *Appl. Syst. Innov.* **2019**, *2*, 18. [CrossRef]
28. Zhang, D.; Zhang, Y.; Zhang, Z.; Ahmed, N.; Zhang, Y.; Li, F.; Belic, M.R.; Xiao, M. On unveiling the link between fractional Schrödinger equation and light propogation in honeycomb lattice. *Ann. Phys.* **2017**, *529*, 1–6. [CrossRef]
29. Zhang, Y.; Zhang, H.; Belic, M.R.; Zhu, Y.; Zhang, W.; Zhang, Y.; Christodoulides, D.N.; Xiao, M. On PT symmetry in a fractional Schrödinger equation. *Laser Photonics Rev.* **2016**, *10*, 526. [CrossRef]
30. Zhang, Y.; Liu, X.; Belic, M.R.; Zhang, W.; Zhang, Y.; Xiao, M. On propogation dynamics of a light beam in a fractional Schrödinger equation. *Phys. Rev. Lett.* **2015**, *115*, 180403. [CrossRef]
31. Jarad, F.; Abdeljawad, T. Generalized fractional derivatives and Laplace transform. *Discrete Contin. Dyn. Syst.* **2018**. [CrossRef]
32. Ameen, R.; Jarad, F.; Abdeljawad, T. Ulam stability of fractional differential equations with a generalized Caputo derivative. *Filomat* **2018**, *32*, 5265–5274. [CrossRef]

33. Chen, H.; Katugampola, U.N. Hermite-Hadamard and Hermite-Hadamard-Fejér type inequalities for generalized fractional integrals. *J. Math. Anal. Appl.* **2017**, *446*, 1274–1291. [CrossRef]
34. Dragomir, S.S.; Gomm, I. Some Hermite-Hadamard type inequalities for functions whose exponentials are convex. *Stud. Univ. Babes-Bolyai Math.* **2015**, *60*, 527–534
35. Dragomir, S.S.; Agarwal, R.P. Two inequalities for differentiable mappings and applications to special means of real numbers and to trapezoidal formula. *Appl. Math. Lett.* **1998**, *11*, 91–95. [CrossRef]
36. Farid, G., Nazeer, W., Saleem, M. S., Mehmood. S., Kang, S. M.: Bounds of Riemann-Liouville fractional integrals in General form via convex functions and their applications. *Mathematics* **2018**, *6*, 248. [CrossRef]
37. Alirezaei, G.; Mathar, R. On exponentially concave functions and their impact in information theory. *J. Inform. Theory Appl.* **2018**, *9*, 265–274
38. Antczak, T. On (p, r)-invex sets and functions. *J. Math. Anal. Appl.* **2001**, *263*, 355–379. [CrossRef]
39. Pal, S.; Wong, T.K.L. On exponentially concave functions and a new information geometry.*Ann. Probab.* **2018**, *46*, 1070–1113. [CrossRef]
40. Pecaric, J.; Jaksetic, J. On exponential convexity, Euler-Radau expansions and stolarsky means, Rad Hrvat. *Matematicke Znanosti* **2013**, *17*, 81–94.
41. Jakestic, J.; Pecaric, J. Exponential convexity method. *J. Convex Anal.* **2012**, *20*, 181–187.
42. Bernstein, S.N. Sur les fonctions absolument monotones. *Acta. Math.* **1929**, *52*, 1–66. [CrossRef]
43. Jarad, F.; Ugurlu, E.; Abdeljawad, T.; Baleanu, D. On a new class of fractional operators. *Adv. Differ. Equ.* **2017**, *2017*, 247. [CrossRef]
44. Khan, T.U.; Khan, M.A. Generalized conformable fractional operators. *J. Comput. Appl. Math.* **2019**, *346*, 378–389. [CrossRef]

9

Stability Results for a Coupled System of Impulsive Fractional Differential Equations

Akbar Zada [1], Shaheen Fatima [1], Zeeshan Ali [1], Jiafa Xu [2] and Yujun Cui [3,*]

1 Department of Mathematics, University of Peshawar, Khyber Pakhtunkhwa 25000, Pakistan; akbarzada@uop.edu.pk (A.Z.); shaheenfatima072@gmail.com (S.F.); zeeshanmaths1@gmail.com (Z.A.)
2 School of Mathematical Sciences, Chongqing Normal University, Chongqing 401331, China; 20150028@cqnu.edu.cn
3 State Key Laboratory of Mining Disaster Prevention and Control Co-founded by Shandong Province and the Ministry of Science and Technology, Shandong University of Science and Technology, Qingdao 266590, China
* Correspondence: cyj720201@sdust.edu.cn

Abstract: In this paper, we establish sufficient conditions for the existence, uniqueness and Ulam–Hyers stability of the solutions of a coupled system of nonlinear fractional impulsive differential equations. The existence and uniqueness results are carried out via Banach contraction principle and Schauder's fixed point theorem. The main theoretical results are well illustrated with the help of an example.

Keywords: caputo fractional derivative; coupled system; impulses; existence theory; stability theory

1. Introduction

Fractional differential equations (FDEs) provide an excellent tool for the description of memory and hereditary properties of different processes and materials. Thus, contrary to the classical derivative, the fractional derivative is nonlocal. Fractional calculus has played a very important role in enhancing the mathematical modeling of several phenomena occurring in engineering and scientific disciplines, such as blood flow systems, control theory, aerodynamics, the nonlinear oscillation of earthquake, the fluid-dynamic traffic model, polymer rheology, regular variation in thermodynamics, etc. FDEs are more accurate than the integer-order derivatives. Therefore, in the last few decades, fractional calculus has received great attention from researchers [1–13]. On the other hand, it is impossible to describe the complicated systems and processes with a single differential equation. Therefore, the coupled systems involving FDEs have also received incredible attention; consequently, many results are devoted to them [14–31].

It is well known that the effects of a pulse cannot be ignored in many processes and phenomena. For example, in biological systems such as heart beats, blood flows, mechanical systems with impact, population dynamical systems and so on. Thus, researchers used differential equations with impulses to describe the aforesaid kinds of phenomena. Therefore, many mathematicians studied impulsive FDEs with different boundary conditions; see [32–40] and references cited therein.

In fields such as numerical analysis, optimization theory, and nonlinear analysis, we mostly deal with the approximate solutions and hence we need to check how close these solutions are to the actual solutions of the related system. For this purpose, many approaches can be used, but the approach of Ulam–Hyers stability is a simple and easy one. The aforesaid stability was first initiated by Ulam in 1940 and then was confirmed by Hyers in 1941 [41,42]. That's why this stability is known as Ulam–Hyers stability. In 1978 [43], Rassias generalized the Ulam–Hyers stability by considering variables. Thereafter, mathematicians extended the work mentioned above to functional, differential, integrals and FDEs; for more information about the topic, the reader is recommended to [44–59].

Inspired from the above discussion, in this article, we study the existence, uniqueness and stability analysis of a coupled system of nonlinear FDEs with impulses of the form:

$$\begin{cases} {}^c\mathbf{D}^{\alpha}x(t) + h(t, {}^c\mathbf{D}^{a}x(t), {}^c\mathbf{D}^{b}y(t)) = 0, \ t \neq t_m, \ m = 1, 2 \dots, n, \\ {}^c\mathbf{D}^{\beta}y(t) + w(t, {}^c\mathbf{D}^{a}x(t), {}^c\mathbf{D}^{b}y(t)) = 0, \ t \neq t_m, \ m = 1, 2 \dots, n, \\ \Delta x \mid_{t=t_m} = M_{1m}(x(t_m)), \ \Delta x' \mid_{t=t_m} = N_{1m}(x(t_m)), \ \Delta x'' \mid_{t=t_m} = O_{1m}(x(t_m)), \\ \Delta y \mid_{t=t_m} = M_{2m}(y(t_m)), \ \Delta y' \mid_{t=t_m} = N_{2m}(y(t_m)), \ \Delta y'' \mid_{t=t_m} = O_{2m}(y(t_m)), \\ x(0) = x'(0) = 0, \ {}^c\mathbf{D}^{\varepsilon}x(\Omega) = x''(1), \\ y(0) = y'(0) = 0, \ {}^c\mathbf{D}^{\rho}y(\Phi) = y''(1), \end{cases} \tag{1}$$

where $t \in J = [0, 1]$, $2 < \alpha, \beta \leqslant 3$, $0 < a, b, \varepsilon, \Omega, \rho, \Phi < 1$. ${}^c\mathbf{D}$ stands for Caputo fractional derivative and $h, w : J \times \mathbb{R}^3 \to \mathbb{R}$ are continuous functions. $M_{1m}, M_{2m}, N_{1m}, N_{2m}, O_{1m}, O_{2m} \in C(\mathbb{R}, \mathbb{R})$ and t_m satisfied $0 = t_0 < t_1 < \cdots < t_n < t_{n+1} = 1$, $\Delta x \mid_{t=t_m} = x(t_m^+) - x(t_m^-)$, $\Delta x' \mid_{t=t_m} = x'(t_m^+) - x'(t_m^-)$, $\Delta x'' \mid_{t=t_m} = x''(t_m^+) - x''(t_m^-)$, $\Delta y \mid_{t=t_m} = y(t_m^+) - y(t_m^-)$, $\Delta y' \mid_{t=t_m} = y'(t_m^+) - y'(t_m^-)$, $\Delta y'' \mid_{t=t_m} = y''(t_m^+) - y''(t_m^-)$, $x(t_m^+), y(t_m^+)$, and $x(t_m^-), y(t_m^-)$ represent the right and left limits of $x(t), y(t)$, respectively, at $t = t_m$.

The remaining article is organized as follows: In Section 2, we give some definitions and lemmas related to fractional calculus. In Section 3, we establish our main results about the existence and uniqueness of solutions for the proposed system (1). In Section 4, we study the Ulam–Hyers stability. In Section 5, we provide an example to support our main results.

2. Background Materials

In this section, we give some basic definitions of fractional calculus that will be used throughout the article.

Definition 1. *(see [60]) If $x : (0, \infty) \to \mathbb{R}$ and $\alpha > 0$, then the Caputo fractional derivative of order α is defined as*

$${}^c\mathbf{D}^{\alpha}x(t) = \frac{1}{\Gamma(n-\alpha)} \int_0^t (t-s)^{n-\alpha-1} x^{(n)}(s) ds, \ n-1 < \alpha < n, \ n = [\alpha] + 1,$$

where $[\alpha]$ denotes the integer part of real number α, provided that the right side is pointwise defined on $(0, \infty)$.

Definition 2. *(see [60]) The Riemann–Liouville fractional integral of order $\alpha > 0$ for a function $x : (0, \infty) \to \mathbb{R}$ is defined as*

$$\mathbf{I}^{\alpha}x(t) = \frac{1}{\Gamma(\alpha)} \int_0^t (t-s)^{\alpha-1} x(s) ds, \quad t > 0,$$

provided that the right side is pointwise defined on $(0, \infty)$, where Γ is the Euler Gamma function.

Lemma 1. *(see [60]) The solution of the differential equations involving Caputo derivative ${}^c\mathbf{D}^{\alpha}x(t) = f(t)$, $t \in J$, has the form:*

$$\mathbf{I}^{\alpha c}\mathbf{D}^{\alpha}x(t) = \mathbf{I}^{\alpha}f(t) + e_0 + e_1 t + \cdots + e_{n-1}t^{n-1},$$

for some $e_i \in \mathbb{R}$, $i = 0, 1, \dots, n-1$, $n = [\alpha] + 1$.

Lemma 2. *(see [60]) If $\alpha, \beta > 0$, $t \in J$, then, for $x(t)$, we have*

$${}^c\mathbf{D}^{\alpha}\mathbf{I}^{\alpha}x(t) = x(t), \quad \mathbf{I}^{\alpha}\mathbf{I}^{\beta}x(t) = \mathbf{I}^{\alpha+\beta}x(t).$$

Lemma 3. *(**Banach contraction principle**, see [59]) If X is real Banach space and $\mathbf{W} : X \longrightarrow X$ is a contraction mapping, then $\mathbf{W}$ has a unique fixed point in X.*

Theorem 1. *(**Schauder fixed point theorem**, see [59]) If ω is a closed bounded convex subset of a Banach space X and $\mathbf{W}:\omega\longrightarrow\omega$ is completely continuous, then $\mathbf{W}$ has at least one fixed point in ω.*

For the sake of convenience, we introduce the Banach space as follows:

Let $J=[0,1],\ J'=J/\{t_1,t_2,\ldots,t_n\}$. Define the set by

$$X=PC(J)=\{x(t):x(t),x'(t),x''(t),{}^{c}\mathbf{D}^{a}x(t),{}^{c}\mathbf{D}^{b}x(t)\in C(J'),x(t_m^+)\text{ and }x(t_m^-)$$
$$\text{exists and satisfying } x(t_m^-)=x(t_m),1\leqslant m\leqslant n\}.$$

It is easy to verify that X is a Banach space equipped with the norm:

$$\|x\|_0=\max\{\sup_{t\in J}|x(t)|,\sup_{t\in J}|x'(t)|,\sup_{t\in J}|x''(t)|,\sup_{t\in J}|{}^{c}\mathbf{D}^{a}x(t)|,\sup_{t\in J}|{}^{c}\mathbf{D}^{b}x(t)|\},\forall\, x(t)\in PC(J).$$

Similarly, we can define a set $Y=PC(J)$, which is a Banach space endowed with the defined norm:

$$\|y\|_0=\max\{\sup_{t\in J}|y(t)|,\sup_{t\in J}|y'(t)|,\sup_{t\in J}|y''(t)|,\sup_{t\in J}|{}^{c}\mathbf{D}^{a}y(t)|,\sup_{t\in J}|{}^{c}\mathbf{D}^{b}y(t)|\},\forall\, y(t)\in PC(J).$$

Furthermore, we define the Banach space $Y'=X\times Y$ with the norms $\|(x,y)\|=\|x\|_0+\|y\|_0$ and $\|(x,y)\|=\max\{\|x\|_0,\|y\|_0\}$.

Definition 3. *A pair of functions $(x(t),y(t))\in Y'$ is called a solution of (1) if $(x(t),y(t))$ satisfy all the equations and boundary value conditions of the system (1).*

Lemma 4. *Assume that $f\in C(J,\mathbb{R})$. A function $x\in PC(J)$ is a solution of the boundary value system*

$$\begin{cases}{}^{c}\mathbf{D}^{\alpha}x(t)+f(t)=0,\ 2<\alpha\leqslant 3,\\ \Delta x\mid_{t=tm}=M_{1m}(x(t_m)),\ m=1,2,\ldots,n,\\ \Delta x'\mid_{t=tm}=N_{1m}(x(t_m)),\ m=1,2,\ldots,n,\\ \Delta x''\mid_{t=tm}=O_{1m}(x(t_m)),\ m=1,2,\ldots,n,\\ x(0)=x'(0)=0,\ {}^{c}\mathbf{D}^{\varepsilon}x(\Omega)=x''(1),\ 0<\varepsilon,\Omega<1,\end{cases}\tag{2}$$

if and only if $x\in PC(J)$ is the solution of integral equation

$$x(t)=\begin{cases}-\frac{1}{\Gamma(\alpha)}\int_0^t(t-s)f(s)ds+e^*t^2,\ t\in[0,t_1],\\ -\frac{1}{\Gamma(\alpha)}\int_{t_m}^t(t-s)^{\alpha-1}f(s)ds-\frac{1}{\Gamma(\alpha)}\sum_{j=1}^m\int_{t_{j-1}}^{t_j}(t_j-s)^{\alpha-1}f(s)ds\\ -\frac{1}{\Gamma(\alpha-1)}\sum_{j=1}^m(t-t_j)\int_{t_{j-1}}^{t_j}(t_j-s)^{\alpha-2}f(s)ds-\frac{1}{2\Gamma(\alpha-2)}\sum_{j=1}^m(t-t_j)^2\\ \times\int_{t_{j-1}}^{t_j}(t_j-s)^{\alpha-3}f(s)ds+\sum_{j=1}^m M_{1j}(x(t_j))+\sum_{j=1}^m(t-t_j)N_{1j}(x(t_j))\\ +\sum_{j=1}^m\frac{(t-t_j)^2}{2}O_{1j}(x(t_j))+e^*t^2,\ t\in(t_m,t_{m+1}],\ 1\leqslant m\leqslant n,\end{cases}\tag{3}$$

where $t\in(t_m,t_{m+1}],1\leqslant m\leqslant n,$

$$
\begin{aligned}
e^{*} =& \frac{\Gamma(3-\varepsilon)}{2(\Gamma(3-\varepsilon)-\Omega^{2-\varepsilon})}\Bigg[-\frac{1}{\Gamma(\alpha-\varepsilon)}\int_{t_j}^{\Omega}(\Omega-s)^{\alpha-\varepsilon-1}f(s)ds-\frac{\Omega^{1-\varepsilon}}{\Gamma(\alpha-1)\Gamma(2-\varepsilon)}\\
&\times\sum_{j=1}^{m}\int_{t_{j-1}}^{t_j}(t_j-s)^{\alpha-2}f(s)ds-\frac{\Omega^{2-\varepsilon}}{\Gamma(\alpha-2)\Gamma(3-\varepsilon)}\sum_{j=1}^{m}\int_{t_{j-1}}^{t_j}(t_j-s)^{\alpha-3}f(s)ds\\
&+\frac{\Omega^{1-\varepsilon}}{\Gamma(\alpha-2)\Gamma(2-\varepsilon)}\sum_{j=1}^{m}t_j\int_{t_{j-1}}^{t_j}(t_j-s)^{\alpha-3}f(s)ds+\frac{\Omega^{1-\varepsilon}}{\Gamma(2-\varepsilon)}\sum_{j=1}^{m}N_{1j}(x(t_j))\\
&+\frac{\Omega^{2-\varepsilon}}{\Gamma(3-\varepsilon)}\sum_{j=1}^{m}O_{1j}(x(t_j))-\frac{\Omega^{1-\varepsilon}}{\Gamma(2-\varepsilon)}\sum_{j=1}^{m}t_jO_{1j}(x(t_j))+\frac{1}{\Gamma(\alpha-2)}\int_{t_m}^{1}(1-s)^{\alpha-3}f(s)ds\\
&+\frac{1}{\Gamma(\alpha-2)}\sum_{j=1}^{m}\int_{t_{j-1}}^{t_j}(t_j-s)^{\alpha-3}f(s)ds-\sum_{j=1}^{m}O_{1j}(x(t_j))\Bigg].
\end{aligned}
$$

Proof. Applying Lemma 1, for some constants $e_0, e_1, e_2 \in \mathbb{R}$, we have

$$
\begin{aligned}
x(t) &= -\mathbf{I}^{\alpha}f(s)ds+e_0+e_1t+e_2t^2\\
&= -\frac{1}{\Gamma(\alpha)}\int_0^t(t-s)^{\alpha-1}f(s)ds+e_0+e_1t+e_2t^2,\ t\in[0,t_1].
\end{aligned}\tag{4}
$$

Then, we obtain

$$
\begin{cases}
x'(t) = -\dfrac{1}{\Gamma(\alpha-1)}\displaystyle\int_0^t(t-s)^{\alpha-2}f(s)ds+e_1+2e_2t,\\
x''(t) = -\dfrac{1}{\Gamma(\alpha-2)}\displaystyle\int_0^t(t-s)^{\alpha-3}f(s)ds+2e_2.
\end{cases}\tag{5}
$$

When $t\in(t_1,t_2)$, we have

$$
\begin{cases}
x(t) = -\dfrac{1}{\Gamma(\alpha)}\displaystyle\int_{t_1}^t(t-s)^{\alpha-1}f(s)ds+e_3+e_4(t-t_1)+e_5(t-t_1)^2,\\
x'(t) = -\dfrac{1}{\Gamma(\alpha-1)}\displaystyle\int_{t_1}^t(t-s)^{\alpha-2}f(s)ds+e_4+2e_5(t-t_1),\\
x''(t) = -\dfrac{1}{\Gamma(\alpha-2)}\displaystyle\int_{t_1}^t(t-s)^{\alpha-3}f(s)ds+2e_5,
\end{cases}\tag{6}
$$

where e_3, e_4, e_5 are arbitrary constants, from (4)–(6), we can find

$$
x(t_1^-) = -\frac{1}{\Gamma(\alpha)}\int_0^{t_1}(t_1-s)^{\alpha-1}f(s)ds+e_0+e_1t_1+e_2t_1^2,\ \ x(t_1^+)=e_3,\tag{7}
$$

$$
x'(t_1^-) = -\frac{1}{\Gamma(\alpha-1)}\int_0^{t_1}(t_1-s)^{\alpha-2}f(s)ds+e_1+2e_2t_1,\ \ x'(t_1^+)=e_4,\tag{8}
$$

$$
x''(t_1^-) = -\frac{1}{\Gamma(\alpha-2)}\int_0^{t_1}(t_1-s)^{\alpha-3}f(s)ds+2e_2,\ \ x''(t_1^+)=2e_5.\tag{9}
$$

Furthermore, $\Delta x\mid_{t=t_1}= M_{11}(x(t_1))$, $\Delta x'\mid_{t=t_1}= N_{11}(x(t_1))$, $\Delta x''\mid_{t=t_1}= O_{11}(x(t_1))$, and (7)–(9) give us:

$$
\begin{cases}
e_3 = -\dfrac{1}{\Gamma(\alpha)}\displaystyle\int_0^{t_1}(t_1-s)^{\alpha-1}f(s)ds+e_0+e_1t_1+e_2t_1^2+M_{11}(x(t_1)),\\
e_4 = -\dfrac{1}{\Gamma(\alpha-1)}\displaystyle\int_0^{t_1}(t_1-s)^{\alpha-2}f(s)ds+e_1+2e_2t_1+N_{11}(x(t_1)),\\
e_5 = -\dfrac{1}{2\Gamma(\alpha-2)}\displaystyle\int_0^{t_1}(t_1-s)^{\alpha-3}f(s)ds+e_2+\dfrac{1}{2}O_{11}(x(t_1)).
\end{cases}
$$

Plugging e_3, e_4, and e_5 into the first equation of (6) for $t \in (t_1, t_2]$, we have

$$x(t) = -\frac{1}{\Gamma(\alpha)}\int_{t_1}^{t}(t-s)^{\alpha-1}f(s)ds - \frac{1}{\Gamma(\alpha)}\int_{0}^{t_1}(t_1-s)f(s)ds - \frac{(t-t_1)}{\Gamma(\alpha-1)}\int_{0}^{t_1}(t_1-s)^{\alpha-2}f(s)ds - \frac{(t-t_1)^2}{2\Gamma(\alpha-2)}\int_{0}^{t_1}(t_1-s)^{\alpha-3}f(s)ds + M_{11}(x(t_1)) + (t-t_1)N_{11}(x(t_1)) + \frac{(t-t_1)^2}{2}O_{11}(x(t_1)) + e_0 + e_1 t + e_2 t^2.$$

Repeating the same process for $t \in (t_m, t_{m+1}]$ such that $(m = 1, 2, \dots, n)$, then we can write

$$\begin{aligned} x(t) =& -\frac{1}{\Gamma(\alpha)}\int_{t_m}^{t}(t-s)^{\alpha-1}f(s)ds - \frac{1}{\Gamma(\alpha)}\sum_{j=1}^{m}\int_{t_{j-1}}^{t_j}(t_j-s)^{\alpha-1}f(s)ds - \frac{1}{\Gamma(\alpha-1)}\sum_{j=1}^{m}(t-t_j) \\ &\times \int_{t_{j-1}}^{t_j}(t_j-s)^{\alpha-2}f(s)ds - \frac{1}{2\Gamma(\alpha-2)}\sum_{j=1}^{m}(t-t_j)^2\int_{t_{j-1}}^{t_j}(t_j-s)^{\alpha-3}f(s)ds + \sum_{j=1}^{m}M_{1j}(x(t_j)) \\ &+ \sum_{j=1}^{m}(t-t_j)N_{1j}(x(t_j)) + \sum_{j=1}^{m}\frac{(t-t_j)^2}{2}O_{1j}(x(t_j)) + e_0 + e_1 t + e_2 t^2. \end{aligned} \tag{10}$$

Furthermore, we have

$$\begin{aligned} x''(t) =& -\frac{1}{\Gamma(\alpha-2)}\int_{t_m}^{t}(t-s)^{\alpha-3}f(s)ds - \frac{1}{\Gamma(\alpha-2)}\sum_{j=1}^{m}\int_{t_{j-1}}^{t_j}(t_j-s)^{\alpha-3}f(s)ds + O_{1j}(x(t_j)) \\ &+2e_2. \end{aligned} \tag{11}$$

By utilizing conditions $x(0) = x'(0) = 0$ in (4), we get $e_0 = e_1 = 0$. In addition, it follows from (11) that

$$\begin{aligned} x''(1) =& -\frac{1}{\Gamma(\alpha-2)}\int_{t_m}^{1}(1-s)^{\alpha-3}f(s)ds - \frac{1}{\Gamma(\alpha-2)}\sum_{j=1}^{m}\int_{t_{j-1}}^{t_j}(t_j-s)^{\alpha-3}f(s)ds + O_{1j}(x(t_j)) \\ &+2e_2. \end{aligned} \tag{12}$$

In view of $j \in \{0, 1, \dots, n\}$ such that $\Omega \in (t_j, t_{j+1}]$, we have

$$\begin{aligned} x(\Omega) =& -\frac{1}{\Gamma(\alpha)}\int_{t_j}^{\Omega}(\Omega-s)^{\alpha-1}f(s)ds - \frac{1}{\Gamma(\alpha)}\sum_{j=1}^{m}\int_{t_{j-1}}^{t_j}(t_j-s)^{\alpha-1}f(s)ds - \frac{1}{\Gamma(\alpha-1)}\sum_{j=1}^{m}(\Omega-t_j) \\ &\times \int_{t_{j-1}}^{t_j}(t_j-s)^{\alpha-2}f(s)ds - \frac{1}{2\Gamma(\alpha-2)}\sum_{j=1}^{m}(\Omega-t_j)^2\int_{t_{j-1}}^{t_j}(t_j-s)^{\alpha-3}f(s)ds \\ &+ \sum_{j=1}^{m}M_{1j}(x(t_j)) + \sum_{j=1}^{m}(\Omega-t_j)N_{1j}(x(t_j)) + \frac{(\Omega-t_j)^2}{2}O_{1j}(x(t_j)) + e_2\Omega^2. \end{aligned} \tag{13}$$

By applying result (5), we get

$$\begin{aligned} {}^cD_{0^+}^{\varepsilon}x(\Omega) =& -\frac{1}{\Gamma(\alpha-\varepsilon)}\int_{t_j}^{\Omega}(\Omega-s)^{\alpha-\varepsilon-1}f(s)ds - \frac{\Omega^{1-\varepsilon}}{\Gamma(\alpha-1)\Gamma(2-\varepsilon)}\sum_{j=1}^{m}\int_{t_{j-1}}^{t_j}(t_j-s)^{\alpha-2}f(s)ds \\ &-\frac{\Omega^{2-\varepsilon}}{\Gamma(\alpha-2)\Gamma(3-\varepsilon)}\sum_{j=1}^{m}\int_{t_{j-1}}^{t_j}(t_j-s)^{\alpha-3}f(s)ds + \frac{\Omega^{1-\varepsilon}}{\Gamma(\alpha-2)\Gamma(2-\varepsilon)}\sum_{j=1}^{m}t_j \\ &\times \int_{t_{j-1}}^{t_j}(t_j-s)^{\alpha-3}f(s)ds + \frac{\Omega^{1-\varepsilon}}{\Gamma(2-\varepsilon)}\sum_{j=1}^{m}N_{1j}(x(t_j)) + \frac{\Omega^{2-\varepsilon}}{\Gamma(3-\varepsilon)}\sum_{j=1}^{m}O_{1j} \\ &-\frac{\Omega^{1-\varepsilon}}{\Gamma(2-\varepsilon)}\sum_{j=1}^{m}t_jO_{1j} + \frac{2\Omega^{2-\varepsilon}}{\Gamma(3-\varepsilon)}e_2. \end{aligned} \tag{14}$$

Since ${}^{c}D^{\varepsilon}_{0+}x(\Omega) = x''(1)$, thus (13) and (14) gives

$$e_2 = \frac{\Gamma(3-\varepsilon)}{2(\Gamma(3-\varepsilon)-\Omega^{2-\varepsilon})}\Bigg[-\frac{1}{\Gamma(\alpha-\varepsilon)}\int_{t_j}^{\Omega}(\Omega-s)^{\alpha-\varepsilon-1}f(s)ds - \frac{\Omega^{1-\varepsilon}}{\Gamma(\alpha-1)\Gamma(2-\varepsilon)}$$

$$\times\sum_{j=1}^{m}\int_{t_{j-1}}^{t_j}(t_j-s)^{\alpha-2}f(s)ds - \frac{\Omega^{2-\varepsilon}}{\Gamma(\alpha-2)\Gamma(3-\varepsilon)}\sum_{j=1}^{m}\int_{t_{j-1}}^{t_j}(t_j-s)^{\alpha-3}f(s)ds$$

$$+\frac{\Omega^{1-\varepsilon}}{\Gamma(\alpha-2)\Gamma(2-\varepsilon)}\sum_{j=1}^{m}t_j\int_{t_{j-1}}^{t_j}(t_j-s)^{\alpha-3}f(s)ds + \frac{\Omega^{1-\varepsilon}}{\Gamma(2-\varepsilon)}\sum_{j=1}^{m}N_{1j}(x(t_j))$$

$$+\frac{\Omega^{2-\varepsilon}}{\Gamma(3-\varepsilon)}\sum_{j=1}^{m}O_{1j} - \frac{\Omega^{1-\varepsilon}}{\Gamma(2-\varepsilon)}\sum_{j=1}^{m}t_jO_{1j} + \frac{1}{\Gamma(\alpha-2)}\int_{t_m}^{1}(1-s)^{\alpha-3}f(s)ds$$

$$+\frac{1}{\Gamma(\alpha-2)}\sum_{j=1}^{m}\int_{t_{j-1}}^{t_j}(t_j-s)^{\alpha-3}f(s)ds - \sum_{j=1}^{m}O_{1j}(x(t_j))\Bigg].$$

Plugging the values of e_0, e_1 and e_2 into (4) and (10), (3) can thus be obtained. Conversely, we consider that $x(t)$ is a solution of (3). Then, it is obvious that (3) satisfies (2). □

Similarly as in Lemma 4, we can prove the following:

Lemma 5. *Let $\vartheta \in C(J, \mathbb{R})$. A function $y \in PC(J)$ is the solution of*

$$\begin{cases} {}^{c}\mathbf{D}^{\beta}y(t) + \vartheta(t) = 0, \ 2 < \beta \leqslant 3, \\ \Delta y\,|_{t=t_m} = M_{2m}(y(t_m)), \ m = 1,2,\ldots,n, \\ \Delta y'\,|_{t=t_m} = N_{2m}(y(t_m)), \ m = 1,2,\ldots,n, \\ \Delta y''\,|_{t=t_m} = O_{2m}(y(t_m)), \ m = 1,2,\ldots,n, \\ y(0) = y'(0) = 0, \ {}^{c}\mathbf{D}^{\rho}_{0+}y(\Phi) = y''(1), \ 0 < \rho, \Phi < 1, \end{cases} \tag{15}$$

if and only if $y \in PC(J)$ is the solution of the integral equation

$$y(t) = \begin{cases} -\frac{1}{\Gamma(\beta)}\int_0^t (t-s)^{\beta-1}\vartheta(s)ds + c^*t^2, \ t \in [0,t_1], \\ -\frac{1}{\Gamma(\beta)}\int_{t_m}^t (t-s)^{\beta-1}\vartheta(s)ds - \frac{1}{\Gamma(\beta)}\sum_{j=1}^{m}\int_{t_{j-1}}^{t_j}(t_j-s)^{\beta-1}\vartheta(s)ds \\ -\frac{1}{\Gamma(\beta-1)}\sum_{j=1}^{m}(t-t_j)\int_{t_{j-1}}^{t_j}(t_j-s)^{\beta-2}\vartheta(s)ds - \frac{1}{2\Gamma(\beta-2)}\sum_{j=1}^{m}(t-t_j)^2 \\ \times\int_{t_{j-1}}^{t_j}(t_j-s)^{\beta-3}\vartheta(s)ds + \sum_{j=1}^{m}M_{2j}(y(t_j)) + \sum_{j=1}^{m}(t-t_j)N_{2j}(y(t_j)) \\ +\sum_{j=1}^{m}\frac{(t-t_j)^2}{2}O_{2j}(y(t_j)) + c^*t^2, \ t \in (t_m, t_{m+1}], \ 1 \leqslant m \leqslant n, \end{cases} \tag{16}$$

where $c \in \{0,1,\ldots,n\}$ such that $\Phi \in (t_c, t_{c+1}]$, and

$$
\begin{aligned}
c^* =& \frac{\Gamma(3-\rho)}{2(\Gamma(3-\rho)-\Phi^{2-\rho})}\Bigg[-\frac{1}{\Gamma(\beta-\rho)}\int_{t_j}^{\Phi}(\Phi-s)^{\beta-\rho-1}\vartheta(s)ds - \frac{\Phi^{1-\rho}}{\Gamma(\beta-1)\Gamma(2-\rho)} \\
&\times\sum_{j=1}^{m}\int_{t_{j-1}}^{t_j}(t_j-s)^{\beta-2}\vartheta(s)ds - \frac{\Phi^{2-\rho}}{\Gamma(\beta-2)\Gamma(3-\rho)}\sum_{j=1}^{m}\int_{t_{j-1}}^{t_j}(t_j-s)^{\beta-3}\vartheta(s)ds \\
&+\frac{\Phi^{1-\rho}}{\Gamma(\beta-2)\Gamma(2-\rho)}\sum_{j=1}^{m}t_j\int_{t_{j-1}}^{t_j}(t_j-s)^{\beta-3}\vartheta(s)ds + \frac{\Phi^{1-\rho}}{\Gamma(2-\rho)}\sum_{j=1}^{m}N_{2j}(y(t_j)) \\
&+\frac{\Phi^{2-\rho}}{\Gamma(3-\rho)}\sum_{j=1}^{m}O_{2j}(y(t_j)) - \frac{\Phi^{1-\rho}}{\Gamma(2-\rho)}\sum_{j=1}^{m}t_jO_{2j}(y(t_j)) + \frac{1}{\Gamma(\beta-2)}\int_{t_m}^{1}(1-s)^{\beta-3}\vartheta(s)ds \\
&+\frac{1}{\Gamma(\beta-2)}\sum_{j=1}^{m}\int_{t_{j-1}}^{t_j}(t_j-s)^{\beta-3}\vartheta(s)ds - \sum_{j=1}^{m}O_{2j}(y(t_j))\Bigg].
\end{aligned}
$$

3. Main Results

In this section, we use fixed point theorems to prove the existence of solutions to problem (1). According to Lemmas 4 and 5, we define operator $\mathbf{W}: Y' \longrightarrow Y'$ by

$$
\mathbf{W}(x,y)(t) = (\mathbf{W}_1(x,y)(t), \mathbf{W}_2(x,y)(t))^T, \forall\,(x,y)\in Y',\ t\in[0,1], \tag{17}
$$

where

$$
\mathbf{W}_1(x,y)(t) = \begin{cases}
-\frac{1}{\Gamma(\alpha)}\int_0^t(t-s)h(s,{}^c\mathbf{D}^a x(s),{}^c\mathbf{D}^b y(s))ds + e^*t^2,\ t\in[0,t_1], \\
-\frac{1}{\Gamma(\alpha)}\int_{t_m}^t(t-s)^{\alpha-1}h(s,{}^c\mathbf{D}^a x(s),{}^c\mathbf{D}^b y(s))ds \\
-\frac{1}{\Gamma(\alpha)}\sum_{j=1}^m\int_{t_{j-1}}^{t_j}(t_j-s)^{\alpha-1}h(s,{}^c\mathbf{D}^a x(s),{}^c\mathbf{D}^b y(s))ds \\
-\frac{1}{\Gamma(\alpha-1)}\sum_{j=1}^m(t-t_j)\int_{t_{j-1}}^{t_j}(t_j-s)^{\alpha-2}h(s,{}^c\mathbf{D}^a x(s),{}^c\mathbf{D}^b y(s))ds \\
-\frac{1}{2\Gamma(\alpha-2)}\sum_{j=1}^m(t-t_j)^2\int_{t_{j-1}}^{t_j}(t_j-s)^{\alpha-3}h(s,{}^c\mathbf{D}^a x(s),{}^c\mathbf{D}^b y(s))ds \\
+\sum_{j=1}^m M_{1j}(x(t_j)) + \sum_{j=1}^m(t-t_j)N_{1j}(x(t_j)) + \sum_{j=1}^m\frac{(t-t_j)^2}{2}O_{1j}(x(t_j)) + e^*t^2, \\
t\in(t_m,t_{m+1}],\ 1\leqslant m\leqslant n,
\end{cases}
$$

and

$$
\mathbf{W}_2(x,y)(t) = \begin{cases}
-\frac{1}{\Gamma(\beta)}\int_0^t(t-s)^{\beta-1}w(s,{}^c\mathbf{D}^a x(s),{}^c\mathbf{D}^b y(s))ds + c^*t^2,\ t\in[0,t_1], \\
-\frac{1}{\Gamma(\beta)}\int_{t_m}^t(t-s)^{\beta-1}w(s,{}^c\mathbf{D}^a x(s),{}^c\mathbf{D}^b y(s))ds \\
-\frac{1}{\Gamma(\beta)}\sum_{j=1}^m\int_{t_{j-1}}^{t_j}(t_j-s)^{\beta-1}w(s,{}^c\mathbf{D}^a x(s),{}^c\mathbf{D}^b y(s))ds \\
-\frac{1}{\Gamma(\beta-1)}\sum_{j=1}^m(t-t_j)\int_{t_{j-1}}^{t_j}(t_j-s)^{\beta-2}w(s,{}^c\mathbf{D}^a x(s),{}^c\mathbf{D}^b y(s))ds \\
-\frac{1}{2\Gamma(\beta-2)}\sum_{j=1}^m(t-t_j)^2\int_{t_{j-1}}^{t_j}(t_j-s)^{\beta-3}w(s,{}^c\mathbf{D}^a x(s),{}^c\mathbf{D}^b y(s))ds \\
+\sum_{j=1}^m M_{2j}(y(t_j)) + \sum_{j=1}^m(t-t_j)N_{2j}(y(t_j)) + \sum_{j=1}^m\frac{(t-t_j)^2}{2}O_{2j}(y(t_j)) + c^*t^2, \\
t\in(t_m,t_{m+1}],\ 1\leqslant m\leqslant n,
\end{cases}
$$

with

$$
\begin{aligned}
e^* =& \mathcal{Z}_\alpha\Bigg[-\frac{1}{\Gamma(\alpha-\varepsilon)}\int_{t_j}^{\Omega}(\Omega-s)^{\alpha-\varepsilon-1}h(s,{}^c\mathbf{D}^a x(s),{}^c\mathbf{D}^b y(s))ds \\
&-\frac{\Omega^{1-\varepsilon}}{\Gamma(\alpha-1)\Gamma(2-\varepsilon)}\sum_{j=1}^{m}\int_{t_{j-1}}^{t_j}(t_j-s)^{\alpha-2}h(s,{}^c\mathbf{D}^a x(s),{}^c\mathbf{D}^b y(s))ds \\
&-\frac{\Omega^{2-\varepsilon}}{\Gamma(\alpha-2)\Gamma(3-\varepsilon)}\sum_{j=1}^{m}\int_{t_{j-1}}^{t_j}(t_j-s)^{\alpha-3}h(s,{}^c\mathbf{D}^a x(s),{}^c\mathbf{D}^b y(s))ds
\end{aligned}
$$

$$+\frac{\Omega^{1-\varepsilon}}{\Gamma(\alpha-2)\Gamma(2-\varepsilon)}\sum_{j=1}^{m}t_j\int_{t_{j-1}}^{t_j}(t_j-s)^{\alpha-3}h(s,{}^c\mathbf{D}^a x(s),{}^c\mathbf{D}^b y(s))ds$$

$$+\frac{\Omega^{1-\varepsilon}}{\Gamma(2-\varepsilon)}\sum_{j=1}^{m}N_{1j}(x(t_j))+\frac{\Omega^{2-\varepsilon}}{\Gamma(3-\varepsilon)}\sum_{j=1}^{m}O_{1j}(x(t_j))-\frac{\Omega^{1-\varepsilon}}{\Gamma(2-\varepsilon)}\sum_{j=1}^{m}t_jO_{1j}(x(t_j))$$

$$+\frac{1}{\Gamma(\alpha-2)}\int_{t_m}^{1}(1-s)^{\alpha-3}h(s,{}^c\mathbf{D}^a x(s),{}^c\mathbf{D}^b y(s))ds$$

$$+\frac{1}{\Gamma(\alpha-2)}\sum_{j=1}^{m}\int_{t_{j-1}}^{t_j}(t_j-s)^{\alpha-3}h(s,{}^c\mathbf{D}^a x(s),{}^c\mathbf{D}^b y(s))ds-\sum_{j=1}^{m}O_{1j}(x(t_j))\Bigg],$$

$$c^* =\mathcal{Z}_\beta\Bigg[-\frac{1}{\Gamma(\beta-\rho)}\int_{t_j}^{\Phi}(\Phi-s)^{\beta-\rho-1}w(s,{}^c\mathbf{D}^a x(s),{}^c\mathbf{D}^b y(s))ds$$

$$-\frac{\Phi^{1-\rho}}{\Gamma(\beta-1)\Gamma(2-\rho)}\sum_{j=1}^{m}\int_{t_{j-1}}^{t_j}(t_j-s)^{\beta-2}w(s,{}^c\mathbf{D}^a x(s),{}^c\mathbf{D}^b y(s))ds$$

$$-\frac{\Phi^{2-\rho}}{\Gamma(\beta-2)\Gamma(3-\rho)}\sum_{j=1}^{m}\int_{t_{j-1}}^{t_j}(t_j-s)^{\beta-3}w(s,{}^c\mathbf{D}^a x(s),{}^c\mathbf{D}^b y(s))ds$$

$$+\frac{\Phi^{1-\rho}}{\Gamma(\beta-2)\Gamma(2-\rho)}\sum_{j=1}^{m}t_j\int_{t_{j-1}}^{t_j}(t_j-s)^{\beta-3}w(s,{}^c\mathbf{D}^a x(s),{}^c\mathbf{D}^b y(s))ds)$$

$$+\frac{\Phi^{1-\rho}}{\Gamma(2-\rho)}\sum_{j=1}^{m}N_{2j}(y(t_j)+\frac{\Phi^{2-\rho}}{\Gamma(3-\rho)}\sum_{j=1}^{m}O_{2j}(y(t_j))-\frac{\Phi^{1-\rho}}{\Gamma(2-\rho)}\sum_{j=1}^{m}t_jO_{2j}(y(t_j))$$

$$+\frac{1}{\Gamma(\beta-2)}\int_{t_m}^{1}(1-s)^{\beta-3}w(s,{}^c\mathbf{D}^a x(s),{}^c\mathbf{D}^b y(s))ds$$

$$+\frac{1}{\Gamma(\beta-2)}\sum_{j=1}^{m}\int_{t_{j-1}}^{t_j}(t_j-s)^{\beta-3}w(s,{}^c\mathbf{D}^a x(s),{}^c\mathbf{D}^b y(s))ds-\sum_{j=1}^{m}O_{2j}(y(t_j))\Bigg],$$

where $\mathcal{Z}_\alpha = \frac{\Gamma(3-\varepsilon)}{2(\Gamma(3-\varepsilon)-\Omega^{2-\varepsilon})}$ and $\mathcal{Z}_\beta = \frac{\Gamma(3-\rho)}{2(\Gamma(3-\rho)-\Phi^{2-\rho})}$. Thus, solving problem (1) is equivalent to obtain a fixed point of the operator $\mathbf{W}$. Next, we have to prove the uniqueness of solutions of problem (1).

Theorem 2. *Let the following conditions $(\mathbf{M}_1)-(\mathbf{M}_3)$ hold, and then the boundary value problem (1) has a unique solution.*

$(\mathbf{M}_1)$: *For all $t \in J$ and $x_j, y_j \in \mathbb{R}$ $(j = 1,2)$ there exists some positive constants μ_j, μ_j' $(j = 1,2)$ such that*

$$\begin{aligned}|h(t,x_1,y_1)-h(t,x_2,y_2)| &\leqslant \mu_1|x_1-x_2|+\mu_2|y_1-y_2|,\\ |w(t,x_1,y_1)-w(t,x_2,y_2)| &\leqslant \mu_1'|x_1-x_2|+\mu_2'|y_1-y_2|.\end{aligned}$$

$(\mathbf{M}_2)$: *For all $x, y \in \mathbb{R}$, there exist some positive constants $I_{jk}, \hat{I}_{jk}, \check{I}_{jk}(j = 1,2;\ m = 1,2,\dots,n)$ such that*

$$\begin{aligned}|M_{jm}(x)-M_{jm}(y)| &\leqslant I_{jm}|x-y|,\\ |N_{jm}(x)-N_{jm}(y)| &\leqslant \hat{I}_{jm}|x-y|,\\ |O_{jm}(x)-O_{jm}(y)| &\leqslant \check{I}_{jm}|x-y|.\end{aligned}$$

$(\mathbf{M}_3)$:

$$\left[(\mu_1+\mu_2)\left(\frac{2}{\Gamma(\alpha+1)}+\frac{1}{\Gamma(\alpha)}+\frac{1}{2\Gamma(\alpha-1)}+\frac{\mathcal{Z}_\alpha}{\Gamma(\alpha-\varepsilon+1)}+\frac{\mathcal{Z}_\alpha\Omega^{1-\varepsilon}}{\Gamma(\alpha)\Gamma(2-\varepsilon)}+\frac{\mathcal{Z}_\alpha\Omega^{2-\varepsilon}}{\Gamma(\alpha-1)\Gamma(3-\varepsilon)}\right.\right.$$
$$\left.+\frac{\mathcal{Z}_\alpha\Omega^{1-\varepsilon}}{\Gamma(\alpha-1)\Gamma(2-\varepsilon)}+\frac{2\mathcal{Z}_\alpha}{\Gamma(\alpha-1)}\right)+\sum_{j=1}^{m}I_{1j}+\sum_{j=1}^{m}\hat{I}_{1j}+\frac{\mathcal{Z}_\alpha\Omega^{1-\varepsilon}}{\Gamma(2-\varepsilon)}\sum_{j=1}^{m}\hat{I}_{1j}+\frac{\mathcal{Z}_\alpha\Omega^{2-\varepsilon}}{\Gamma(3-\varepsilon)}\sum_{j=1}^{m}\check{I}_{1j}$$
$$\left.+\frac{\mathcal{Z}_\alpha\Omega^{1-\varepsilon}}{\Gamma(2-\varepsilon)}\sum_{j=1}^{m}\check{I}_{1j}+\mathcal{Z}_\alpha\sum_{j=1}^{m}\check{I}_{1j}+\frac{1}{2}\sum_{j=1}^{m}\check{I}_{1j}\right]<1$$

and

$$\left[(\mu_1'+\mu_2')\left(\frac{2}{\Gamma(\beta+1)}+\frac{1}{\Gamma(\beta)}+\frac{1}{2\Gamma(\beta-1)}+\frac{\mathcal{Z}_\beta}{\Gamma(\beta-\rho+1)}+\frac{\mathcal{Z}_\beta\Phi^{1-\rho}}{\Gamma(\beta)\Gamma(2-\rho)}+\frac{\mathcal{Z}_\beta\Phi^{2-\rho}}{\Gamma(\beta-1)\Gamma(3-\rho)}\right.\right.$$
$$\left.+\frac{\mathcal{Z}_\beta\Phi^{1-\rho}}{\Gamma(\beta-1)\Gamma(2-\rho)}+\frac{2\mathcal{Z}_\beta}{\Gamma(\beta-1)}\right)+\sum_{j=1}^{m}I_{2j}+\sum_{j=1}^{m}\hat{I}_{2j}+\frac{\mathcal{Z}_\beta\Phi^{1-\rho}}{\Gamma(2-\rho)}\sum_{j=1}^{m}\hat{I}_{2j}+\frac{\mathcal{Z}_\beta\Phi^{2-\rho}}{\Gamma(3-\rho)}\sum_{j=1}^{m}\check{I}_{2j}$$
$$\left.+\frac{\mathcal{Z}_\beta\Phi^{1-\rho}}{\Gamma(2-\rho)}\sum_{j=1}^{m}\check{I}_{2j}+\mathcal{Z}_\beta\sum_{j=1}^{m}\check{I}_{2j}+\frac{1}{2}\sum_{j=1}^{m}\check{I}_{2j}\right]<1.$$

Proof. By using the Banach contraction principle, we can prove that $\mathbf{W}$, defined by (17), has a fixed point. Before proving the main result first, we will prove the contraction. When $t\in J$, from (17) and conditions $(\mathbf{M}_1)-(\mathbf{M}_2)$, for all $(x_1,y_1),(x_2,y_2)\in Y'$, we have

$$\begin{aligned}
&\left|\mathbf{W}_1(x_1,y_1)(t)-\mathbf{W}_1(x_2,y_2)(t)\right|\\
\leqslant&\frac{1}{\Gamma(\alpha)}\int_0^t(t-s)^{\alpha-1}\left|h(s,{}^c\mathbf{D}^a x_1(s),{}^c\mathbf{D}^b y_1(s))-h(s,{}^c\mathbf{D}^a x_2(s),{}^c\mathbf{D}^b y_2(s))\right|ds\\
&+\frac{\mathcal{Z}_\alpha t^2}{\Gamma(\alpha-\varepsilon)}\int_{t_j}^{\Omega}(\Omega-s)^{\alpha-\varepsilon-1}\left|h(s,{}^c\mathbf{D}^a x_1(s),{}^c\mathbf{D}^b y_1(s))-h(s,{}^c\mathbf{D}^a x_2(s),{}^c\mathbf{D}^b y_2(s))\right|ds\\
&+\frac{\mathcal{Z}_\alpha t^2\Omega^{1-\varepsilon}}{\Gamma(\alpha-1)\Gamma(2-\varepsilon)}\sum_{j=1}^{m}\int_{t_{j-1}}^{t_j}(t_j-s)^{\alpha-2}\left|h(s,{}^c\mathbf{D}^a x_1(s),{}^c\mathbf{D}^b y_1(s))-h(s,{}^c\mathbf{D}^a x_2(s),{}^c\mathbf{D}^b y_2(s))\right|ds\\
&+\frac{\mathcal{Z}_\alpha t^2\Omega^{2-\varepsilon}}{\Gamma(\alpha-2)\Gamma(3-\varepsilon)}\sum_{j=1}^{m}\int_{t_{j-1}}^{t_j}(t_j-s)^{\alpha-3}\left|h(s,{}^c\mathbf{D}^a x_1(s),{}^c\mathbf{D}^b y_1(s))-h(s,{}^c\mathbf{D}^a x_2(s),{}^c\mathbf{D}^b y_2(s))\right|ds\\
&+\frac{\mathcal{Z}_\alpha t^2\Omega^{1-\varepsilon}}{\Gamma(\alpha-2)\Gamma(2-\varepsilon)}\sum_{j=1}^{m}t_j\int_{t_{j-1}}^{t_j}(t_j-s)^{\alpha-3}\left|h(s,{}^c\mathbf{D}^a x_1(s),{}^c\mathbf{D}^b y_1(s))-h(s,{}^c\mathbf{D}^a x_2(s),{}^c\mathbf{D}^b y_2(s))\right|ds\\
&+\frac{\mathcal{Z}_\alpha t^2\Omega^{1-\varepsilon}}{\Gamma(2-\varepsilon)}\sum_{j=1}^{m}\left|N_{1j}(x_1(t_j))-N_{1j}(x_2(t_j))\right|+\frac{\mathcal{Z}_\alpha t^2\Omega^{2-\varepsilon}}{\Gamma(3-\varepsilon)}\sum_{j=1}^{m}\left|O_{1j}(x_1(t_j))-O_{1j}(x_2(t_j))\right|\\
&+\frac{\mathcal{Z}_\alpha t^2}{\Gamma(\alpha-2)}\int_{t_m}^{1}(1-s)^{\alpha-3}\left|h(s,{}^c\mathbf{D}^a x_1(s),{}^c\mathbf{D}^b y_1(s))-h(s,{}^c\mathbf{D}^a x_2(s),{}^c\mathbf{D}^b y_2(s))\right|ds\\
&+\frac{\mathcal{Z}_\alpha t^2}{\Gamma(\alpha-2)}\sum_{j=1}^{m}\int_{t_{j-1}}^{t_j}(t_j-s)^{\alpha-3}\left|h(s,{}^c\mathbf{D}^a x_1(s),{}^c\mathbf{D}^b y_1(s))-h(s,{}^c\mathbf{D}^a x_2(s),{}^c\mathbf{D}^b y_2(s))\right|ds\\
&+\frac{\mathcal{Z}_\alpha t^2\Omega^{1-\varepsilon}}{\Gamma(2-\varepsilon)}\sum_{j=1}^{m}|t_j|\left|O_{1j}(x_1(t_j))-O_{1j}(x_2(t_j))\right|+\mathcal{Z}_\alpha t^2\sum_{j=1}^{m}\left|O_{1j}(x_1(t_j))-O_{1j}(x_2(t_j))\right|,\\
\leqslant&\frac{1}{\Gamma(\alpha+1)}\left[\mu_1\left|{}^c\mathbf{D}^a x_1(s)-{}^c\mathbf{D}^a x_2(s)\right|+\mu_2\left|{}^c\mathbf{D}^b y_1(s)-{}^c\mathbf{D}^b y_2(s)\right|\right]
\end{aligned}$$

$$
\begin{aligned}
&+\frac{\mathcal{Z}_\alpha}{\Gamma(\alpha-\varepsilon+1)}\left[\mu_1\left|{}^c\mathbf{D}^a x_1(s)-{}^c\,\mathbf{D}^a x_2(s)\right|+\mu_2\left|{}^c\mathbf{D}^b y_1(s)-{}^c\,\mathbf{D}^b y_2(s)\right|\right]\\
&+\frac{\mathcal{Z}_\alpha\Omega^{1-\varepsilon}}{\Gamma(\alpha)\Gamma(2-\varepsilon)}\left[\mu_1\left|{}^c\mathbf{D}^a x_1(s)-{}^c\,\mathbf{D}^a x_2(s)\right|+\mu_2\left|{}^c\mathbf{D}^b y_1(s)-{}^c\,\mathbf{D}^b y_2(s)\right|\right]\\
&+\frac{\mathcal{Z}_\alpha\Omega^{2-\varepsilon}}{\Gamma(\alpha-1)\Gamma(3-\varepsilon)}\left[\mu_1\left|{}^c\mathbf{D}^a x_1(s)-{}^c\,\mathbf{D}^a x_2(s)\right|+\mu_2\left|{}^c\mathbf{D}^a y_1(s)-{}^c\,\mathbf{D}^a y_2(s)\right|\right]\\
&+\frac{\mathcal{Z}_\alpha\Omega^{1-\varepsilon}}{\Gamma(\alpha-1)\Gamma(2-\varepsilon)}\left[\mu_1\left|{}^c\mathbf{D}^a x_1(s)-{}^c\,\mathbf{D}^a x_2(s)\right|+\mu_2\left|{}^c\mathbf{D}^b y_1(s)-{}^c\,\mathbf{D}^b y_2(s)\right|\right]\\
&+\frac{\mathcal{Z}_\alpha\Omega^{1-\varepsilon}}{\Gamma(2-\varepsilon)}\sum_{j=1}^{m}\left|N_{1j}(x_1(t_j))-N_{1j}(x_2(t_j))\right|+\frac{\mathcal{Z}_\alpha\Omega^{2-\varepsilon}}{\Gamma(3-\varepsilon)}\sum_{j=1}^{m}\left|O_{1j}(x_1(t_j))-O_{1j}(x_2(t_j))\right|\\
&+\frac{\mathcal{Z}_\alpha}{\Gamma(\alpha-1)}\left[\mu_1\left|{}^c\mathbf{D}^a x_1(s)-{}^c\,\mathbf{D}^a x_2(s)\right|+\mu_2\left|{}^c\mathbf{D}^b y_1(s)-{}^c\,\mathbf{D}^b y_2(s)\right|\right]\\
&+\frac{\mathcal{Z}_\alpha}{\Gamma(\alpha-1)}\left[\mu_1\left|{}^c\mathbf{D}^a x_1(s)-{}^c\,\mathbf{D}^a x_2(s)\right|+\mu_2\left|{}^c\mathbf{D}^b y_1(s)-{}^c\,\mathbf{D}^b y_2(s)\right|\right]\\
&+\frac{\mathcal{Z}_\alpha\Omega^{1-\varepsilon}}{\Gamma(2-\varepsilon)}\sum_{j=1}^{m}t_j\left|O_{1j}(x_1(t_j))-O_{1j}(x_2(t_j))\right|+\mathcal{Z}_\alpha\sum_{j=1}^{m}\left|O_{1j}(x_1(t_j))-O_{1j}(x_2(t_j))\right|,\\
\leqslant&\frac{1}{\Gamma(\alpha+1)}\left[\mu_1\|x_1-x_2\|_0+\mu_2\|y_1-y_2\|_0\right]+\frac{\mathcal{Z}_\alpha}{\Gamma(\alpha-\varepsilon+1)}\left[\mu_1\|x_1-x_2\|_0+\mu_2\|y_1-y_2\|_0\right]\\
&+\frac{\mathcal{Z}_\alpha\Omega^{1-\varepsilon}}{\Gamma(\alpha)\Gamma(2-\varepsilon)}\left[\mu_1\|x_1-x_2\|_0+\mu_2\|y_1-y_2\|_0\right]+\frac{\mathcal{Z}_\alpha\Omega^{2-\varepsilon}}{\Gamma(\alpha-1)\Gamma(3-\varepsilon)}\left[\mu_1\|x_1-x_2\|_0+\mu_2\|y_1-y_2\|_0\right]\\
&+\tfrac{\mathcal{Z}_\alpha\Omega^{1-\varepsilon}}{\Gamma(\alpha-1)\Gamma(2-\varepsilon)}\left[\mu_1\|x_1-x_2\|_0+\mu_2\|y_1-y_2\|_0\right]+\tfrac{\mathcal{Z}_\alpha\Omega^{1-\varepsilon}}{\Gamma(2-\varepsilon)}\textstyle\sum_{j=1}^{m}\hat{I}_{1j}\|x_1-x_2\|_0+\tfrac{\mathcal{Z}_\alpha\Omega^{2-\varepsilon}}{\Gamma(3-\varepsilon)}\textstyle\sum_{j=1}^{m}\check{I}_{1j} \qquad (18)\\
&\times\|x_1-x_2\|_0+\frac{\mathcal{Z}_\alpha}{\Gamma(\alpha-1)}\left[\mu_1\|x_1-x_2\|_0+\mu_2\|y_1-y_2\|_0\right]+\frac{\mathcal{Z}_\alpha}{\Gamma(\alpha-1)}\left[\mu_1\|x_1-x_2\|_0+\mu_2\|y_1-y_2\|_0\right]\\
&+\frac{\mathcal{Z}_\alpha\Omega^{1-\varepsilon}}{\Gamma(2-\varepsilon)}\sum_{j=1}^{m}\check{I}_{1j}\|x_1-x_2\|_0+\mathcal{Z}_\alpha\sum_{j=1}^{m}\check{I}_{1j}\|x_1-x_2\|_0,\\
\leqslant&\frac{1}{\Gamma(\alpha+1)}[\mu_1+\mu_2]\|(x_1-x_2,y_1-y_2)\|+\frac{\mathcal{Z}_\alpha}{\Gamma(\alpha-\varepsilon+1)}[\mu_1+\mu_2]\|(x_1-x_2,y_1-y_2)\|\\
&+\frac{\mathcal{Z}_\alpha\Omega^{1-\varepsilon}}{\Gamma(\alpha)\Gamma(2-\varepsilon)}[\mu_1+\mu_2]\|(x_1-x_2,y_1-y_2)\|+\frac{\mathcal{Z}_\alpha\Omega^{2-\varepsilon}}{\Gamma(\alpha-1)\Gamma(3-\varepsilon)}[\mu_1+\mu_2]\|(x_1-x_2,y_1-y_2)\|\\
&+\frac{\mathcal{Z}_\alpha\Omega^{1-\varepsilon}}{\Gamma(\alpha-1)\Gamma(2-\varepsilon)}[\mu_1+\mu_2]\|(x_1-x_2,y_1-y_2)\|+\frac{\mathcal{Z}_\alpha\Omega^{1-\varepsilon}}{\Gamma(2-\varepsilon)}\sum_{j=1}^{m}\hat{I}_{1j}\|(x_1-x_2,y_1-y_2)\|\\
&+\frac{\mathcal{Z}_\alpha\Omega^{2-\varepsilon}}{\Gamma(3-\varepsilon)}\sum_{j=1}^{m}\check{I}_{1j}\|(x_1-x_2,y_1-y_2)\|+\frac{\mathcal{Z}_\alpha}{\Gamma(\alpha-1)}[\mu_1+\mu_2]\|(x_1-x_2,y_1-y_2)\|\\
&+\frac{\mathcal{Z}_\alpha}{\Gamma(\alpha-1)}[\mu_1+\mu_2]\|(x_1-x_2,y_1-y_2)\|+\frac{\mathcal{Z}_\alpha\Omega^{1-\varepsilon}}{\Gamma(2-\varepsilon)}\sum_{j=1}^{m}\check{I}_{1j}\|(x_1-x_2,y_1-y_2)\|\\
&+\mathcal{Z}_\alpha\sum_{j=1}^{m}\check{I}_{1j}\|(x_1-x_2,y_1-y_2)\|,\\
\leqslant&\mathcal{Z}_\alpha\Bigg[(\mu_1+\mu_2)\Bigg(\frac{1}{\mathcal{Z}_\alpha\Gamma(\alpha+1)}+\frac{1}{\Gamma(\alpha-\varepsilon+1)}+\frac{\Omega^{1-\varepsilon}}{\Gamma(\alpha)\Gamma(2-\varepsilon)}+\frac{\Omega^{2-\varepsilon}}{\Gamma(\alpha-1)\Gamma(3-\varepsilon)}+\frac{\Omega^{1-\varepsilon}}{\Gamma(\alpha-1)\Gamma(2-\varepsilon)}\\
&+\frac{2}{\Gamma(\alpha-1)}\Bigg)+\frac{\Omega^{1-\varepsilon}}{\Gamma(2-\varepsilon)}\sum_{j=1}^{m}\hat{I}_{1j}+\frac{\Omega^{2-\varepsilon}}{\Gamma(3-\varepsilon)}\sum_{j=1}^{m}\check{I}_{1j}+\frac{\Omega^{1-\varepsilon}}{\Gamma(2-\varepsilon)}\sum_{j=1}^{m}\check{I}_{1j}+\sum_{j=1}^{m}\check{I}_{1j}\Bigg]\\
&\times\|(x_1-x_2,y_1-y_2)\|,\ t\in[0,t_1].
\end{aligned}
$$

When $t \in (t_m, t_{m+1}]$, then

$$
\begin{aligned}
&\left|\mathbf{W}_1(x_1,y_1)(t)-\mathbf{W}_1(x_2,y_2)(t)\right| \\
\leqslant&\frac{1}{\Gamma(\alpha)}\int_{t_m}^{t}(t-s)^{\alpha-1}\left|h(s,{}^c\mathbf{D}^a x_1(s),{}^c\mathbf{D}^b y_1(s))-h(s,{}^c\mathbf{D}^a x_2(s),{}^c\mathbf{D}^b y_2(s))\right|ds \\
&+\frac{1}{\Gamma(\alpha)}\sum_{j=1}^{m}\int_{t_{j-1}}^{t_j}(t_j-s)^{\alpha-1}\left|h(s,{}^c\mathbf{D}^a x_1(s),{}^c\mathbf{D}^b y_1(s))-h(s,{}^c\mathbf{D}^a x_2(s),{}^c\mathbf{D}^b y_2(s))\right|ds \\
&+\frac{1}{\Gamma(\alpha-1)}\sum_{j=1}^{m}\left|(t-t_j)\right|\int_{t_{j-1}}^{t_j}(t_j-s)^{\alpha-2}\left|h(s,{}^c\mathbf{D}^a x_1(s),{}^c\mathbf{D}^b y_1(s))-h(s,{}^c\mathbf{D}^a x_2(s),{}^c\mathbf{D}^b y_2(s))\right|ds \\
&+\frac{1}{2\Gamma(\alpha-2)}\sum_{j=1}^{m}\left|(t-t_j)^2\right|\int_{t_{j-1}}^{t_j}(t_j-s)^{\alpha-3}\left|h(s,{}^c\mathbf{D}^a x_1(s),{}^c\mathbf{D}^b y_1(s))-h(s,{}^c\mathbf{D}^a x_2(s),{}^c\mathbf{D}^b y_2(s))\right|ds \\
&+\sum_{j=1}^{m}\left|M_{1j}(x_1(t_j))-M_{1j}(x_2(t_j))\right|+\sum_{j=1}^{m}\left|(t-t_j)\right|\left|N_{1j}(x_1(t_j))-N_{1j}(x_2(t_j))\right| \\
&+\frac{\mathcal{Z}_\alpha t^2}{\Gamma(\alpha-\varepsilon)}\int_{t_j}^{\Omega}(\Omega-s)^{\alpha-\varepsilon-1}\left|h(s,{}^c\mathbf{D}^a x_1(s),{}^c\mathbf{D}^b y_1(s))-h(s,{}^c\mathbf{D}^a x_2(s),{}^c\mathbf{D}^b y_2(s))\right|ds \\
&+\frac{\mathcal{Z}_\alpha t^2\Omega^{1-\varepsilon}}{\Gamma(\alpha-1)\Gamma(2-\varepsilon)}\textstyle\sum_{j=1}^{m}\int_{t_{j-1}}^{t_j}(t_j-s)^{\alpha-2}\left|h(s,{}^c\mathbf{D}^a x_1(s),{}^c\mathbf{D}^b y_1(s))-h(s,{}^c\mathbf{D}^a x_2(s),{}^c\mathbf{D}^b y_2(s))\right|ds \qquad (19) \\
&+\frac{\mathcal{Z}_\alpha t^2\Omega^{2-\varepsilon}}{\Gamma(\alpha-2)\Gamma(3-\varepsilon)}\textstyle\sum_{j=1}^{m}\int_{t_{j-1}}^{t_j}(t_j-s)^{\alpha-3}\left|h(s,{}^c\mathbf{D}^a x_1(s),{}^c\mathbf{D}^b y_1(s))-h(s,{}^c\mathbf{D}^a x_2(s),{}^c\mathbf{D}^b y_2(s))\right|ds \\
&+\frac{\mathcal{Z}_\alpha t^2\Omega^{1-\varepsilon}}{\Gamma(\alpha-2)\Gamma(2-\varepsilon)}\textstyle\sum_{j=1}^{m}|t_j|\int_{t_{j-1}}^{t_j}(t_j-s)^{\alpha-3}\left|h(s,{}^c\mathbf{D}^a x_1(s),{}^c\mathbf{D}^b y_1(s))-h(s,{}^c\mathbf{D}^a x_2(s),{}^c\mathbf{D}^b y_2(s))\right|ds \\
&+\frac{\mathcal{Z}_\alpha t^2\Omega^{1-\varepsilon}}{\Gamma(2-\varepsilon)}\sum_{j=1}^{m}\left|N_{1j}(x_1(t_j))-N_{1j}(x_2(t_j))\right|+\frac{\mathcal{Z}_\alpha t^2\Omega^{2-\varepsilon}}{\Gamma(3-\varepsilon)}\sum_{j=1}^{m}\left|O_{1j}(x_1(t_j))-O_{1j}(x_2(t_j))\right| \\
&+\frac{\mathcal{Z}_\alpha t^2}{\Gamma(\alpha-2)}\int_{t_m}^{1}(1-s)^{\alpha-3}\left|h(s,{}^c\mathbf{D}^a x_1(s),{}^c\mathbf{D}^b y_1(s))-h(s,{}^c\mathbf{D}^a x_2(s),{}^c\mathbf{D}^b y_2(s))\right|ds \\
&+\frac{\mathcal{Z}_\alpha t^2}{\Gamma(\alpha-2)}\sum_{j=1}^{m}\int_{t_{j-1}}^{t_j}(t_j-s)^{\alpha-3}\left|h(s,{}^c\mathbf{D}^a x_1(s),{}^c\mathbf{D}^b y_1(s))-h(s,{}^c\mathbf{D}^a x_2(s),{}^c\mathbf{D}^b y_2(s))\right|ds \\
&+\frac{\mathcal{Z}_\alpha t^2\Omega^{1-\varepsilon}}{\Gamma(2-\varepsilon)}\sum_{j=1}^{m}|t_j|\left|O_{1j}(x_1(t_j))-O_{1j}(x_2(t_j))\right|+\mathcal{Z}_\alpha t^2\sum_{j=1}^{m}\left|O_{1j}(x_1(t_j))-O_{1j}(x_2(t_j))\right| \\
&+\sum_{j=1}^{m}\frac{\left|(t-t_j)^2\right|}{2}\left|O_{1j}(x_1(t_j))-O_{1j}(x_2(t_j))\right|.
\end{aligned}
$$

Utilizing $(\mathbf{M}_1)$ and $(\mathbf{M}_2)$ in (19) and taking the maximum, we get

$$
\begin{aligned}
\leqslant&\frac{1}{\Gamma(\alpha+1)}\left[\mu_1\|x_1-x_2\|_0+\mu_2\|y_1-y_2\|_0\right]+\frac{1}{\Gamma(\alpha+1)}\left[\mu_1\|x_1-x_2\|_0+\mu_2\|y_1-y_2\|_0\right] \\
&+\frac{1}{\Gamma(\alpha)}\left[\mu_1\|x_1-x_2\|_0+\mu_2\|y_1-y_2\|_0\right]+\frac{1}{2\Gamma(\alpha-1)}\left[\mu_1\|x_1-x_2\|_0+\mu_2\|y_1-y_2\|_0\right] \\
&+\sum_{j=1}^{m}I_{1j}\|x_1-x_2\|_0+\sum_{j=1}^{m}\hat{I}_{1j}\|x_1-x_2\|_0+\frac{\mathcal{Z}_\alpha}{\Gamma(\alpha-\varepsilon+1)}\left[\mu_1\|x_1-x_2\|_0+\mu_2\|y_1-y_2\|_0\right] \\
&+\frac{\mathcal{Z}_\alpha\Omega^{1-\varepsilon}}{\Gamma(\alpha)\Gamma(2-\varepsilon)}\left[\mu_1\|x_1-x_2\|_0+\mu_2\|y_1-y_2\|_0\right]+\frac{\mathcal{Z}_\alpha\Omega^{2-\varepsilon}}{\Gamma(\alpha-1)\Gamma(3-\varepsilon)}\left[\mu_1\|x_1-x_2\|_0+\mu_2\|y_1-y_2\|_0\right]
\end{aligned}
$$

$$
\begin{aligned}
&+\frac{Z_\alpha \Omega^{1-\varepsilon}}{\Gamma(\alpha-1)\Gamma(2-\varepsilon)}\left[\mu_1\|x_1-x_2\|_0+\mu_2\|y_1-y_2\|_0\right]+\frac{Z_\alpha \Omega^{1-\varepsilon}}{\Gamma(2-\varepsilon)}\sum_{j=1}^{m}\hat{I}_{1j}\|x_1-x_2\|_0\\
&+\frac{Z_\alpha \Omega^{2-\varepsilon}}{\Gamma(3-\varepsilon)}\sum_{j=1}^{m}\check{I}_{1j}\|x_1-x_2\|_0+\frac{Z_\alpha}{\Gamma(\alpha-1)}\left[\mu_1\|x_1-x_2\|_0+\mu_2\|y_1-y_2\|_0\right]\\
&+\frac{Z_\alpha}{\Gamma(\alpha-1)}\left[\mu_1\|x_1-x_2\|_0+\mu_2\|y_1-y_2\|_0\right]+\frac{Z_\alpha \Omega^{1-\varepsilon}}{\Gamma(2-\varepsilon)}\sum_{j=1}^{m}\check{I}_{1j}\|x_1-x_2\|_0+Z_\alpha\sum_{j=1}^{m}\check{I}_{1j}\|x_1-x_2\|_0\\
&+\frac{1}{2}\sum_{j=1}^{m}\check{I}_{1j}\|x_1-x_2\|_0,\\
&\leqslant\frac{1}{\Gamma(\alpha+1)}[\mu_1+\mu_2]\|(x_1-x_2,y_1-y_2)\|+\frac{1}{\Gamma(\alpha+1)}[\mu_1+\mu_2]\|(x_1-x_2,y_1-y_2)\|\\
&+\frac{1}{\Gamma(\alpha)}[\mu_1+\mu_2]\|(x_1-x_2,y_1-y_2)\|+\frac{1}{2\Gamma(\alpha-1)}[\mu_1+\mu_2]\|(x_1-x_2,y_1-y_2)\|\\
&+\textstyle\sum_{j=1}^{m}I_{1j}\|(x_1-x_2,y_1-y_2)\|+\sum_{j=1}^{m}\hat{I}_{1j}\|(x_1-x_2,y_1-y_2)\|+\frac{Z_\alpha}{\Gamma(\alpha-\varepsilon+1)}[\mu_1+\mu_2]\|(x_1-x_2,y_1-y_2)\|\\
&+\tfrac{Z_\alpha\Omega^{1-\varepsilon}}{\Gamma(\alpha)\Gamma(2-\varepsilon)}[\mu_1+\mu_2]\|(x_1-x_2,y_1-y_2)\|+\tfrac{Z_\alpha\Omega^{2-\varepsilon}}{\Gamma(\alpha-1)\Gamma(3-\varepsilon)}[\mu_1+\mu_2]\|(x_1-x_2,y_1-y_2)\| \qquad (20)\\
&+\frac{Z_\alpha \Omega^{1-\varepsilon}}{\Gamma(\alpha-1)\Gamma(2-\varepsilon)}[\mu_1+\mu_2]\|(x_1-x_2,y_1-y_2)\|+\frac{Z_\alpha \Omega^{1-\varepsilon}}{\Gamma(2-\varepsilon)}\sum_{j=1}^{m}\hat{I}_{1j}\|(x_1-x_2,y_1-y_2)\|\\
&+\frac{Z_\alpha \Omega^{2-\varepsilon}}{\Gamma(3-\varepsilon)}\sum_{j=1}^{m}\check{I}_{1j}\|(x_1-x_2,y_1-y_2)\|+\frac{Z_\alpha}{\Gamma(\alpha-1)}[\mu_1+\mu_2]\|(x_1-x_2,y_1-y_2)\|\\
&+\frac{Z_\alpha}{\Gamma(\alpha-1)}[\mu_1+\mu_2]\|(x_1-x_2,y_1-y_2)\|+\frac{Z_\alpha \Omega^{1-\varepsilon}}{\Gamma(2-\varepsilon)}\sum_{j=1}^{m}\check{I}_{1j}\|(x_1-x_2,y_1-y_2)\|\\
&+Z_\alpha\sum_{j=1}^{m}\check{I}_{1j}\|(x_1-x_2,y_1-y_2)\|+\frac{1}{2}\sum_{j=1}^{m}\check{I}_{1j}\|(x_1-x_2,y_1-y_2)\|\\
&\leqslant\Bigg[(\mu_1+\mu_2)\Bigg(\frac{2}{\Gamma(\alpha+1)}+\frac{1}{\Gamma(\alpha)}+\frac{1}{2\Gamma(\alpha-1)}+\frac{Z_\alpha}{\Gamma(\alpha-\varepsilon+1)}+\frac{Z_\alpha\Omega^{1-\varepsilon}}{\Gamma(\alpha)\Gamma(2-\varepsilon)}+\frac{Z_\alpha\Omega^{2-\varepsilon}}{\Gamma(\alpha-1)\Gamma(3-\varepsilon)}\\
&+\frac{Z_\alpha\Omega^{1-\varepsilon}}{\Gamma(\alpha-1)\Gamma(2-\varepsilon)}+\frac{2Z_\alpha}{\Gamma(\alpha-1)}\Bigg)+\sum_{j=1}^{m}I_{1j}+\sum_{j=1}^{m}\hat{I}_{1j}+\frac{Z_\alpha\Omega^{1-\varepsilon}}{\Gamma(2-\varepsilon)}\sum_{j=1}^{m}\hat{I}_{1j}+\frac{Z_\alpha\Omega^{2-\varepsilon}}{\Gamma(3-\varepsilon)}\sum_{j=1}^{m}\check{I}_{1j}\\
&+\frac{Z_\alpha\Omega^{1-\varepsilon}}{\Gamma(2-\varepsilon)}\sum_{j=1}^{m}\check{I}_{1j}+Z_\alpha\sum_{j=1}^{m}\check{I}_{1j}+\frac{1}{2}\sum_{j=1}^{m}\check{I}_{1j}\Bigg]\|(x_1-x_2,y_1-y_2)\|.
\end{aligned}
$$

In the same fashion, we can obtain

$$
\begin{aligned}
&|\mathbf{W}_2(x_1,y_1)(t)-\mathbf{W}_2(x_2,y_2)(t)|\\
\leqslant\ & Z_\beta\Bigg[(\mu_1'+\mu_2')\Bigg(\tfrac{1}{Z_\beta\Gamma(\beta+1)}+\tfrac{1}{\Gamma(\beta-\rho+1)}+\tfrac{\Phi^{1-\rho}}{\Gamma(\beta)\Gamma(2-\rho)}+\tfrac{\Phi^{2-\rho}}{\Gamma(\beta-1)\Gamma(3-\rho)}+\tfrac{\Phi^{1-\rho}}{\Gamma(\beta-1)\Gamma(2-\rho)}\\
&+\tfrac{2}{\Gamma(\beta-1)}\Bigg)+\tfrac{\Phi^{1-\rho}}{\Gamma(2-\rho)}\textstyle\sum_{j=1}^{m}\hat{I}_{2j}+\tfrac{\Phi^{2-\rho}}{\Gamma(3-\rho)}\sum_{j=1}^{m}\check{I}_{2j}+\tfrac{\Phi^{1-\rho}}{\Gamma(2-\rho)}\sum_{j=1}^{m}\check{I}_{2j}+\sum_{j=1}^{m}\check{I}_{2j}\Bigg] \qquad (21)\\
&\times\|(x_1-x_2,y_1-y_2)\|,\ t\in[0,t_1],
\end{aligned}
$$

and

$$\begin{aligned}&\left|\mathbf{W}_2(x_1,y_1)(t)-\mathbf{W}_2(x_2,y_2)(t)\right|\\ \leqslant\ &\Bigg[(\mu_1'+\mu_2')\Big(\tfrac{2}{\Gamma(\beta+1)}+\tfrac{1}{\Gamma(\beta)}+\tfrac{1}{2\Gamma(\beta-1)}+\tfrac{\mathcal{Z}_\beta}{\Gamma(\beta-\rho+1)}+\tfrac{\mathcal{Z}_\beta\Phi^{1-\rho}}{\Gamma(\beta)\Gamma(2-\rho)}+\tfrac{\mathcal{Z}_\beta\Phi^{2-\rho}}{\Gamma(\beta-1)\Gamma(3-\rho)}\\ &+\tfrac{\mathcal{Z}_\beta\Phi^{1-\rho}}{\Gamma(\beta-1)\Gamma(2-\rho)}+\tfrac{2\mathcal{Z}_\beta}{\Gamma(\beta-1)}\Big)+\textstyle\sum_{j=1}^m \mathrm{I}_{2j}+\sum_{j=1}^m \hat{\mathrm{I}}_{2j}+\tfrac{\mathcal{Z}_\beta\Phi^{1-\rho}}{\Gamma(2-\rho)}\sum_{j=1}^m \hat{\mathrm{I}}_{2j}+\tfrac{\mathcal{Z}_\beta\Phi^{2-\rho}}{\Gamma(3-\rho)}\sum_{j=1}^m \check{\mathrm{I}}_{2j}\\ &+\tfrac{\mathcal{Z}_\beta\Phi^{1-\rho}}{\Gamma(2-\rho)}\textstyle\sum_{j=1}^m \check{\mathrm{I}}_{2j}+\mathcal{Z}_\beta\sum_{j=1}^m \check{\mathrm{I}}_{2j}+\frac{1}{2}\sum_{j=1}^m \check{\mathrm{I}}_{2j}\Bigg]\|(x_1-x_2,y_1-y_2)\|,\ t\in(t_m,t_{m+1}].\end{aligned}\tag{22}$$

Thus, from (18)–(22) and $(\mathbf{M}_3)$, we infer that $\mathbf{W}$ is a contraction mapping. According to Lemma 3, $\mathbf{W}$ has a fixed point $(x^*(t),y^*(t))\in Y'$, which is unique. Therefore, problem (1) has a unique solution $(x^*(t),y^*(t))$. □

Theorem 3. *Let* $(\mathbf{M}_1)-(\mathbf{M}_2)$, $(\mathbf{M}_4)$ *and for all* $t\in J$ *such that* $h(t,0,0)=w(t,0,0)=0$, $M_{ik}=N_{ik}=O_{ik}=0$, $(i=1,2;\ k=1,2,\ldots,n)$ *hold. Then,* (1) *has at least one solution* $(x^*(t),y^*(t))$.

Proof. For the sake of simplicity, let us denote

$$\begin{aligned}\varpi=&\Bigg[(\mu_1+\mu_2)\left(\frac{3}{\Gamma(\alpha+1)}+\frac{1}{\Gamma(\alpha)}+\frac{1}{2\Gamma(\alpha-1)}+\frac{\mathcal{Z}_\alpha}{\Gamma(\alpha-\varepsilon+1)}+\frac{\mathcal{Z}_\alpha\Omega^{1-\varepsilon}}{\Gamma(\alpha)\Gamma(2-\varepsilon)}+\frac{\mathcal{Z}_\alpha\Omega^{2-\varepsilon}}{\Gamma(\alpha-1)\Gamma(3-\varepsilon)}\right.\\ &\left.+\frac{\mathcal{Z}_\alpha\Omega^{1-\varepsilon}}{\Gamma(\alpha-1)\Gamma(2-\varepsilon)}+\frac{2\mathcal{Z}_\alpha}{\Gamma(\alpha-1)}\right)+\sum_{j=1}^m \mathrm{I}_{1j}+\sum_{j=1}^m \hat{\mathrm{I}}_{1j}+\frac{\mathcal{Z}_\alpha\Omega^{1-\varepsilon}}{\Gamma(2-\varepsilon)}\sum_{j=1}^m \hat{\mathrm{I}}_{1j}+\frac{\mathcal{Z}_\alpha\Omega^{2-\varepsilon}}{\Gamma(3-\varepsilon)}\sum_{j=1}^m \check{\mathrm{I}}_{1j}\\ &+\frac{\mathcal{Z}_\alpha\Omega^{1-\varepsilon}}{\Gamma(2-\varepsilon)}\sum_{j=1}^m \check{\mathrm{I}}_{1j}+\mathcal{Z}_\alpha\sum_{j=1}^m \check{\mathrm{I}}_{1j}+\frac{1}{2}\sum_{j=1}^m \check{\mathrm{I}}_{1j}\Bigg],\\ \xi=&\Bigg[(\mu_1'+\mu_2')\left(\frac{3}{\Gamma(\beta+1)}+\frac{1}{\Gamma(\beta)}+\frac{1}{2\Gamma(\beta-1)}+\frac{\mathcal{Z}_\beta}{\Gamma(\beta-\rho+1)}+\frac{\mathcal{Z}_\beta\Phi^{1-\rho}}{\Gamma(\beta)\Gamma(2-\rho)}+\frac{\mathcal{Z}_\beta\Phi^{2-\rho}}{\Gamma(\beta-1)\Gamma(3-\rho)}\right.\\ &\left.+\frac{\mathcal{Z}_\beta\Phi^{1-\rho}}{\Gamma(\beta-1)\Gamma(2-\rho)}+\frac{2\mathcal{Z}_\beta}{\Gamma(\beta-1)}\right)+\sum_{j=1}^m \mathrm{I}_{2j}+\sum_{j=1}^m \hat{\mathrm{I}}_{2j}+\frac{\mathcal{Z}_\beta\Phi^{1-\rho}}{\Gamma(2-\rho)}\sum_{j=1}^m \hat{\mathrm{I}}_{2j}+\frac{\mathcal{Z}_\beta\Phi^{2-\rho}}{\Gamma(3-\rho)}\sum_{j=1}^m \check{\mathrm{I}}_{2j}\\ &+\frac{\mathcal{Z}_\beta\Phi^{1-\rho}}{\Gamma(2-\rho)}\sum_{j=1}^m \check{\mathrm{I}}_{2j}+\mathcal{Z}_\beta\sum_{j=1}^m \check{\mathrm{I}}_{2j}+\frac{1}{2}\sum_{j=1}^m \check{\mathrm{I}}_{2j}\Bigg],\end{aligned}$$

and $\mathcal{R}_\upsilon=\max\{(\frac{1}{\varpi}+1,\frac{1}{\xi}+1)\}$. Define the operator $\mathbf{W}$, as in (17), and a closed ball of Banach space Y' as follows:

$$\upsilon=\{(x,y)\in Y':\|(x,y)\|\leqslant\mathcal{R}_\upsilon\}.\tag{23}$$

Similar to (18)–(22), we easily show that $\mathbf{W}(\upsilon)\subset\upsilon$ by applying $(\mathbf{M}_4)$. $\mathbf{W}(\upsilon)\subset\upsilon$ indicates that $\mathbf{W}(\upsilon)$ is uniformly bounded in Y'. The continuity of the operator $\mathbf{W}$ is follows from the continuity of h,w,M_{im},N_{im} and O_{im}. Now, we need to prove that $\mathbf{W}:\upsilon\to\upsilon$ is equicontinuous. Let $(x,y)\in\upsilon$ and $\ell_1,\ell_2\in[0,1]$ with $\ell_1<\ell_2$. When $0\leqslant\ell_1<\ell_2\leqslant t_1$, similar to Equation (18), we have

$$\begin{aligned}&\left|\mathbf{W}_1(x,y)(\ell_2)-\mathbf{W}_1(x,y)(\ell_1)\right|\\ =&\Bigg|\frac{1}{\Gamma(\alpha)}\int_0^{\ell_1}[(\ell_2-s)^{\alpha-1}-(\ell_1-s)^{\alpha-1}]h(s,{}^c\mathbf{D}^a x(s),{}^c\mathbf{D}^b y(s))ds+\frac{1}{\Gamma(\alpha)}\int_{\ell_1}^{\ell_2}(\ell_2-s)^{\alpha-1}\\ &\times h(s,{}^c\mathbf{D}^a x(s),{}^c\mathbf{D}^b y(s))ds-e^*(\ell_2-\ell_1)^2\Bigg|,\end{aligned}$$

$$
\begin{aligned}
&\leqslant \frac{1}{\Gamma(\alpha)}\int_0^{\ell_1}\left[(\ell_2-s)^{\alpha-1}-(\ell_1-s)^{\alpha-1}\right]\left|h(s,{}^c\mathbf{D}^a x(s),{}^c\mathbf{D}^b y(s))\right|ds+\frac{1}{\Gamma(\alpha)}\int_{\ell_1}^{\ell_2}(\ell_2-s)^{\alpha-1}\\
&\quad\times\left|h(s,{}^c\mathbf{D}^a x(s),{}^c\mathbf{D}^b y(s))\right|ds+\left|e^*\right|(\ell_2-\ell_1)^2,\\
&\leqslant \frac{\mu_1+\mu_2}{\Gamma(\alpha)}\|(x,y)\|\int_0^{\ell_1}\left[(\ell_2-s)^{\alpha-1}-(\ell_1-s)^{\alpha-1}\right]ds+\frac{\mu_1+\mu_2}{\Gamma(\alpha+1)}\|(x,y)\|(\ell_2-\ell_1)^{\alpha}+\left|e^*\right|(\ell_2-\ell_1)^2,\\
&\leqslant \frac{\mu_1+\mu_2}{\Gamma(\alpha+1)}\|(x,y)\|\left[(\ell_2-\ell_1)^{\alpha}-(\ell_2^{\alpha}-\ell_1^{\alpha})\right]+\frac{\mu_1+\mu_2}{\Gamma(\alpha+1)}\|(x,y)\|(\ell_2-\ell_1)^{\alpha}+|e^*|(\ell_2-\ell_1)^2, \qquad (24)\\
&\leqslant \mathcal{Z}_\alpha\Bigg[(\mu_1+\mu_2)\Bigg(\tfrac{2}{\mathcal{Z}_\alpha\Gamma(\alpha+1)}+\tfrac{1}{\Gamma(\alpha-\varepsilon+1)}+\tfrac{\Omega^{1-\varepsilon}}{\Gamma(\alpha)\Gamma(2-\varepsilon)}+\tfrac{\Omega^{2-\varepsilon}}{\Gamma(\alpha-1)\Gamma(3-\varepsilon)}+\tfrac{\Omega^{1-\varepsilon}}{\Gamma(\alpha-1)\Gamma(2-\varepsilon)}\\
&\quad+\tfrac{2}{\Gamma(\alpha-1)}\Bigg)+\tfrac{\Omega^{1-\varepsilon}}{\Gamma(2-\varepsilon)}\textstyle\sum_{j=1}^{m}\hat{I}_{1j}+\tfrac{\Omega^{2-\varepsilon}}{\Gamma(3-\varepsilon)}\sum_{j=1}^{m}\check{I}_{1j}+\tfrac{\Omega^{1-\varepsilon}}{\Gamma(2-\varepsilon)}\sum_{j=1}^{m}\check{I}_{1j}+\sum_{j=1}^{m}\check{I}_{1j}\Bigg]\|x,y\|(\ell_2-\ell_1)^2,\\
&\leqslant \varpi\mathcal{R}_\upsilon(\ell_2-\ell_1)^2.
\end{aligned}
$$

In the same fashion, we obtain

$$
\left|\mathbf{W}_2(x,y)(\ell_2)-\mathbf{W}_2(x,y)(\ell_1)\right|\leqslant \xi\mathcal{R}_\upsilon(\ell_2-\ell_1)^2. \qquad (25)
$$

In addition, we obtain the same result when $t_m<\ell_1<\ell_2\leqslant t_{m+1}$, $1\leqslant m\leqslant n$, similar to (20)

$$
\left|\mathbf{W}_1(x,y)(\ell_2)-\mathbf{W}_1(x,y)(\ell_1)\right|\leqslant \varpi\mathcal{R}_\upsilon(\ell_2-\ell_1)^2 \qquad (26)
$$

and

$$
\left|\mathbf{W}_2(x,y)(\ell_2)-\mathbf{W}_2(x,y)(\ell_1)\right|\leqslant \xi\mathcal{R}_\upsilon(\ell_2-\ell_1)^2. \qquad (27)
$$

Thus, it follows from (24)–(27) that, for any $\epsilon>0$, there exists a positive constant $\sigma=\frac{\epsilon}{\mathcal{R}_\upsilon}\min\{\frac{1}{\varpi},\frac{1}{\xi}\}$ independent of ℓ_1,ℓ_2 and (x,y) such that $\|\mathbf{W}(x,y)(\ell_2)-\mathbf{W}(x,y)(\ell_1)\|<\epsilon$, whenever $|\ell_2-\ell_1|\leqslant\sigma$. Thereby, $\mathbf{W}:Y'\to Y'$ is equicontinuous. By the Arzela–Ascoli theorem, we know that $\mathbf{W}:Y'\to Y'$ is completely continuous. In view of Theorem 1, $\mathbf{W}$ has a unique fixed point $(x^*(t),y^*(t))\in\overline{\upsilon}$, which is a solution of system (1). □

4. Ulam–Hyers Stability

In this section, we are interested in Ulam–Hyers stability and its types for the solution of (1).

Definition 4. *[61] Problem* (1) *is Ulam–Hyers stable if there exists a constant $\mathbf{K}_{\alpha,\beta}=(\mathbf{K}_\alpha,\mathbf{K}_\beta)>0$ such that, for any $\epsilon=(\epsilon_\alpha,\epsilon_\beta)>0$, and $m=1,2,\dots,n$, there exists a solution $(x,y)\in Y'$ of:*

$$
\begin{cases}
\left|{}^c\mathbf{D}^{\alpha}x(t)-h(t,{}^c\mathbf{D}^a x(t),{}^c\mathbf{D}^b y(t))\right|\leqslant\epsilon_\alpha,\\
\left|\Delta x(t_m)-M_{1m}(x(t_m))\right|\leqslant\epsilon_\alpha,\\
\left|\Delta x'(t_m)-N_{1m}(x(t_m))\right|\leqslant\epsilon_\alpha,\\
\left|\Delta x''(t_m)-O_{1m}(x(t_m))\right|\leqslant\epsilon_\alpha,\\
\left|{}^c\mathbf{D}^{\beta}y(t)-w(t,{}^c\mathbf{D}^a x(t),{}^c\mathbf{D}^b y(t))\right|\leqslant\epsilon_\beta,\\
\left|\Delta y(t_m)-M_{2m}(y(t_m))\right|\leqslant\epsilon_\beta,\\
\left|\Delta y'(t_m)-N_{2m}(y(t_m))\right|\leqslant\epsilon_\beta,\\
\left|\Delta y''(t_m)-O_{2m}(y(t_m))\right|\leqslant\epsilon_\beta,
\end{cases} \qquad (28)
$$

corresponding to a solution $(\zeta,\chi)\in Y'$ of (1) *such that*

$$
\left|(x,y)(t)-(\zeta,\chi)(t)\right|\leqslant\mathbf{K}_{\alpha,\beta}\,\epsilon.
$$

Definition 5. *[61] Problem* (1) *is generalized Ulam–Hyers stable if there exists a function* $\Theta_{\alpha,\beta} \in C(\mathbb{R}_+, \mathbb{R}_+)$, $\Theta_{\alpha,\beta}(0) = 0$ *for each* $\epsilon > 0$, *such that for every solution* $(x, y) \in Y'$ *of the inequality* (28). *there is a solution* $(\zeta, \chi) \in Y'$ *of* (1) *such that*

$$\left|(x,y)(t) - (\zeta,\chi)(t)\right| \leqslant \Theta_{\alpha,\beta}(\epsilon).$$

Definition 6. *[61] Problem* (1) *is Ulam–Hyers–Rassias stable with respect to* $(\Psi_{\alpha,\beta}, \varphi_{\alpha,\beta})$, *where* $\Psi_{\alpha,\beta} = (\Psi_\alpha, \Psi_\beta) \in C(J, \mathbb{R})$ *and* $\varphi_{\alpha,\beta} = (\varphi_\alpha, \varphi_\beta) \in C(J, \mathbb{R})$, *if, for every* $\epsilon = (\epsilon_\alpha, \epsilon_\beta) > 0$, *there exists a real number* $\mathbf{K}_{\Psi,\varphi} > 0$, *such that for* $m = 1, 2, \ldots, n$ *and for a solution* $(x, y) \in Y'$ *of:*

$$\begin{cases} \left|{}^c\mathbf{D}^\alpha x(t) - h(t, {}^c\mathbf{D}^a x(t), {}^c\mathbf{D}^b y(t))\right| \leqslant \Psi_\alpha(t), \\ \left|\Delta x(t_m) - M_{1m}(x(t_m))\right| \leqslant \varphi_\alpha, \\ \left|\Delta x'(t_m) - N_{1m}(x(t_m))\right| \leqslant \varphi_\alpha, \\ \left|\Delta x''(t_m) - O_{1m}(x(t_m))\right| \leqslant \varphi_\alpha, \\ \left|{}^c\mathbf{D}^\beta y(t) - w(t, {}^c\mathbf{D}^a x(t), {}^c\mathbf{D}^b y(t))\right| \leqslant \Psi_\beta(t), \\ \left|\Delta y(t_m) - M_{2m}(y(t_m))\right| \leqslant \varphi_\beta, \\ \left|\Delta y'(t_m) - N_{2m}(y(t_m))\right| \leqslant \varphi_\beta, \\ \left|\Delta y''(t_m) - O_{2m}(y(t_m))\right| \leqslant \varphi_\beta, \end{cases} \tag{29}$$

there is a solution $(\zeta, \chi) \in Y'$ *of* (1) *such that*

$$\left|(x,y)(t) - (\zeta,\chi)(t)\right| \leqslant \mathbf{K}_{\Psi,\varphi}(\Psi_{\alpha,\beta}(t) + \varphi_{\alpha,\beta})\epsilon.$$

Definition 7. *[61] Problem* (1) *is generalized Ulam–Hyers–Rassias stable with respect to* $(\Psi_{\alpha,\beta}, \varphi_{\alpha,\beta}) \in C(J, \mathbb{R})$, *if there exists a real number* $\mathbf{K}_{\Psi,\varphi} > 0$, *such that for* $m = 1, 2, \ldots, n$ *and for every solution* $(x, y) \in Y'$ *of the following:*

$$\begin{cases} \left|{}^c\mathbf{D}^\alpha x(t) - h(t, {}^c\mathbf{D}^a x(t), {}^c\mathbf{D}^b y(t))\right| \leqslant \Psi_\alpha(t)\epsilon_\alpha, \\ \left|\Delta x(t_m) - M_{1m}(x(t_m))\right| \leqslant \varphi_\alpha\epsilon_\alpha, \\ \left|\Delta x'(t_m) - N_{1m}(x(t_m))\right| \leqslant \varphi_\alpha\epsilon_\alpha, \\ \left|\Delta x''(t_m) - O_{1m}(x(t_m))\right| \leqslant \varphi_\alpha\epsilon_\alpha, \\ \left|{}^c\mathbf{D}^\beta y(t) - w(t, {}^c\mathbf{D}^a x(t), {}^c\mathbf{D}^b y(t))\right| \leqslant \Psi_\beta(t)\epsilon_\beta, \\ \left|\Delta y(t_m) - M_{2m}(y(t_m))\right| \leqslant \varphi_\beta\epsilon_\beta, \\ \left|\Delta y'(t_m) - N_{2m}(y(t_m))\right| \leqslant \varphi_\beta\epsilon_\beta, \\ \left|\Delta y''(t_m) - O_{2m}(y(t_m))\right| \leqslant \varphi_\beta\epsilon_\beta, \end{cases} \tag{30}$$

there is a solution $(\zeta, \chi) \in Y'$ *of* (1) *such that*

$$\left|(x,y)(t) - (\zeta,\chi)(t)\right| \leqslant \mathbf{K}_{\Psi,\varphi}(\Psi_{\alpha,\beta}(t) + \varphi_{\alpha,\beta}).$$

Remark 1. *A function* $(x, y) \in Y'$ *is a solution of the inequality* (28), *if and only if there exist functions* $F_h, \hat{F}_w \in Y'$ *and a sequence* $F_m, \hat{F}_m$, $m = 1, 2, \ldots, n$ *depending on* (x, y), *such that*

- $|F_h(t)| \leqslant \epsilon_\alpha$, $|\hat{F}_w(t)| \leqslant \epsilon_\beta$, $|F_m| \leqslant \epsilon_\alpha$, $|\hat{F}_m| \leqslant \epsilon_\beta$, $t \in J_m$, $m = 1, \ldots, n$;
- ${}^cD^\alpha x(t) = -h(t, {}^cD^a x(t), {}^cD^b y(t)) + F_h(t)$;
- $\Delta x \mid_{t=t_m} = M_{1m}(x(t_m)) + F_m$;
- $\Delta x \mid_{t=t_m} = N_{1m}(x(t_m)) + F_m$;
- $\Delta x \mid_{t=t_m} = O_{1m}(x(t_m)) + F_m$;
- ${}^cD^\beta x(t) = -w(t, {}^cD^a x(t), {}^cD^b y(t)) + \hat{F}_w(t)$;
- $\Delta y \mid_{t=t_m} = M_{2m}(y(t_m)) + \hat{F}_m$;
- $\Delta y \mid_{t=t_m} = N_{2m}(y(t_m)) + \hat{F}_m$;
- $\Delta y \mid_{t=t_m} = O_{2m}(y(t_m)) + \hat{F}_m$.

Remark 2. *A function* $(x,y) \in Y'$ *is a solution of the inequality* (29), *if and only if there exist functions* $\digamma_h, \hat{\digamma}_w \in Y'$ *and a sequence* $\digamma_m, \hat{\digamma}_m$, $m = 1,2,\ldots,n$ *depending on* (x,y), *such that*

- $|\digamma_h(t)| \leqslant \Psi_\alpha$, $|\hat{\digamma}_w(t)| \leqslant \Psi_\beta$, $|\digamma_m| \leqslant \varphi_\alpha$, $|\digamma_m| \leqslant \varphi_\beta$, $t \in J_m$, $m = 1,\ldots,n$;
- ${}^cD^\alpha x(t) = -h(t, {}^cD^a x(t), {}^cD^b y(t)) + \digamma_h(t)$;
- $\Delta x \mid_{t=t_m} = M_{1m}(x(t_m)) + \digamma_m$;
- $\Delta x \mid_{t=t_m} = N_{1m}(x(t_m)) + \digamma_m$;
- $\Delta x \mid_{t=t_m} = O_{1m}(x(t_m)) + \digamma_m$;
- ${}^cD^\beta x(t) = -w(t, {}^cD^a x(t), {}^cD^b y(t)) + \hat{\digamma}_w(t)$;
- $\Delta y \mid_{t=t_m} = M_{2m}(y(t_m)) + \hat{\digamma}_m$;
- $\Delta y \mid_{t=t_m} = N_{2m}(y(t_m)) + \hat{\digamma}_m$;
- $\Delta y \mid_{t=t_m} = O_{2m}(y(t_m)) + \hat{\digamma}_m$.

Similarly, one can easily state such a remark for the inequality (30).

Theorem 4. *If the assumptions* $(\mathbf{M}_1) - (\mathbf{M}_2)$ *hold with*

$$\Lambda_0 = 1 - \frac{\Lambda_2^*\Lambda_4^*}{(1-\Lambda_1^*)(1-\Lambda_3^*)} > 0, \tag{31}$$

then (1) *is Ulam–Hyers and generalized Ulam–Hyers stable.*

Proof. Let $(x,y) \in Y'$ be any solution of the inequality (28) and let $(\zeta,\chi) \in Y'$ be the unique solution of the following:

$$\begin{cases} {}^c\mathbf{D}^\alpha \zeta(t) + h(t, {}^c\mathbf{D}^a \zeta(t), {}^c\mathbf{D}^b \chi(t)) = 0, \ t \neq t_m, \ m = 1,2,\ldots,n, \\ {}^c\mathbf{D}^\beta \chi(t) + w(t, {}^c\mathbf{D}^a \zeta(t), {}^c\mathbf{D}^b \chi(t)) = 0, \ t \neq t_m, \ m = 1,2,\ldots,n, \\ \Delta\zeta \mid_{t=t_m} = M_{1m}(\zeta(t_m)), \ \Delta\zeta' \mid_{t=t_m} = N_{1m}(\zeta(t_m)), \ \Delta\zeta'' \mid_{t=t_m} = O_{1m}(\zeta(t_m)), \\ \Delta\chi \mid_{t=t_m} = M_{2m}(\chi(t_m)), \ \Delta\chi' \mid_{t=t_m} = N_{2m}(\chi(t_m)), \ \Delta\chi'' \mid_{t=t_m} = O_{2m}(\chi(t_m)), \\ \zeta(0) = \zeta'(0) = 0, \ {}^cD^\varepsilon \zeta(\Omega) = \zeta''(1), \\ \chi(0) = \chi'(0) = 0, \ {}^cD^\rho \chi(\Phi) = \chi''(1). \end{cases} \tag{32}$$

By Lemma 2.4, we have

$$\begin{cases} {}^c\mathbf{D}^\alpha x(t) + h(t, {}^c\mathbf{D}^a x(t), {}^c\mathbf{D}^b y(t)) = \digamma_h(t), \ t \neq t_m, \ m = 1,2\ldots,n, \\ {}^c\mathbf{D}^\beta y(t) + w(t, {}^c\mathbf{D}^a x(t), {}^c\mathbf{D}^b y(t)) = \digamma_w(t), \ t \neq t_m, \ m = 1,2\ldots,n, \\ \Delta x \mid_{t=t_m} = M_{1m}(x(t_m)) + \digamma_m, \ \Delta x' \mid_{t=t_m} = N_{1m}(x(t_m)) + \digamma_m, \ \Delta x'' \mid_{t=t_m} = O_{1m}(x(t_m)) + \digamma_m, \\ \Delta y \mid_{t=t_m} = M_{2m}(y(t_m)) + \hat{\digamma}_m, \ \Delta y' \mid_{t=t_m} = N_{2m}(y(t_m)) + \hat{\digamma}_m, \ \Delta y'' \mid_{t=t_m} = O_{2m}(y(t_m)) + \hat{\digamma}_m, \\ x(0) = x'(0) = 0, \ {}^c\mathbf{D}^\varepsilon x(\Omega) = x''(1), \\ y(0) = y'(0) = 0, \ {}^c\mathbf{D}^\rho y(\Phi) = y''(1). \end{cases} \tag{33}$$

Since (x, y) is a solution of the inequality (28) and $t \in J$; hence, by Remark 1, we obtain

$$x(t) = \begin{cases} -\frac{1}{\Gamma(\alpha)} \int_0^t (t-s)\left[h(s,{}^c\mathbf{D}^a x(s),{}^c\mathbf{D}^b y(s)) + \mathcal{F}_h(s)\right] ds + e^* t^2, \ t \in [0, t_1], \\ -\frac{1}{\Gamma(\alpha)} \int_{t_m}^t (t-s)^{\alpha-1}\left[h(s,{}^c\mathbf{D}^a x(s),{}^c\mathbf{D}^b y(s)) + \mathcal{F}_h(s)\right] ds \\ -\frac{1}{\Gamma(\alpha)} \sum_{j=1}^m \int_{t_{j-1}}^{t_j} (t_j-s)^{\alpha-1}\left[h(s,{}^c\mathbf{D}^a x(s),{}^c\mathbf{D}^b y(s)) + \mathcal{F}_h(s)\right] ds \\ -\frac{1}{\Gamma(\alpha-1)} \sum_{j=1}^m (t-t_j) \int_{t_{j-1}}^{t_j} (t_j-s)^{\alpha-2}\left[h(s,{}^c\mathbf{D}^a x(s),{}^c\mathbf{D}^b y(s)) + \mathcal{F}_h(s)\right] ds \\ -\frac{1}{2\Gamma(\alpha-2)} \sum_{j=1}^m (t-t_j)^2 \int_{t_{j-1}}^{t_j} (t_j-s)^{\alpha-3}\left[h(s,{}^c\mathbf{D}^a x(s),{}^c\mathbf{D}^b y(s)) + \mathcal{F}_h(s)\right] ds \\ +\sum_{j=1}^m \left[M_{1j}(x(t_j)) + \mathcal{F}_m\right] + \sum_{j=1}^m (t-t_j)\left[N_{1j}(x(t_j)) + \mathcal{F}_m\right] + \sum_{j=1}^m \frac{(t-t_j)^2}{2} \\ \times\left[O_{1j}(x(t_j)) + \mathcal{F}_m\right] + e^* t^2, \ t \in (t_m, t_{m+1}], \ 1 \leqslant m \leqslant n. \end{cases}$$

$$y(t) = \begin{cases} -\frac{1}{\Gamma(\beta)} \int_0^t (t-s)^{\beta-1}\left[w(s,{}^c\mathbf{D}^a x(s),{}^c\mathbf{D}^b y(s)) + \mathcal{F}_w(s)\right] ds + c^* t^2, \ t \in [0, t_1], \\ -\frac{1}{\Gamma(\beta)} \int_{t_m}^t (t-s)^{\beta-1}\left[w(s,{}^c\mathbf{D}^a x(s),{}^c\mathbf{D}^b y(s)) + \mathcal{F}_w(s)\right] ds \\ -\frac{1}{\Gamma(\beta)} \sum_{j=1}^m \int_{t_{j-1}}^{t_j} (t_j-s)^{\beta-1} w(s,{}^c\mathbf{D}^a x(s),{}^c\mathbf{D}^b y(s)) ds \\ -\frac{1}{\Gamma(\beta-1)} \sum_{j=1}^m (t-t_j) \int_{t_{j-1}}^{t_j} (t_j-s)^{\beta-2} w(s,{}^c\mathbf{D}^a x(s),{}^c\mathbf{D}^b y(s)) ds \\ -\frac{1}{2\Gamma(\beta-2)} \sum_{j=1}^m (t-t_j)^2 \int_{t_{j-1}}^{t_j} (t_j-s)^{\beta-3}\left[w(s,{}^c\mathbf{D}^a x(s),{}^c\mathbf{D}^b y(s)) + \mathcal{F}_w(s)\right] ds \\ +\sum_{j=1}^m \left[M_{2j}(y(t_j)) + \hat{\mathcal{F}}_m\right] + \sum_{j=1}^m (t-t_j)\left[M_{2j}(y(t_j)) + \hat{\mathcal{F}}_m\right] + \sum_{j=1}^m \frac{(t-t_j)^2}{2} \\ \times\left[O_{2j}(y(t_j)) + \hat{\mathcal{F}}_m\right] + c^* t^2, \ t \in (t_m, t_{m+1}], \ 1 \leqslant m \leqslant n, \end{cases}$$

where

$$\begin{aligned} e^* = \mathcal{Z}_\alpha \Bigg[& -\frac{1}{\Gamma(\alpha-\varepsilon)} \int_{t_j}^{\Omega} (\Omega-s)^{\alpha-\varepsilon-1}\left[h(s,{}^c\mathbf{D}^a x(s),{}^c\mathbf{D}^b y(s)) + \mathcal{F}_h(s)\right] ds \\ & -\frac{\Omega^{1-\varepsilon}}{\Gamma(\alpha-1)\Gamma(2-\varepsilon)} \sum_{j=1}^m \int_{t_{j-1}}^{t_j} (t_j-s)^{\alpha-2}\left[h(s,{}^c\mathbf{D}^a x(s),{}^c\mathbf{D}^b y(s)) + \mathcal{F}_h(s)\right] ds \\ & -\frac{\Omega^{2-\varepsilon}}{\Gamma(\alpha-2)\Gamma(3-\varepsilon)} \sum_{j=1}^m \int_{t_{j-1}}^{t_j} (t_j-s)^{\alpha-3}\left[h(s,{}^c\mathbf{D}^a x(s),{}^c\mathbf{D}^b y(s)) + \mathcal{F}_h(s)\right] ds \\ & +\frac{\Omega^{1-\varepsilon}}{\Gamma(\alpha-2)\Gamma(2-\varepsilon)} \sum_{j=1}^m t_j \int_{t_{j-1}}^{t_j} (t_j-s)^{\alpha-3}\left[h(s,{}^c\mathbf{D}^a x(s),{}^c\mathbf{D}^b y(s)) + \mathcal{F}_h(s)\right] ds \\ & +\frac{\Omega^{1-\varepsilon}}{\Gamma(2-\varepsilon)} \sum_{j=1}^m \left[N_{1j}(x(t_j)) + \mathcal{F}_m\right] + \frac{\Omega^{2-\varepsilon}}{\Gamma(3-\varepsilon)} \sum_{j=1}^m \left[O_{1j}(x(t_j)) + \mathcal{F}_m\right] - \frac{\Omega^{1-\varepsilon}}{\Gamma(2-\varepsilon)} \sum_{j=1}^m t_j \\ & \times \left[O_{1j}(x(t_j)) + \mathcal{F}_m\right] + \frac{1}{\Gamma(\alpha-2)} \int_{t_m}^1 (1-s)^{\alpha-3}\left[h(s,{}^c\mathbf{D}^a x(s),{}^c\mathbf{D}^b y(s)) + \mathcal{F}_h(s)\right] ds \\ & +\frac{1}{\Gamma(\alpha-2)} \sum_{j=1}^m \int_{t_{j-1}}^{t_j} (t_j-s)^{\alpha-3}\left[h(s,{}^c\mathbf{D}^a x(s),{}^c\mathbf{D}^b y(s)) + \mathcal{F}_h(s)\right] ds - \sum_{j=1}^m \left[O_{1j}(x(t_j)) + \mathcal{F}_m\right] \Bigg], \\ c^* = \mathcal{Z}_\beta \Bigg[& -\frac{1}{\Gamma(\beta-\rho)} \int_{t_j}^{\Phi} (\Phi-s)^{\beta-\rho-1}\left[w(s,{}^c\mathbf{D}^a x(s),{}^c\mathbf{D}^b y(s)) + \mathcal{F}_w(s)\right] ds \\ & -\frac{\Phi^{1-\rho}}{\Gamma(\beta-1)\Gamma(2-\rho)} \sum_{j=1}^m \int_{t_{j-1}}^{t_j} (t_j-s)^{\beta-2}\left[w(s,{}^c\mathbf{D}^a x(s),{}^c\mathbf{D}^b y(s)) + \mathcal{F}_w(s)\right] ds \\ & -\frac{\Phi^{2-\rho}}{\Gamma(\beta-2)\Gamma(3-\rho)} \sum_{j=1}^m \int_{t_{j-1}}^{t_j} (t_j-s)^{\beta-3}\left[w(s,{}^c\mathbf{D}^a x(s),{}^c\mathbf{D}^b y(s)) + \mathcal{F}_w(s)\right] ds \\ & +\frac{\Phi^{1-\rho}}{\Gamma(\beta-2)\Gamma(2-\rho)} \sum_{j=1}^m t_j \int_{t_{j-1}}^{t_j} (t_j-s)^{\beta-3}\left[w(s,{}^c\mathbf{D}^a x(s),{}^c\mathbf{D}^b y(s)) + \mathcal{F}_w(s)\right] ds \end{aligned}$$

$$+\frac{\Phi^{1-\rho}}{\Gamma(2-\rho)}\sum_{j=1}^{m}\left[N_{2j}(y(t_j))+\mathcal{F}_m\right]+\frac{\Phi^{2-\rho}}{\Gamma(3-\rho)}\sum_{j=1}^{m}\left[O_{2j}(y(t_j))+\mathcal{F}_m\right]-\frac{\Phi^{1-\rho}}{\Gamma(2-\rho)}\sum_{j=1}^{m}t_j$$
$$\times\left[O_{2j}(y(t_j))+\mathcal{F}_m\right]+\frac{1}{\Gamma(\beta-2)}\int_{t_m}^{1}(1-s)^{\beta-3}\left[w(s,{}^c\mathbf{D}^a x(s),{}^c\mathbf{D}^b y(s))+\mathcal{F}_w(s)\right]ds$$
$$+\frac{1}{\Gamma(\beta-2)}\sum_{j=1}^{m}\int_{t_{j-1}}^{t_j}(t_j-s)^{\beta-3}\left[w(s,{}^c\mathbf{D}^a x(s),{}^c\mathbf{D}^b y(s))+\mathcal{F}_w(s)\right]ds-\sum_{j=1}^{m}\left[O_{2j}(y(t_j))+\mathcal{F}_m\right]\Bigg].$$

For $t\in[0,t_1]$, we have

$$\begin{aligned}x(t)=&\frac{1}{\Gamma(\alpha)}\int_0^t(t-s)^{\alpha-1}\left[h(s,{}^c\mathbf{D}^a x_1(s),{}^c\mathbf{D}^b y_1(s))+\mathcal{F}_h(s)\right]ds\\&+\frac{\mathcal{Z}_\alpha t^2}{\Gamma(\alpha-\varepsilon)}\int_{t_j}^{\Omega}(\Omega-s)^{\alpha-\varepsilon-1}\left[h(s,{}^c\mathbf{D}^a x_1(s),{}^c\mathbf{D}^b y_1(s))+\mathcal{F}_h(s)\right]ds\\&+\frac{\mathcal{Z}_\alpha t^2\Omega^{1-\varepsilon}}{\Gamma(\alpha-1)\Gamma(2-\varepsilon)}\sum_{j=1}^{m}\int_{t_{j-1}}^{t_j}(t_j-s)^{\alpha-2}\left[h(s,{}^c\mathbf{D}^a x_1(s),{}^c\mathbf{D}^b y_1(s))+\mathcal{F}_h(s)\right]ds\\&+\frac{\mathcal{Z}_\alpha t^2\Omega^{2-\varepsilon}}{\Gamma(\alpha-2)\Gamma(3-\varepsilon)}\sum_{j=1}^{m}\int_{t_{j-1}}^{t_j}(t_j-s)^{\alpha-3}\left[h(s,{}^c\mathbf{D}^a x_1(s),{}^c\mathbf{D}^b y_1(s))+\mathcal{F}_h(s)\right]ds\\&+\frac{\mathcal{Z}_\alpha t^2\Omega^{1-\varepsilon}}{\Gamma(\alpha-2)\Gamma(2-\varepsilon)}\sum_{j=1}^{m}t_j\int_{t_{j-1}}^{t_j}(t_j-s)^{\alpha-3}\left[h(s,{}^c\mathbf{D}^a x_1(s),{}^c\mathbf{D}^b y_1(s))+\mathcal{F}_h(s)\right]ds\\&+\frac{\mathcal{Z}_\alpha t^2\Omega^{1-\varepsilon}}{\Gamma(2-\varepsilon)}\sum_{j=1}^{m}\left[N_{1j}(x(t_j))+\mathcal{F}_m\right]+\frac{\mathcal{Z}_\alpha t^2\Omega^{2-\varepsilon}}{\Gamma(3-\varepsilon)}\sum_{j=1}^{m}\left[O_{1j}(x(t_j))+\mathcal{F}_m\right]\\&+\frac{\mathcal{Z}_\alpha t^2}{\Gamma(\alpha-2)}\int_{t_m}^{1}(1-s)^{\alpha-3}\left[h(s,{}^c\mathbf{D}^a x_1(s),{}^c\mathbf{D}^b y_1(s))+\mathcal{F}_h(s)\right]ds\\&+\frac{\mathcal{Z}_\alpha t^2}{\Gamma(\alpha-2)}\sum_{j=1}^{m}\int_{t_{j-1}}^{t_j}(t_j-s)^{\alpha-3}\left[h(s,{}^c\mathbf{D}^a x_1(s),{}^c\mathbf{D}^b y_1(s))+\mathcal{F}_h(s)\right]ds\\&+\frac{\mathcal{Z}_\alpha t^2\Omega^{1-\varepsilon}}{\Gamma(2-\varepsilon)}\sum_{j=1}^{m}t_j\left[O_{1j}(x(t_j))+\mathcal{F}_m\right]+\mathcal{Z}_\alpha t^2\sum_{j=1}^{m}\left[O_{1j}(x(t_j))+\mathcal{F}_m\right].\end{aligned}\tag{34}$$

For computational convenience, we use $s(t)$ for the sum of terms which are free of $\mathcal{F}$; then, (34) becomes

$$\begin{aligned}&|x(t)-s_1(t)|\\\leqslant&\frac{1}{\Gamma(\alpha)}\int_0^t(t-s)^{\alpha-1}|\mathcal{F}_h(s)|ds+\frac{\mathcal{Z}_\alpha}{\Gamma(\alpha-\varepsilon)}\int_{t_j}^{\Omega}(\Omega-s)^{\alpha-\varepsilon-1}|\mathcal{F}_h(s)|ds\\&+\frac{\mathcal{Z}_\alpha\Omega^{1-\varepsilon}}{\Gamma(\alpha-1)\Gamma(2-\varepsilon)}\sum_{j=1}^{m}\int_{t_{j-1}}^{t_j}(t_j-s)^{\alpha-2}|\mathcal{F}_h(s)|ds+\frac{\mathcal{Z}_\alpha\Omega^{2-\varepsilon}}{\Gamma(\alpha-2)\Gamma(3-\varepsilon)}\sum_{j=1}^{m}\int_{t_{j-1}}^{t_j}(t_j-s)^{\alpha-3}|\mathcal{F}_h(s)|ds\\&+\frac{\mathcal{Z}_\alpha\Omega^{1-\varepsilon}}{\Gamma(\alpha-2)\Gamma(2-\varepsilon)}\sum_{j=1}^{m}|t_j|\int_{t_{j-1}}^{t_j}(t_j-s)^{\alpha-3}|\mathcal{F}_h(s)|ds+\frac{\mathcal{Z}_\alpha\Omega^{1-\varepsilon}}{\Gamma(2-\varepsilon)}\sum_{j=1}^{m}|\mathcal{F}_m|+\frac{\mathcal{Z}_\alpha\Omega^{2-\varepsilon}}{\Gamma(3-\varepsilon)}\sum_{j=1}^{m}|\mathcal{F}_m|\\&+\frac{\mathcal{Z}_\alpha}{\Gamma(\alpha-2)}\int_{t_m}^{1}(1-s)^{\alpha-3}|\mathcal{F}_h(s)|ds+\frac{\mathcal{Z}_\alpha}{\Gamma(\alpha-2)}\sum_{j=1}^{m}\int_{t_{j-1}}^{t_j}(t_j-s)^{\alpha-3}|\mathcal{F}_h(s)|ds\\&+\frac{\mathcal{Z}_\alpha\Omega^{1-\varepsilon}}{\Gamma(2-\varepsilon)}\sum_{j=1}^{m}|t_j||\mathcal{F}_m|+\mathcal{Z}_\alpha\sum_{j=1}^{m}|\mathcal{F}_m|.\end{aligned}$$

By utilizing Remark 1, we get

$$
\begin{aligned}
&\left|x(t)-s_1(t)\right| \\
\leqslant&\Big[\frac{1}{\Gamma(\alpha+1)}+\frac{\mathcal{Z}_\alpha}{\Gamma(\alpha-\varepsilon+1)}+\frac{\mathcal{Z}_\alpha\Omega^{1-\varepsilon}}{\Gamma(\alpha)\Gamma(2-\varepsilon)}+\frac{\mathcal{Z}_\alpha\Omega^{2-\varepsilon}}{\Gamma(\alpha-1)\Gamma(3-\varepsilon)}+\frac{\mathcal{Z}_\alpha\Omega^{1-\varepsilon}}{\Gamma(\alpha-1)\Gamma(2-\varepsilon)}+\frac{2\mathcal{Z}_\alpha\Omega^{1-\varepsilon}}{\Gamma(2-\varepsilon)} \\
&+\frac{\mathcal{Z}_\alpha\Omega^{2-\varepsilon}}{\Gamma(3-\varepsilon)}+\frac{2\mathcal{Z}_\alpha}{\Gamma(\alpha-1)}+\mathcal{Z}_\alpha\Big]\epsilon_\alpha.
\end{aligned} \tag{35}
$$

Let

$$
\begin{aligned}
\mathcal{Q}_1=&\frac{1}{\Gamma(\alpha+1)}+\frac{\mathcal{Z}_\alpha}{\Gamma(\alpha-\varepsilon+1)}+\frac{\mathcal{Z}_\alpha\Omega^{1-\varepsilon}}{\Gamma(\alpha)\Gamma(2-\varepsilon)}+\frac{\mathcal{Z}_\alpha\Omega^{2-\varepsilon}}{\Gamma(\alpha-1)\Gamma(3-\varepsilon)}+\frac{\mathcal{Z}_\alpha\Omega^{1-\varepsilon}}{\Gamma(\alpha-1)\Gamma(2-\varepsilon)}+\frac{2\mathcal{Z}_\alpha\Omega^{1-\varepsilon}}{\Gamma(2-\varepsilon)} \\
&+\frac{\mathcal{Z}_\alpha\Omega^{2-\varepsilon}}{\Gamma(3-\varepsilon)}+\frac{2\mathcal{Z}_\alpha}{\Gamma(\alpha-1)}+\mathcal{Z}_\alpha.
\end{aligned}
$$

Thus, (35) becomes

$$
\left|x(t)-s_1(t)\right| \leqslant \mathcal{Q}_1\epsilon_\alpha. \tag{36}
$$

Let

$$
\left|x(t)-\zeta(t)\right|=\left|x(t)-s_1(t)+s_1(t)-\zeta(t)\right| \leqslant \left|x(t)-s_1(t)\right|+\left|s_1(t)-\zeta(t)\right|. \tag{37}
$$

Using (36) in (37), we have

$$
\begin{aligned}
&\left|x(t)-\zeta(t)\right| \\
\leqslant&\mathcal{Q}_1\epsilon_\alpha+\frac{1}{\Gamma(\alpha)}\int_0^t (t-s)^{\alpha-1}\left|h(s,{}^c\mathbf{D}^a x(s),{}^c\mathbf{D}^b y(s))-h(s,{}^c\mathbf{D}^a\zeta(s),{}^c\mathbf{D}^b\chi(s))\right|ds \\
&+\frac{\mathcal{Z}_\alpha t^2}{\Gamma(\alpha-\varepsilon)}\int_{t_j}^{\Omega}(\Omega-s)^{\alpha-\varepsilon-1}\left|h(s,{}^c\mathbf{D}^a x(s),{}^c\mathbf{D}^b y(s))-h(s,{}^c\mathbf{D}^a\zeta(s),{}^c\mathbf{D}^b\chi(s))\right|ds \\
&+\frac{\mathcal{Z}_\alpha t^2\Omega^{1-\varepsilon}}{\Gamma(\alpha-1)\Gamma(2-\varepsilon)}\sum_{j=1}^{m}\int_{t_{j-1}}^{t_j}(t_j-s)^{\alpha-2}\left|h(s,{}^c\mathbf{D}^a x(s),{}^c\mathbf{D}^b y(s))-h(s,{}^c\mathbf{D}^a\zeta(s),{}^c\mathbf{D}^b\chi(s))\right|ds \\
&+\frac{\mathcal{Z}_\alpha t^2\Omega^{2-\varepsilon}}{\Gamma(\alpha-2)\Gamma(3-\varepsilon)}\sum_{j=1}^{m}\int_{t_{j-1}}^{t_j}(t_j-s)^{\alpha-3}\left|h(s,{}^c\mathbf{D}^a x(s),{}^c\mathbf{D}^b y(s))-h(s,{}^c\mathbf{D}^a\zeta(s),{}^c\mathbf{D}^b\chi(s))\right|ds \\
&+\frac{\mathcal{Z}_\alpha t^2\Omega^{1-\varepsilon}}{\Gamma(\alpha-2)\Gamma(2-\varepsilon)}\sum_{j=1}^{m}t_j\int_{t_{j-1}}^{t_j}(t_j-s)^{\alpha-3}\left|h(s,{}^c\mathbf{D}^a x(s),{}^c\mathbf{D}^b y(s))-h(s,{}^c\mathbf{D}^a\zeta(s),{}^c\mathbf{D}^b\chi(s))\right|ds \\
&+\frac{\mathcal{Z}_\alpha t^2\Omega^{1-\varepsilon}}{\Gamma(2-\varepsilon)}\sum_{j=1}^{m}\left|N_{1j}(x(t_j))-N_{1j}(\zeta(t_j))\right|+\frac{\mathcal{Z}_\alpha t^2\Omega^{2-\varepsilon}}{\Gamma(3-\varepsilon)}\sum_{j=1}^{m}\left|O_{1j}(x(t_j))-O_{1j}(\zeta(t_j))\right| \\
&+\frac{\mathcal{Z}_\alpha t^2}{\Gamma(\alpha-2)}\int_{t_m}^{1}(1-s)^{\alpha-3}\left|h(s,{}^c\mathbf{D}^a x(s),{}^c\mathbf{D}^b y(s))-h(s,{}^c\mathbf{D}^a\zeta(s),{}^c\mathbf{D}^b\chi(s))\right|ds \\
&+\frac{\mathcal{Z}_\alpha t^2}{\Gamma(\alpha-2)}\sum_{j=1}^{m}\int_{t_{j-1}}^{t_j}(t_j-s)^{\alpha-3}\left|h(s,{}^c\mathbf{D}^a x(s),{}^c\mathbf{D}^b y(s))-h(s,{}^c\mathbf{D}^a\zeta(s),{}^c\mathbf{D}^b\chi(s))\right|ds \\
&+\frac{\mathcal{Z}_\alpha t^2\Omega^{1-\varepsilon}}{\Gamma(2-\varepsilon)}\sum_{j=1}^{m}\left|t_j\right|\left|O_{1j}(x(t_j))-O_{1j}(\zeta(t_j))\right|+\mathcal{Z}_\alpha t^2\sum_{j=1}^{m}\left|O_{1j}(x(t_j))-O_{1j}(\zeta(t_j))\right|.
\end{aligned}
$$

Utilizing $(\mathbf{M}_1)$ and $(\mathbf{M}_2)$, we get

$$
\begin{aligned}
&\leqslant \mathcal{Q}_1\epsilon_\alpha + \frac{1}{\Gamma(\alpha+1)}\left[\mu_1\|x-\zeta\|_0+\mu_2\|y-\chi\|_0\right] + \frac{\mathcal{Z}_\alpha}{\Gamma(\alpha-\varepsilon+1)}\left[\mu_1\|x-\zeta\|_0+\mu_2\|y-\chi\|_0\right] \\
&+ \frac{\mathcal{Z}_\alpha\Omega^{1-\varepsilon}}{\Gamma(\alpha)\Gamma(2-\varepsilon)}\left[\mu_1\|x-\zeta\|_0+\mu_2\|y-\chi\|_0\right] + \frac{\mathcal{Z}_\alpha\Omega^{2-\varepsilon}}{\Gamma(\alpha-1)\Gamma(3-\varepsilon)}\left[\mu_1\|x-\zeta\|_0+\mu_2\|y-\chi\|_0\right] \\
&+ \frac{\mathcal{Z}_\alpha\Omega^{1-\varepsilon}}{\Gamma(\alpha-1)\Gamma(2-\varepsilon)}\left[\mu_1\|x-\zeta\|_0+\mu_2\|y-\chi\|_0\right] + \frac{\mathcal{Z}_\alpha\Omega^{1-\varepsilon}}{\Gamma(2-\varepsilon)}\sum_{j=1}^{m}\hat{I}_{1j}\|x-\zeta\|_0 \\
&+ \frac{\mathcal{Z}_\alpha\Omega^{2-\varepsilon}}{\Gamma(3-\varepsilon)}\sum_{j=1}^{m}\check{I}_{1j}\|x-\zeta\|_0 + \frac{\mathcal{Z}_\alpha}{\Gamma(\alpha-1)}\left[\mu_1\|x-\zeta\|_0+\mu_2\|y-\chi\|_0\right] + \frac{\mathcal{Z}_\alpha}{\Gamma(\alpha-1)} \\
&\times\left[\mu_1\|x-\zeta\|_0+\mu_2\|y-\chi\|_0\right] + \frac{\mathcal{Z}_\alpha\Omega^{1-\varepsilon}}{\Gamma(2-\varepsilon)}\sum_{j=1}^{m}\check{I}_{1j}\|x-\zeta\|_0 + \mathcal{Z}_\alpha\sum_{j=1}^{m}\check{I}_{1j}\|x-\zeta\|_0.
\end{aligned}
\tag{38}
$$

After some calculation and rearrangement in (38), we get

$$
\|x-\zeta\|_0 - \frac{\Lambda_2}{(1-\Lambda_1)}\|y-\chi\|_0 \leqslant \frac{\mathcal{Q}_1\epsilon_\alpha}{(1-\Lambda_1)}, \tag{39}
$$

where

$$
\begin{aligned}
\Lambda_1 =& \mathcal{Z}_\alpha\Bigg[\mu_1\Bigg(\frac{1}{\mathcal{Z}_\alpha\Gamma(\alpha+1)} + \frac{1}{\Gamma(\alpha-\varepsilon+1)} + \frac{\Omega^{1-\varepsilon}}{\Gamma(\alpha)\Gamma(2-\varepsilon)} + \frac{\Omega^{2-\varepsilon}}{\Gamma(\alpha-1)\Gamma(3-\varepsilon)} + \frac{\Omega^{1-\varepsilon}}{\Gamma(\alpha-1)\Gamma(2-\varepsilon)} \\
&+ \frac{2}{\Gamma(\alpha-1)}\Bigg) + \frac{\Omega^{1-\varepsilon}}{\Gamma(2-\varepsilon)}\sum_{j=1}^{m}\hat{I}_{1j} + \frac{\Omega^{2-\varepsilon}}{\Gamma(3-\varepsilon)}\sum_{j=1}^{m}\check{I}_{1j} + \frac{\Omega^{1-\varepsilon}}{\Gamma(2-\varepsilon)}\sum_{j=1}^{m}\check{I}_{1j} + \sum_{j=1}^{m}\check{I}_{1j}\Bigg], \\
\Lambda_2 =& \mathcal{Z}_\alpha\Bigg[\mu_2\Bigg(\frac{1}{\mathcal{Z}_\alpha\Gamma(\alpha+1)} + \frac{1}{\Gamma(\alpha-\varepsilon+1)} + \frac{\Omega^{1-\varepsilon}}{\Gamma(\alpha)\Gamma(2-\varepsilon)} + \frac{\Omega^{2-\varepsilon}}{\Gamma(\alpha-1)\Gamma(3-\varepsilon)} + \frac{\Omega^{1-\varepsilon}}{\Gamma(\alpha-1)\Gamma(2-\varepsilon)} \\
&+ \frac{2}{\Gamma(\alpha-1)}\Bigg)\Bigg].
\end{aligned}
$$

In addition, for $t \in (t_m, t_{m+1}]$, we have

$$
\begin{aligned}
x(t) =& \frac{1}{\Gamma(\alpha)}\int_{t_m}^{t}(t-s)^{\alpha-1}\left[h(s,{}^c\mathbf{D}^a x(s),{}^c\mathbf{D}^b y(s)) + \mathcal{F}_h(s)\right]ds \\
&+ \frac{1}{\Gamma(\alpha)}\sum_{j=1}^{m}\int_{t_{j-1}}^{t_j}(t_j-s)^{\alpha-1}\left[h(s,{}^c\mathbf{D}^a x(s),{}^c\mathbf{D}^b y(s)) + \mathcal{F}_h(s)\right]ds \\
&+ \frac{1}{\Gamma(\alpha-1)}\sum_{j=1}^{m}(t-t_j)\int_{t_{j-1}}^{t_j}(t_j-s)^{\alpha-2}\left[h(s,{}^c\mathbf{D}^a x(s),{}^c\mathbf{D}^b y(s)) + \mathcal{F}_h(s)\right]ds \\
&+ \frac{1}{2\Gamma(\alpha-2)}\sum_{j=1}^{m}(t-t_j)^2\int_{t_{j-1}}^{t_j}(t_j-s)^{\alpha-3}\left[h(s,{}^c\mathbf{D}^a x(s),{}^c\mathbf{D}^b y(s)) + \mathcal{F}_h(s)\right]ds \\
&+ \sum_{j=1}^{m}\left[M_{1j}x(t_j) + \mathcal{F}_m\right] + \sum_{j=1}^{m}(t-t_j)\left[N_{1j}(x(t_j)) + \mathcal{F}_m\right] \\
&+ \frac{\mathcal{Z}_\alpha t^2}{\Gamma(\alpha-\varepsilon)}\int_{t_j}^{\Omega}(\Omega-s)^{\alpha-\varepsilon-1}\left[h(s,{}^c\mathbf{D}^a x(s),{}^c\mathbf{D}^b y(s)) + \mathcal{F}_h(s)\right]ds \\
&+ \frac{\mathcal{Z}_\alpha t^2\Omega^{1-\varepsilon}}{\Gamma(\alpha-1)\Gamma(2-\varepsilon)}\sum_{j=1}^{m}\int_{t_{j-1}}^{t_j}(t_j-s)^{\alpha-2}\left[h(s,{}^c\mathbf{D}^a x(s),{}^c\mathbf{D}^b y(s)) + \mathcal{F}_h(s)\right]ds
\end{aligned}
$$

$$
\begin{aligned}
&+\frac{\mathcal{Z}_\alpha t^2\Omega^{2-\varepsilon}}{\Gamma(\alpha-2)\Gamma(3-\varepsilon)}\sum_{j=1}^{m}\int_{t_{j-1}}^{t_j}(t_j-s)^{\alpha-3}\left[h(s,{}^c\mathbf{D}^a x(s),{}^c\mathbf{D}^b y(s))+\mathcal{F}_h(s)\right]ds\\
&+\frac{\mathcal{Z}_\alpha t^2\Omega^{1-\varepsilon}}{\Gamma(\alpha-2)\Gamma(2-\varepsilon)}\sum_{j=1}^{m}t_j\int_{t_{j-1}}^{t_j}(t_j-s)^{\alpha-3}\left[h(s,{}^c\mathbf{D}^a x(s),{}^c\mathbf{D}^b y(s))+\mathcal{F}_h(s)\right]ds\\
&+\frac{\mathcal{Z}_\alpha t^2\Omega^{1-\varepsilon}}{\Gamma(2-\varepsilon)}\sum_{j=1}^{m}\left[N_{1j}(x(t_j))+\mathcal{F}_m\right]+\frac{\mathcal{Z}_\alpha t^2\Omega^{2-\varepsilon}}{\Gamma(3-\varepsilon)}\sum_{j=1}^{m}\left[O_{1j}(x(t_j))+\mathcal{F}_m\right]\\
&+\frac{\mathcal{Z}_\alpha t^2}{\Gamma(\alpha-2)}\int_{t_m}^{1}(1-s)^{\alpha-3}\left[h(s,{}^c\mathbf{D}^a x(s),{}^c\mathbf{D}^b y(s))+\mathcal{F}_h(s)\right]ds\\
&+\frac{\mathcal{Z}_\alpha t^2}{\Gamma(\alpha-2)}\sum_{j=1}^{m}\int_{t_{j-1}}^{t_j}(t_j-s)^{\alpha-3}\left[h(s,{}^c\mathbf{D}^a x(s),{}^c\mathbf{D}^b y(s))+\mathcal{F}_h(s)\right]ds\\
&+\frac{\mathcal{Z}_\alpha t^2\Omega^{1-\varepsilon}}{\Gamma(2-\varepsilon)}\sum_{j=1}^{m}t_j\left[O_{1j}(x(t_j))+\mathcal{F}_m\right]+\mathcal{Z}_\alpha t^2\sum_{j=1}^{m}\left[O_{1j}(x(t_j))+\mathcal{F}_m\right]\\
&+\sum_{j=1}^{m}\frac{(t-t_j)^2}{2}\left[O_{1j}(x(t_j))+\mathcal{F}_m\right].
\end{aligned}
$$

For computational convenience, we use $s_1^*(t)$ for the sum of terms which are free of $\mathcal{F}$, so we have

$$
\begin{aligned}
&|x(t)-s_1^*(t)|\\
\leqslant&\frac{1}{\Gamma(\alpha)}\int_{t_m}^{t}(t-s)^{\alpha-1}|\mathcal{F}_h(s)|ds+\frac{1}{\Gamma(\alpha)}\sum_{j=1}^{m}\int_{t_{j-1}}^{t_j}(t_j-s)^{\alpha-1}|\mathcal{F}_h(s)|ds\\
&+\frac{1}{\Gamma(\alpha-1)}\sum_{j=1}^{m}|(t-t_j)|\int_{t_{j-1}}^{t_j}(t_j-s)^{\alpha-2}|\mathcal{F}_h(s)|ds+\frac{1}{2\Gamma(\alpha-2)}\sum_{j=1}^{m}|(t-t_j)^2|\int_{t_{j-1}}^{t_j}(t_j-s)^{\alpha-3}\\
&\times|\mathcal{F}_h(s)|ds+\sum_{j=1}^{m}|\mathcal{F}_m|+\sum_{j=1}^{m}|(t-t_j)||\mathcal{F}_m|+\frac{\mathcal{Z}_\alpha}{\Gamma(\alpha-\varepsilon)}\int_{t_j}^{\Omega}(\Omega-s)^{\alpha-\varepsilon-1}|\mathcal{F}_h(s)|ds\\
&+\frac{\mathcal{Z}_\alpha\Omega^{1-\varepsilon}}{\Gamma(\alpha-1)\Gamma(2-\varepsilon)}\sum_{j=1}^{m}\int_{t_{j-1}}^{t_j}(t_j-s)^{\alpha-2}|\mathcal{F}_h(s)|ds+\frac{\mathcal{Z}_\alpha\Omega^{2-\varepsilon}}{\Gamma(\alpha-2)\Gamma(3-\varepsilon)}\sum_{j=1}^{m}\int_{t_{j-1}}^{t_j}(t_j-s)^{\alpha-3}|\mathcal{F}_h(s)|ds\\
&+\frac{\mathcal{Z}_\alpha\Omega^{1-\varepsilon}}{\Gamma(\alpha-2)\Gamma(2-\varepsilon)}\sum_{j=1}^{m}t_j\int_{t_{j-1}}^{t_j}(t_j-s)^{\alpha-3}|\mathcal{F}_h(s)|ds+\frac{\mathcal{Z}_\alpha\Omega^{1-\varepsilon}}{\Gamma(2-\varepsilon)}\sum_{j=1}^{m}|\mathcal{F}_m|+\frac{\mathcal{Z}_\alpha\Omega^{2-\varepsilon}}{\Gamma(3-\varepsilon)}\sum_{j=1}^{m}|\mathcal{F}_m|\\
&+\frac{\mathcal{Z}_\alpha}{\Gamma(\alpha-2)}\int_{t_m}^{1}(1-s)^{\alpha-3}|\mathcal{F}_h(s)|ds+\frac{\mathcal{Z}_\alpha}{\Gamma(\alpha-2)}\sum_{j=1}^{m}\int_{t_{j-1}}^{t_j}(t_j-s)^{\alpha-3}|\mathcal{F}_h(s)|ds\\
&+\frac{\mathcal{Z}_\alpha\Omega^{1-\varepsilon}}{\Gamma(2-\varepsilon)}\sum_{j=1}^{m}|t_j||\mathcal{F}_m|+\mathcal{Z}_\alpha\sum_{j=1}^{m}|\mathcal{F}_m|+\sum_{j=1}^{m}\frac{|(t-t_j)^2|}{2}|\mathcal{F}_m|.
\end{aligned}
$$

By utilizing Remark 1, we get

$$
\begin{aligned}
&|x(t)-s_1^*(t)|\\
\leqslant&\bigg(\tfrac{2}{\Gamma(\alpha+1)}+\tfrac{1}{\Gamma(\alpha)}+\tfrac{4\mathcal{Z}_\alpha+1}{2\Gamma(\alpha-1)}+\tfrac{3\mathcal{Z}_\alpha+4}{2}+\tfrac{\mathcal{Z}_\alpha}{\Gamma(\alpha-\varepsilon+1)}+\tfrac{\mathcal{Z}_\alpha\Omega^{1-\varepsilon}}{\Gamma(\alpha)\Gamma(2-\varepsilon)}+\tfrac{2\mathcal{Z}_\alpha\Omega^{1-\varepsilon}}{\Gamma(2-\varepsilon)}+\tfrac{\mathcal{Z}_\alpha\Omega^{2-\varepsilon}}{\Gamma(3-\varepsilon)}\\
&+\frac{\mathcal{Z}_\alpha\Omega^{2-\varepsilon}}{\Gamma(\alpha-1)\Gamma(3-\varepsilon)}+\frac{\mathcal{Z}_\alpha\Omega^{1-\varepsilon}}{\Gamma(\alpha-1)\Gamma(2-\varepsilon)}\bigg)\epsilon_\alpha.
\end{aligned}
\tag{40}
$$

For computational convenience, let

$$\mathcal{Q}_\alpha = \frac{2}{\Gamma(\alpha+1)} + \frac{1}{\Gamma(\alpha)} + \frac{4\mathcal{Z}_\alpha+1}{2\Gamma(\alpha-1)} + \frac{3\mathcal{Z}_\alpha+4}{2} + \frac{\mathcal{Z}_\alpha}{\Gamma(\alpha-\varepsilon+1)} + \frac{\mathcal{Z}_\alpha\Omega^{1-\varepsilon}}{\Gamma(\alpha)\Gamma(2-\varepsilon)} + \frac{2\mathcal{Z}_\alpha\Omega^{1-\varepsilon}}{\Gamma(2-\varepsilon)} + \frac{\mathcal{Z}_\alpha\Omega^{2-\varepsilon}}{\Gamma(3-\varepsilon)} + \frac{\mathcal{Z}_\alpha\Omega^{2-\varepsilon}}{\Gamma(\alpha-1)\Gamma(3-\varepsilon)} + \frac{\mathcal{Z}_\alpha\Omega^{1-\varepsilon}}{\Gamma(\alpha-1)\Gamma(2-\varepsilon)}.$$

Thus, (40) becomes

$$\left|x(t) - s_1^*(t)\right| \leqslant \mathcal{Q}_\alpha \epsilon_\alpha. \tag{41}$$

Let

$$\left|x(t) - \zeta(t)\right| = \left|x(t) - s_1^*(t) + s_1^*(t) - \zeta(t)\right| \leqslant \left|x(t) - s_1^*(t)\right| + \left|s_1^*(t) - \zeta(t)\right|. \tag{42}$$

Using (41) in (42), we get

$$\begin{aligned}
&\left|x(t) - \zeta(t)\right| \\
&\leqslant \mathcal{Q}_\alpha \epsilon_\alpha + \frac{1}{\Gamma(\alpha)}\int_{t_m}^{t}(t-s)^{\alpha-1}\left|h(s,{}^c\mathbf{D}^a x(s),{}^c\mathbf{D}^b y(s)) - h(s,{}^c\mathbf{D}^a\zeta(s),{}^c\mathbf{D}^b\chi(s))\right|ds \\
&+ \frac{1}{\Gamma(\alpha)}\sum_{j=1}^{m}\int_{t_{j-1}}^{t_j}(t_j-s)^{\alpha-1}\left|h(s,{}^c\mathbf{D}^a x(s),{}^c\mathbf{D}^b y(s)) - h(s,{}^c\mathbf{D}^a\zeta(s),{}^c\mathbf{D}^b\chi(s))\right|ds \\
&+ \frac{1}{\Gamma(\alpha-1)}\sum_{j=1}^{m}\left|(t-t_j)\right|\int_{t_{j-1}}^{t_j}(t_j-s)^{\alpha-2}\left|h(s,{}^c\mathbf{D}^a x(s),{}^c\mathbf{D}^b y(s)) - h(s,{}^c\mathbf{D}^a\zeta(s),{}^c\mathbf{D}^b\chi(s))\right|ds \\
&+ \frac{1}{2\Gamma(\alpha-2)}\sum_{j=1}^{m}\left|(t-t_j)^2\right|\int_{t_{j-1}}^{t_j}(t_j-s)^{\alpha-3}\left|h(s,{}^c\mathbf{D}^a x(s),{}^c\mathbf{D}^b y(s)) - h(s,{}^c\mathbf{D}^a\zeta(s),{}^c\mathbf{D}^b\chi(s))\right|ds \\
&+ \sum_{j=1}^{m}\left|M_{1j}(x(t_j)) - M_{1j}(\zeta(t_j))\right| + \sum_{j=1}^{m}\left|(t-t_j)\right|\left|N_{1j}(x(t_j)) - N_{1j}(\zeta(t_j))\right| \\
&+ \frac{\mathcal{Z}_\alpha t^2}{\Gamma(\alpha-\varepsilon)}\int_{t_j}^{\Omega}(\Omega-s)^{\alpha-\varepsilon-1}\left|h(s,{}^c\mathbf{D}^a x(s),{}^c\mathbf{D}^b y(s)) - h(s,{}^c\mathbf{D}^a\zeta(s),{}^c\mathbf{D}^b\chi(s))\right|ds \\
&+ \frac{\mathcal{Z}_\alpha t^2\Omega^{1-\varepsilon}}{\Gamma(\alpha-1)\Gamma(2-\varepsilon)}\sum_{j=1}^{m}\int_{t_{j-1}}^{t_j}(t_j-s)^{\alpha-2}\left|h(s,{}^c\mathbf{D}^a x(s),{}^c\mathbf{D}^b y(s)) - h(s,{}^c\mathbf{D}^a\zeta(s),{}^c\mathbf{D}^b\chi(s))\right|ds \\
&+ \frac{\mathcal{Z}_\alpha t^2\Omega^{2-\varepsilon}}{\Gamma(\alpha-2)\Gamma(3-\varepsilon)}\sum_{j=1}^{m}\int_{t_{j-1}}^{t_j}(t_j-s)^{\alpha-3}\left|h(s,{}^c\mathbf{D}^a x(s),{}^c\mathbf{D}^b y(s)) - h(s,{}^c\mathbf{D}^a\zeta(s),{}^c\mathbf{D}^b\chi(s))\right|ds \\
&+ \frac{\mathcal{Z}_\alpha t^2\Omega^{1-\varepsilon}}{\Gamma(\alpha-2)\Gamma(2-\varepsilon)}\sum_{j=1}^{m}\left|t_j\right|\int_{t_{j-1}}^{t_j}(t_j-s)^{\alpha-3}\left|h(s,{}^c\mathbf{D}^a x(s),{}^c\mathbf{D}^b y(s)) - h(s,{}^c\mathbf{D}^a\zeta(s),{}^c\mathbf{D}^b\chi(s))\right|ds \\
&+ \frac{\mathcal{Z}_\alpha t^2\Omega^{1-\varepsilon}}{\Gamma(2-\varepsilon)}\sum_{j=1}^{m}\left|N_{1j}(x(t_j)) - N_{1j}(\zeta(t_j))\right| + \frac{\mathcal{Z}_\alpha t^2\Omega^{2-\varepsilon}}{\Gamma(3-\varepsilon)}\sum_{j=1}^{m}\left|O_{1j}(x(t_j)) - O_{1j}(\zeta(t_j))\right| \\
&+ \frac{\mathcal{Z}_\alpha t^2}{\Gamma(\alpha-2)}\int_{t_m}^{1}(1-s)^{\alpha-3}\left|h(s,{}^c\mathbf{D}^a x(s),{}^c\mathbf{D}^b y(s)) - h(s,{}^c\mathbf{D}^a\zeta(s),{}^c\mathbf{D}^b\chi(s))\right|ds \\
&+ \frac{\mathcal{Z}_\alpha t^2}{\Gamma(\alpha-2)}\sum_{j=1}^{m}\int_{t_{j-1}}^{t_j}(t_j-s)^{\alpha-3}\left|h(s,{}^c\mathbf{D}^a x(s),{}^c\mathbf{D}^b y(s)) - h(s,{}^c\mathbf{D}^a\zeta(s),{}^c\mathbf{D}^b\chi(s))\right|ds
\end{aligned}$$

$$+\frac{\mathcal{Z}_\alpha t^2\Omega^{1-\varepsilon}}{\Gamma(2-\varepsilon)}\sum_{j=1}^{m}|t_j||O_{1j}(x(t_j))-O_{1j}(\zeta(t_j))|+\mathcal{Z}_\alpha t^2\sum_{j=1}^{m}|O_{1j}(x(t_j))-O_{1j}(\zeta(t_j))|$$
$$+\sum_{j=1}^{m}\frac{|(t-t_j)^2|}{2}|O_{1j}(x(t_j))-O_{1j}(\zeta(t_j))|.$$

Using $(\mathbf{M}_1)$ and $(\mathbf{M}_2)$, we have

$$\begin{aligned}
&\leqslant \mathcal{Q}_\alpha\epsilon_\alpha+\tfrac{1}{\Gamma(\alpha+1)}\left[\mu_1\|x-\zeta\|_0+\mu_2\|y-\chi\|_0\right]+\tfrac{1}{\Gamma(\alpha+1)}\left[\mu_1\|x-\zeta\|_0+\mu_2\|y-\chi\|_0\right]\\
&+\tfrac{1}{\Gamma(\alpha)}\left[\mu_1\|x-\zeta\|_0+\mu_2\|y-\chi\|_0\right]+\tfrac{1}{2\Gamma(\alpha-1)}\left[\mu_1\|x-\zeta\|_0+\mu_2\|y-\chi\|_0\right]\\
&+\textstyle\sum_{j=1}^{m}I_{1j}\|x-\zeta\|_0+\sum_{j=1}^{m}\hat{I}_{1j}\|x-\zeta\|_0+\tfrac{\mathcal{Z}_\alpha}{\Gamma(\alpha-\varepsilon+1)}\left[\mu_1\|x-\zeta\|_0+\mu_2\|y-\chi\|_0\right]\\
&+\tfrac{\mathcal{Z}_\alpha\Omega^{1-\varepsilon}}{\Gamma(\alpha)\Gamma(2-\varepsilon)}\left[\mu_1\|x-\zeta\|_0+\mu_2\|y-\chi\|_0\right]+\tfrac{\mathcal{Z}_\alpha\Omega^{2-\varepsilon}}{\Gamma(\alpha-1)\Gamma(3-\varepsilon)}\left[\mu_1\|x-\zeta\|_0+\mu_2\|y-\chi\|_0\right]\\
&+\tfrac{\mathcal{Z}_\alpha\Omega^{1-\varepsilon}}{\Gamma(\alpha-1)\Gamma(2-\varepsilon)}\left[\mu_1\|x-\zeta\|_0+\mu_2\|y-\chi\|_0\right]+\tfrac{\mathcal{Z}_\alpha\Omega^{1-\varepsilon}}{\Gamma(2-\varepsilon)}\textstyle\sum_{j=1}^{m}\hat{I}_{1j}\|x-\zeta\|_0\\
&+\tfrac{\mathcal{Z}_\alpha\Omega^{2-\varepsilon}}{\Gamma(3-\varepsilon)}\textstyle\sum_{j=1}^{m}\hat{I}_{1j}\|x-\zeta\|_0+\tfrac{\mathcal{Z}_\alpha}{\Gamma(\alpha-1)}\left[\mu_1\|x-\zeta\|_0+\mu_2\|y-\chi\|_0\right]\\
&+\tfrac{\mathcal{Z}_\alpha}{\Gamma(\alpha-1)}\left[\mu_1\|x-\zeta\|_0+\mu_2\|y-\chi\|_0\right]+\tfrac{\mathcal{Z}_\alpha\Omega^{1-\varepsilon}}{\Gamma(2-\varepsilon)}\textstyle\sum_{j=1}^{m}\check{I}_{1j}\|x-\zeta\|_0+\mathcal{Z}_\alpha\sum_{j=1}^{m}\check{I}_{1j}\|x-\zeta\|_0\\
&+\tfrac{1}{2}\textstyle\sum_{j=1}^{m}\check{I}_{1j}\|x-\zeta\|_0.
\end{aligned}\tag{43}$$

After some calculation and rearrangement in (43), we get

$$\|x-\zeta\|_0-\frac{\Lambda_2^*}{(1-\Lambda_1^*)}\|y-\chi\|_0\leqslant\frac{\mathcal{Q}_\alpha\epsilon_\alpha}{(1-\Lambda_1^*)},\tag{44}$$

where

$$\begin{aligned}
\Lambda_1^*=&\Bigg[\mu_1\Bigg(\frac{2}{\Gamma(\alpha+1)}+\frac{1}{\Gamma(\alpha)}+\frac{1}{2\Gamma(\alpha-1)}+\frac{\mathcal{Z}_\alpha}{\Gamma(\alpha-\varepsilon+1)}+\frac{\mathcal{Z}_\alpha\Omega^{1-\varepsilon}}{\Gamma(\alpha)\Gamma(2-\varepsilon)}+\frac{\mathcal{Z}_\alpha\Omega^{2-\varepsilon}}{\Gamma(\alpha-1)\Gamma(3-\varepsilon)}\\
&+\frac{\mathcal{Z}_\alpha\Omega^{1-\varepsilon}}{\Gamma(\alpha-1)\Gamma(2-\varepsilon)}+\frac{2\mathcal{Z}_\alpha}{\Gamma(\alpha-1)}\Bigg)+\sum_{j=1}^{m}I_{1j}+\sum_{j=1}^{m}\hat{I}_{1j}+\frac{\mathcal{Z}_\alpha\Omega^{1-\varepsilon}}{\Gamma(2-\varepsilon)}\sum_{j=1}^{m}\hat{I}_{1j}+\frac{\mathcal{Z}_\alpha\Omega^{2-\varepsilon}}{\Gamma(3-\varepsilon)}\sum_{j=1}^{m}\hat{I}_{1j}\\
&+\frac{\mathcal{Z}_\alpha\Omega^{1-\varepsilon}}{\Gamma(2-\varepsilon)}\sum_{j=1}^{m}\check{I}_{1j}+\mathcal{Z}_\alpha\sum_{j=1}^{m}\check{I}_{1j}+\frac{1}{2}\sum_{j=1}^{m}\check{I}_{1j}\Bigg],\\
\Lambda_2^*=&\mu_2\Bigg(\frac{2}{\Gamma(\alpha+1)}+\frac{1}{\Gamma(\alpha)}+\frac{1}{2\Gamma(\alpha-1)}+\frac{\mathcal{Z}_\alpha}{\Gamma(\alpha-\varepsilon+1)}+\frac{\mathcal{Z}_\alpha\Omega^{1-\varepsilon}}{\Gamma(\alpha)\Gamma(2-\varepsilon)}+\frac{\mathcal{Z}_\alpha\Omega^{2-\varepsilon}}{\Gamma(\alpha-1)\Gamma(3-\varepsilon)}\\
&+\frac{\mathcal{Z}_\alpha\Omega^{1-\varepsilon}}{\Gamma(\alpha-1)\Gamma(2-\varepsilon)}+\frac{2\mathcal{Z}_\alpha}{\Gamma(\alpha-1)}\Bigg).
\end{aligned}$$

On the similar fashion, for $t\in[0,t_1]$, and utilizing $(\mathbf{M}_1)-(\mathbf{M}_2)$, we can find

$$|y(t)-s_2(t)|\leqslant\mathcal{Q}_2\epsilon_\beta,\tag{45}$$

where $s_2(t)$ are those terms which are free of $\mathcal{F}$ and

$$\begin{aligned}
\mathcal{Q}_2=&\frac{1}{\Gamma(\beta+1)}+\frac{\mathcal{Z}_\beta}{\Gamma(\beta-\rho+1)}+\frac{\mathcal{Z}_\beta\Phi^{1-\rho}}{\Gamma(\beta)\Gamma(2-\rho)}+\frac{\mathcal{Z}_\beta\Phi^{2-\rho}}{\Gamma(\beta-1)\Gamma(3-\rho)}+\frac{\mathcal{Z}_\beta\Phi^{1-\rho}}{\Gamma(\beta-1)\Gamma(2-\rho)}+\frac{2\mathcal{Z}_\beta\Phi^{1-\rho}}{\Gamma(2-\rho)}\\
&+\frac{\mathcal{Z}_\beta\Phi^{2-\rho}}{\Gamma(3-\rho)}+\frac{2\mathcal{Z}_\beta}{\Gamma(\beta-1)}+\mathcal{Z}_\beta,
\end{aligned}$$

and

$$\|y-\chi\|_0 - \frac{\Lambda_4}{(1-\Lambda_3)}\|x-\zeta\|_0 \leqslant \frac{\mathcal{Q}_2\epsilon_\beta}{(1-\Lambda_3)}, \tag{46}$$

where

$$\begin{aligned}
\Lambda_3 =& \mathcal{Z}_\beta\Bigg[\mu_1'\Bigg(\frac{1}{\mathcal{Z}_\beta\Gamma(\beta+1)} + \frac{1}{\Gamma(\beta-\rho+1)} + \frac{\Phi^{1-\rho}}{\Gamma(\beta)\Gamma(2-\rho)} + \frac{\Phi^{2-\rho}}{\Gamma(\beta-1)\Gamma(3-\rho)} + \frac{\Phi^{1-\rho}}{\Gamma(\beta-1)\Gamma(2-\rho)} \\
&+ \frac{2}{\Gamma(\beta-1)}\Bigg) + \frac{\Phi^{1-\rho}}{\Gamma(2-\rho)}\sum_{j=1}^{m}\hat{I}_{2j} + \frac{\Phi^{2-\rho}}{\Gamma(3-\rho)}\sum_{j=1}^{m}\check{I}_{2j} + \frac{\Phi^{1-\rho}}{\Gamma(2-\rho)}\sum_{j=1}^{m}\check{I}_{2j} + \sum_{j=1}^{m}\check{I}_{2j}\Bigg], \\
\Lambda_4 =& \mathcal{Z}_\beta\Bigg[\mu_2'\Bigg(\frac{1}{\mathcal{Z}_\beta\Gamma(\beta+1)} + \frac{1}{\Gamma(\beta-\rho+1)} + \frac{\Phi^{1-\rho}}{\Gamma(\beta)\Gamma(2-\rho)} + \frac{\Phi^{2-\rho}}{\Gamma(\beta-1)\Gamma(3-\rho)} + \frac{\Phi^{1-\rho}}{\Gamma(\beta-1)\Gamma(2-\rho)} \\
&+ \frac{2}{\Gamma(\beta-1)}\Bigg)\Bigg].
\end{aligned}$$

In addition, for $t \in (t_m, t_{m+1}]$, $1 \leqslant m \leqslant n$, we can get

$$\left|y(t) - s_2^*(t)\right| \leqslant \mathcal{Q}_\beta\epsilon_\beta, \tag{47}$$

where $s_2^*(t)$ are those terms which are free of $\mathcal{F}$ and

$$\begin{aligned}
\mathcal{Q}_\beta =& \frac{2}{\Gamma(\beta+1)} + \frac{1}{\Gamma(\beta)} + \frac{4\mathcal{Z}_\beta+1}{2\Gamma(\beta-1)} + \frac{3\mathcal{Z}_\beta+4}{2} + \frac{\mathcal{Z}_\beta}{\Gamma(\beta-\rho+1)} + \frac{\mathcal{Z}_\beta\Phi^{1-\rho}}{\Gamma(\beta)\Gamma(2-\rho)} + \frac{2\mathcal{Z}_\beta\Phi^{1-\rho}}{\Gamma(2-\rho)} + \frac{\mathcal{Z}_\beta\Phi^{2-\rho}}{\Gamma(3-\rho)} \\
&+ \frac{\mathcal{Z}_\beta\Phi^{2-\rho}}{\Gamma(\beta-1)\Gamma(3-\rho)} + \frac{\mathcal{Z}_\beta\Phi^{1-\rho}}{\Gamma(\beta-1)\Gamma(2-\rho)}.
\end{aligned}$$

In addition,

$$\|y-\chi\|_0 - \frac{\Lambda_4^*}{(1-\Lambda_3^*)}\|x-\zeta\|_0 \leqslant \frac{\mathcal{Q}_\beta\epsilon_\beta}{(1-\Lambda_3^*)}, \tag{48}$$

where

$$\begin{aligned}
\Lambda_3^* =& \Bigg[\mu_2'\Bigg(\frac{2}{\Gamma(\beta+1)} + \frac{1}{\Gamma(\beta)} + \frac{1}{2\Gamma(\beta-1)} + \frac{\mathcal{Z}_\beta}{\Gamma(\beta-\rho+1)} + \frac{\mathcal{Z}_\beta\Phi^{1-\rho}}{\Gamma(\beta)\Gamma(2-\rho)} + \frac{\mathcal{Z}_\beta\Phi^{2-\rho}}{\Gamma(\beta-1)\Gamma(3-\rho)} \\
&+ \frac{\mathcal{Z}_\beta\Phi^{1-\rho}}{\Gamma(\beta-1)\Gamma(2-\rho)} + \frac{2\mathcal{Z}_\beta}{\Gamma(\beta-1)}\Bigg) + \sum_{j=1}^{m} I_{2j} + \sum_{j=1}^{m}\hat{I}_{2j} + \frac{\mathcal{Z}_\beta\Phi^{1-\rho}}{\Gamma(2-\rho)}\sum_{j=1}^{m}\hat{I}_{2j} + \frac{\mathcal{Z}_\beta\Phi^{2-\rho}}{\Gamma(3-\rho)}\sum_{j=1}^{m}\check{I}_{2j} \\
&+ \frac{\mathcal{Z}_\beta\Phi^{1-\rho}}{\Gamma(2-\rho)}\sum_{j=1}^{m}\check{I}_{2j} + \mathcal{Z}_\beta\sum_{j=1}^{m}\check{I}_{2j} + \frac{1}{2}\sum_{j=1}^{m}\check{I}_{2j}\Bigg], \\
\Lambda_4^* =& \mu_1'\Bigg(\frac{2}{\Gamma(\beta+1)} + \frac{1}{\Gamma(\beta)} + \frac{1}{2\Gamma(\beta-1)} + \frac{\mathcal{Z}_\beta}{\Gamma(\beta-\rho+1)} + \frac{\mathcal{Z}_\beta\Phi^{1-\rho}}{\Gamma(\beta)\Gamma(2-\rho)} + \frac{\mathcal{Z}_\beta\Phi^{2-\rho}}{\Gamma(\beta-1)\Gamma(3-\rho)} \\
&+ \frac{\mathcal{Z}_\beta\Phi^{1-\rho}}{\Gamma(\beta-1)\Gamma(2-\rho)} + \frac{2\mathcal{Z}_\beta}{\Gamma(\beta-1)}\Bigg).
\end{aligned}$$

The equivalent matrix of Equations (44) and (48) is given as:

$$\begin{bmatrix} 1 & -\frac{\Lambda_2^*}{(1-\Lambda_1^*)} \\ -\frac{\Lambda_4^*}{(1-\Lambda_3^*)} & 1 \end{bmatrix}\begin{bmatrix} \|x-\zeta\|_0 \\ \|y-\chi\|_0 \end{bmatrix} \leqslant \begin{bmatrix} \frac{\mathcal{Q}_\alpha\epsilon_\alpha}{(1-\Lambda_1^*)} \\ \frac{\mathcal{Q}_\beta\epsilon_\beta}{(1-\Lambda_3^*)} \end{bmatrix}.$$

Solving the above inequality, we get

$$\begin{bmatrix} \|x-\zeta\|_0 \\ \|y-\chi\|_0 \end{bmatrix} \leqslant \begin{bmatrix} \frac{1}{\Lambda_0} & \frac{\Lambda_2^*}{\Lambda_0(1-\Lambda_1^*)} \\ \frac{\Lambda_4^*}{\Lambda_0(1-\Lambda_3^*)} & \frac{1}{\Lambda_0} \end{bmatrix} \begin{bmatrix} \frac{\Omega_\alpha \epsilon_\alpha}{(1-\Lambda_1^*)} \\ \frac{\Omega_\beta \epsilon_\beta}{(1-\Lambda_3^*)} \end{bmatrix},$$

where

$$\Lambda_0 = 1 - \frac{\Lambda_2^*\Lambda_4^*}{(1-\Lambda_1^*)(1-\Lambda_3^*)} > 0.$$

Further simplification of the above system gives

$$\|x-\zeta\|_0 \leqslant \frac{\Omega_\alpha \epsilon_\alpha}{\Lambda_0(1-\Lambda_1^*)} + \frac{\Lambda_2^*\Omega_\beta \epsilon_\beta}{\Lambda_0(1-\Lambda_1^*)(1-\Lambda_3^*)},$$
$$\|y-\chi\|_0 \leqslant \frac{\Omega_\beta \epsilon_\beta}{\Lambda_0(1-\Lambda_3^*)} + \frac{\Lambda_4^*\Omega_\alpha \epsilon_\alpha}{\Lambda_0(1-\Lambda_1^*)(1-\Lambda_3^*)},$$

from which we have

$$\|x-\zeta\|_0 + \|y-\chi\|_0 \leqslant \frac{\Omega_\alpha \epsilon_\alpha}{\Lambda_0(1-\Lambda_1^*)} + \frac{\Omega_\beta \epsilon_\beta}{\Lambda_0(1-\Lambda_3^*)} + \frac{\Lambda_4^*\Omega_\alpha \epsilon_\alpha}{\Lambda_0(1-\Lambda_1^*)(1-\Lambda_3^*)} + \frac{\Lambda_2^*\Omega_\beta \epsilon_\beta}{\Lambda_0(1-\Lambda_1^*)(1-\Lambda_3^*)}. \quad (49)$$

Let $\max\{\epsilon_\alpha, \epsilon_\beta\} = \epsilon$; then, from (49), we get

$$\|(x-\zeta, y-\chi)\| \leqslant \mathbf{K}_{\alpha,\beta}\,\epsilon,$$

which implies that

$$\|(x,y)-(\zeta,\chi)\| \leqslant \mathbf{K}_{\alpha,\beta}\,\epsilon, \quad (50)$$

where

$$\mathbf{K}_{\alpha,\beta} = \left[\frac{\Omega_\alpha}{\Lambda_0(1-\Lambda_1^*)} + \frac{\Omega_\beta}{\Lambda_0(1-\Lambda_3^*)} + \frac{\Lambda_4^*\Omega_\alpha}{\Lambda_0(1-\Lambda_1^*)(1-\Lambda_3^*)} + \frac{\Lambda_2^*\Omega_\beta}{\Lambda_0(1-\Lambda_1^*)(1-\Lambda_3^*)}\right].$$

Hence, problem (1) is Ulam–Hyers stable. Moreover, if we set $\Theta(\epsilon) = \mathbf{K}_{x,y}\,\epsilon$; $\Theta(0) = 0$ in (50), then problem (1) is generalized Ulam–Hyers stable. □

($\mathbf{M}_5$): Let $\Psi_\alpha, \Psi_\beta \in PC(J, \mathbb{R}^+)$ be nondecreasing functions; then, for $t \in J$, there are $\lambda_\alpha, \lambda_\beta > 0$ such that

$$\mathbf{I}^\alpha\Psi_\alpha(t) \leqslant \lambda_\alpha\Psi_\alpha(t), \;\; \mathbf{I}^{\alpha-1}\Psi_\alpha(t) \leqslant \lambda_\alpha\Psi_\alpha(t),$$
$$\mathbf{I}^{\alpha-2}\Psi_\alpha(t) \leqslant \lambda_\alpha\Psi_\alpha(t), \;\; \mathbf{I}^{\alpha-\varepsilon}\Psi_\alpha(t) \leqslant \lambda_\alpha\Psi_\alpha(t).$$

Similarly,

$$\mathbf{I}^\beta\Psi_\beta(t) \leqslant \lambda_\beta\Psi_\beta(t), \;\; \mathbf{I}^{\beta-1}\Psi_\beta(t) \leqslant \lambda_\beta\Psi_\beta(t),$$
$$\mathbf{I}^{\beta-2}\Psi_\beta(t) \leqslant \lambda_\beta\Psi_\beta(t), \;\; \mathbf{I}^{\beta-\rho}\Psi_\beta(t) \leqslant \lambda_\beta\Psi_\beta(t).$$

Theorem 5. *Assume that* $(\mathbf{M}_1) - (\mathbf{M}_2)$ *and* $(\mathbf{M}_5)$ *are satisfied; then, by Definition 6 and Definition 7, Problem (1) is Ulam–Hyers–Rassias stable with respect to* $(\Psi_{\alpha,\beta}, \varphi_{\alpha,\beta})$*, as well as generalized Ulam–Hyers–Rassias stable.*

5. Example

To substantiate the aforemention demonstrated theory, we supply the following problem:

$$\begin{cases} {}^{c}\mathbf{D}^{\alpha}x(t)+h(t,{}^{c}\mathbf{D}^{a}x(t),{}^{c}\mathbf{D}^{b}y(t))=0, t\neq t_m, \\ {}^{c}\mathbf{D}^{\beta}y(t)+w(t,{}^{c}\mathbf{D}^{a}x(t),{}^{c}\mathbf{D}^{b}y(t))=0, \\ \Delta x\mid_{t=t_m}=M_{1m}x(t_m),\ \Delta x'\mid_{t=t_m}=N_{1m}x(t_m),\ \Delta x''\mid_{t=t_m}=O_{1m}x(t_m), \\ \Delta y\mid_{t=t_m}=M_{2m}y(t_m),\ \Delta y'\mid_{t=t_m}=N_{2m}y(t_m),\ \Delta y''\mid_{t=t_m}=O_{2m}y(t_m), \\ x(t)='x(t)=0,\ {}^{c}\mathbf{D}^{\varepsilon}x(\Omega)=x''(1),\ y(0)=y'(0)=0,\ {}^{c}\mathbf{D}^{\rho}y(\Phi)=y''(1), \\ 0<\varepsilon,\Omega<1,\ 0<\rho,\Phi<1. \end{cases} \tag{51}$$

Take $J=[0,1]$, $\alpha=\frac{5}{2}$, $\beta=\sqrt{5}$, $\varepsilon=\frac{1}{2}$, $\Omega=\frac{1}{3}$, $a=\frac{1}{2}$, $b=\frac{1}{8}$, $t_1=\frac{1}{2}$, $\rho=\frac{1}{3}$, $\Phi=\frac{1}{5}$, $h(t,x,y)=e^{t}+\frac{|x|+|y|}{40(1+|x|+|y|)}$, $w(t,x,y)=\frac{\cos t+|x|+|y|}{40(1+|x|+|y|)}$, $M_{11}(x)=M_{21}(x)=\frac{|x|}{5+|x|}$, $N_{11}(x)=N_{21}(x)=\frac{|x|}{10+|x|}$, $O_{11}(x)=O_{21}(x)=\frac{|x|}{15+|x|}$. By direct computation, we have $\mu_1=\mu_2=\mu_1'=\mu_2'=\frac{1}{40}$, $I_{11}=I_{21}=\frac{1}{5}$, $\hat{I}_{11}=\hat{I}_{21}=\frac{1}{10}$, $\check{I}_{11}=\check{I}_{21}=\frac{1}{15}$,

$$\begin{aligned}&\Bigg[(\mu_1+\mu_2)\bigg(\frac{2}{\Gamma(\alpha+1)}+\frac{1}{\Gamma(\alpha)}+\frac{1}{2\Gamma(\alpha-1)}+\frac{Z_\alpha}{\Gamma(\alpha-\varepsilon+1)}+\frac{Z_\alpha\Omega^{1-\varepsilon}}{\Gamma(\alpha)\Gamma(2-\varepsilon)}+\frac{Z_\alpha\Omega^{2-\varepsilon}}{\Gamma(\alpha-1)\Gamma(3-\varepsilon)}\\&+\frac{Z_\alpha\Omega^{1-\varepsilon}}{\Gamma(\alpha-1)\Gamma(2-\varepsilon)}+\frac{2Z_\alpha}{\Gamma(\alpha-1)}\bigg)+\sum_{j=1}^{m}I_{1j}+\sum_{j=1}^{m}\hat{I}_{1j}+\frac{Z_\alpha\Omega^{1-\varepsilon}}{\Gamma(2-\varepsilon)}\sum_{j=1}^{m}\hat{I}_{1j}+\frac{Z_\alpha\Omega^{2-\varepsilon}}{\Gamma(3-\varepsilon)}\sum_{j=1}^{m}\check{I}_{1j}\\&+\frac{Z_\alpha\Omega^{1-\varepsilon}}{\Gamma(2-\varepsilon)}\sum_{j=1}^{m}\check{I}_{1j}+Z_\alpha\sum_{j=1}^{m}\check{I}_{1j}+\frac{1}{2}\sum_{j=1}^{m}\check{I}_{1j}\Bigg]\approx 0.680277<1.\end{aligned}$$

Similarly,

$$\begin{aligned}&\Bigg[(\mu_1'+\mu_2')\bigg(\frac{2}{\Gamma(\beta+1)}+\frac{1}{\Gamma(\beta)}+\frac{1}{2\Gamma(\beta-1)}+\frac{Z_\beta}{\Gamma(\beta-\rho+1)}+\frac{Z_\beta\Phi^{1-\rho}}{\Gamma(\beta)\Gamma(2-\rho)}+\frac{Z_\beta\Phi^{2-\rho}}{\Gamma(\beta-1)\Gamma(3-\rho)}\\&+\frac{Z_\beta\Phi^{1-\rho}}{\Gamma(\beta-1)\Gamma(2-\rho)}+\frac{2Z_\beta}{\Gamma(\beta-1)}\bigg)+\sum_{j=1}^{m}I_{2j}+\sum_{j=1}^{m}\hat{I}_{2j}+\frac{Z_\beta\Phi^{1-\rho}}{\Gamma(2-\rho)}\sum_{j=1}^{m}\hat{I}_{2j}+\frac{Z_\beta\Phi^{2-\rho}}{\Gamma(3-\rho)}\sum_{j=1}^{m}\check{I}_{2j}\\&+\frac{Z_\beta\Phi^{1-\rho}}{\Gamma(2-\rho)}\sum_{j=1}^{m}\check{I}_{2j}+Z_\beta\sum_{j=1}^{m}\check{I}_{2j}+\frac{1}{2}\sum_{j=1}^{m}\check{I}_{2j}\Bigg]\approx 0.427420<1.\end{aligned}$$

Thus, by Theorem 2, Problem (51) has a unique solution. Furthermore, all of the assumptions are satisfied, so the Problem (51) is Ulam–Hyers, generalized Ulam–Hyers, Ulam–Hyers–Rassias and generalized Ulam–Hyers–Rassias stable.

6. Conclusions

In the above study, we have successfully built up existence theory for the solutions of system (1). The required analysis has been developed with the help of the Banach contraction principle and Schauder fixed point theorem. We found that the fractional order coupled system is additionally complicated and challenging as compared to the single FDEs. We also concluded that, if we increase the order or boundary conditions, then the end result turns into extra accurate. Our results are new and fascinating. Our methods can be used to study the existence of solutions for the high order or multiple-point boundary value systems of a nonlinear coupled system of FDEs. Furthermore, we have presented different kinds of Ulam–Hyers stability results for the solution of the considered system (1). In addition, we have presented our main theoretical results with the help of an example. In the future, this concept can be extended to more applied and complicated problems of applied nature. The obtained results can be used in fields like numerical analysis and managerial sciences including business mathematics and economics, etc.

Author Contributions: S.F. and Z.A. contributed equally in writing this article; supervision, A.Z., J.X. and Y.C. All authors read and approved the final manuscript.

References

1. Brikaa, M. Existence results for a couple system of nonlinear fractional differential equation with three point boundary conditions. *J. Fract. Calc. Appl.* **2015**, *3*, 1–10.
2. Henderson, J.; Luca, R. Positive solutions for a system of fractional differential equations with coupled integral boundary conditions. *Appl. Math. Comput.* **2014**, *249*, 182–197. [CrossRef]
3. Kilbas, A.; Srivastava, H.; Trujillo, J. Theory and Applications of Fractional Differential Equations. In *North–Holland Mathematics Studies*; Elsevier: Amsterdam, The Netherlands, 2006; Volume 204.
4. Klafter, J.; Lim, S.C. *Fractional Dynamics in Physics*; Metzler, R., Ed.; World Scientific: Singapore, 2011.
5. Podlubny, I. *Fractional Differential Equations*; Academic Press: San Diego, CA, USA, 1999.
6. Yang, W. Positive solutions for a coupled system of nonlinear fractional differential equations with integral boundary conditions. *Comput. Math. Appl.* **2012**, *63*, 288–297. [CrossRef]
7. Zhang, K.; Fu, Z. Solutions for a class of Hadamard fractional boundary value problems with sign-changing nonlinearity. *J. Funct. Spaces* **2019**, *2019*, 9046472. [CrossRef]
8. Zhang, K.; Wang, J.; Ma, W. Solutions for integral boundary value problems of nonlinear Hadamard fractional differential equations. *J. Funct. Spaces* **2018**, *2018*, 2193234. [CrossRef]
9. Zou, Y.; He, G. On the uniqueness of solutions for a class of fractional differential equations. *Appl. Math. Lett.* **2017**, *74*, 68–73. [CrossRef]
10. Yue, Z.; Zou, Y. New uniqueness results for fractional differential equation with dependence on the first order derivative. *Adv. Differ. Equ.* **2019**, *2019*, 38. [CrossRef]
11. Fu, Z.; Bai, S.; O'Regan, D.; Xu, J. Nontrivial solutions for an integral boundary value problem involving Riemann–Liouville fractional derivatives. *J. Inequal. Appl.* **2019**, *2019*, 104. [CrossRef]
12. Zhang, K.; O'Regan, D.; Xu, J.; Fu, Z. Nontrivial solutions for a higher order nonlinear fractional boundary value problem involving Riemann–Liouville fractional derivatives. *J. Funct. Spaces* **2019**, *2019*, 2381530. [CrossRef]
13. Pu, R.; Zhang, X.; Cui, Y.; Li, P.; Wang, W. Positive solutions for singular semipositone fractional differential equation subject to multipoint boundary conditions. *J. Funct. Spaces* **2017**, *2017*, 5892616. [CrossRef]
14. Zhang, Y. Existence results for a coupled system of nonlinear fractional multi-point boundary value problems at resonance. *J. Inequal. Appl.* **2018**, *2018*, 198. [CrossRef] [PubMed]
15. Zhang, Y.; Bai, Z.; Feng, T. Existence results for a coupled system of nonlinear fractional three-point boundary value problems at resonance. *Comput. Math. Appl.* **2011**, *61*, 1032–1047. [CrossRef]
16. Qi, T.; Liu, Y.; Zou, Y. Existence result for a class of coupled fractional differential systems with integral boundary value conditions. *J. Nonlinear Sci. Appl.* **2017**, *10*, 4034–4045. [CrossRef]
17. Qi, T.; Liu, Y.; Cui, Y. Existence of solutions for a class of coupled fractional differential systems with nonlocal boundary conditions. *J. Funct. Spaces* **2017**, *2017*, 6703860. [CrossRef]
18. Xu, J.; Goodrich, C.S.; Cui, Y. Positive solutions for a system of first-order discrete fractional boundary value problems with semipositone nonlinearities. *Revista de la Real Academia de Ciencias Exactas Físicas y Naturales Serie A Matemáticas* **2019**, *113*, 1343–1358. [CrossRef]
19. Cheng, W.; Xu, J.; Cui, Y.; Ge, Q. Positive solutions for a class of fractional difference systems with coupled boundary conditions. *Adv. Differ. Equ.* **2019**, *2019*, 249. [CrossRef]
20. Cheng, W.; Xu, J.; Cui, Y. Positive solutions for a system of nonlinear semipositone fractional q-difference equations with q-integral boundary conditions. *J. Nonlinear Sci. Appl.* **2017**, *10*, 4430–4440. [CrossRef]
21. Wang, F.; Cui, Y. Positive solutions for an infinite system of fractional order boundary value problems. *Adv. Differ. Equ.* **2019**, *2019*, 169. [CrossRef]
22. Qiu, X.; Xu, J.; O'Regan, D.; Cui, Y. Positive solutions for a system of nonlinear semipositone boundary value problems with Riemann–Liouville fractional derivatives. *J. Funct. Spaces* **2018**, *2018*, 7351653. [CrossRef]

23. Hao, X.; Wang, H.; Liu, L.; Cui, Y. Positive solutions for a system of nonlinear fractional nonlocal boundary value problems with parameters and p-Laplacian operator. *Bound. Value Probl.* **2017**, *2017*, 182. [CrossRef]
24. Zhang, X.; Liu, L.; Zou, Y. Fixed-point theorems for systems of operator equations and their applications to the fractional differential equations. *J. Funct. Spaces* **2018**, *2018*, 7469868. [CrossRef]
25. Zhang, X.; Liu, L.; Wu, Y.; Zou, Y. Existence and uniqueness of solutions for systems of fractional differential equations with Riemann-Stieltjes integral boundary condition. *Adv. Differ. Equ.* **2018**, *2018*, 204. [CrossRef]
26. Li, H.; Zhang, J. Positive solutions for a system of fractional differential equations with two parameters. *J. Funct. Spaces* **2018**, *2018*, 1462505. [CrossRef]
27. Zhao, Y.; Hou, X.; Sun, Y.; Bai, Z. Solvability for some class of multi-order nonlinear fractional systems. *Adv. Differ. Equ.* **2019**, *2019*, 23. [CrossRef]
28. Zhao, Y.; Sun, Y.; Wang, Y.; Bai, Z. Asymptotical stabilization of the nonlinear upper triangular fractional-order systems. *Adv. Differ. Equ.* **2019**, *2019*, 157. [CrossRef]
29. Cui, Y. Multiplicity results for positive solutions to differential systems of singular coupled integral boundary value problems. *Math. Probl. Eng.* **2017**, *2017*, 3608352. [CrossRef]
30. Jiang, J.; O'Regan, D.; Xu, J.; Fu, Z. Positive solutions for a system of nonlinear Hadamard fractional differential equations involving coupled integral boundary conditions. *J. Inequal. Appl.* **2019**, *2019*, 204. [CrossRef]
31. Zhai, C.; Wang, W.; Li, H. A uniqueness method to a new Hadamard fractional differential system with four-point boundary conditions. *J. Inequal. Appl.* **2018**, *2018*, 207. [CrossRef]
32. Wang, J.; Fečkan, M.; Zhou, Y. A survey on impulsive fractional differential equations. *Fract. Calc. Appl. Anal.* **2016**, *19*, 806–831. [CrossRef]
33. Yukunthorn, W.; Ahmad, B.; Ntouyas, S.; Tariboon, J. On Caputo-Hadamard type fractional impulsive hybrid systems with nonlinear fractional integral conditions. *Nonlinear Anal. Hybrid Syst.* **2016**, *19*, 77–92. [CrossRef]
34. Zhao, K.; Liang, J. Solvability of triple-point integral boundary value problems for a class of impulsive fractional differential equations. *Adv. Differ. Equ.* **2017**, *2017*, 50. [CrossRef]
35. Fu, X.; Bao, X. Some existence results for nonlinear fractional differential equations with impulsive and fractional integral boundary conditions. *Adv. Differ. Equ.* **2014**, *2014*, 129. [CrossRef]
36. Zhao, K. Multiple positive solutions of integral BVPs for high-order nonlinear fractional differential equations with impulses and distributed delays. *Dyn. Syst.* **2015**, *30*, 208–223. [CrossRef]
37. Zhao, K. Impulsive boundary value problems for two classes of fractional differential equation with two different Caputo fractional derivatives. *Mediterr. J. Math.* **2016**, *13*, 1033–1050. [CrossRef]
38. Wang, Y.; Liu, Y.; Cui, Y. Infinitely many solutions for impulsive fractional boundary value problem with p-Laplacian. *Bound. Value Probl.* **2018**, *2018*, 94. [CrossRef]
39. Zuo, M.; Hao, X.; Liu, L.; Cui, Y. Existence results for impulsive fractional integro-differential equation of mixed type with constant coefficient and antiperiodic boundary conditions. *Bound. Value Probl.* **2017**, *2017*, 161. [CrossRef]
40. Bai, Z.; Dong, X.; Yin, C. Existence results for impulsive nonlinear fractional differential equation with mixed boundary conditions. *Bound. Value Probl.* **2016**, *2016*, 63. [CrossRef]
41. Ulam, S.M. *Problems in Modern Mathematics*; John Wiley and Sons: New York, NY, USA, 1940.
42. Ulam, S.M. *A Collection of Mathematical Problems*; Interscience: New York, NY, USA, 1960.
43. Rassias, T.M. On the stability of the linear mapping in Banach spaces. *Proc. Am. Math. Soc.* **1978**, *72*, 297–300. [CrossRef]
44. Ali, Z.; Zada, A.; Shah, K. Ulam stability to a toppled systems of nonlinear implicit fractional order boundary value problem. *Bound. Value Probl.* **2018**, *2018*, 175. [CrossRef]
45. Ali, Z.; Zada, A.; Shah, K. On Ulam's stability for a coupled systems of nonlinear implicit fractional differential equations. *Bull. Malays. Math. Sci. Soc.* **2019**, *42*, 2681–2699. [CrossRef]
46. Ahmad, N.; Ali, Z.; Shah, K.; Zada, A.; Rahman, G. Analysis of implicit type nonlinear dynamical problem of impulsive fractional differential equations. *Complexity* **2018**, *2018*, 6423974. [CrossRef]
47. Zada, A.; Shah, S.O.; Shah, R. Hyers–Ulam stability of non-autonomous systems in terms of boundedness of Cauchy problem. *Appl. Math. Comput.* **2015**, *271*, 512–518. [CrossRef]
48. Shah, S.O.; Zada, A. Existence, uniqueness and stability of solution to mixed integral dynamic systems

with instantaneous and noninstantaneous impulses on time scales. *Appl. Math. Comput.* **2019**, *359*, 202–213. [CrossRef]
49. Zada, A.; Mashal, A. Stability analysis of nth order nonlinear impulsive differential equations in Quasi-Banach space. *Numer. Funct. Anal. Optim.* **2019**. [CrossRef]
50. Ali, A.; Rabiei, F.; Shah, K. On Ulam's type stability for a class of impulsive fractional differential equations with nonlinear integral boundary conditions. *J. Nonlinear Sci. Appl.* **2017**, *10*, 4760–4775. [CrossRef]
51. Riaz, U.; Zada, A.; Ali, Z.; Ahmad, M.; Xu, J.; Fu, Z. Analysis of nonlinear coupled systems of impulsive fractional differential equations with Hadamard derivatives. *Math. Probl. Eng.* **2019**, *2019*, 5093572. [CrossRef]
52. Riaz, U.; Zada, A.; Ali, Z.; Cui, Y.; Xu, J. Analysis of coupled systems of implicit impulsive fractional differential equations involving Hadamard derivatives. *Adv. Differ. Equ.* **2019**, *2019*, 226. [CrossRef]
53. Ali, Z.; Kumam, P.; Shah, K.; Zada, A. Investigation of Ulam stability results of a coupled system of nonlinear implicit fractional differential equations. *Mathematics* **2019**, *7*, 341. [CrossRef]
54. Wang, J.; Zada, A.; Waheed, H. Stability analysis of a coupled system of nonlinear implicit fractional anti-periodic boundary value problem. *Math. Meth. App. Sci.* **2019**. [CrossRef]
55. Shah, K.; Shah, L.; Ahmad, S.; Rassias, J.M.; Li, Y. Monotone iterative techniques together with Hyers-Ulam-Rassias stability. *Math. Meth. Appl. Sci.* **2019**, 1–18. [CrossRef]
56. Zada, A.; Ali, W.; Park, C. Ulam's type stability of higher order nonlinear delay differential equations via integral inequality of Grönwall–Bellman–Bihari's type. *Appl. Math. Comput.* **2019**, *350*, 60–65.
57. Jung, S.M. *Hyers–Ulam–Rassias Stability of Functional Equations in Nonlinear Analysis*; Springer: New York, NY, USA, 2011.
58. Kumam, P.; Ali, A.; Shah, K.; Khan, R.A. Existence results and Hyers–Ulam stability to a class of nonlinear arbitrary order differential equations. *J. Nonlinear Sci. Appl.* **2017**, *10*, 2986–2997. [CrossRef]
59. Guo, D.; Lakshmikantham, V. *Nonlinear Problems in Abstract Cone*; Academic Press: Orlando, FL, USA, 1988.
60. Miller, K.; Ross, B. *An Introduction to the Fractional Calculus and Fractional Differential Equations*; Wiley: New York, NY, USA, 1993.
61. Rus, I.A. Ulam stabilities of ordinary differential equations in a Banachspace. *Carpathian J. Math.* **2010**, *26*, 103–107.

10

Existence and Iterative Method for some Riemann Fractional Nonlinear Boundary Value Problems

Imed Bachar [1,*], Habib Mâagli [2,3] and Hassan Eltayeb [1]

1 Mathematics Department, College of Science, King Saud University, P.O. Box 2455, Riyadh 11451, Saudi Arabia; hgadain@ksu.edu.sa
2 Department of Mathematics, College of Sciences and Arts, King Abdulaziz University, Rabigh Campus, P.O. Box 344, Rabigh 21911, Saudi Arabia; abobaker@kau.edu.sa or habib.maagli@fst.rnu.tn
3 Analyse Harmonique et Théorie du Potentiel, LR10ES09 Modélisation Mathématique, Faculté des Sciences de Tunis, Université de Tunis El Manar, Tunis 2092, Tunisie
* Correspondence: abachar@ksu.edu.sa;

Abstract: In this paper, we prove the existence and uniqueness of solution for some Riemann–Liouville fractional nonlinear boundary value problems. The positivity of the solution and the monotony of iterations are also considered. Some examples are presented to illustrate the main results. Our results generalize those obtained by Wei et al (Existence and iterative method for some fourth order nonlinear boundary value problems. Appl. Math. Lett. 2019, 87, 101–107.) to the fractional setting.

Keywords: fractional differential equation; Green's function; existence and uniqueness of solution; positivity of solution; iterative method

---

1. Introduction

Forth-order boundary value problems, can be used to model the deformation of the elastic beam, which is considered to be one of the most used elements in structures such as bridges, buildings and aircraft (see, for instance, [1,2]).

In the literature problems of the form

$$u^{(4)}(x) = f(x, u(x), u''(x)),\ x \in (0,1), \tag{1}$$

subject to different types of boundary conditions have been extensively studied (see, for example, [1–11] and the references therein).

Under adequate conditions imposed on f and using different approach, the existence, uniqueness and qualitative properties of solutions have been considered.

In [1], Aftabizadeh considered Equation (1) together with the boundary conditions:

$$u(0) = u(1) = u''(0) = u''(1) = 0, \tag{2}$$

where $f : [0,1] \times \mathbb{R}^2 \to \mathbb{R}$ is continuous. Under adequate conditions imposed on f he proved that problem (1)–(2) has a unique solution. To do this, he transforms Equation (1) into a second-order integro-differential equation and apply the Schauder's fixed point theorem.

In [4], by using the method of lower and upper solutions for a fourth-order equation and some restrictive conditions on f, Bai established an existence result to problem (1)–(2).

In [7], Dang et al., to prove the existence and uniqueness of a solution of the problem (1)–(2), they reduced the problem to an operator equation for the right-hand side function and proved the contraction of the operator under some convenient conditions on f. The positivity of the solution and the monotony of iterations are also considered. This idea was also used by Dang and Qey for cantilever beam equation [12].

Recently, in [11], Wei et al. considered the following problem

$$\begin{cases} u^{(4)}(x) = f(x, u(x), u'(x)), & t \in (0,1), \\ u(0) = u'(0) = u'(1) = u''(1) = 0. \end{cases} \tag{3}$$

Observe that problem (3) cannot be reduced to two second-order problems. Nevertheless, following the idea developed in [7], they proved the existence and uniqueness of this problem.

Motivated by the mentioned works, in this paper, we generalize the results obtained in [11] to the fractional setting.

More precisely, we are concerned with the following problem

$$\begin{cases} D^{\beta}(D^{\alpha}u)(x) = f(x, u(x), D^{\alpha}u(x)), \quad x \in (0,1), \\ u(0) = D^{\alpha}u(0) = D^{\alpha}u(1) = (D^{\alpha}u)'(1) = 0, \end{cases} \tag{4}$$

where $0 < \alpha \leq 1$, $2 < \beta \leq 3$, and $f : [0,1] \times \mathbb{R}^2 \to \mathbb{R}$ continuous function satisfying some adequate assumptions. Here D^{α} (resp. D^{β}) denotes the Riemann–Liouville fractional derivative of order α (resp. β).

It is worth mentioning that many authors studied fractional differential equations which were applied in many fields such as physics, mechanics, chemistry, and engineering; (see, for instance [13–32] and the references therein).

Following a different approach, they addressed the question of existence and uniqueness of positive continuous solution.

In [31], the authors considered the two-dimensional fractional Schrödinger equation (FSE) without potential

$$i\frac{\partial \psi}{\partial z} - \left(-\frac{\partial^2}{\partial x^2} - \frac{\partial^2}{\partial y^2}\right)^{\frac{\alpha}{2}} \psi = 0, \tag{5}$$

for the slowly varying envelope ψ of the optical field and $1 < \alpha \leq 2$.

They transformed Equation (5) into a Dirac–Weyl-like equation, which is used to establish a link with light propagation in the honeycomb lattice (HCL). They discovered a very similar behavior—the conical diffraction. This similarity in behavior is broken if an additional potential is brought into system.

Our paper is organized as follows. In Section 2, we establish some estimates on the Green's function and we prove appropriate inequalities on some integral operators involving the Green' function. In Section 3, under adequate conditions imposed on function f, we prove the existence and uniqueness of a solution of problem (4). Our approach is based on the Banach contraction principle. The positivity of the solution and the monotony of iterations are also considered. Some examples are given to illustrate our existence results.

Throughout this paper, we denote by $C([0,1])$ the set of continuous functions in $[0,1]$. We recall that the space $C([0,1])$ equipped with the uniform norm $\|u\| := \max_{x \in [0,1]} |u(x)|$ is a Banach space.

2. Preliminary Results

2.1. Fractional Calculus

We recall in this section some basic definitions on fractional calculus (see [33–36]).

Definition 1. *The Riemann–Liouville fractional integral of order $\gamma > 0$ for a measurable function $f : (0, \infty) \to \mathbb{R}$ is defined as*

$$I^{\gamma} f(x) = \frac{1}{\Gamma(\gamma)} \int_0^x (x - t)^{\gamma - 1} f(t)\, dt, \quad x > 0,$$

provided that the right-hand side is pointwise defined on $(0,\infty)$. Here Γ is the Euler Gamma function.

Definition 2. *The Riemann–Liouville fractional derivative of order $\gamma > 0$ for a measurable function $f : (0,\infty) \to \mathbb{R}$ is defined as*

$$D^{\gamma} f(x) = \frac{1}{\Gamma(n-\gamma)} \left(\frac{d}{dx}\right)^{n} \int_{0}^{x} (x-t)^{n-\gamma-1} f(t)\,dt = \left(\frac{d}{dx}\right)^{n} I^{n-\gamma} f(x),$$

provided that the right-hand side is pointwise defined on $(0,\infty)$. Here $n = [\gamma]+1$, where $[\gamma]$ denotes the integer part of γ.

Please note that if $\gamma = m \in \mathbb{N}\backslash\{0\}$, then we obtain the classical derivative of order m.

Lemma 1. *Let $\gamma > 0$ and $u \in C(0,1) \cap L^1(0,1)$. Then we have*

(i) *For $0 < \gamma < \delta$, $D^{\gamma} I^{\delta} u = I^{\delta-\gamma} u$ and $D^{\gamma} I^{\gamma} u = u$.*

(ii) *$D^{\gamma} u(x) = 0$ if and only if $u(x) = c_1 x^{\gamma-1} + c_2 x^{\gamma-2} + \ldots + c_n x^{\gamma-m}$,*

where m is the smallest integer greather than or equal to γ and $c_i \in \mathbb{R}$ $(i = 1,\ldots,m)$ are arbitrary constants.

(iii) *Assume that $D^{\gamma} u \in C(0,1) \cap L^1(0,1)$, then*

$$I^{\gamma} D^{\gamma} u(x) = u(x) + c_1 x^{\gamma-1} + c_2 x^{\gamma-2} + \ldots + c_m x^{\gamma-m},$$

where m is the smallest integer greather than or equal to γ and $c_i \in \mathbb{R}$ $(i = 1,\ldots,m)$ are arbitrary constants.

Proof. For the convenience of the reader, we provide the proof of property (ii) which plays an important role in the rest of the paper.

The property is clear if $\gamma = m \in \mathbb{N}\backslash\{0\}$. Next we assume that $m-1 < \gamma < m$.

We claim that for $i = 1,2,\ldots,m$,

$$D^{\gamma}(t^{\gamma-i})(x) = 0.$$

Indeed, by elementary calculus, we have

$$I^{m-\gamma}(t^{\gamma-i})(x) = \frac{1}{\Gamma(m-\gamma)} \int_{0}^{x} (x-t)^{m-\gamma-1} t^{\gamma-i} dt = \frac{\Gamma(\gamma+1-i)}{\Gamma(m-i+1)} x^{m-i}.$$

Hence

$$D^{\gamma}(t^{\gamma-i})(x) = \left(\frac{d}{dx}\right)^{m} \left(I^{m-\gamma}(t^{\gamma-i})\right)(x) = 0.$$

Therefore, if $u(x) = \sum_{i=1}^{m} c_i x^{\gamma-i}$, then $D^{\gamma} u(x) = 0$.

Conversely, assume that $D^{\gamma} u(x) = 0$.

From Definition 2, we obtain

$$I^{m-\gamma} u(x) = a_0 + a_1 x + \ldots + a_{m-1} x^{m-1},$$

where $a_i \in \mathbb{R}(i = 0,1,\ldots,m-1)$ are arbitrary constants.

Using property (i), we deduce that

$$\begin{aligned} u(x) &= D^{m-\gamma}(I^{m-\gamma}u)(x) \\ &= \sum_{i=0}^{m-1} a_i D^{m-\gamma}(t^i)(x) \\ &= \sum_{i=0}^{m-1} a_i \frac{\Gamma(1+i)}{\Gamma(1+i-m+\gamma)} x^{i-m+\gamma} \\ &= \sum_{i=1}^{m} c_i x^{\gamma-i}, \end{aligned}$$

where $c_i \in \mathbb{R}(i = 1, ..., m)$ are arbitrary constants. □

2.2. Estimates on the Green's Function

Lemma 2. *Let $2 < \beta \leq 3$ and $\varphi \in C([0,1])$, then the boundary-value problem,*

$$\begin{cases} D^{\beta} v(x) = \varphi(x) \text{ in } (0,1), \\ v(0) = v(1) = v'(1) = 0, \end{cases} \tag{6}$$

has a unique solution

$$v(x) = \int_0^1 G_{\beta}(x,t)\, \varphi(t)\, dt, \tag{7}$$

where for $x, t \in [0,1]$,

$$\begin{aligned} G_{\beta}(x,t) &= \frac{1}{\Gamma(\beta)} \begin{cases} G(x,t), & \text{for } 0 \leq x \leq t \leq 1, \\ G(x,t) + (x-t)^{\beta-1}, & \text{for } 0 \leq t \leq x \leq 1, \end{cases} \\ &= \frac{1}{\Gamma(\beta)} (G(x,t) + (\max(x-t,0))^{\beta-1}), \end{aligned} \tag{8}$$

with

$$G(x,t) \ := x^{\beta-2}(1-t)^{\beta-2}\left[(\beta-1)(t-x) + (\beta-2)\, x\, (1-t)\right] \tag{9}$$

$$= (\beta-1)t(1-x)x^{\beta-2}(1-t)^{\beta-2} - x^{\beta-1}(1-t)^{\beta-1}. \tag{10}$$

$G_{\beta}(x,t)$ is called Green's function of boundary-value problem (6).

Proof. By means of Lemma 1, we can reduce equation $D^{\beta} v(x) = \varphi(x)$ to an equivalent integral equation

$$v(x) = c_1 x^{\beta-1} + c_2 x^{\beta-2} + c_3 x^{\beta-3} + I^{\beta}\varphi(x), \tag{11}$$

where $(c_1, c_2, c_3) \in \mathbb{R}^3$.

The boundary condition $v(0) = 0$ implies that $c_3 = 0$, while the condition $v(1) = 0$, gives

$$c_1 + c_2 + I^{\beta}\varphi(1) = 0. \tag{12}$$

On the other hand, since $v'(1) = 0$, we obtain

$$(\beta-1)c_1 + (\beta-2)c_2 + I^{\beta-1}\varphi(1) = 0.$$

Hence

$$c_1 = (\beta-2)\, I^{\beta}\varphi(1) - I^{\beta-1}\varphi(1) \text{ and } c_2 = I^{\beta-1}\varphi(1) - (\beta-1)\, I^{\beta}\varphi(1).$$

Therefore the unique solution of problem (6) is

$$
\begin{aligned}
v(x) &= \frac{(\beta-2)}{\Gamma(\beta)}\int_0^1 x^{\beta-1}(1-t)^{\beta-1}\varphi(t)\,dt - \frac{1}{\Gamma(\beta-1)}\int_0^1 x^{\beta-1}(1-t)^{\beta-2}\varphi(t)\,dt \\
&+(\frac{1}{\Gamma(\beta-1)}\int_0^1 x^{\beta-2}(1-t)^{\beta-2}\varphi(t)\,dt - \frac{(\beta-1)}{\Gamma(\beta)}\int_0^1 x^{\beta-2}(1-t)^{\beta-1}\varphi(t)\,dt \\
&+\frac{1}{\Gamma(\beta)}\int_0^x (x-t)^{\beta-1}\varphi(t)\,dt \\
&= \frac{1}{\Gamma(\beta)}\int_0^1 x^{\beta-2}(1-t)^{\beta-2}((\beta-1)(t-x)+(\beta-2)x(1-t))\varphi(t)\,dt \\
&+\frac{1}{\Gamma(\beta)}\int_0^x (x-t)^{\beta-1}\varphi(t)\,dt \\
&= \int_0^1 G_\beta(x,t)\,\varphi(t)\,dt.
\end{aligned}
$$

The proof is completed. □

In the following, for some values of β we give the representation of the Green function $G_\beta(x,t)$ with the contours and the projections on some coordinate planes (see Figures 1–3). These details give an immediate idea of the behavior of these functions.

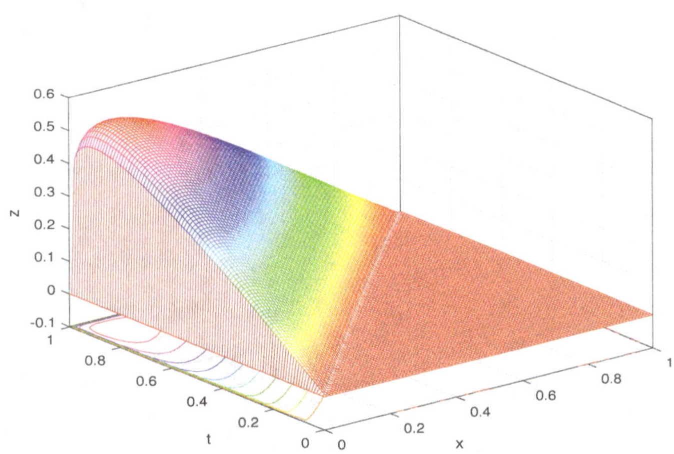

(**a**) $G_\beta(x,t)$ and contours.

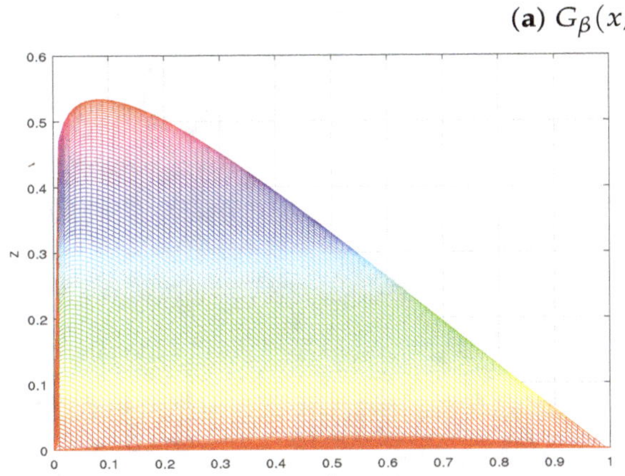

(**b**) Projection of graph of $G_\beta(x,t)$ on the plane xz.

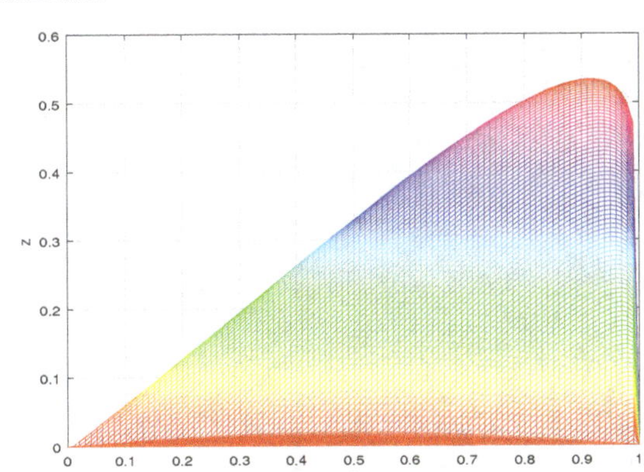

(**c**) Projection of graph of $G_\beta(x,t)$ on the plane tz.

Figure 1. The Green function for $\beta = 2.1$.

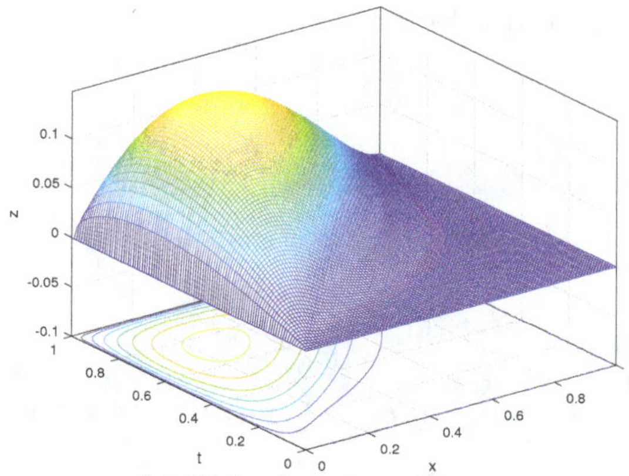

(**a**) $G_\beta(x,t)$ and contours.

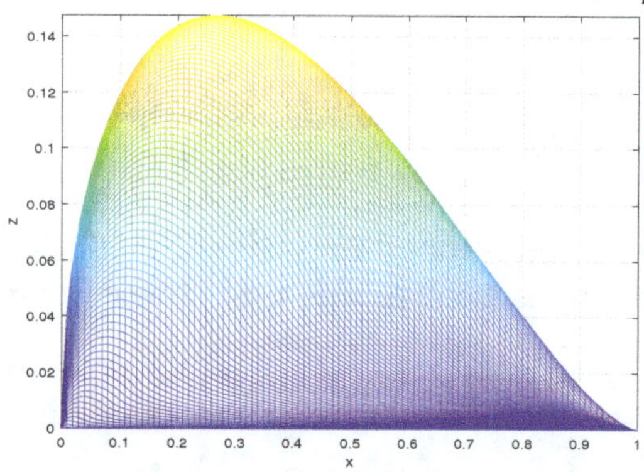

(**b**) Projection of graph of $G_\beta(x,t)$ on the plane xz.

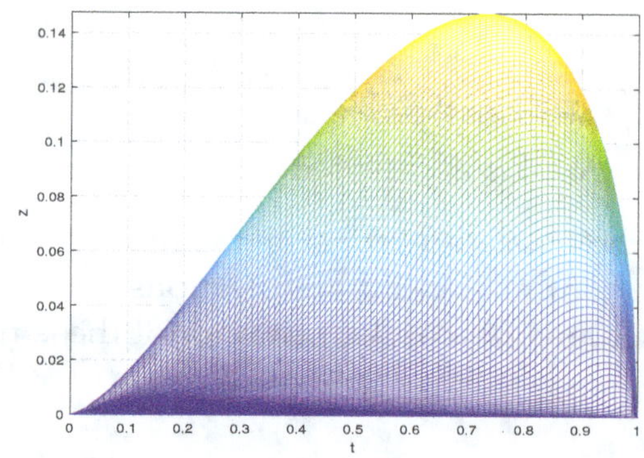

(**c**) Projection of graph of $G_\beta(x,t)$ on the plane tz.

Figure 2. The Green function for $\beta = 5/2$.

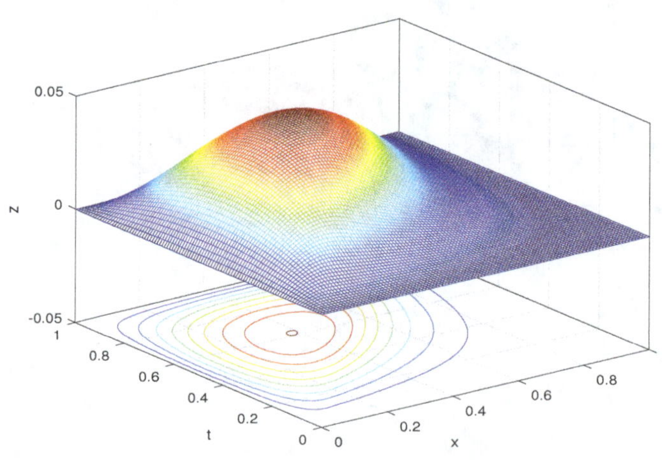

(**a**) $G_\beta(x,t)$ and contours.

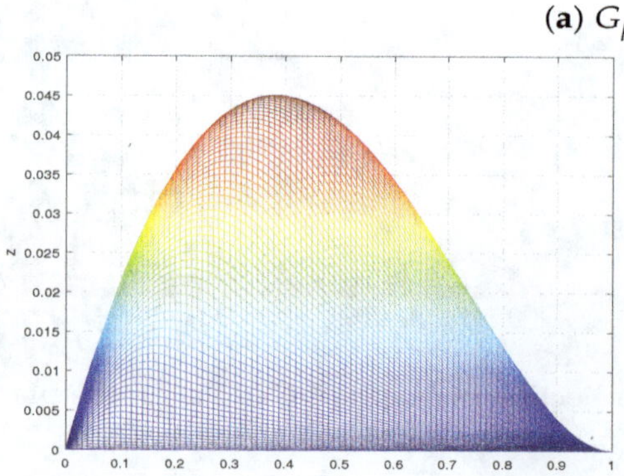

(**b**) Projection of graph of $G_\beta(x,t)$ on the plane xz.

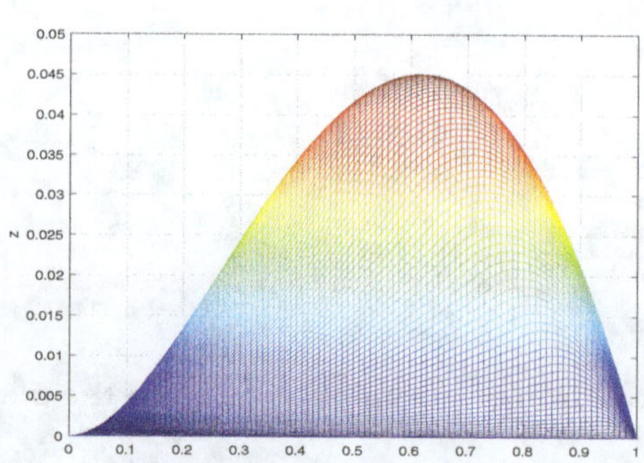

(**c**) Projection of graph of $G_\beta(x,t)$ on the plane tz.

Figure 3. The Green function for $\beta = 3$.

Proposition 1. *Let $2 < \beta \leq 3$. The Green function $G_\beta(x,t)$ satisfies the following properties.*

(i) $(x,t) \to G_\beta(x,t)$ *is continuous on* $[0,1] \times [0,1]$.

(ii) *For* $0 \leq x \leq t \leq 1$, *we have*

$$(\beta - 2)\, H(x,t) \leq \Gamma(\beta)\, G_\beta(x,t) \leq (\beta - 1)\, H(x,t),$$

where $H(x,t) := t\,(1-x)\,x^{\beta-2}\,(1-t)^{\beta-2}$.

(iii) *For* $0 \leq t \leq x \leq 1$, *we have*

$$(\beta - 2)\, \overline{H}(x,t) \leq 2\Gamma(\beta-1)\, G_\beta(x,t) \leq \overline{H}(x,t),$$

where $\overline{H}(x,t) := t^2\,(1-x)^2\,x^{\beta-3}\,(1-t)^{\beta-3}$.

Proof. (i) It is clear.
(ii) Assume that $0 \leq x \leq t \leq 1$. From (8) and (9) we have

$$\begin{aligned}
\Gamma(\beta)\, G_\beta(x,t) &= x^{\beta-2}\,(1-t)^{\beta-2}\,[(\beta-1)(t-x) + (\beta-2)\,x\,(1-t)] \\
&\leq (\beta-1)x^{\beta-2}\,(1-t)^{\beta-2}\,[(t-x) + x\,(1-t)] \\
&\leq (\beta-1)H(x,t).
\end{aligned}$$

On the other hand, since $t - x \geq 0$, we get

$$\Gamma(\beta)\, G_\beta(x,t) \geq (\beta-2)H(x,t).$$

(iii) Now, assume that $0 \leq t \leq x \leq 1$.
Since

$$x^{\beta-1}\,(1-t)^{\beta-1} - (x-t)^{\beta-1} = (\beta-1)t(1-x)\int_0^1 (x - t + st(1-x))^{\beta-2}ds,$$

it follows from (8) and (10) that

$$G_\beta(x,t) = \frac{1}{\Gamma(\beta-1)} t(1-x)x^{\beta-2}\,(1-t)^{\beta-2}\int_0^1 (1 - (\frac{x-t+st(1-x)}{x(1-t)})^{\beta-2})ds. \tag{13}$$

Now, using the fact that

$$(\beta-2)(1 - (\frac{x-t+st(1-x)}{x(1-t)})) \leq 1 - (\frac{x-t+st(1-x)}{x(1-t)})^{\beta-2} \leq 1 - (\frac{x-t+st(1-x)}{x(1-t)}), \tag{14}$$

we deduce from (13) that

$$\begin{aligned}
\Gamma(\beta-1)\, G_\beta(x,t) &\leq t(1-x)x^{\beta-3}\,(1-t)^{\beta-3}\int_0^1 t(1-x)(1-s)ds \\
&\leq \frac{1}{2}t^2(1-x)^2x^{\beta-3}\,(1-t)^{\beta-3}.
\end{aligned}$$

Similarly, using again (13) and (14), we obtain

$$\Gamma(\beta-1)\, G_\beta(x,t) \geq \frac{(\beta-2)}{2}t^2(1-x)^2x^{\beta-3}\,(1-t)^{\beta-3}.$$

□

Throughout this paper, for $2 < \beta \leq 3$ and $\varphi \in C([0,1])$, we denote by

$$G_\beta \varphi(x) = \int_0^1 G_\beta(x,t)\,\varphi(t)dt, \text{ for } x \in [0,1], \tag{15}$$

where $G_\beta(x,t)$ is given by (8).

Lemma 3. *Let* $0 < \alpha \leq 1, 2 < \beta \leq 3$ *and* $\varphi \in C([0,1])$. *Then the following assertions hold:*

$$\|G_\beta \varphi\| \leq K_\beta \|\varphi\| \text{ and } \|I^\alpha(G_\beta \varphi)\| \leq M_{\alpha,\beta} \|\varphi\|, \tag{16}$$

where

$$K_\beta := \frac{4}{\beta^2 \Gamma(\beta+1)} \left(\frac{\beta-2}{\beta}\right)^{\beta-2}, \tag{17}$$

and

$$M_{\alpha,\beta} := \frac{\omega^{\alpha+\beta-2}}{\beta^2 \Gamma(\alpha+\beta+1)} \left(\left(1+\sqrt{\frac{(\alpha+\beta-1)(1-\alpha)}{\beta-1}}\right)^2 + \frac{\alpha(\alpha+\beta)}{\beta-1}\right), \tag{18}$$

with $\omega := \frac{\alpha+\beta-1}{\beta} - \frac{1}{\beta}\sqrt{\frac{(\alpha+\beta-1)(1-\alpha)}{\beta-1}}$.

Proof. Let $\varphi \in C([0,1])$. By (15), we have for $x \in [0,1]$

$$|G_\beta \varphi(x)| \leq \|\varphi\| \int_0^1 G_\beta(x,t)\,dt. \tag{19}$$

Using Lemma 2, we obtain

$$\begin{aligned}
\int_0^1 G_\beta(x,t)\,dt &= \frac{1}{\Gamma(\beta)} \int_0^1 G(x,t)\,dt + \frac{1}{\Gamma(\beta)} \int_0^x (x-t)^{\beta-1}\,dt \\
&= \frac{(\beta-1)}{\Gamma(\beta)} (1-x) x^{\beta-2} \int_0^1 t(1-t)^{\beta-2}\,dt \\
&\quad -\frac{1}{\Gamma(\beta)} x^{\beta-1} \int_0^1 (1-t)^{\beta-1}\,dt + \frac{1}{\Gamma(\beta)} \int_0^x (x-t)^{\beta-1}\,dt \\
&= \frac{1}{\Gamma(\beta+1)} ((1-x)x^{\beta-2} - x^{\beta-1} + x^\beta) \\
&= \frac{1}{\Gamma(\beta+1)} (1-x)^2 x^{\beta-2} := \theta(x). \qquad (20)
\end{aligned}$$

By simple computation we obtain

$$\|\theta\| = \max_{x \in [0,1]} |\theta(x)| = \theta\left(\frac{\beta-2}{\beta}\right) = K_\beta. \tag{21}$$

Hence from (19) and (21), we get the first inequality in (16).

Now, using Definition 1 and (20), we obtain for $x \in [0,1]$

$$\begin{aligned} \left|I^{\alpha}(G_{\beta}\varphi)(x)\right| &\leq \frac{\|\varphi\|}{\Gamma(\alpha)\Gamma(\beta+1)}\int_0^x (x-t)^{\alpha-1}(1-t)^2 t^{\beta-2}dt \\ &= \frac{\|\varphi\|}{\Gamma(\alpha)\Gamma(\beta+1)}\int_0^x (x-t)^{\alpha-1}(t^{\beta}-2t^{\beta-1}+t^{\beta-2})dt \\ &= \frac{\|\varphi\|}{\Gamma(\beta+1)}\Big(\frac{\Gamma(\beta+1)}{\Gamma(\alpha+\beta+1)}x^{\alpha+\beta}-2\frac{\Gamma(\beta)}{\Gamma(\alpha+\beta)}x^{\alpha+\beta-1} \\ &\quad +\frac{\Gamma(\beta-1)}{\Gamma(\alpha+\beta-1)}x^{\alpha+\beta-2}\Big) \\ &= \frac{\|\varphi\|}{\Gamma(\alpha+\beta)}\psi(x), \end{aligned} \tag{22}$$

where

$$\psi(x) = \frac{1}{\alpha+\beta}x^{\alpha+\beta} - \frac{2}{\beta}x^{\alpha+\beta-1} + \frac{\alpha+\beta-1}{\beta(\beta-1)}x^{\alpha+\beta-2}.$$

Observe that

$$\begin{aligned} \psi'(x) &= x^{\alpha+\beta-3}\Big(x^2 - 2\frac{(\alpha+\beta-1)}{\beta}x + \frac{(\alpha+\beta-1)(\alpha+\beta-2)}{\beta(\beta-1)}\Big) \\ &= x^{\alpha+\beta-3}(x-\omega)(x-\overline{\omega}), \end{aligned}$$

where $\omega = \frac{(\alpha+\beta-1)}{\beta} - \frac{1}{\beta}\sqrt{\frac{(\alpha+\beta-1)(1-\alpha)}{\beta-1}}$ and $\overline{\omega} = \frac{(\alpha+\beta-1)}{\beta} + \frac{1}{\beta}\sqrt{\frac{(\alpha+\beta-1)(1-\alpha)}{\beta-1}}$.
Since $\omega \in (0,1]$ and $\overline{\omega} \geq 1$, it follows that $\psi'(x) \geq 0$ on $[0,\omega]$ and $\psi'(x) \leq 0$ on $[\omega,1]$.
Hence

$$\|\psi\| = \psi(\omega). \tag{23}$$

By combining (22) and (23), we obtain the second inequality in (16). □

3. Main Results

Let $0 < \alpha \leq 1$ and $2 < \beta \leq 3$. For each real number $M > 0$, denote by

$$\mathcal{D}_M = \{(x,u,v) \in \mathbb{R}^3 : 0 \leq x \leq 1, \ |u| \leq MM_{\alpha,\beta}, \ |v| \leq MK_{\beta}\},$$

where K_β and $M_{\alpha,\beta}$ are respectively given by (17) and (18).
By $B[O,M]$, we denote the closed ball centered at O with radius M in the space $C([0,1])$.

3.1. Existence and Uniqueness of a Solution

Theorem 1. *Let $f : [0,1] \times \mathbb{R}^2 \to \mathbb{R}$ be a continuous function and assume that there exist numbers $M, L_1, L_2 \geq 0$ such that*

(i) $|f(x,u,v)| \leq M$ *for any* $(x,u,v) \in \mathcal{D}_M$.
(ii) $|f(x,u_2,v_2) - f(x,u_1,v_1)| \leq L_1|u_2-u_1| + L_2|v_2-v_1|$,
for any $(x,u_i,v_i) \in \mathcal{D}_M$, $i = 1,2$.
(iii) $q := L_1M_{\alpha,\beta} + L_2K_\beta < 1$.

Then the boundary value problem (4) has a unique solution $u \in C([0,1])$ satisfying

$$\|u\| \leq MM_{\alpha,\beta} \text{ and } \|D^{\alpha}u\| \leq MK_{\beta}. \tag{24}$$

Proof. Consider the operator $T : C([0,1]) \to C([0,1])$ defined for $\varphi \in C([0,1])$ by

$$T\varphi(x) = f(x, I^{\alpha}(G_{\beta}\varphi)(x), G_{\beta}\varphi(x)), \ x \in [0,1], \tag{25}$$

where $G_\beta\varphi$ is defined by (15) and I^α is the Riemann–Liouville fractional integral operator given by Definition 1.

We shall investigate problem (4) via the operator equation (25).

Observe that if φ is a fixed point of the operator T, then by Lemma 1, (15) and Lemma 2,

$$u(x) := I^\alpha(G_\beta\varphi)(x), \tag{26}$$

is a solution of problem (4) and vice versa.

We claim that T is a contraction operator from $B[O, M]$ into itself.

First, we show that the operator T maps $B[O, M]$ into itself.

Indeed, since φ is continuous and by Proposition 1 (i) the Green' s function $G_\beta(x, t)$ is continuous on $[0, 1] \times [0, 1]$, it is not difficult to check that $T\varphi$ is continuous on $[0, 1]$.

Now, for any $\varphi \in B[O, M]$, we have by Lemma 3

$$\|G_\beta\varphi\| \leq MK_\beta \text{ and } \|I^\alpha(G_\beta\varphi)\| \leq MM_{\alpha,\beta}. \tag{27}$$

Hence, for $x \in [0, 1]$, we have $(x, I^\alpha(G_\beta\varphi)(x), G_\beta\varphi(x)) \in \mathcal{D}_M$. Therefore, from assumption (i), it follows that $\|T\varphi\| \leq M$. Therefore, the operator T maps $B[O, M]$ into itself.

Secondly, we prove that $T : B[O, M] \to B[O, M]$ is a contraction operator. Indeed, for any $\varphi_1, \varphi_2 \in B[O, M]$, by using assumption (ii) and Lemma 3, we obtain for $x \in [0, 1]$,

$$\begin{aligned} |T\varphi_2(x) - T\varphi_1(x)| &= |f(x, I^\alpha(G_\beta\varphi_2)(x), G_\beta\varphi_2(x)) - f(x, I^\alpha(G_\beta\varphi_1)(x), G_\beta\varphi_1(x))| \\ &\leq L_1 \|I^\alpha(G_\beta\varphi_2) - I^\alpha(G_\beta\varphi_1)\| + L_2 \|G_\beta\varphi_2 - G_\beta\varphi_1)\| \\ &= L_1 \|I^\alpha(G_\beta(\varphi_2 - \varphi_1))\| + L_2 \|G_\beta(\varphi_2 - \varphi_1)\| \\ &\leq L_1 M_{\alpha,\beta} \|\varphi_2 - \varphi_1\| + L_2 K_\beta \|\varphi_2 - \varphi_1\| \\ &= q \|\varphi_2 - \varphi_1\|, \end{aligned}$$

where q is defined in assumption (iii).

Therefore, T is a contraction operator in $B[O, M]$. Hence, it has a unique fixed point φ in $B[O, M]$.

Therefore, problem (4) has a unique solution $u \in C([0, 1])$ given by (26). The estimates (24) follow from Lemma 3 and the fact that $\|\varphi\| \leq M$.

The the proof is completed. □

Next, we present a particular case of Theorem 1. To this end, denote

$$\mathcal{D}_M^+ = \{(x, u, v) \in \mathbb{R}^3 : 0 \leq x \leq 1,\ 0 \leq u \leq MM_{\alpha,\beta},\ 0 \leq v \leq MK_\beta\}.$$

Corollary 1. *Let $f : [0, 1] \times \mathbb{R}^2 \to \mathbb{R}$ be a continuous function and assume that there exists numbers $M, L_1, L_2 \geq 0$ such that*

(i) $0 \leq f(x, u, v) \leq M$ *for any* $(x, u, v) \in \mathcal{D}_M^+$.

(ii) $|f(x, u_2, v_2) - f(x, u_1, v_1)| \leq L_1 |u_2 - u_1| + L_2 |v_2 - v_1|$,

for any $(x, u_i, v_i) \in \mathcal{D}_M^+$, $i = 1, 2$.

(iii) $q := L_1 M_{\alpha,\beta} + L_2 K_\beta < 1$.

Then the boundary value problem (4) has a unique nonnegative solution $u \in C([0, 1])$ satisfying

$$0 \leq u(x) \leq MM_{\alpha,\beta} \text{ and } 0 \leq D^\alpha u \leq MK_\beta. \tag{28}$$

3.2. Iterative Method and Examples

Consider the following iterative process.

$$\begin{cases} \text{Let } \varphi_0 \in B[O,M], \\ \varphi_{k+1}(x) := T\varphi_k(x) = f(x, I^\alpha(G_\beta\varphi_k)(x), G_\beta\varphi_k(x)), \text{ for } k = 0,1,...;\ x \in [0,1]. \end{cases} \tag{29}$$

Theorem 2. *Assume that hypotheses of Theorem 1 are satisfied. The sequence $(\varphi_k)_{k\geq 0}$ converges with the rate of geometric progression and we have*

$$\|I^\alpha(G_\beta\varphi_k) - u\| \leq M_{\alpha,\beta}\frac{q^k}{1-q}\|\varphi_1 - \varphi_0\|, \tag{30}$$

where u is the exact solution of problem (4) and q is given in assumption (iii) in Theorem 1.

Proof. It is known by the Banach contracting mapping principle that the sequence $(\varphi_k)_{k\geq 0}$ converges with the rate of geometric progression and we have

$$\|\varphi_k - \varphi\| \leq \frac{q^k}{1-q}\|\varphi_1 - \varphi_0\|, \tag{31}$$

where φ is the unique fixed point of the operator T in $B[O,M]$.

Using this fact and Lemma 3, we obtain

$$\begin{aligned} \|I^\alpha(G_\beta\varphi_k) - u\| &= \|I^\alpha(G_\beta\varphi_k) - I^\alpha(G_\beta\varphi)\| \\ &= \|I^\alpha(G_\beta(\varphi_k - \varphi))\| \\ &\leq M_{\alpha,\beta}\|\varphi_k - \varphi\| \\ &\leq M_{\alpha,\beta}\frac{q^k}{1-q}\|\varphi_1 - \varphi_0\|. \end{aligned}$$

The proof is completed. □

Proposition 2. (*Monotony*)*Assume that hypotheses of Theorem 1 are satisfied. In addition, we assume that the function $f(x,u,v)$ is nondecreasing in u and v for any $(x,u,v) \in \mathcal{D}_M$. Let $\varphi_0, \psi_0 \in B[O,M]$ be initial approximations such that $\varphi_0(x) \leq \psi_0(x)$, for all $x \in [0,1]$. Then*

(i) for all $k \in \mathbb{N}$ and $x \in [0,1]$,

$$I^\alpha(G_\beta\varphi_k)(x) \leq I^\alpha(G_\beta\psi_k)(x). \tag{32}$$

(ii) Suppose further that for all $(x,u,v) \in \mathcal{D}_M$

$$\varphi_0(x) \leq f(x,u,v) \leq \psi_0(x). \tag{33}$$

Then the sequences $(I^\alpha(G_\beta\varphi_k))_{k\geq 0}$ and $(I^\alpha(G_\beta\psi_k))_{k\geq 0}$ converge to the unique solution u of problem (4) and

$$I^\alpha(G_\beta\varphi_k) \leq I^\alpha(G_\beta\varphi_{k+1}) \leq u \leq I^\alpha(G_\beta\psi_{k+1}) \leq I^\alpha(G_\beta\psi_k). \tag{34}$$

In particular, if $\varphi_0 \geq 0$ (resp. $\psi_0 \leq 0$), then u is nonnegative (resp. nonpositive) solution.

Proof. (i) We claim that for all $k \in \mathbb{N}$, we have

$$\varphi_k(x) \leq \psi_k(x), \text{ on } [0,1]. \tag{35}$$

We proceed by induction. From hypothesis, the inequality is clear for $k = 0$. For a given $k \in \mathbb{N}$, assume that $\varphi_k(x) \leq \psi_k(x)$.

Since the Green function is nonnegative, we deduce from (15) and Definition 1 that

$$G_\beta \varphi_k \leq G_\beta \psi_k \text{ and } I^\alpha(G_\beta \varphi_k) \leq I^\alpha(G_\beta \psi_k).$$

Combining this fact and that the function $f(x, u, v)$ is nondecreasing in u and v, we obtain

$$\varphi_{k+1}(x) := f(x, I^\alpha(G_\beta \varphi_k)(x), G_\beta \varphi_k(x)) \leq f(x, I^\alpha(G_\beta \psi_k)(x), G_\beta \psi_k(x)) = \psi_{k+1}(x).$$

So our claim is proved.

Using (35), (15) and Definition 1 we get inequality in (32)

(ii) From Theorem 2, we know that the sequences $(I^\alpha(G_\beta \varphi_k))_{k \geq 0}$ and $(I^\alpha(G_\beta \psi_k))_{k \geq 0}$ converge to the unique solution u of problem (4).

We claim that the sequence $(\varphi_k)_{k \geq 0}$ is nondecreasing.

Indeed, since for $x \in [0, 1]$, we have $(x, I^\alpha(G_\beta \varphi_0)(x), G_\beta \varphi_0(x)) \in \mathcal{D}_M$, we deduce from (33) that

$$\varphi_0(x) \leq f(x, I^\alpha(G_\beta \varphi_0)(x), G_\beta \varphi_0(x)) = \varphi_1(x).$$

Assume that $\varphi_k(x) \leq \varphi_{k+1}(x)$. From (15), Definition 1 and the monotony of the function f, we deduce that

$$\varphi_{k+1}(x) = f(x, I^\alpha(G_\beta \varphi_k)(x), G_\beta \varphi_k(x)) \leq f(x, I^\alpha(G_\beta \varphi_{k+1})(x), G_\beta \varphi_{k+1}(x)) = \varphi_{k+2}(x).$$

Hence the sequence $(\varphi_k)_{k \geq 0}$ is nondecreasing.

Therefore, by using again (15) and Definition 1, it follows that the sequence $(I^\alpha(G_\beta \varphi_k))_{k \geq 0}$ is nondecreasing.

Since the sequence $(I^\alpha(G_\beta \varphi_k))_{k \geq 0}$ converges to u, we obtain

$$I^\alpha(G_\beta \varphi_k) \leq I^\alpha(G_\beta \varphi_{k+1}) \leq u$$

Similarly, we prove that the sequence $(I^\alpha(G_\beta \psi_k))_{k \geq 0}$ is nonincreasing and that

$$u \leq I^\alpha(G_\beta \psi_{k+1}) \leq I^\alpha(G_\beta \psi_k).$$

So inequalities in (34) are proved.

Finally, from (34), we have

$$I^\alpha(G_\beta \varphi_0) \leq u \leq I^\alpha(G_\beta \psi_0).$$

This implies that if $\varphi_0 \geq 0$ (resp. $\psi_0 \leq 0$), then u is nonnegative (resp. nonpositive) solution.

This completes the proof. □

Example 1. *Consider the following boundary value problem:*

$$\begin{cases} D^{\frac{5}{2}}(D^{\frac{1}{2}}u)(x) = xu(x) + x^2(D^{\frac{1}{2}}u(x))^2 + 2x + 1, & x \in (0,1), \\ u(0) = D^{\frac{1}{2}}u(0) = D^{\frac{1}{2}}u(1) = (D^{\frac{1}{2}}u)'(1) = 0. \end{cases} \tag{36}$$

In this case $K_{\frac{5}{2}} = 8.6123 \times 10^{-2}$, $M_{\frac{1}{2},\frac{5}{2}} = 5.4279 \times 10^{-2}$ *and* $f(x, u, v) = xu + x^2v^2 + 2x + 1$. *So condition (i) in Theorem 1 will be satisfied if we choose* $M > 0$ *such that*

$$MM_{\frac{1}{2},\frac{5}{2}} + M^2 K_{\frac{5}{2}}^2 + 3 \leq M.$$

It is easy to verify that $M = 4$ *is an example of suitable choice.*

Since

$$f'_u = x \text{ and } f'_v = 2x^2 v,$$

it follows that for any $(x,u,v) \in \mathcal{D}_4 = \{(x,u,v), 0 \le x \le 1, |u| \le 4M_{\frac{1}{2},\frac{5}{2}}, |v| \le 4K_{\frac{5}{2}}\}$,

$$|f'_u| \le 1 \text{ and } |f'_v| \le 8K_{\frac{5}{2}} \le 1.$$

Hence, $L_1 = 1$ *and* $L_2 = 1$ *satisfy the condition (ii) in Theorem 1. Also, we have* $q := L_1 M_{\frac{1}{2},\frac{5}{2}} + L_2 K_{\frac{5}{2}} = M_{\frac{1}{2},\frac{5}{2}} + K_{\frac{5}{2}} < 1$.

Thus by Theorem 1, problem (36) has a unique solution, and the iterative method converges.

In Figure 4, we present the approximation of the unique solution of problem (36) with $u_k(x) := I^{\frac{1}{2}}(G_{\frac{5}{2}}\varphi_k)(x)$ *and* $\varphi_0(x) := 2x + 1$.

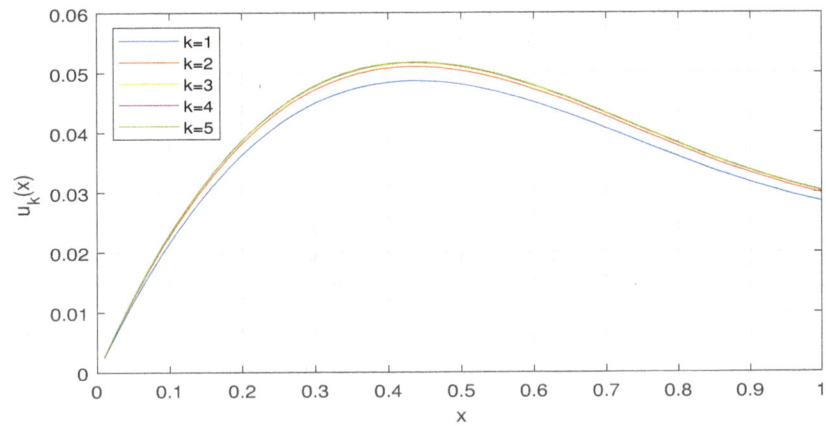

Figure 4. The approximation of the solution of problem (36).

Example 2. *Consider the following boundary value problem:*

$$\begin{cases} D^{\frac{8}{3}}(u')(x) = -3u^2(u'(x))^2 + 3u(x) + 4u'(x) + \sin(\pi x), & x \in (0,1), \\ u(0) = u'(0) = u'(1) = u''(1) = 0. \end{cases} \tag{37}$$

In this example, $K_{\frac{8}{3}} = 5.5637 \times 10^{-2}$, $M_{1,\frac{8}{3}} = 2.1030 \times 10^{-2}$ *and* $f(x,u,v) = -3u^2v^2 + 3u + 4v + \sin(\pi x)$.

As in Example 1, we verify that all conditions of Theorem 1 are satisfied with $M = 3$, $L_1 = 4$ *and* $L_2 = 5$. *Hence problem (37) has a unique solution, and the iterative method converges. Moreover, since in* $\mathcal{D}_3$ *we have* $f'_u \ge 0$ *and* $f'_v \ge 0$, *the function* $f(x,u,v)$ *is nondecreasing in both* u *and* v. *Take the initial approximation* $\varphi_0 = f(x,0,0) = \sin(\pi x) \ge 0, 0 \le x \le 1$. *By the positivity of the Green's function and Lemma 3, we have*

$$0 \le v_0 := G_{\frac{8}{3}}\varphi_0 \le K_{\frac{8}{3}} \text{ and } 0 \le u_0 := I^1(G_{\frac{8}{3}}\varphi_0) \le M_{1,\frac{8}{3}}.$$

Therefore form the iterative process (29), we obtain

$$\begin{aligned} \varphi_1(x) &= f(x, u_0(x), v_0(x)) \\ &= -3u_0^2 v_0^2 + 3u_0 + 4v_0 + \sin(\pi x) \\ &= 3u_0(1 - u_0 v_0^2) + 4v_0 + \sin(\pi x) \\ &\ge \sin(\pi x) = \varphi_0. \end{aligned}$$

By Proposition 2, $(u_k := I^\alpha(G_\beta \varphi_k))_{k\ge 0}$ *is a nonnegative increasing sequence which converges to the unique nonnegative solution* u. *Some iterations are depicted in Figure 5.*

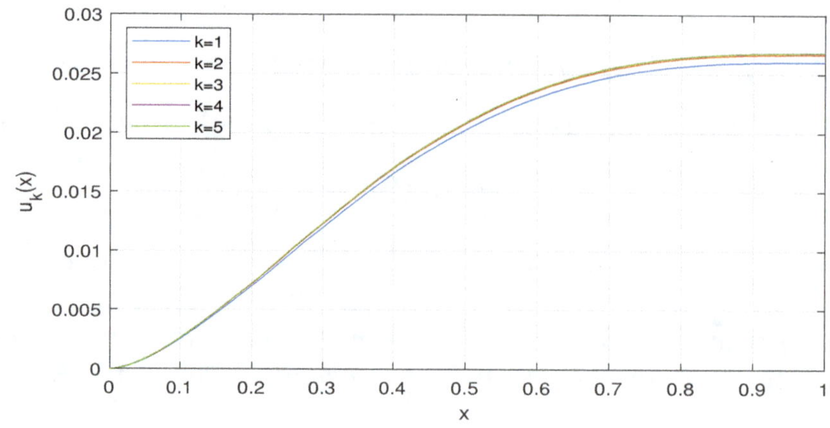

Figure 5. The approximation of the solution of problem (37).

Author Contributions: Investigation, I.B., H.M. and H.E.; Writing–review and editing, I.B. and H.M. All authors contributed equally to the writing of this paper. All authors read and approved the final manuscript.

Acknowledgments: The authors would like to extend their sincere appreciation to the Deanship of Scientific Research at King Saud University for its funding this Research group NO (RG-1435-043). We also would like to thank the referees for their careful reading of the manuscript and for helpful suggestions which improved the quality of the paper.

References

1. Aftabizadeh, A.R. Existence and uniqueness theorems for fourth-order boundary value problems. *J. Math. Anal. Appl.* **1986**, *116*, 415–426. [CrossRef]
2. Li, Y. Existence of positive solutions for the cantilever beam equations with fully nonlinear terms. *Nonlinear Anal.* **2016**, *27*, 221–237. [CrossRef]
3. Alsaedi, R.S. Existence and global behavior of positive solutions for some fourth order boundary value problems. *Abstr. Appl. Anal.* **2014**, *2014*, 657926. [CrossRef]
4. Bai, Z. The method of lower and upper solutions for a bending of an elastic beam equation. *J. Math. Anal. Appl.* **2010**, *248*, 195–202. [CrossRef]
5. Bai, Z. Positive solutions of some nonlocal fourth-order boundary value problem. *Appl. Math. Comput.* **2010**, *215*, 4191–4197. [CrossRef]
6. Dang, Q.A. Iterative method for solving the Neumann boundary value problem for biharmonic type equation. *J. Comput. Appl. Math.* **2006**, *196*, 634–643. [CrossRef]
7. Dang, Q.A.; Dang, Q.L.; Ngo, T.K.Q. A novel efficient method for nonlinear boundary value problems. *Numer. Algorithms* **2017**, *76*, 427–439.
8. Li, Y. A monotone iterative technique for solving the bending elastic beam equations. *Appl. Math. Comput.* **2010**, *2017*, 2200–2208. [CrossRef]
9. Ma, T.F.; da Silva, J. Iterative solutions for a beam equation with nonlinear boundary conditions of third order. *Appl. Math. Comput.* **2004**, *159*, 11–18. [CrossRef]
10. Ma, R.; Tisdell, C.C. Positive solutions of singular sublinear fourth order boundary value problems. *Appl. Anal.* **2005**, *12*, 1199–1220. [CrossRef]
11. Wei, Y.; Song, Q.; Bai, Z. Existence and iterative method for some fourth order nonlinear boundary value problems. *Appl. Math. Lett.* **2019**, *87*, 101–107. [CrossRef]
12. Dang, Q.; Quy, N. Existence results and iterative method for solving the cantilever beam equation with fully nonlinear term. *Nonlinear Anal. RWA* **2017**, *36*, 56–68. [CrossRef]
13. Agarwal, R.P.; O'Regan, O.; Staněk, S. Positive solutions for Dirichlet problems of singular nonlinear fractional differential equations. *J. Math. Anal. Appl.* **2010**, *371*, 57–68. [CrossRef]

14. Bachar, I.; Mâagli, M.; Toumi, F.; Zine el Abidine, Z. Existence and Global Asymptotic Behavior of Positive Solutions for Sublinear and Superlinear Fractional Boundary Value Problems. *Chin. Ann. Math. Ser. B* **2016**, *37*, 1–28. [CrossRef]
15. Bai, Z.; Lü, H. Positive solutions for boundary value problem of nonlinear fractional differential equation. *J. Math. Anal. Appl.* **2005**, *311*, 495–505. [CrossRef]
16. Diethelm, K.; Freed, A.D. On the solution of nonlinear fractional order differential equations used in the modeling of viscoplasticity. In *Scientific Computing in Chemical Engineering. II-Computational Fluid Dynamics, Reaction Engineering and Molecular Properties*; Keil, F., Mackens, W., Voss, H., Werther, J., Eds.; Springer: Heidelberg, Germany, 1999; pp. 217–224.
17. Gaul, L.; Klein, P.; Kempfle, S. Damping description involving fractional operators. *Mech. Syst. Signal Process.* **1991**, *5*, 81–88. [CrossRef]
18. Glockle, W.G.; Nonnenmacher, T.F. A fractional calculus approach of self-similar protein dynamics. *Biophys. J.* **1995**, *68*, 46–53. [CrossRef]
19. Graef, J.R.; Kong, L.; Kong, Q.; Wang, M. Existence and uniqueness of solutions for a fractional boundary value problem with Dirichlet boundary condition. *Electron. J. Qual. Theory Differ. Equ.* **2013**, *55*, 1–11. [CrossRef]
20. Hilfer, R. *Applications of Fractional Calculus in Physics*; World Scientific: Singapore, 2000.
21. Kaufmann, E.R.; Mboumi, E. Positive solutions of a boundary value problem for a nonlinear fractional differential equation. *Electron. J. Qual. Theory Differ. Equ.* **2008**, *3*, 1–11. [CrossRef]
22. Liang, S.; Zhang, J. Positive solutions for boundary value problems of nonlinear fractional differential equation. *Nonlinear Anal.* **2009**, *71*, 5545–5550. [CrossRef]
23. Mâagli, H.; Mhadhebi, N.; Zeddini, N. Existence and Estimates of Positive Solutions for Some Singular Fractional Boundary Value Problems. *Abstr. Appl. Anal.* **2009**, *2009*, 120781. [CrossRef]
24. Mâagli, H.; Mhadhebi, N.; Zeddini, N. Existence and Exact Asymptotic Behavior of Positive Solutions for a Fractional Boundary Value Problem. *Abstr. Appl. Anal.* **2013**, *2013*, 420514. [CrossRef]
25. Scher, H.; Montroll, E. Anomalous transit-time dispersion in amorphous solids. *Phys. Rev. B.* **1975**, *12*, 2455–2477. [CrossRef]
26. Timoshenko, S.; Gere, J.M. *Theory of Elastic Stability*; McGraw-Hill: New York, NY, USA, 1961.
27. Xu, X.; Jiang, D.; Yuan, C. Multiple positive solutions for the boundary value problem of a nonlinear fractional differenteial equation. *Nonlinear Anal.* **2009**, *71*, 4676–4688. [CrossRef]
28. Xu, X.; Jiang, D.; Yuan, C. Singular Positone and Semipositone Boundary Value Problems of Nonlinear Fractional Differential Equations. *Math. Probl. Eng.* **2009**, *2009*, 535209.
29. Zhang, Y.; Liu, X.; Belić, M.R.; Zhong, W.; Zhang, Y.; Xiao, M. Propagation Dynamics of a Light Beam in a Fractional Schrödinger Equation. *Phys. Rev. Lett.* **2015**, *115*, 180403. [CrossRef]
30. Zhang, X.; Liu, L.; Wu, Y. Multiple positive solutions of a singular fractional differential equation with negatively perturbed term. *Math. Comput. Model.* **2012**, *55*, 1263–1274. [CrossRef]
31. Zhang, D.; Zhang, Y.; Zhang, Z.; Ahmed, N.; Zhang,Y.; Li, F.; Belić, M.R.; Xiao, M. Unveiling the Link Between Fractional Schrödinger Equation and Light Propagation in Honeycomb Lattice. *Ann. Phys. (Berlin)* **2017**, *529*, 1700149. [CrossRef]
32. Zhang, Y.; Zhong, H.; Belić, M.R.; Zhu, Y.; Zhong, W.; Zhang, Y.; Christodoulides, D.N.; Xiao, M. $\mathcal{PT}$ symmetry in a fractional Schrödinger equation. *Laser Photonics Rev.* **2016**, *10*, 526–531. [CrossRef]
33. Kilbas, A.; Srivastava, H.; Trujillo, J. Theory and Applications of Fractional Differential Equations. In *North-Holland Mathematics Studies Vol. 204*; Elsevier: Amsterdam, The Netherlands, 2006.
34. Miller, K.; Ross, B. *An Introduction to the Fractional Calculus and Fractional Differential Equations*; Wiley and Sons: New York, NY, USA, 1993.
35. Podlubny, I. Fractional Differential Equations. In *Mathematics in Science and Engineering*; Academic Press: New York, NY, USA, 1999.
36. Samko, S.; Kilbas, A.; Marichev, O. *Fractional Integrals and Derivative. Theory and Applications*; Gordon and Breach: Yverdon, Switzerland, 1993.

11

On a Generalized Langevin type Nonlocal Fractional Integral Multivalued Problem

Ahmed Alsaedi [1], Bashir Ahmad [1,*], Madeaha Alghanmi [1] and Sotiris K. Ntouyas [1,2]

1 Nonlinear Analysis and Applied Mathematics (NAAM)-Research Group, Department of Mathematics, Faculty of Science, King Abdulaziz University, P.O. Box 80203, Jeddah 21589, Saudi Arabia; aalsaedi@hotmail.com (A.A.); madeaha@hotmail.com (M.A.); sntouyas@uoi.gr (S.K.N.)

2 Department of Mathematics, University of Ioannina, 451 10 Ioannina, Greece

* Correspondence: bmuhammed@kau.edu.sa

Abstract: We establish sufficient criteria for the existence of solutions for a nonlinear generalized Langevin-type nonlocal fractional-order integral multivalued problem. The convex and non-convex cases for the multivalued map involved in the given problem are considered. Our results rely on Leray–Schauder nonlinear alternative for multivalued maps and Covitz and Nadler's fixed point theorem. Illustrative examples for the main results are included.

Keywords: differential inclusions; Caputo-type fractional derivative; fractional integral; existence; fixed point

---

1. Introduction

Fractional calculus is the extension of classical calculus which deals with differential and integral operators of fractional order. It has evolved into a significant and popular branch of mathematical analysis owing to its extensive applications in the mathematical modeling of applied and technical problems. The literature on fractional calculus is now much enriched and covers a wide range of interesting results, for instance [1–6]. For a comprehensive treatment of Hadamard-type fractional differential equations and inclusions, we refer the reader to the text [7].

The Langevin equation is found to be an effective tool to describe stochastic problems in fluctuating situations. A modified type of this equation is used in various functional approaches for fractal media. A variety of boundary value problems involving the Langevin equation have been investigated by several authors. In [8], existence and uniqueness results for a nonlinear Langevin equation involving two fractional orders supplemented with three-point boundary conditions were obtained. An impulsive boundary value problem for a nonlinear Langevin equation involving two different fractional derivatives was investigated in [9]. Some existing results for Langevin fractional differential inclusions with two indices were derived in [10]. In [11], the authors proved the existence of and uniqueness results for an anti-periodic boundary value problem of a system of Langevin fractional differential equations. In [12], the authors investigated a nonlinear fractional Langevin equation with anti-periodic boundary conditions by applying coupled fixed point theorems. In a recent work [13], the authors obtained some existence results for a fractional Langevin equation with nonlinearity depending on Riemann–Liouville fractional integral, and complemented with nonlocal multi-point and multi-strip boundary conditions.

In the present paper, we study the existence of solutions for a nonlinear generalized Langevin type nonlocal fractional-order integral multivalued problem given by

$$\begin{cases} {}^{\rho}_{c}D^{\alpha}_{a+}({}^{\rho}_{c}D^{\beta}_{a+}+\lambda)x(t) \in F(t,x(t)), \;\; t \in J := [a,T], \;\; \lambda \in \mathbb{R}, \\ x(a)=0, \;\; x(\eta)=0, \;\; x(T)=\mu\, {}^{\rho}I^{\gamma}_{a+}x(\xi), \quad a<\eta<\xi<T, \mu \in \mathbb{R}, \end{cases} \tag{1}$$

where ${}^{\rho}_{c}D^{\alpha}_{a+}$, ${}^{\rho}_{c}D^{\beta}_{a+}$ denote the Caputo-type generalized fractional differential operators of order $1 < \alpha \le 2$, $0 < \beta < 1$, $\rho > 0$, respectively, $F : [a, T] \times \mathbb{R} \to \mathcal{P}(\mathbb{R})$ is a multi-valued map ($\mathcal{P}(\mathbb{R})$ is the family of all nonempty subsets of $\mathbb{R}$), ${}^{\rho}I^{\gamma}_{a+}$ is the generalized fractional integral operator of order $\gamma > 0$ and $\rho > 0$. Here we emphasize that the single-valued analogue of the problem (1) was discussed in [14].

The rest of the paper is arranged as follows. The background material related to our work is outlined in Section 3. The existence results for the problem (1) are presented in Section 3. The first result for the problem (1), associated with the convex valued mutivalued map, is derived with the aid of Leray–Schauder nonlinear alternative for multivalued maps, while the result for non-convex valued map for the problem (1) is proved by applying a fixed point theorem due to Covitz and Nadler. Section 4 contains the illustrative examples for the main results. We summarize the work established in this paper, and its implications, in the last section.

2. Preliminaries

Define by $X^p_c(a,b)$ the space of all complex-valued Lebesgue measurable functions ϕ on (a,b) equipped with the norm:

$$\|\phi\|_{X^p_c} = \Big(\int_a^b |x^c\phi(x)|^p \frac{dx}{x}\Big)^{1/p} < \infty, \;\; c \in \mathbb{R}, 1 \le p \le \infty.$$

Let $AC^n_\delta[a,b]$ denote the class of all absolutely continuous functions g possessing δ^{n-1}-derivative ($\delta^{n-1}g \in AC([a,b],\mathbb{R})$), endowed with the norm $\|g\|_{AC^n_\delta} = \sum_{k=0}^{n-1} \|\delta^k g\|_C$.

Definition 1. *The left-sided and right-sided generalized fractional integrals for $g \in X^p_c(a,b)$ of order $\beta > 0$ and $\rho > 0$, denoted by ${}^{\rho}I^{\beta}_{a+}g$ and ${}^{\rho}I^{\beta}_{b-}g$ respectively, are defined by [15]*

$$({}^{\rho}I^{\beta}_{a+}g)(t) = \frac{\rho^{1-\beta}}{\Gamma(\beta)} \int_a^t \frac{s^{\rho-1}}{(t^\rho - s^\rho)^{1-\beta}} g(s)ds, \;\; -\infty < a < t < b < \infty, \tag{2}$$

$$({}^{\rho}I^{\beta}_{b-}g)(t) = \frac{\rho^{1-\alpha}}{\Gamma(\beta)} \int_t^b \frac{s^{\rho-1}}{(s^\rho - t^\rho)^{1-\beta}} g(s)ds, \;\; -\infty < a < t < b < \infty. \tag{3}$$

Definition 2. *Let $\beta > 0$, $n = [\beta] + 1$ and $\rho > 0$. We define the generalized fractional derivatives, associated with the generalized fractional integrals (2) and (3), for $0 \le a < t < b < \infty$, as follows [16]:*

$$({}^{\rho}D^{\beta}_{a+}g)(t) = \Big(t^{1-\rho}\frac{d}{dt}\Big)^n ({}^{\rho}I^{n-\beta}_{a+}g)(t) = \frac{\rho^{\beta-n+1}}{\Gamma(n-\beta)} \Big(t^{1-\rho}\frac{d}{dt}\Big)^n \int_a^t \frac{s^{\rho-1}}{(t^\rho - s^\rho)^{\beta-n+1}} g(s)ds, \tag{4}$$

$$({}^{\rho}D^{\beta}_{b-}g)(t) = \Big(-t^{1-\rho}\frac{d}{dt}\Big)^n ({}^{\rho}I^{n-\beta}_{b-}g)(t) = \frac{\rho^{\beta-n+1}}{\Gamma(n-\beta)} \Big(-t^{1-\rho}\frac{d}{dt}\Big)^n \int_t^b \frac{s^{\rho-1}}{(s^\rho - t^\rho)^{\beta-n+1}} g(s)ds, \tag{5}$$

provided the integrals in the above expressions exist.

Definition 3. *Let $g \in AC^n_\delta[a,b]$ and $\beta > 0, n = [\beta] + 1$. Then the Caputo-type generalized fractional derivatives ${}^{\rho}_{c}D^{\beta}_{a+}g$ and ${}^{\rho}_{c}D^{\beta}_{b-}g$ are respectively defined via (4) and (5) by [17]*

$${}^{\rho}_{c}D^{\beta}_{a+}g(x) = {}^{\rho}D^{\beta}_{a+}\Big[g(t) - \sum_{k=0}^{n-1} \frac{\delta^k g(a)}{k!}\Big(\frac{t^\rho - a^\rho}{\rho}\Big)^k\Big](x), \;\; \delta = x^{1-\rho}\frac{d}{dx}, \tag{6}$$

$${}^{\rho}_{c}D^{\beta}_{b-}g(x) = {}^{\rho}D^{\beta}_{b-}\Big[g(t) - \sum_{k=0}^{n-1} \frac{(-1)^k\delta^k g(b)}{k!}\Big(\frac{b^\rho - t^\rho}{\rho}\Big)^k\Big](x), \;\; \delta = x^{1-\rho}\frac{d}{dx}. \tag{7}$$

Remark 1. *The left and right generalized Caputo derivatives of order β for $g \in AC^n_\delta[a,b]$, are respectively given by [17]*

$$ {}^\rho_c D^\beta_{a+} g(t) = \frac{1}{\Gamma(n-\beta)} \int_a^t \Big(\frac{t^\rho - s^\rho}{\rho}\Big)^{n-\beta-1} \frac{(\delta^n g)(s) ds}{s^{1-\rho}}, \tag{8}$$

$$ {}^\rho_c D^\beta_{b-} g(t) = \frac{1}{\Gamma(n-\beta)} \int_t^b \Big(\frac{s^\rho - t^\rho}{\rho}\Big)^{n-\alpha-1} \frac{(-1)^n(\delta^n g)(s) ds}{s^{1-\rho}}. \tag{9}$$

Lemma 1. *Let $g \in AC^n_\delta[a,b]$ or $C^n_\delta[a,b]$. Then, for $\beta \in \mathbb{R}$, the following results hold [17]:*

$$ {}^\rho I^\beta_{a+} {}^\rho_c D^\beta_{a+} g(x) = g(x) - \sum_{k=0}^{n-1} \frac{(\delta^k g)(a)}{k!} \Big(\frac{x^\rho - a^\rho}{\rho}\Big)^k,$$

$$ {}^\rho I^\beta_{b-} {}^\rho_c D^\beta_{b-} g(x) = g(x) - \sum_{k=0}^{n-1} \frac{(-1)^k(\delta^k g)(a)}{k!} \Big(\frac{b^\rho - x^\rho}{\rho}\Big)^k.$$

In particular, for $0 < \beta \le 1$, we have

$$ {}^\rho I^\beta_{a+} {}^\rho_c D^\beta_{a+} g(x) = g(x) - g(a), \quad {}^\rho I^\beta_{b-} {}^\rho_c D^\beta_{b-} g(x) = g(x) - g(b).$$

We need the following known lemma [14] in the sequel.

Lemma 2. *Let $h \in C([a,T],\mathbb{R})$ and $x \in AC^3_\delta(J)$. Then the unique solution of linear problem:*

$$\begin{cases} {}^\rho_c D^\alpha_{a+}({}^\rho_c D^\beta_{a+} + \lambda)x(t) = h(t), \ \ t \in J := [a,T], \\ x(a) = 0, \ \ x(\eta) = 0, \ \ x(T) = \mu^\rho I^\gamma_{a+} x(\xi), \quad a < \eta < \xi < T, \end{cases} \tag{10}$$

is given by:

$$\begin{aligned} x(t) \ =\ & {}^\rho I^{\alpha+\beta}_{a+} h(t) - \lambda {}^\rho I^\beta_{a+} x(t) + \frac{(t^\rho - a^\rho)^\beta(\eta^\rho - t^\rho)}{\rho^{\beta+1}\Gamma(\beta+2)\Omega} \Big\{ {}^\rho I^{\alpha+\beta}_{a+} h(T) - \lambda {}^\rho I^\beta_{a+} x(T) \\ & - \mu^\rho I^{\alpha+\beta+\gamma}_{a+} h(\xi) + \mu\lambda^\rho I^{\beta+\gamma}_{a+} x(\xi) \Big\} - \frac{(t^\rho - a^\rho)^\beta}{\Omega(\eta^\rho - a^\rho)^\beta} \Big(\frac{(T^\rho - a^\rho)^\beta (T^\rho - t^\rho)}{\rho^{\beta+1}\Gamma(\beta+2)} \\ & - \frac{\mu(\xi^\rho - a^\rho)^{\beta+\gamma}[(\beta+1)(\xi^\rho - t^\rho) - \gamma(t^\rho - a^\rho)]}{\rho^{\beta+\gamma+1}\Gamma(\beta+\gamma+2)(\beta+1)} \Big) \Big\{ {}^\rho I^{\alpha+\beta}_{a+} h(\eta) - \lambda^\rho I^\beta_{a+} x(\eta) \Big\}, \end{aligned} \tag{11}$$

where it is assumed that

$$\Omega = \Big[\frac{(T^\rho - a^\rho)^\beta (T^\rho - \eta^\rho)}{\rho^{\beta+1}\Gamma(\beta+2)} - \frac{\mu(\xi^\rho - a^\rho)^{\beta+\gamma}[(\beta+1)(\xi^\rho - \eta^\rho) - \gamma(\eta^\rho - a^\rho)]}{\rho^{\beta+\gamma+1}\Gamma(\beta+\gamma+2)(\beta+1)} \Big] \neq 0. \tag{12}$$

3. Main Results

We begin this section with the definition of a solution for the multi-valued problem (1).

Definition 4. *A function $x \in C(J,\mathbb{R})$ is called a solution of the problem (1) if we can find a function $v \in L^1(J,\mathbb{R})$ with $v(t) \in F(t,x)$ a.e. on J such that $x(a) = 0, x(\eta) = 0, x(T) = \mu^\rho I^\gamma_{a+} x(\xi)$ and*

$$\begin{aligned} x(t) \ =\ & {}^\rho I^{\alpha+\beta}_{a+} v(t) - \lambda^\rho I^\beta_{a+} x(t) + \frac{(t^\rho - a^\rho)^\beta(\eta^\rho - t^\rho)}{\rho^{\beta+1}\Gamma(\beta+2)\Omega} \Big\{ {}^\rho I^{\alpha+\beta}_{a+} v(T) - \lambda^\rho I^\beta_{a+} x(T) \\ & - \mu^\rho I^{\alpha+\beta+\gamma}_{a+} v(\xi) + \mu\lambda^\rho I^{\beta+\gamma}_{a+} x(\xi) \Big\} - \frac{(t^\rho - a^\rho)^\beta}{\Omega(\eta^\rho - a^\rho)^\beta} \Big(\frac{(T^\rho - a^\rho)^\beta (T^\rho - t^\rho)}{\rho^{\beta+1}\Gamma(\beta+2)} \\ & - \frac{\mu(\xi^\rho - a^\rho)^{\beta+\gamma}[(\beta+1)(\xi^\rho - t^\rho) - \gamma(t^\rho - a^\rho)]}{\rho^{\beta+\gamma+1}\Gamma(\beta+\gamma+2)(\beta+1)} \Big) \Big\{ {}^\rho I^{\alpha+\beta}_{a+} v(\eta) - \lambda^\rho I^\beta_{a+} x(\eta) \Big\}. \end{aligned} \tag{13}$$

For the sake of computational convenience, we set

$$\begin{aligned}\Lambda_1 &= \frac{(T^\rho - a^\rho)^{\alpha+\beta}}{\rho^{\alpha+\beta}\Gamma(\alpha+\beta+1)}\Big[1+\frac{\zeta_1}{\rho^{\beta+1}\Gamma(\beta+2)|\Omega|}\Big] + \frac{|\mu|(\xi^\rho - a^\rho)^{\alpha+\beta+\gamma}\zeta_1}{\rho^{\alpha+2\beta+\gamma+1}\Gamma(\alpha+\beta+\gamma+1)\Gamma(\beta+2)|\Omega|}\\ &\quad + \frac{(\eta^\rho - a^\rho)^\alpha \zeta_2}{\rho^{\alpha+\beta}\Gamma(\alpha+\beta+1)|\Omega|},\end{aligned} \tag{14}$$

$$\begin{aligned}\Lambda_2 &= \frac{|\lambda|(T^\rho - a^\rho)^{\beta}}{\rho^{\beta}\Gamma(\beta+1)}\Big[1+\frac{\zeta_1}{\rho^{\beta+1}\Gamma(\beta+2)|\Omega|}\Big] + \frac{|\mu||\lambda|(\xi^\rho - a^\rho)^{\beta+\gamma}\zeta_1}{\rho^{2\beta+\gamma+1}\Gamma(\beta+\gamma+1)\Gamma(\beta+2)|\Omega|}\\ &\quad + \frac{|\lambda|\zeta_2}{\rho^{\beta}\Gamma(\beta+1)|\Omega|},\end{aligned} \tag{15}$$

where

$$\zeta_1 := \max_{t\in[a,T]} \Big|(t^\rho - a^\rho)^\beta(\eta^\rho - t^\rho)\Big|, \tag{16}$$

$$\zeta_2 := \max_{t\in[a,T]} \Big|(t^\rho - a^\rho)^\beta\Big[\frac{(T^\rho - a^\rho)^\beta(T^\rho - t^\rho)}{\rho^{\beta+1}\Gamma(\beta+2)} - \frac{\mu(\xi^\rho - a^\rho)^{\beta+\gamma}[(\beta+1)(\xi^\rho - t^\rho) - \gamma(t^\rho - a^\rho)]}{\rho^{\beta+\gamma+1}\Gamma(\beta+\gamma+2)(\beta+1)}\Big]\Big|. \tag{17}$$

We define the set of selections of F by $S_{F,x} := \{y \in L^1(J,\mathbb{R}) : y(t) \in F(t,x(t)) \text{ on } J\}$ for each $x \in C(J,\mathbb{R})$.

3.1. *The Upper Semicontinuous Case*

In the following result, we assume that the multivalued map F is convex-valued and apply Leray–Schauder nonlinear alternative for multivalued maps [18] to prove the existence of solutions for the problem at hand.

Theorem 1. *Assume that:*

- (A_1) *$F : J \times \mathbb{R} \to \mathcal{P}_{cp,c}(\mathbb{R})$ is L^1-Carathéodory, where $\mathcal{P}_{cp,c}(\mathbb{R}) = \{\mathcal{Y} \in \mathcal{P}(\mathbb{R}) : \mathcal{Y}$ is compact and convex$\}$;*
- (A_2) *there exist a function $P \in C(J,\mathbb{R}^+)$ and a continuous nondecreasing function $Q : [0,\infty) \to (0,\infty)$ such that $\|F(t,x)\|_{\mathcal{P}} := \sup\{|y| : y \in F(t,x)\} \le P(t)Q(|x|)$ for each $(t,x) \in J \times \mathbb{R}$;*
- (A_3) *there exists a constant $M > 0$ such that*

$$\frac{(1-\Lambda_2)M}{\Lambda_1\|P\|Q(M)} > 1, \quad \Lambda_2 < 1,$$

where Λ_1 and Λ_2 are respectively given by (14) and (15).

Then the problem (1) has at least one solution on J.

Proof. Let us first convert the problem (1) into a fixed point problem by introducing a multivalued map: $N : C(J,\mathbb{R}) \to \mathcal{P}(C(J,\mathbb{R}))$ as

$$N(x) = \left\{\begin{array}{l} h \in C(J,\mathbb{R}) : \\ h(t) = \left\{\begin{array}{l} {}^\rho I_{a+}^{\alpha+\beta} v(t) - \lambda\, {}^\rho I_{a+}^{\beta} x(t) \\ + \dfrac{(t^\rho - a^\rho)^\beta(\eta^\rho - t^\rho)}{\rho^{\beta+1}\Gamma(\beta+2)\Omega}\Big\{{}^\rho I_{a+}^{\alpha+\beta} v(T) - \lambda\, {}^\rho I_{a+}^{\beta} x(T) \\ -\mu\, {}^\rho I_{a+}^{\alpha+\beta+\gamma} v(\xi) + \mu\lambda\, {}^\rho I_{a+}^{\beta+\gamma} x(\xi)\Big\} - \dfrac{(t^\rho - a^\rho)^\beta}{\Omega(\eta^\rho - a^\rho)^\beta}\Big(\dfrac{(T^\rho - a^\rho)^\beta(T^\rho - t^\rho)}{\rho^{\beta+1}\Gamma(\beta+2)} \\ - \dfrac{\mu(\xi^\rho - a^\rho)^{\beta+\gamma}[(\beta+1)(\xi^\rho - t^\rho) - \gamma(t^\rho - a^\rho)]}{\rho^{\beta+\gamma+1}\Gamma(\beta+\gamma+2)(\beta+1)}\Big)\Big\{{}^\rho I_{a+}^{\alpha+\beta} v(\eta) - \lambda\, {}^\rho I_{a+}^{\beta} x(\eta)\Big\}, \end{array}\right. \end{array}\right\}$$

for $v \in S_{F,x}$.

It is clear that fixed points of N are solutions of problem (1). So we need to verify that the operator N satisfies all the conditions of Leray–Schauder nonlinear alternative [18]. This will be done in several steps.

Step 1. *$N(x)$ is convex for each $x \in C(J, \mathbb{R})$.*

Indeed, if h_1, h_2 belongs to $N(x)$, then there exist $v_1, v_2 \in S_{F,x}$ such that, for each $t \in J$, we have

$$\begin{aligned} h_i(t) &= {}^\rho I_{a+}^{\alpha+\beta} v_i(t) - \lambda^\rho I_{a+}^{\beta} x(t) + \frac{(t^\rho - a^\rho)^\beta (\eta^\rho - t^\rho)}{\rho^{\beta+1}\Gamma(\beta+2)\Omega} \Big\{ {}^\rho I_{a+}^{\alpha+\beta} v_i(T) - \lambda^\rho I_{a+}^{\beta} x(T) \\ &\quad - \mu^\rho I_{a+}^{\alpha+\beta+\gamma} v_i(\xi) + \mu\lambda^\rho I_{a+}^{\beta+\gamma} x(\xi) \Big\} - \frac{(t^\rho - a^\rho)^\beta}{\Omega(\eta^\rho - a^\rho)^\beta} \Big(\frac{(T^\rho - a^\rho)^\beta (T^\rho - t^\rho)}{\rho^{\beta+1}\Gamma(\beta+2)} \\ &\quad - \frac{\mu(\xi^\rho - a^\rho)^{\beta+\gamma}[(\beta+1)(\xi^\rho - t^\rho) - \gamma(t^\rho - a^\rho)]}{\rho^{\beta+\gamma+1}\Gamma(\beta+\gamma+2)(\beta+1)} \Big) \Big\{ {}^\rho I_{a+}^{\alpha+\beta} v_i(\eta) - \lambda^\rho I_{a+}^{\beta} x(\eta) \Big\}, i = 1, 2. \end{aligned}$$

Let $t \in J$ and $\theta \in (0,1)$. Then

$$\begin{aligned} & [\theta h_1 + (1-\theta)h_2](t) \\ = {}& {}^\rho I_{a+}^{\alpha+\beta} [\theta v_1(s) + (1-\theta)v_2(s)](t) - \lambda^\rho I_{a+}^{\beta} x(t) \\ & + \frac{(t^\rho - a^\rho)^\beta (\eta^\rho - t^\rho)}{\rho^{\beta+1}\Gamma(\beta+2)\Omega} \Big\{ {}^\rho I_{a+}^{\alpha+\beta} [\theta v_1(s) + (1-\theta)v_2(s)](T) - \lambda^\rho I_{a+}^{\beta} x(T) \\ & - \mu^\rho I_{a+}^{\alpha+\beta+\gamma} [\theta v_1(s) + (1-\theta)v_2(s)](\xi) + \mu\lambda^\rho I_{a+}^{\beta+\gamma} x(\xi) \Big\} - \frac{(t^\rho - a^\rho)^\beta}{\Omega(\eta^\rho - a^\rho)^\beta} \Big(\frac{(T^\rho - a^\rho)^\beta (T^\rho - t^\rho)}{\rho^{\beta+1}\Gamma(\beta+2)} \\ & - \frac{\mu(\xi^\rho - a^\rho)^{\beta+\gamma}[(\beta+1)(\xi^\rho - t^\rho) - \gamma(t^\rho - a^\rho)]}{\rho^{\beta+\gamma+1}\Gamma(\beta+\gamma+2)(\beta+1)} \Big) \Big\{ {}^\rho I_{a+}^{\alpha+\beta} [\theta v_1(s) + (1-\theta)v_2(s)](\eta) - \lambda^\rho I_{a+}^{\beta} x(\eta) \Big\}. \end{aligned}$$

Since F has convex values ($S_{F,x}$ is convex), therefore, $\theta h_1 + (1-\theta)h_2 \in N(x)$.

Step 2. *$N(x)$ maps bounded sets (balls) into bounded sets in $C(J, \mathbb{R})$.*

Let $B_r = \{x \in C(J, \mathbb{R}) : \|x\| \le r\}$ be a bounded ball in $C(J, \mathbb{R})$, where r is a positive number. Then, for each $h \in N(x), x \in B_r$, there exists $v \in S_{F,x}$ such that

$$\begin{aligned} h(t) &= {}^\rho I_{a+}^{\alpha+\beta} v(t) - \lambda^\rho I_{a+}^{\beta} x(t) + \frac{(t^\rho - a^\rho)^\beta (\eta^\rho - t^\rho)}{\rho^{\beta+1}\Gamma(\beta+2)\Omega} \Big\{ {}^\rho I_{a+}^{\alpha+\beta} v(T) - \lambda^\rho I_{a+}^{\beta} x(T) \\ &\quad - \mu^\rho I_{a+}^{\alpha+\beta+\gamma} v(\xi) + \mu\lambda^\rho I_{a+}^{\beta+\gamma} x(\xi) \Big\} - \frac{(t^\rho - a^\rho)^\beta}{\Omega(\eta^\rho - a^\rho)^\beta} \Big(\frac{(T^\rho - a^\rho)^\beta (T^\rho - t^\rho)}{\rho^{\beta+1}\Gamma(\beta+2)} \\ &\quad - \frac{\mu(\xi^\rho - a^\rho)^{\beta+\gamma}[(\beta+1)(\xi^\rho - t^\rho) - \gamma(t^\rho - a^\rho)]}{\rho^{\beta+\gamma+1}\Gamma(\beta+\gamma+2)(\beta+1)} \Big) \Big\{ {}^\rho I_{a+}^{\alpha+\beta} v(\eta) - \lambda^\rho I_{a+}^{\beta} x(\eta) \Big\}. \end{aligned}$$

In view of (H_2), for each $t \in J$, we find that

$$\begin{aligned}
|h(t)| &\leq {}^{\rho}I_{a+}^{\alpha+\beta}|v(t)| + |\lambda|\,{}^{\rho}I_{a+}^{\beta}|x(t)| + \frac{|(t^\rho - a^\rho)^\beta(\eta^\rho - t^\rho)|}{\rho^{\beta+1}\Gamma(\beta+2)|\Omega|}\Big\{{}^{\rho}I_{a+}^{\alpha+\beta}|v(T)| + \lambda^{\rho}I_{a+}^{\beta}|x(T)| \\
&+|\mu|\,{}^{\rho}I_{a+}^{\alpha+\beta+\gamma}|v(\xi)| + |\mu\lambda|\,{}^{\rho}I_{a+}^{\beta+\gamma}|x(\xi)|\Big\} + \Big|\frac{(t^\rho - a^\rho)^\beta}{\Omega(\eta^\rho - a^\rho)^\beta}\Big(\frac{(T^\rho - a^\rho)^\beta(T^\rho - t^\rho)}{\rho^{\beta+1}\Gamma(\beta+2)} \\
&+\frac{\mu(\xi^\rho - a^\rho)^{\beta+\gamma}[(\beta+1)(\xi^\rho - t^\rho) - \gamma(t^\rho - a^\rho)]}{\rho^{\beta+\gamma+1}\Gamma(\beta+\gamma+2)(\beta+1)}\Big)\Big|\Big\{{}^{\rho}I_{a+}^{\alpha+\beta}|v(\eta)| + |\lambda|\,{}^{\rho}I_{a+}^{\beta}|x(\eta)|\Big\} \\
&\leq \|P\|Q(\|x\|)\Big(\frac{(T^\rho - a^\rho)^{\alpha+\beta}}{\rho^{\alpha+\beta}\Gamma(\alpha+\beta+1)}\Big[1 + \frac{\zeta_1}{\rho^{\beta+1}\Gamma(\beta+2)|\Omega|}\Big] \\
&+\frac{|\mu|(\xi^\rho - a^\rho)^{\alpha+\beta+\gamma}\zeta_1}{\rho^{\alpha+2\beta+\gamma+1}\Gamma(\alpha+\beta+\gamma+1)\Gamma(\beta+2)|\Omega|} + \frac{(\eta^\rho - a^\rho)^{\alpha}\zeta_2}{\rho^{\alpha+\beta}\Gamma(\alpha+\beta+1)|\Omega|}\Big) \\
&+\|x\|\Big(\frac{|\lambda|(T^\rho - a^\rho)^\beta}{\rho^\beta\Gamma(\beta+1)}\Big[1 + \frac{\zeta_1}{\rho^{\beta+1}\Gamma(\beta+2)|\Omega|}\Big] \\
&+\frac{|\mu||\lambda|(\xi^\rho - a^\rho)^{\beta+\gamma}\zeta_1}{\rho^{2\beta+\gamma+1}\Gamma(\beta+\gamma+1)\Gamma(\beta+2)|\Omega|} + \frac{|\lambda|\zeta_2}{\rho^\beta\Gamma(\beta+1)|\Omega|}\Big) \\
&= \Lambda_1\|P\|Q(\|x\|) + \Lambda_2\|x\|,
\end{aligned}$$

which leads to $\|h\| \leq \Lambda_1\|P\|Q(r) + \Lambda_2 r$.

Step 3. *$N(x)$ maps bounded sets into equicontinuous sets of $C(J,\mathbb{R})$.*

Let x be any element in B_r and $h \in N(x)$. Then there exists a function $v \in S_{F,x}$ such that, for each $t \in J$ we have

$$\begin{aligned}
h(t) &= {}^{\rho}I_{a+}^{\alpha+\beta}v(t) - \lambda^{\rho}I_{a+}^{\beta}x(t) + \frac{(t^\rho - a^\rho)^\beta(\eta^\rho - t^\rho)}{\rho^{\beta+1}\Gamma(\beta+2)\Omega}\Big\{{}^{\rho}I_{a+}^{\alpha+\beta}v(T) - \lambda^{\rho}I_{a+}^{\beta}x(T) \\
&-\mu^{\rho}I_{a+}^{\alpha+\beta+\gamma}v(\xi) + \mu\lambda^{\rho}I_{a+}^{\beta+\gamma}x(\xi)\Big\} - \frac{(t^\rho - a^\rho)^\beta}{\Omega(\eta^\rho - a^\rho)^\beta}\Big(\frac{(T^\rho - a^\rho)^\beta(T^\rho - t^\rho)}{\rho^{\beta+1}\Gamma(\beta+2)} \\
&-\frac{\mu(\xi^\rho - a^\rho)^{\beta+\gamma}[(\beta+1)(\xi^\rho - t^\rho) - \gamma(t^\rho - a^\rho)]}{\rho^{\beta+\gamma+1}\Gamma(\beta+\gamma+2)(\beta+1)}\Big)\Big\{{}^{\rho}I_{a+}^{\alpha+\beta}v(\eta) - \lambda^{\rho}I_{a+}^{\beta}x(\eta)\Big\}.
\end{aligned}$$

Let $\tau_1,\ \tau_2 \in J,\ \tau_1 < \tau_2$. Then

$$
\begin{aligned}
&|h(t_2) - h(t_1)| \\
\leq\ & \Big|\frac{\rho^{1-(\alpha+\beta)}}{\Gamma(\alpha+\beta)}\Big[\int_0^{t_1} s^{\rho-1}[(t_2^\rho - s^\rho)^{\alpha+\beta-1} - (t_1^\rho - s^\rho)^{\alpha+\beta-1}]v(s)ds + \int_{t_1}^{t_2} s^{\rho-1}(t_2^\rho - s^\rho)^{\alpha+\beta-1}v(s)ds\Big] \\
&+\Big[\frac{(t_2^\rho - a^\rho)^\beta(\eta^\rho - t_2^\rho)}{\rho^{\beta+1}\Gamma(\beta+2)\Omega} - \frac{(t_1^\rho - a^\rho)^\beta(\eta^\rho - t_1^\rho)}{\rho^{\beta+1}\Gamma(\beta+2)\Omega}\Big]\Big\{{}^\rho I_{a+}^{\alpha+\beta}v(T) - \mu{}^\rho I_{a+}^{\alpha+\beta+\gamma}v(\xi)\Big\} \\
&-\Big[\frac{(t_2^\rho - a^\rho)^\beta}{\Omega(\eta^\rho - a^\rho)^\beta}\Big[\frac{(T^\rho - a^\rho)^\beta(T^\rho - t_2^\rho)}{\rho^{\beta+1}\Gamma(\beta+2)} - \frac{\mu(\xi^\rho - a^\rho)^{\beta+\gamma}[(\beta+1)(\xi^\rho - t_2^\rho) - \gamma(t^\rho - a^\rho)]}{\rho^{\beta+\gamma+1}\Gamma(\beta+\gamma+2)(\beta+1)}\Big] \\
&-\frac{(t_1^\rho - a^\rho)^\beta}{\Omega(\eta^\rho - a^\rho)^\beta}\Big[\frac{(T^\rho - a^\rho)^\beta(T^\rho - t_1^\rho)}{\rho^{\beta+1}\Gamma(\beta+2)} - \frac{\mu(\xi^\rho - a^\rho)^{\beta+\gamma}[(\beta+1)(\xi^\rho - t_1^\rho) - \gamma(t_1^\rho - a^\rho)]}{\rho^{\beta+\gamma+1}\Gamma(\beta+\gamma+2)(\beta+1)}\Big]\Big] \times \\
&\times {}^\rho I_{a+}^{\alpha+\beta}v(\eta)\Big| \\
&+\Big|\frac{-\lambda\rho^{1-\beta}}{\Gamma(\beta)}\Big[\int_0^{t_1} s^{\rho-1}[(t_2^\rho - s^\rho)^{\beta-1} - (t_1^\rho - s^\rho)^{\beta-1}]x(s)ds + \int_{t_1}^{t_2} s^{\rho-1}(t_2^\rho - s^\rho)^{\beta-1}x(s)ds\Big] \\
&+\Big[\frac{(t_2^\rho - a^\rho)^\beta(\eta^\rho - t_2^\rho)}{\rho^{\beta+1}\Gamma(\beta+2)\Omega} - \frac{(t_1^\rho - a^\rho)^\beta(\eta^\rho - t_1^\rho)}{\rho^{\beta+1}\Gamma(\beta+2)\Omega}\Big]\Big\{-\lambda{}^\rho I_{a+}^{\beta}x(T) + \mu\lambda{}^\rho I_{a+}^{\beta+\gamma}x(\xi)\Big\} \\
&+\Big[\frac{\lambda(t_2^\rho - a^\rho)^\beta}{\Omega(\eta^\rho - a^\rho)^\beta}\Big[\frac{(T^\rho - a^\rho)^\beta(T^\rho - t_2^\rho)}{\rho^{\beta+1}\Gamma(\beta+2)} - \frac{\mu(\xi^\rho - a^\rho)^{\beta+\gamma}[(\beta+1)(\xi^\rho - t_2^\rho) - \gamma(t^\rho - a^\rho)]}{\rho^{\beta+\gamma+1}\Gamma(\beta+\gamma+2)(\beta+1)}\Big] \\
&-\frac{\lambda(t_1^\rho - a^\rho)^\beta}{\Omega(\eta^\rho - a^\rho)^\beta}\Big[\frac{(T^\rho - a^\rho)^\beta(T^\rho - t_1^\rho)}{\rho^{\beta+1}\Gamma(\beta+2)} - \frac{\mu(\xi^\rho - a^\rho)^{\beta+\gamma}[(\beta+1)(\xi^\rho - t_1^\rho) - \gamma(t_1^\rho - a^\rho)]}{\rho^{\beta+\gamma+1}\Gamma(\beta+\gamma+2)(\beta+1)}\Big]\Big] \times \\
&\times {}^\rho I_{a+}^{\beta}x(\eta)\Big| \\
\leq\ & \frac{\|P\|Q(r)}{\rho^{\alpha+\beta}\Gamma(\alpha+\beta+1)}\Big\{|t_2^{\rho(\alpha+\beta)} - t_1^{\rho(\alpha+\beta)}| + 2(t_2^\rho - t_1^\rho)^{\alpha+\beta}\Big\} \\
&+\Big|\frac{(t_2^\rho - a^\rho)^\beta(\eta^\rho - t_2^\rho)}{\rho^{\beta+1}\Gamma(\beta+2)\Omega} - \frac{(t_1^\rho - a^\rho)^\beta(\eta^\rho - t_1^\rho)}{\rho^{\beta+1}\Gamma(\beta+2)\Omega}\Big| \\
&\times\|P\|Q(r)\Big(\frac{(T^\rho - a^\rho)^{\alpha+\beta}}{\rho^{\alpha+\beta}\Gamma(\alpha+\beta+1)} + |\mu|\frac{(\xi^\rho - a^\rho)^{\alpha+\beta+\gamma}}{\rho^{\alpha+\beta+\gamma}\Gamma(\alpha\beta+\gamma+1)}\Big) \\
&+\Big|\frac{(t_2^\rho - a^\rho)^\beta}{\Omega(\eta^\rho - a^\rho)^\beta}\Big[\frac{(T^\rho - a^\rho)^\beta(T^\rho - t_2^\rho)}{\rho^{\beta+1}\Gamma(\beta+2)} - \frac{\mu(\xi^\rho - a^\rho)^{\beta+\gamma}[(\beta+1)(\xi^\rho - t_2^\rho) - \gamma(t_2^\rho - a^\rho)]}{\rho^{\beta+\gamma+1}\Gamma(\beta+\gamma+2)(\beta+1)}\Big] \\
&-\frac{(t_1^\rho - a^\rho)^\beta}{\Omega(\eta^\rho - a^\rho)^\beta}\Big[\frac{(T^\rho - a^\rho)^\beta(T^\rho - t_1^\rho)}{\rho^{\beta+1}\Gamma(\beta+2)} - \frac{\mu(\xi^\rho - a^\rho)^{\beta+\gamma}[(\beta+1)(\xi^\rho - t_1^\rho) - \gamma(t_1^\rho - a^\rho)]}{\rho^{\beta+\gamma+1}\Gamma(\beta+\gamma+2)(\beta+1)}\Big]\Big| \times \\
&\times\|P\|Q(r)\frac{(\eta^\rho - a^\rho)^{\alpha+\beta}}{\rho^{\alpha+\beta}\Gamma(\alpha+\beta+1)} + \frac{r}{\rho^\beta\Gamma(\beta+1)}\Big\{|t_2^{\rho\beta} - t_1^{\rho\beta}| + 2(t_2^\rho - t_1^\rho)^\beta\Big\} \\
&+\Big|\Big[\frac{(t_2^\rho - a^\rho)^\beta(\eta^\rho - t_2^\rho)}{\rho^{\beta+1}\Gamma(\beta+2)\Omega} - \frac{(t_1^\rho - a^\rho)^\beta(\eta^\rho - t_1^\rho)}{\rho^{\beta+1}\Gamma(\beta+2)\Omega}\Big]\Big||\lambda|r\Big(\frac{(T^\rho - a^\rho)^\beta}{\rho^\beta\Gamma(\beta+1)} + |\mu|\frac{(\xi^\rho - a^\rho)^{\beta+\gamma}}{\rho^{\beta+\gamma}\Gamma(\beta+\gamma+1)}\Big) \\
&+\Big|\frac{\lambda(t_2^\rho - a^\rho)^\beta}{\Omega(\eta^\rho - a^\rho)^\beta}\Big[\frac{(T^\rho - a^\rho)^\beta(T^\rho - t_2^\rho)}{\rho^{\beta+1}\Gamma(\beta+2)} - \frac{\mu(\xi^\rho - a^\rho)^{\beta+\gamma}[(\beta+1)(\xi^\rho - t_2^\rho) - \gamma(t^\rho - a^\rho)]}{\rho^{\beta+\gamma+1}\Gamma(\beta+\gamma+2)(\beta+1)}\Big] \\
&-\frac{\lambda(t_1^\rho - a^\rho)^\beta}{\Omega(\eta^\rho - a^\rho)^\beta}\Big[\frac{(T^\rho - a^\rho)^\beta(T^\rho - t_1^\rho)}{\rho^{\beta+1}\Gamma(\beta+2)} - \frac{\mu(\xi^\rho - a^\rho)^{\beta+\gamma}[(\beta+1)(\xi^\rho - t_1^\rho) - \gamma(t_1^\rho - a^\rho)]}{\rho^{\beta+\gamma+1}\Gamma(\beta+\gamma+2)(\beta+1)}\Big]\Big| \times \\
&\times\frac{(\eta^\rho - a^\rho)^\beta}{\rho^\beta\Gamma(\beta+1)} \to 0 \text{ when } t_1 \to t_2, \text{ independently of } x \in B_r.
\end{aligned}
$$

Combining the outcome of Steps 1–3 with Arzelá-Ascoli theorem leads to the conclusion that $N : C(J,\mathbb{R}) \to \mathcal{P}(C(J,\mathbb{R}))$ is completely continuous.

Next, we show that N has a closed graph. Then it will follow by Proposition 1.2 in [19] that the operator N is u.s.c.

Step 4. *N has a closed graph.*

Suppose that there exists $x_n \to x_*$, $h_n \in N(x_n)$ and $h_n \to h_*$. Then we have to establish that $h_* \in N(x_*)$. Since $h_n \in N(x_n)$, there exists $v_n \in S_{F,x_n}$. In consequence, for each $t \in J$, we get

$$\begin{aligned} h_n(t) &= {}^\rho I_{a+}^{\alpha+\beta} v_n(t) - \lambda^\rho I_{a+}^{\beta} x_n(t) + \frac{(t^\rho - a^\rho)^\beta(\eta^\rho - t^\rho)}{\rho^{\beta+1}\Gamma(\beta+2)\Omega}\Big\{{}^\rho I_{a+}^{\alpha+\beta} v_n(T) - \lambda^\rho I_{a+}^{\beta} x_n(T) \\ &\quad -\mu^\rho I_{a+}^{\alpha+\beta+\gamma} v_n(\xi) + \mu\lambda^\rho I_{a+}^{\beta+\gamma} x_n(\xi)\Big\} - \frac{(t^\rho - a^\rho)^\beta}{\Omega(\eta^\rho - a^\rho)^\beta}\Big(\frac{(T^\rho - a^\rho)^\beta(T^\rho - t^\rho)}{\rho^{\beta+1}\Gamma(\beta+2)} \\ &\quad - \frac{\mu(\xi^\rho - a^\rho)^{\beta+\gamma}[(\beta+1)(\xi^\rho - t^\rho) - \gamma(t^\rho - a^\rho)]}{\rho^{\beta+\gamma+1}\Gamma(\beta+\gamma+2)(\beta+1)}\Big)\Big\{{}^\rho I_{a+}^{\alpha+\beta} v_n(\eta) - \lambda^\rho I_{a+}^{\beta} x_n(\eta)\Big\}. \end{aligned}$$

Next we show that there exists $v_* \in S_{F,x_*}$ such that, for each $t \in J$,

$$\begin{aligned} h_*(t) &= {}^\rho I_{a+}^{\alpha+\beta} v_*(t) - \lambda^\rho I_{a+}^{\beta} x_*(t) + \frac{(t^\rho - a^\rho)^\beta(\eta^\rho - t^\rho)}{\rho^{\beta+1}\Gamma(\beta+2)\Omega}\Big\{{}^\rho I_{a+}^{\alpha+\beta} v_*(T) - \lambda^\rho I_{a+}^{\beta} x_*(T) \\ &\quad -\mu^\rho I_{a+}^{\alpha+\beta+\gamma} v_*(\xi) + \mu\lambda^\rho I_{a+}^{\beta+\gamma} x_*(\xi)\Big\} - \frac{(t^\rho - a^\rho)^\beta}{\Omega(\eta^\rho - a^\rho)^\beta}\Big(\frac{(T^\rho - a^\rho)^\beta(T^\rho - t^\rho)}{\rho^{\beta+1}\Gamma(\beta+2)} \\ &\quad - \frac{\mu(\xi^\rho - a^\rho)^{\beta+\gamma}[(\beta+1)(\xi^\rho - t^\rho) - \gamma(t^\rho - a^\rho)]}{\rho^{\beta+\gamma+1}\Gamma(\beta+\gamma+2)(\beta+1)}\Big)\Big\{{}^\rho I_{a+}^{\alpha+\beta} v_*(\eta) - \lambda^\rho I_{a+}^{\beta} x_*(\eta)\Big\}. \end{aligned}$$

Consider the continuous linear operator $\Theta : L^1(J,\mathbb{R}) \to C(J,\mathbb{R})$ given by

$$\begin{aligned} v \to \Theta(v)(t) &= {}^\rho I_{a+}^{\alpha+\beta} v(t) - \lambda^\rho I_{a+}^{\beta} x(t) + \frac{(t^\rho - a^\rho)^\beta(\eta^\rho - t^\rho)}{\rho^{\beta+1}\Gamma(\beta+2)\Omega}\Big\{{}^\rho I_{a+}^{\alpha+\beta} v(T) - \lambda^\rho I_{a+}^{\beta} x(T) \\ &\quad -\mu^\rho I_{a+}^{\alpha+\beta+\gamma} v(\xi) + \mu\lambda^\rho I_{a+}^{\beta+\gamma} x(\xi)\Big\} - \frac{(t^\rho - a^\rho)^\beta}{\Omega(\eta^\rho - a^\rho)^\beta}\Big(\frac{(T^\rho - a^\rho)^\beta(T^\rho - t^\rho)}{\rho^{\beta+1}\Gamma(\beta+2)} \\ &\quad - \frac{\mu(\xi^\rho - a^\rho)^{\beta+\gamma}[(\beta+1)(\xi^\rho - t^\rho) - \gamma(t^\rho - a^\rho)]}{\rho^{\beta+\gamma+1}\Gamma(\beta+\gamma+2)(\beta+1)}\Big)\Big\{{}^\rho I_{a+}^{\alpha+\beta} v(\eta) - \lambda^\rho I_{a+}^{\beta} x(\eta)\Big\}. \end{aligned}$$

Notice that $\|h_n(t) - h_*(t)\| \to 0$ as $n \to \infty$. So we deduce by a closed graph result obtained in [20] that $\Theta \circ S_{F,x}$ is a closed graph operator. Furthermore, $h_n \in \Theta(S_{F,x_n})$. Since $x_n \to x_*$, therefore we have

$$\begin{aligned} h_*(t) &= {}^\rho I_{a+}^{\alpha+\beta} v_*(t) - \lambda^\rho I_{a+}^{\beta} x_*(t) + \frac{(t^\rho - a^\rho)^\beta(\eta^\rho - t^\rho)}{\rho^{\beta+1}\Gamma(\beta+2)\Omega}\Big\{{}^\rho I_{a+}^{\alpha+\beta} v_*(T) - \lambda^\rho I_{a+}^{\beta} x_*(T) \\ &\quad -\mu^\rho I_{a+}^{\alpha+\beta+\gamma} v_*(\xi) + \mu\lambda^\rho I_{a+}^{\beta+\gamma} x_*(\xi)\Big\} - \frac{(t^\rho - a^\rho)^\beta}{\Omega(\eta^\rho - a^\rho)^\beta}\Big(\frac{(T^\rho - a^\rho)^\beta(T^\rho - t^\rho)}{\rho^{\beta+1}\Gamma(\beta+2)} \\ &\quad - \frac{\mu(\xi^\rho - a^\rho)^{\beta+\gamma}[(\beta+1)(\xi^\rho - t^\rho) - \gamma(t^\rho - a^\rho)]}{\rho^{\beta+\gamma+1}\Gamma(\beta+\gamma+2)(\beta+1)}\Big)\Big\{{}^\rho I_{a+}^{\alpha+\beta} v_*(\eta) - \lambda^\rho I_{a+}^{\beta} x_*(\eta)\Big\}, \end{aligned}$$

for some $v_* \in S_{F,x_*}$

Step 5. *There exists an open set $\mathcal{V} \subseteq C(J,\mathbb{R})$ with $x \notin \theta N(x)$ for any $\theta \in (0,1)$ and all $x \in \partial\mathcal{V}$.*

Take $\theta \in (0,1)$, $x \in \theta N(x)$ and $t \in J$. Then we show that there exists $v \in L^1(J,\mathbb{R})$ with $v \in S_{F,x}$ such that

$$\begin{aligned} x(t) &= \theta^\rho I_{a+}^{\alpha+\beta} v(t) - \theta\lambda^\rho I_{a+}^{\beta} x(t) + \theta\frac{(t^\rho - a^\rho)^\beta(\eta^\rho - t^\rho)}{\rho^{\beta+1}\Gamma(\beta+2)\Omega}\Big\{{}^\rho I_{a+}^{\alpha+\beta} v_*(T) - \lambda^\rho I_{a+}^{\beta} x(T) \\ &\quad -\mu^\rho I_{a+}^{\alpha+\beta+\gamma} v(\xi) + \mu\lambda^\rho I_{a+}^{\beta+\gamma} x(\xi)\Big\} - \theta\frac{(t^\rho - a^\rho)^\beta}{\Omega(\eta^\rho - a^\rho)^\beta}\Big(\frac{(T^\rho - a^\rho)^\beta(T^\rho - t^\rho)}{\rho^{\beta+1}\Gamma(\beta+2)} \\ &\quad - \frac{\mu(\xi^\rho - a^\rho)^{\beta+\gamma}[(\beta+1)(\xi^\rho - t^\rho) - \gamma(t^\rho - a^\rho)]}{\rho^{\beta+\gamma+1}\Gamma(\beta+\gamma+2)(\beta+1)}\Big)\Big\{{}^\rho I_{a+}^{\alpha+\beta} v(\eta) - \lambda^\rho I_{a+}^{\beta} x(\eta)\Big\}. \end{aligned}$$

Using the computations done in Step 2, for each $t \in J$, we get

$$|x(t)| \leq \Lambda_1 \|P\| Q(\|x\|) + \Lambda_2 \|x\|,$$

which yields

$$\frac{(1-\Lambda_2)\|x\|}{\Lambda_1 \|P\| Q(\|x\|)} \leq 1.$$

By (A_3), there exists M such that $\|x\| \neq M$. Define a set

$$\mathcal{V} = \{x \in C(J, \mathbb{R}) : \|x\| < M\}.$$

Observe that the operator $N : \overline{\mathcal{V}} \to \mathcal{P}(C(J, \mathbb{R}))$ is a compact multivalued map, u.s.c. with convex closed values. With the given choice of $\mathcal{V}$, it is not possible to find $x \in \partial\mathcal{V}$ satisfying $x \in \theta N(x)$ for some $\theta \in (0,1)$. Consequently, by the nonlinear alternative of Leray–Schauder type [18], the operator N has a fixed point $x \in \overline{\mathcal{V}}$, which corresponds to a solution of the problem (1). This finishes the proof. □

3.2. The Lipschitz Case

Let $(\mathcal{X}, d)$ denote a metric space induced from the normed space $(\mathcal{X}; \|\cdot\|)$. Let $H_d : \mathcal{P}(\mathcal{X}) \times \mathcal{P}(\mathcal{X}) \to \mathbb{R} \cup \{\infty\}$ be defined by $H_d(A_1, A_2) = \max\{\sup_{a_1 \in A_1} d(a_1, A_2), \sup_{a_2 \in A_2} d(A_1, a_2)\}$, where $d(A_1, a_2) = \inf_{a_1 \in A_1} d(a_1; a_2)$ and $d(a_1, A_2) = \inf_{a_2 \in A_2} d(a_1; a_2)$. Then $(\mathcal{P}_{b,cl}(\mathcal{X}), H_d)$ is a metric space (see [21]), where $\mathcal{P}_{b,cl}(\mathcal{X}) = \{\mathcal{Y} \in \mathcal{P}(\mathcal{X}) : \mathcal{Y} \text{ is bounded and closed}\}$,

The following result deals with the non-convex valued case of the problem (1) and is based on Covitz and Nadler's fixed point theorem [22]: "If $N : \mathcal{X} \to \mathcal{P}_{cl}(\mathcal{X})$ is a contraction, then $FixN \neq \emptyset$, where $\mathcal{P}_{cl}(\mathcal{X}) = \{\mathcal{Y} \in \mathcal{P}(\mathcal{X}) : \mathcal{Y} \text{ is closed}\}$".

Theorem 2. *Assume that*

(A_4) *$F : J \times \mathbb{R} \to \mathcal{P}_{cp}(\mathbb{R})$ is such that $F(\cdot, x) : J \to \mathcal{P}_{cp}(\mathbb{R})$ is measurable for each $x \in \mathbb{R}$, where $\mathcal{P}_{cp}(\mathbb{R}) = \{\mathcal{Y} \in \mathcal{P}(\mathbb{R}) : \mathcal{Y} \text{ is compact}\}$;*

(A_5) *$H_d(F(t,x), F(t,\widehat{x})) \leq \varpi(t)|x - \widehat{x}|$ for almost all $t \in J$ and $x, \widehat{x} \in \mathbb{R}$ with $\varpi \in C(J, \mathbb{R}^+)$ and $d(0, F(t,0)) \leq \varpi(t)$ for almost all $t \in J$.*

Then the problem (1) has at least one solution on J if

$$\|\varpi\|\Lambda_1 + \Lambda_2 < 1, \tag{18}$$

where Λ_1 and Λ_2 are respectively given by (14) and (15).

Proof. Let us verify that the operator $N : C(J, \mathbb{R}) \to \mathcal{P}(C(J, \mathbb{R}))$, defined in the proof of the last theorem, satisfies the hypothesis of Covitz and Nadler fixed point theorem [22]. We establish it in two steps.

Step I. $N(x)$ is nonempty and closed for every $v \in S_{F,x}$.

Since the set-valued map $F(\cdot, x(\cdot))$ is measurable, it admits a measurable selection $v : J \to \mathbb{R}$ by the measurable selection theorem ([23], Theorem III.6). By (A_5), we have

$$|v(t)| \leq \varpi(t)(1 + |x(t)|),$$

that is, $v \in L^1(J, \mathbb{R})$. So F is integrably bounded. Therefore, $S_{F,x} \neq \emptyset$.

Now we establish that $N(x)$ is closed for each $x \in C(J,\mathbb{R})$. Let $\{u_n\}_{n\geq 0} \in N(x)$ be such that $u_n \to u$ as $n \to \infty$ in $C(J,\mathbb{R})$. Then $u \in C(J,\mathbb{R})$ and we can find $v_n \in S_{F,x_n}$ such that, for each $t \in J$,

$$\begin{aligned} u_n(t) &= {}^\rho I_{a+}^{\alpha+\beta} v_n(t) - \lambda^\rho I_{a+}^{\beta} x_n(t) + \frac{(t^\rho - a^\rho)^\beta(\eta^\rho - t^\rho)}{\rho^{\beta+1}\Gamma(\beta+2)\Omega}\Big\{{}^\rho I_{a+}^{\alpha+\beta} v(T) - \lambda^\rho I_{a+}^{\beta} x_n(T) \\ &\quad - \mu^\rho I_{a+}^{\alpha+\beta+\gamma} v_n(\xi) + \mu\lambda^\rho I_{a+}^{\beta+\gamma} x_n(\xi)\Big\} - \frac{(t^\rho - a^\rho)^\beta}{\Omega(\eta^\rho - a^\rho)^\beta}\Big(\frac{(T^\rho - a^\rho)^\beta(T^\rho - t^\rho)}{\rho^{\beta+1}\Gamma(\beta+2)} \\ &\quad - \frac{\mu(\xi^\rho - a^\rho)^{\beta+\gamma}[(\beta+1)(\xi^\rho - t^\rho) - \gamma(t^\rho - a^\rho)]}{\rho^{\beta+\gamma+1}\Gamma(\beta+\gamma+2)(\beta+1)}\Big)\Big\{{}^\rho I_{a+}^{\alpha+\beta} v_n(\eta) - \lambda^\rho I_{a+}^{\beta} x_n(\eta)\Big\}. \end{aligned}$$

As F has compact values, we can pass onto a subsequence (if necessary) to obtain that v_n converges to v in $L^1(J,\mathbb{R})$. So $v \in S_{F,x}$. Then, for each $t \in J$, we get

$$\begin{aligned} u_n(t) \to v(t) &= {}^\rho I_{a+}^{\alpha+\beta} v(t) - \lambda^\rho I_{a+}^{\beta} x(t) + \frac{(t^\rho - a^\rho)^\beta(\eta^\rho - t^\rho)}{\rho^{\beta+1}\Gamma(\beta+2)\Omega}\Big\{{}^\rho I_{a+}^{\alpha+\beta} v(T) - \lambda^\rho I_{a+}^{\beta} x(T) \\ &\quad - \mu^\rho I_{a+}^{\alpha+\beta+\gamma} v(\xi) + \mu\lambda^\rho I_{a+}^{\beta+\gamma} x(\xi)\Big\} - \frac{(t^\rho - a^\rho)^\beta}{\Omega(\eta^\rho - a^\rho)^\beta}\Big(\frac{(T^\rho - a^\rho)^\beta(T^\rho - t^\rho)}{\rho^{\beta+1}\Gamma(\beta+2)} \\ &\quad - \frac{\mu(\xi^\rho - a^\rho)^{\beta+\gamma}[(\beta+1)(\xi^\rho - t^\rho) - \gamma(t^\rho - a^\rho)]}{\rho^{\beta+\gamma+1}\Gamma(\beta+\gamma+2)(\beta+1)}\Big)\Big\{{}^\rho I_{a+}^{\alpha+\beta} v(\eta) - \lambda^\rho I_{a+}^{\beta} x(\eta)\Big\}, \end{aligned}$$

which implies that $u \in N(x)$.

Step II. We establish that there exists $0 < \hat{\theta} < 1$ $(\hat{\theta} = \Lambda_1\|\varpi\| + \Lambda_2)$ satisfying

$$H_d(N(x), N(\widehat{x})) \leq \hat{\theta}\|x - \widehat{x}\| \quad \text{for each} \quad x, \widehat{x} \in C(J,\mathbb{R}).$$

Let us take $x, \widehat{x} \in C(J,\mathbb{R})$ and $h_1 \in N(x)$. Then there exists $v_1(t) \in F(t, x(t))$ such that, for each $t \in J$,

$$\begin{aligned} h_1(t) &= {}^\rho I_{a+}^{\alpha+\beta} v_1(t) - \lambda^\rho I_{a+}^{\beta} x(t) + \frac{(t^\rho - a^\rho)^\beta(\eta^\rho - t^\rho)}{\rho^{\beta+1}\Gamma(\beta+2)\Omega}\Big\{{}^\rho I_{a+}^{\alpha+\beta} v_1(T) - \lambda^\rho I_{a+}^{\beta} x(T) \\ &\quad - \mu^\rho I_{a+}^{\alpha+\beta+\gamma} v_1(\xi) + \mu\lambda^\rho I_{a+}^{\beta+\gamma} x(\xi)\Big\} - \frac{(t^\rho - a^\rho)^\beta}{\Omega(\eta^\rho - a^\rho)^\beta}\Big(\frac{(T^\rho - a^\rho)^\beta(T^\rho - t^\rho)}{\rho^{\beta+1}\Gamma(\beta+2)} \\ &\quad - \frac{\mu(\xi^\rho - a^\rho)^{\beta+\gamma}[(\beta+1)(\xi^\rho - t^\rho) - \gamma(t^\rho - a^\rho)]}{\rho^{\beta+\gamma+1}\Gamma(\beta+\gamma+2)(\beta+1)}\Big)\Big\{{}^\rho I_{a+}^{\alpha+\beta} v_1(\eta) - \lambda^\rho I_{a+}^{\beta} x(\eta)\Big\}. \end{aligned}$$

By (A_5), we have that $H_d(F(t,x), F(t,\widehat{x})) \leq \varpi(t)|x(t) - \widehat{x}(t)|$. So, there exists $w(t) \in F(t, \widehat{x}(t))$ satisfying $|v_1(t) - w| \leq \varpi(t)|x(t) - \widehat{x}(t)|, \ t \in J$.

Define $\mathcal{W} : J \to \mathcal{P}(\mathbb{R})$ by

$$\mathcal{W}(t) = \{w \in \mathbb{R} : |v_1(t) - w| \leq \varpi(t)|x(t) - \widehat{x}(t)|\}.$$

As the multivalued operator $\mathcal{W}(t) \cap F(t, \widehat{x}(t))$ is measurable by Proposition III.4 [23], we can find a function $v_2(t)$ which is a measurable selection for $\mathcal{W}$. So $v_2(t) \in F(t, \widehat{x}(t))$ and for each $t \in J$, we have $|v_1(t) - v_2(t)| \leq \varpi(t)|x(t) - \widehat{x}(t)|$. For each $t \in J$, we define

$$\begin{aligned} h_2(t) &= {}^\rho I_{a+}^{\alpha+\beta} v_2(t) - \lambda^\rho I_{a+}^{\beta} \widehat{x}(t) + \frac{(t^\rho - a^\rho)^\beta(\eta^\rho - t^\rho)}{\rho^{\beta+1}\Gamma(\beta+2)\Omega}\Big\{{}^\rho I_{a+}^{\alpha+\beta} v_2(T) - \lambda^\rho I_{a+}^{\beta} \widehat{x}(T) \\ &\quad - \mu^\rho I_{a+}^{\alpha+\beta+\gamma} v_2(\xi) + \mu\lambda^\rho I_{a+}^{\beta+\gamma} \widehat{x}(\xi)\Big\} - \frac{(t^\rho - a^\rho)^\beta}{\Omega(\eta^\rho - a^\rho)^\beta}\Big(\frac{(T^\rho - a^\rho)^\beta(T^\rho - t^\rho)}{\rho^{\beta+1}\Gamma(\beta+2)} \\ &\quad - \frac{\mu(\xi^\rho - a^\rho)^{\beta+\gamma}[(\beta+1)(\xi^\rho - t^\rho) - \gamma(t^\rho - a^\rho)]}{\rho^{\beta+\gamma+1}\Gamma(\beta+\gamma+2)(\beta+1)}\Big)\Big\{{}^\rho I_{a+}^{\alpha+\beta} v_2(\eta) - \lambda^\rho I_{a+}^{\beta} \widehat{x}(\eta)\Big\}. \end{aligned}$$

As a result, we get

$$
\begin{aligned}
&|h_1(t)-h_2(t)| \\
= \; & \Big|{}^{\rho}I_{a+}^{\alpha+\beta}[v_2(s)-v_1(s)](t)-\lambda^{\rho}I_{a+}^{\beta}[x(t)-\widehat{x}(t)] \\
& +\frac{(t^{\rho}-a^{\rho})^{\beta}(\eta^{\rho}-t^{\rho})}{\rho^{\beta+1}\Gamma(\beta+2)\Omega}\Big\{{}^{\rho}I_{a+}^{\alpha+\beta}[v_2(s)-v_1(s)](T)-\lambda^{\rho}I_{a+}^{\beta}[x(T)-\widehat{x}(T)] \\
& -\mu^{\rho}I_{a+}^{\alpha+\beta+\gamma}[v_2(s)-v_1(s)](\xi)+\mu\lambda^{\rho}I_{a+}^{\beta+\gamma}[x(\xi)-\widehat{x}(\xi)]\Big\}-\frac{(t^{\rho}-a^{\rho})^{\beta}}{\Omega(\eta^{\rho}-a^{\rho})^{\beta}}\Big(\frac{(T^{\rho}-a^{\rho})^{\beta}(T^{\rho}-t^{\rho})}{\rho^{\beta+1}\Gamma(\beta+2)} \\
& -\frac{\mu(\xi^{\rho}-a^{\rho})^{\beta+\gamma}[(\beta+1)(\xi^{\rho}-t^{\rho})-\gamma(t^{\rho}-a^{\rho})]}{\rho^{\beta+\gamma+1}\Gamma(\beta+\gamma+2)(\beta+1)}\Big)\Big\{{}^{\rho}I_{a+}^{\alpha+\beta}[v_2(s)-v_1(s)](\eta)-\lambda^{\rho}I_{a+}^{\beta}[x(\eta)-\widehat{x}(\eta)]\Big\}\Big| \\
\leq \; & \|\varpi\|\|x-\widehat{x}\|\Big(\frac{(T^{\rho}-a^{\rho})^{\alpha+\beta}}{\rho^{\alpha+\beta}\Gamma(\alpha+\beta+1)}\Big[1+\frac{\zeta_1}{\rho^{\beta+1}\Gamma(\beta+2)|\Omega|}\Big] \\
& +\frac{|\mu|(\xi^{\rho}-a^{\rho})^{\alpha+\beta+\gamma}\zeta_1}{\rho^{\alpha+2\beta+\gamma+1}\Gamma(\alpha+\beta+\gamma+1)\Gamma(\beta+2)|\Omega|} \\
& +\frac{(\eta^{\rho}-a^{\rho})^{\alpha}\zeta_2}{\rho^{\alpha+\beta}\Gamma(\alpha+\beta+1)|\Omega|}\Big)+\|x-\widehat{x}\|\Big(\frac{|\lambda|(T^{\rho}-a^{\rho})^{\beta}}{\rho^{\beta}\Gamma(\beta+1)}\Big[1+\frac{\zeta_1}{\rho^{\beta+1}\Gamma(\beta+2)|\Omega|}\Big] \\
& +\frac{|\mu||\lambda|(\xi^{\rho}-a^{\rho})^{\beta+\gamma}\zeta_1}{\rho^{2\beta+\gamma+1}\Gamma(\beta+\gamma+1)\Gamma(\beta+2)|\Omega|}+\frac{|\lambda|\zeta_2}{\rho^{\beta}\Gamma(\beta+1)|\Omega|}\Big) \\
= \; & (\|\varpi\|\Lambda_1+\Lambda_2)\|x-\widehat{x}\|.
\end{aligned}
$$

Hence

$$\|h_1-h_2\|\leq(\|\varpi\|\Lambda_1+\Lambda_2)\|x-\widehat{x}\|.$$

Analogously, we can interchange the roles of x and $\widehat{x}$ to get

$$H_d(N(x),N(\widehat{x}))\leq(\|\varpi\|\Lambda_1+\Lambda_2)\|x-\widehat{x}\|,$$

which implies that N is a contraction by the condition (18). Hence, by the conclusion of Covitz and Nadler fixed point theorem [22], N has a fixed point x, which corresponds to a solution of (1). This finishes the proof. □

4. Examples

We illustrate our main results by presenting a numerical example.

Example 1. *Consider the following problem*

$$
\begin{cases}
{}^{1/3}_{\ c}D^{5/4}\left({}^{1/3}_{\ c}D^{1/4}+1/5\right)x(t)\in F(t,x(t)), \quad t\in J:=[1,2], \\
x(1)=0, \quad x(3/2)=0, \quad x(2)=2/7\,{}^{1/3}I^{3/4}x(7/4).
\end{cases} \tag{19}
$$

Here $\rho=1/3$, $\alpha=5/4$, $\beta=1/4$, $\lambda=1/5$, $a=1$, $T=2$ $\eta=3/2$, $\mu=2/7$, $\gamma=3/4$, $\xi=7/4$. Using the given data, we find that $\zeta_1\approx 0.082260$, $\zeta_2\approx 0.232036$, $|\Omega|\approx 0.293634$, $\Lambda_1\approx 1.336009$ and $\Lambda_2\approx 0.673563$, where ζ_1, ζ_2, Λ_1 and Λ_2 are given by (16), (17), (14) and (15) respectively.

(i) Let us consider the function

$$F(t,x(t))=\Big[\frac{1}{\sqrt{t^2+63}}\Big(\frac{|x(t)|}{3}\Big(\frac{|x(t)|}{|x(t)|+1}+2\Big)+1\Big),\frac{e^{-t}}{9t+8}\Big(\sin x(t)+\frac{1}{80}\Big)\Big]. \tag{20}$$

We note that $|F(t,x(t))|\leq P(t)Q(\|x\|)$, where $P(t)=\frac{1}{\sqrt{t^2+63}}$, $Q(\|x\|)=\|x\|+1$. So the assumption (A_2) holds. Moreover, there exists $M>1.047447394$ satisfying (A_3). Thus the hypothesis of Theorem 1 holds true and hence there exists at least one solution for the problem (19) with $F(t,x)$ given by (20) on $[1,2]$.

(ii) To illustrate Theorem 2 we consider the function

$$F(t,x) = \left[\frac{e^{-t}}{19+t}, \frac{1}{(t+4)^2}\left(x + \tan^{-1}(x) + \frac{1}{15}\right)\right]. \tag{21}$$

Clearly $H_d(F(t,x), F(t,\bar{x})) \leq \varpi(t)|x - \bar{x}|$, where $\varpi(t) = \frac{2}{(t+4)^2}$. Also $d(0, F(t,0)) \leq \varpi(t)$ for almost all $t \in [0,1]$ and $\Lambda_1\|\varpi\| + \Lambda_2 \approx 0.7804433180 < 1$. As the hypothesis of Theorem 2 is satisfied, therefore we conclude that the multivalued problem (19) with $F(t,x)$ given by (21) has at least one solution on $[1,2]$.

5. Conclusions

We have introduced a new class of multivalued (inclusions) boundary value problems on an arbitrary domain containing Caputo-type generalized fractional differential operators of different orders and a generalized integral operator. We have considered convex as well as non-convex valued cases for the multi-valued map involved in the given problem. Leray–Schauder nonlinear alternative for multivalued maps plays a central role in proving the existence of solutions for convex valued case of the given problem, while the existence result for the non-convex valued case is based on Covitz and Nadler fixed point theorem. The work presented in this paper is not only new in the given configuration, but will also lead to some new results as special cases. For example, fixing $\mu = 0$ in the obtained results, we obtain the ones for nonlocal three-point boundary conditions: $x(a) = 0, x(\eta) = 0, x(T) = 0, 0 < \eta < T$. For $\rho = 1$, our results specialize to the ones for Liouville–Caputo type fractional differential inclusions complemented with nonlocal generalized integral boundary conditions on an arbitrary domain.

Author Contributions: Conceptualization, B.A. and M.A.; methodology, A.A. and S.K.N.; validation, A.A., B.A., M.A. and S.K.N.; formal analysis, A.A., B.A., M.A. and S.K.N.; writing—original draft preparation, M.A.; writing—review and editing, A.A., B.A. and S.K.N.; project administration, B.A.; funding acquisition, A.A.

Acknowledgments: This project was funded by the Deanship of Scientific Research (DSR), King Abdulaziz University, Jeddah, under grant no. (KEP-PhD-70-130-38). The authors, therefore, acknowledge with thanks DSR technical and financial support.

References

1. Zaslavsky, G.M. *Hamiltonian Chaos and Fractional Dynamics*; Oxford University Press: Oxford, UK, 2005.
2. Magin, R.L. *Fractional Calculus in Bioengineering*; Begell House Publishers: Danbury, CT, USA, 2006.
3. Kilbas, A.A.; Srivastava, H.M.; Trujillo, J.J. *Theory and Applications of Fractional Differential Equations*; North-Holland Mathematics Studies 204; Elsevier Science B.V.: Amsterdam, The Netherlands, 2006.
4. Diethelm, K. *The Analysis of Fractional Differential Equations. An Application-Oriented Exposition Using Differential Operators of Caputo Type*; Lecture Notes in Mathematics 2004; Springer: Berlin, Germany, 2010.
5. Javidi, M.; Ahmad, B. Dynamic analysis of time fractional order phytoplankton-toxic phytoplankton–zooplankton system. *Ecol. Model.* **2015**, *318*, 8–18. [CrossRef]
6. Fallahgoul, H.A.; Focardi, S.M.; Fabozzi, F.J. *Fractional Calculus and Fractional Processes with Applications to Financial Economics. Theory and Application*; Elsevier/Academic Press: London, UK, 2017.
7. Ahmad, B.; Alsaedi, A.; Ntouyas, S.K.; Tariboon, J. *Hadamard-Type Fractional Differential Equations, Inclusions and Inequalities*; Springer: Cham, Switzerland, 2017.
8. Ahmad, B.; Nieto, J.J.; Alsaedi, A.; El-Shahed, M. A study of nonlinear Langevin equation involving two fractional orders in different intervals. *Nonlinear Anal. Real World Appl.* **2012**, *13*, 599–606. [CrossRef]

9. Wang, G.; Zhang, L.; Song, G. Boundary value problem of a nonlinear Langevin equation with two different fractional orders and impulses. *Fixed Point Theory Appl.* **2012**, *2012*, 200. [CrossRef]
10. Ahmad, B.; Ntouyas, S.K. New existence results for differential inclusions involving Langevin equation with two indices. *J. Nonlinear Convex Anal.* **2013**, *14*, 437–450.
11. Muensawat, T.; Ntouyas, S.K.; Tariboon, J. Systems of generalized Sturm-Liouville and Langevin fractional differential equations. *Adv. Differ. Equ.* **2017**, *2017*, 63. [CrossRef]
12. Fazli, H.; Nieto, J.J. Fractional Langevin equation with anti-periodic boundary conditions. *Chaos Solitons Fractals* **2018**, *114*, 332–337. [CrossRef]
13. Ahmad, B.; Alsaedi, A.; Salem, S. On a nonlocal integral boundary value problem of nonlinear Langevin equation with different fractional orders. *Adv. Differ. Equ.* **2019**, *2019*, 57. [CrossRef]
14. Ahmad, B.; Alghanmi, M.; Alsaedi, A.; Srivastava, H.; Ntouyas, K. The Langevin equation in terms of generalized Liouville-Caputo derivatives with nonlocal boundary conditions involving a generalized fractional integral. *Mathematics* **2019**, *7*, 533. [CrossRef]
15. Katugampola, U.N. New Approach to a generalized fractional integral. *Appl. Math. Comput.* **2015**, *218*, 860–865. [CrossRef]
16. Katugampola, U.N. A new approach to generalized fractional derivatives. *Bull. Math. Anal. Appl.* **2014**, *6*, 1–15.
17. Jarad, F.; Abdeljawad, T.; Baleanu, D. On the generalized fractional derivatives and their caputo modification. *J. Nonlinear Sci. Appl.* **2017**, *10*, 2607–2619. [CrossRef]
18. Granas, A.; Dugundji, J. *Fixed Point Theory*; Springer: New York, NY, USA, 2005.
19. Deimling, K. *Multivalued Differential Equations*; De Gruyter: Berlin, Germany, 1992.
20. Lasota, A.; Opial, Z. An application of the Kakutani-Ky Fan theorem in the theory of ordinary differential equations, *Bull. Acad. Pol. Sci. Ser. Sci. Math. Astronom. Phys.* **1965**, *13*, 781–786.
21. Kisielewicz, M. *Differential Inclusions and Optimal Control*; Kluwer: Dordrecht, The Netherlands, 1991.
22. Covitz, H.; Nadler, S.B., Jr. Multivalued contraction mappings in generalized metric spaces. *ISRAEL J. Math.* **1970**, *8*, 5–11. [CrossRef]
23. Castaing, C.; Valadier, M. *Convex Analysis and Measurable Multifunctions*; Lecture Notes in Mathematics 580; Springer: Berlin/Heidelberg, Germany; New York, NY, USA, 1977.

12

Generalized Integral Inequalities for Hermite–Hadamard-Type Inequalities via *s*-Convexity on Fractal Sets

Ohud Almutairi [1,†] and Adem Kılıçman [2,*,†]

1 Department of Mathematics, University of Hafr Al-Batin, Hafr Al-Batin 31991, Saudi Arabia; AhoudbAlmutairi@gmail.com
2 Department of Mathematics and Institute for Mathematical Research, University Putra Malaysia, Serdang 43400, Malaysia
* Correspondence: akilic@upm.edu.my;
† These authors contributed equally to this work.

Abstract: In this article, we establish new Hermite–Hadamard-type inequalities via Riemann–Liouville integrals of a function ψ taking its value in a fractal subset of $\mathbb{R}$ and possessing an appropriate generalized s-convexity property. It is shown that these fractal inequalities give rise to a generalized s-convexity property of ψ. We also prove certain inequalities involving Riemann–Liouville integrals of a function ψ provided that the absolute value of the first or second order derivative of ψ possesses an appropriate fractal s-convexity property.

Keywords: s-convex function; Hermite–Hadamard inequalities; Riemann–Liouville fractional integrals; fractal space

---

1. Introduction

Convexity is considered to be an important property in mathematical analysis. The applications of convex functions can be found in many fields of studies including economics, engineering and optimization (see for example [1,2]). A well-known result which was identified as Hermite–Hadamard inequalities is the reformulation through convexity. These inequalities, widely reported in the literature, can be defined as follows:

Theorem 1. *Suppose that $\psi : [u,v] \subset \mathbb{R} \to \mathbb{R}$ is a convex function on $[u,v]$ with $u < v$, then*

$$\psi\left(\frac{u+v}{2}\right) \leq \frac{1}{v-u}\int_u^v \psi(x)dx \leq \frac{\psi(u)+\psi(v)}{2}. \tag{1}$$

These two inequalities, which are refinement of convexity, can be held in reverse order as concave. Following this, many refinements of convex functions using Hermite–Hadamard inequalities have been continuously studied [3–6]. Given the variation of Hermite–Hadamard inequalities, Dragomir and Fitzpatrick [7] established a new generalization of s-convex functions in the second sense.

Theorem 2. *Suppose that $\psi : \mathbb{R}_+ \to \mathbb{R}_+$ is a s-convex function in the second sense, where $0 < s \leq 1$, $u, v \in \mathbb{R}_+$ and $u < v$. If $\psi \in L^1([u,v])$, then*

$$2^{s-1}\psi\left(\frac{u+v}{2}\right) \leq \frac{1}{v-u}\int_u^v \psi(x)dx \leq \frac{\psi(u)+\psi(v)}{s+1}. \tag{2}$$

Though the Hermite–Hadamard inequalities were established for classical integrals [8], the inequalities can also hold for fractional calculus, such as Riemann–Liouville [9–11], Katugampola [12] and local fractional integrals [13]. Some of these were studied through Mittag–Leffler function [14,15]. Other important generalizations include the work of Sarikaya et al. [16], who proved the Hermite–Hadamard inequalities through fractional integrals as follows:

Theorem 3. *Suppose that $\psi : [u,v] \to \mathbb{R}$ is a non-negative function with $0 \leq u < v$ and $\psi \in L_1[u,v]$. If ψ is convex function on $[u,v]$, we have:*

$$\psi\left(\frac{u+v}{2}\right) \leq \frac{\Gamma(\alpha+1)}{2(v-u)^\alpha}\left[J^\alpha_{u+}\psi(v) + J^\alpha_{v-}\psi(u)\right] \leq \frac{\psi(u)+\psi(v)}{2},$$

where $0 < \alpha \leq 1$.

The s-convexity mentioned in Hudzik and Maligranda [2] was also given as the generalization on fractal sets.

Definition 1 ([17]). *A function $\psi : \mathbb{R}_+ \to \mathbb{R}^\alpha$ is called generalized s-convex in the second sense if*

$$\psi(\gamma_1 u + \gamma_2 v) \leq \gamma_1^{\alpha s}\psi(u) + \gamma_2^{\alpha s}\psi(v), \tag{3}$$

holds for all $u, v \in \mathbb{R}_+$, $\gamma_1, \gamma_2 \geq 0$, with $\gamma_1 + \gamma_2 = 1$ and for some fixed $s \in (0,1]$. The symbol GK^2_s denotes the class of this functions.

The Riemann–Liouville fractional integral is introduced here due to its importance.

Definition 2 ([18]). *Suppose that $\psi \in L_1[u,v]$. The Riemann–Liouville integrals $J^\alpha_{u+}\psi$ and $J^\alpha_{v-}\psi$ of order $\alpha \in \mathbb{R}_+$ are defined by*

$$J^\alpha_{u+}\psi(x) = \frac{1}{\Gamma(\alpha)}\int_u^x (x-\gamma)^{\alpha-1}\psi(\gamma)d\gamma, \quad x > u,$$

and

$$J^\alpha_{v-}\psi(x) = \frac{1}{\Gamma(\alpha)}\int_x^v (\gamma-x)^{\alpha-1}\psi(\gamma)d\gamma, \quad x < v,$$

respectively.

The following lemma for differentiable function is given by Sarikaya et al. [16].

Lemma 1. *Let $\psi : [u,v] \to \mathbb{R}$ be a differentiable function on (u,v) with $u < v$. If $\psi' \in L^1[u,v]$, then we have:*

$$\frac{\psi(u)+\psi(v)}{2} - \frac{\Gamma(\alpha+1)}{2(v-u)^\alpha}\left[J^\alpha_{u+}\psi(v) + J^\alpha_{v-}\psi(u)\right] = \frac{v-u}{2}\int_0^1 \left[(1-\gamma)^\alpha - \gamma^\alpha\right]\psi'(\gamma u + (1-\gamma)v)d\gamma.$$

Wang et al. [9] extended Lemma 1 to include two cases, one of which involves the second derivative of Riemann–Liouville fractional integrals.

Lemma 2. *Let $\psi : [u, v] \to \mathbb{R}$ be a twice-differentiable function on (u, v) with $u < v$. If $\psi'' \in L^1[u, v]$, then*

$$\frac{\psi(u)+\psi(v)}{2} - \frac{\Gamma(\alpha+1)}{2(v-u)^\alpha}\left[J^\alpha_{u+}\psi(v) + J^\alpha_{v-}\psi(u)\right]$$
$$= \frac{(v-u)^2}{2}\int_0^1 \frac{1-(1-\gamma)^{\alpha+1}-\gamma^{\alpha+1}}{\alpha+1}\psi''(\gamma u + (1-\gamma)v)d\gamma,$$

holds.

Even though studies were conducted on generalized Hermite–Hadamard inequality via Riemann–Liouville fractional integrals for s-convexity [16,19–21], inequalities of this type for generalized s-convexity are lacking. Therefore, this paper is aimed at establishing some new integral inequalities via generalized s-convexity on fractal sets. We show that the newly established inequalities are generalizations of Theorem 2. The new Hermite–Hadamard-type inequalities in the class of functions with derivatives in absolute values are shown to be s-convex function on fractal sets. This was achieved using Riemann–Liouville fractional integrals inequalities.

2. Main Results

Our first main result is obtained in the following theorem.

Theorem 4. *Suppose that $\psi : [u, v] \subseteq \mathbb{R}_+ \to \mathbb{R}^\alpha_+$ is a generalized s-convex on $[u, v]$, where $0 < s < 1$, $u, v \in \mathbb{R}_+$ and $u \le v$. If $\psi \in L^1[u, v]$, then we obtain*

$$2^{\alpha(s-1)}\psi\left(\frac{u+v}{2}\right) \le \frac{\Gamma(\alpha+1)}{2(v-u)^\alpha}\left[J^\alpha_{u+}\psi(v) + J^\alpha_{v-}\psi(u)\right] \le \left[\frac{1}{s+1} + \frac{\Gamma(\alpha s+1)\Gamma(\alpha+1)}{\Gamma(\alpha(s+1)+1)}\right]\frac{[\psi(u)+\psi(v)]}{2}. \tag{4}$$

Proof. Since $\psi \in GK^2_s$, we get

$$\psi\left(\frac{x+y}{2}\right) \le \frac{\psi(x)+\psi(y)}{2^{\alpha s}}. \tag{5}$$

Substituting $x = \gamma u + (1-\gamma)v$ and $y = (1-\gamma)u + \gamma v$ with $\gamma \in [0, 1]$ in inequality (5), we obtain

$$2^{\alpha s}\psi\left(\frac{x+y}{2}\right) \le \psi(\gamma u + (1-\gamma)v) + \psi(\gamma v + (1-\gamma)u). \tag{6}$$

Multiplying both sides of (6) by $\gamma^{\alpha-1}$ and integrating the resulting inequality with respect to γ over $[0, 1]$ yields

$$\begin{aligned}\frac{2^{\alpha s}}{\alpha}\psi\left(\frac{x+y}{2}\right) &\le \int_0^1 \gamma^{\alpha-1}\psi(\gamma u + (1-\gamma)v) + \int_0^1 \gamma^{\alpha-1}\psi(\gamma v + (1-\gamma)u)\\ &= \frac{\Gamma(\alpha)}{(v-u)^\alpha}\left[J^\alpha_{u+}\psi(v) + J^\alpha_{v-}\psi(u)\right].\end{aligned} \tag{7}$$

Then the first inequality in (4) is proved.

To prove the second inequality in (4), since $\psi \in GK^2_s$, we get

$$\psi(\gamma u + (1-\gamma)v) \le \gamma^{\alpha s}\psi(u) + (1-\gamma)^{\alpha s}\psi(v), \tag{8}$$

and

$$\psi(\gamma v + (1-\gamma)u) \le \gamma^{\alpha s}\psi(v) + (1-\gamma)^{\alpha s}\psi(u). \tag{9}$$

Combining the inequalities (8) and (9), we obtain

$$\begin{aligned}\psi(\gamma u+(1-\gamma)v)+\psi((1-\gamma)u+\gamma v) &\leq \gamma^{\alpha s}\psi(u)+(1-\gamma)^{\alpha s}\psi(v)+\gamma^{\alpha s}\psi(u)+(1-\gamma)^{\alpha s}\psi(u)\\ &=[\gamma^{\alpha s}+(1-\gamma)^{\alpha s}][\psi(u)+\psi(v)].\end{aligned} \tag{10}$$

A similar technique used in (6) is applied to inequality (10) to get the following:

$$\begin{aligned}\frac{\Gamma(\alpha)}{(v-u)^{\alpha}}[J_{u+}^{\alpha}\psi(v)+J_{v-}^{\alpha}\psi(u)] &\leq \int_0^1 \gamma^{\alpha-1}[\gamma^{\alpha s}+(1-\gamma)^{\alpha s}][\psi(u)+\psi(v)]d\gamma\\ &\leq \left[\frac{1}{\alpha(s+1)}+\frac{\Gamma(\alpha s+1)\Gamma(\alpha)}{\Gamma(\alpha(s+1)+1)}\right][\psi(u)+\psi(v),\end{aligned} \tag{11}$$

where

$$\int_0^1 \gamma^{\alpha s+s-1}d\gamma=\frac{1}{\alpha s+\alpha},$$

and

$$\int_0^1 \gamma^{\alpha-1}(1-\gamma)^{\alpha s}d\gamma=\frac{\Gamma(\alpha s+1)\Gamma(\alpha)}{\Gamma(\alpha s+\alpha+1)}.$$

Using inequalities (7) and (11), we prove Theorem 4. □

Remark 1. *In the second inequality of Theorem 4, the expression* $\frac{1}{2}\left[\frac{1}{s+1}+\frac{\Gamma(\alpha s+1)\Gamma(\alpha+1)}{\Gamma(\alpha(s+1)+1)}\right]$ *for* $0<s\leq 1$ *is the best possible. The map* $\psi:[0,1]\to[0^{\alpha},1^{\alpha}]$ *given by* $\psi(z)=z^{s\alpha}$ *is generalized s-convex in the second sense, and it satisfies the following equalities:*

$$\begin{aligned}\frac{\Gamma(\alpha+1)}{2}[J_{u+}^{\alpha}\psi(1)+J_{v-}^{\alpha}\psi(0)] &= \frac{\Gamma(\alpha+1)}{2}\left[\frac{1}{\Gamma(\alpha)}\left(\frac{1}{\alpha(s+1)}+\frac{\Gamma(\alpha s+1)\Gamma(\alpha)}{\Gamma(\alpha s+\alpha+1)}\right)\right]\\ &= \frac{\Gamma(\alpha+1)}{2}\left[\frac{1}{\alpha\Gamma(\alpha)(s+1)}+\frac{\Gamma(\alpha s+1)\Gamma(\alpha)}{\Gamma(\alpha)\Gamma(\alpha s+\alpha+1)}\right]\\ &= \frac{1}{2}\left[\frac{1}{s+1}+\frac{\Gamma(\alpha s+1)\Gamma(\alpha+1)}{\Gamma(\alpha(s+1)+1)}\right],\end{aligned}$$

and

$$\left[\frac{1}{s+1}+\frac{\Gamma(\alpha s+1)\Gamma(\alpha+1)}{\Gamma(\alpha(s+1)+1)}\right]\frac{[\psi(0)+\psi(1)]}{2}=\frac{1}{2}\left[\frac{1}{s+1}+\frac{\Gamma(\alpha s+1)\Gamma(\alpha+1)}{\Gamma(\alpha(s+1)+1)}\right].$$

Corollary 1. *By taking* $\alpha=1$ *in Theorem 4, the inequalities in (2) of Theorem 2 are recovered.*

This result is the same as Theorem 2.1 in Dragomir and Fitzpatrick [7].

Remark 2. *The equality*

$$\beta(u,v)=\frac{\Gamma(u)\Gamma(v)}{\Gamma(u+v)}$$

implies

$$2^{\alpha s-1}\psi(\frac{u+v}{2}) \leq \frac{\Gamma(\alpha+1)}{2(v-u)^{\alpha}}[J^{\alpha}_{u^+}\psi(v)+J^{\alpha}_{v^-}\psi(u)] \leq \left[\frac{1}{s+1}+\frac{\Gamma(\alpha s+1)\Gamma(\alpha+1)}{\Gamma(\alpha(s+1)+1)}\right]\frac{[\psi(u)+\psi(v)]}{2}$$
$$\leq \left[\frac{1}{s+1}+\beta(\alpha s+1,\alpha+1)(\alpha(s+1)+1)\right] \times \frac{[\psi(u)+\psi(v)]}{2}.$$

Theorem 5. *Suppose that $M:[0,1]\to\mathbb{R}^{\alpha}$ is the mapping given by*

$$M(\gamma)=\frac{\Gamma(\alpha+1)}{(v-u)^{\alpha}}\left[J^{\alpha}_{u^+}\psi\left(\gamma v+(1-\gamma)\frac{u+v}{2}\right)+J^{\alpha}_{v^-}\psi\left(\gamma u+(1-\gamma)\frac{u+v}{2}\right)\right],\gamma(0,1),$$

where $\psi:[u,v]\to\mathbb{R}^{\alpha}$ belongs to GK^2_s, $s\in(0,1]$, $u,v\in\mathbb{R}_+, u<v$ and $\psi\in L^1([u,v])$. Then

(i) $M\in GK^2_s$ *on* $[0,1]$.
(ii) *We have the following inequality:*

$$M(\gamma)\geq 2^{\alpha s}\psi\left(\frac{u+v}{2}\right). \tag{12}$$

(iii) *We have the following inequality:*

$$M(\gamma)\leq \min\{M_1(\gamma),M_2(\gamma)\},\gamma\in[0,1], \tag{13}$$

where

$$M_1(\gamma)=\gamma^{\alpha s}\frac{\Gamma(\alpha+1)}{(v-u)^{\alpha}}[J^{\alpha}_{u^+}\psi(v)+J^{\alpha}_{v^-}\psi(u)]+(1-\gamma)^{\alpha s}\psi\left(\frac{u+v}{2}\right),$$

and

$$M_2(\gamma)=\left[\frac{1}{s+1}+\frac{\Gamma(\alpha s+1)\Gamma(\alpha+1)}{\Gamma(\alpha(s+1)+2)}\right]\left[\psi\left(\gamma u+(1-\gamma)\frac{u+v}{2}\right)+\psi\left(\gamma v+(1-\gamma)\frac{u+v}{2}\right)\right].$$

(iv) *If $\tilde{M}=\max\{M_1(\gamma),M_2(\gamma)\},\gamma\in[0,1]$, then we have*

$$\tilde{M}\leq\left[\frac{1}{s+1}+\frac{\Gamma(\alpha s+1)\Gamma(\alpha+1)}{\Gamma(\alpha(s+1)+1)}\right]\left[\gamma^{\alpha s}[\psi(u)+\psi(v)]+2^{\alpha s}(1-\gamma)^{\alpha s}\psi\left(\frac{u+v}{2}\right)\right].$$

Proof.

(i) Let $\gamma_1,\gamma_2\in[0,1]$ and $\mu_1,\mu_2\geq 0$ with $\mu_1+\mu_2=1$, then

$$\begin{aligned}
M(\mu_1\gamma_1+\mu_2\gamma_2) &= \tfrac{\Gamma(\alpha+1)}{(v-u)^{\alpha}}\Big[J^{\alpha}_{u^+}\psi((\mu_1\gamma_1+\mu_2\gamma_2)\tfrac{u+v}{2}+(1-(\mu_1\gamma_1+\mu_2\gamma_2))v)\\
&\quad +J^{\alpha}_{v^-}\psi\Big((\mu_1\gamma_1+\mu_2\gamma_2)u+(1-(\mu_1\gamma_1+\mu_2\gamma_2))\tfrac{u+v}{2}\Big)\Big]\\
&\leq \tfrac{\Gamma(\alpha+1)}{(v-u)^{\alpha}}\Big[\mu_1^{\alpha s}\Big(J^{\alpha}_{u^+}\psi\Big(\gamma_1\tfrac{u+v}{2}+(1-\gamma_1)v\Big)+J^{\alpha}_{v^-}\psi\Big(\gamma_1 u+(1-\gamma_1)\tfrac{u+v}{2}\Big)\Big)\\
&\quad +\mu_2^{\alpha s}\Big(J^{\alpha}_{u^+}\psi\Big(\gamma_2\tfrac{u+v}{2}+(1-\gamma_2)v\Big)+J^{\alpha}_{v^-}\psi\Big(\gamma_2 u+(1-\gamma_2)\tfrac{u+v}{2}\Big)\Big)\Big]\\
&= \mu_1^{\alpha s}M(\gamma_1)+\mu_2^{\alpha s}M(\gamma_2).
\end{aligned}$$

(ii) Assume that $\gamma \in (0,1]$. Then by the change of variables $q = \gamma v + (1-\gamma)\frac{u+v}{2}$ and $p = \gamma u + (1-\gamma)\frac{u+v}{2}$, we have

$$\begin{aligned} M(\gamma) &= \frac{\Gamma(\alpha+1)}{\gamma^\alpha (v-u)^\alpha}\left[J^\alpha_{(\gamma v+(1-\gamma)\frac{u+v}{2})+}\psi\left(\gamma u + (1-\gamma)\frac{u+v}{2}\right)\right. \\ &\left.+ J^\alpha_{(\gamma u+(1-\gamma)\frac{u+v}{2})-}\psi\left(\gamma v + (1-\gamma)\frac{u+v}{2}\right)\right] \\ &= \frac{\Gamma(\alpha+1)}{(p-q)^\alpha}[J^\alpha_{q+}\psi(p) + J^\alpha_{p-}\psi(q)]. \end{aligned}$$

Applying the first generalized Hermite–Hadamard inequality, we obtain

$$\frac{\Gamma(\alpha+1)}{(p-q)^\alpha}[J^\alpha_{q+}\psi(p) + J^\alpha_{p-}\psi(q)] \geq 2^{\alpha s}\psi\left(\frac{q+p}{2}\right) = 2^{\alpha s}\psi\left(\frac{u+v}{2}\right),$$

and inequality (12) is obtained.
If $\gamma = 0$, the inequality

$$\psi\left(\frac{u+v}{2}\right) \geq 2^{\alpha s-1}\psi\left(\frac{u+v}{2}\right),$$

also holds.

(iii) Applying the second generalized Hermite–Hadamard inequality, we obtain

$$\begin{aligned} \frac{\Gamma(\alpha+1)}{(v-u)^\alpha}[J^\alpha_{u+}\psi(v) + J^\alpha_{v-}\psi(u)] &\leq \left[\frac{1}{s+1} + \frac{\Gamma(\alpha s+1)\Gamma(\alpha+1)}{\Gamma(\alpha(s+1)+1)}\right][\psi(u)+\psi(v)] \\ &= \left[\frac{1}{s+1} + \frac{\Gamma(\alpha s+1)\Gamma(\alpha+1)}{\Gamma(\alpha(s+1)+1)}\right]\left[\psi\left(\gamma u + (1-\gamma)\frac{u+v}{2}\right)\right. \\ &\left.+\psi\left(\gamma v + (1-\gamma)\frac{u+v}{2}\right)\right] \\ &= A_2(\gamma), \forall \gamma \in [0,1]. \end{aligned}$$

Please note that if $\gamma = 0$, then the inequality

$$\psi\left(\frac{u+v}{2}\right) = M(0) \leq M_2(0) = 2^\alpha\left[\frac{1}{s+1} + \frac{\Gamma(\alpha s+1)\Gamma(\alpha+1)}{\Gamma(\alpha(s+1)+1)}\right]\psi\left(\frac{u+v}{2}\right),$$

holds as it is equivalent to

$$\left(\left[\frac{\Gamma(\alpha s+\alpha+1)(s+1)}{\Gamma(\alpha s+1)\Gamma(\alpha+1)}\right] - 2^\alpha\right)\psi\left(\frac{u+v}{2}\right) \leq 0^\alpha,$$

which is known to hold for $s \in (0,1]$.

Since for all $\gamma \in [0,1]$ and $x \in [u,v]$ the inequalities

$$\psi\left(\gamma u + (1-\gamma)\frac{u+v}{2}\right) \leq \gamma^{\alpha s}\psi(u) + (1-\gamma)^{\alpha s}\psi\left(\frac{u+v}{2}\right),$$

and

$$\psi\left(\gamma v + (1-\gamma)\frac{u+v}{2}\right) \leq \gamma^{\alpha s}\psi(v) + (1-\gamma)^{\alpha s}\psi\left(\frac{u+v}{2}\right).$$

are true, we obtain

$$\begin{aligned}M(\gamma) =& \frac{\Gamma(\alpha+1)}{(v-u)^\alpha}\left[J^\alpha_{u+}\psi\left(\gamma v+(1-\gamma)\frac{u+v}{2}\right)\right.\\&\left.+J^\alpha_{v-}\psi\left(\gamma u+(1-\gamma)\frac{u+v}{2}\right)\right]\\&\leqslant \gamma^{\alpha s}\frac{\Gamma(\alpha+1)}{(v-u)^\alpha}\left[J^\alpha_{u+}\psi(v)+J^\alpha_{v-}\psi(u)\right]+(1-\gamma)^{\alpha s}\psi\left(\frac{u+v}{2}\right)\\=&M_1(\gamma)\end{aligned}$$

and the inequality (13) is proved.

(iv) We have

$$\begin{aligned}M_2(\gamma) =& \left[\frac{1}{s+1}+\frac{\Gamma(\alpha s+1)\Gamma(\alpha+1)}{\Gamma(\alpha(s+1)+2)}\right]\left[\psi\left(\gamma u+(1-\gamma)\frac{u+v}{2}\right)\right.\\&\left.+\psi\left(\gamma v+(1-\gamma)\frac{u+v}{2}\right)\right]\\&\leq\left[\frac{1}{s+1}+\frac{\Gamma(\alpha s+1)\Gamma(\alpha+1)}{\Gamma(\alpha(s+1)+1)}\right]\left[\gamma^{\alpha s}\psi(u)+(1-\gamma)^{\alpha s}\psi\left(\frac{u+v}{2}\right)+\gamma^{\alpha s}\psi(v)\right.\\&\left.+(1-\gamma)^{\alpha s}\psi\left(\frac{u+v}{2}\right)\right]\\&=\left[\frac{1}{s+1}+\frac{\Gamma(\alpha s+1)\Gamma(\alpha+1)}{\Gamma(\alpha(s+1)+1)}\right]\left[\gamma^{\alpha s}[\psi(u)+\psi(v)]+2^{\alpha s}(1-\gamma)^{\alpha s}\psi\left(\frac{u+v}{2}\right)\right].\end{aligned}$$

Since

$$\frac{\Gamma(\alpha+1)}{2(v-u)^\alpha}\left[J^\alpha_{u+}\psi(v)+J^\alpha_{v-}\psi(u)\right]\leq\left[\frac{1}{s+1}+\frac{\Gamma(\alpha s+1)\Gamma(\alpha+1)}{\Gamma(\alpha(s+1)+1)}\right]\frac{[\psi(u)+\psi(v)]}{2},$$

and

$$(1-\gamma)^{\alpha s}\psi\left(\frac{u+v}{2}\right)\leq 2^\alpha(1-\gamma)^{\alpha s}\left[\frac{1}{s+1}+\frac{\Gamma(\alpha s+1)\Gamma(\alpha+1)}{\Gamma(\alpha(s+1)+1)}\right]\psi\left(\frac{u+v}{2}\right),$$

then

$$\begin{aligned}M_1(\gamma)\leq&\gamma^{\alpha s}\left[\frac{1}{s+1}+\frac{\Gamma(\alpha s+1)\Gamma(\alpha+1)}{\Gamma(\alpha(s+1)+1)}\right](\psi(u)+\psi(v))\\&+2^\alpha(1-\gamma)^{\alpha s}\left[\frac{1}{s+1}+\frac{\Gamma(\alpha s+1)\Gamma(\alpha+1)}{\Gamma(\alpha(s+1)+1)}\right]\psi\left(\frac{u+v}{2}\right),\end{aligned}$$

and the proof of Theorem 5 is complete.

□

Corollary 2. *Choosing $s=1$ in Theorem 5, we have*

(i)

$$\frac{\Gamma(\alpha+1)}{(v-u)^{\alpha}}\left[J_{u^+}^{\alpha}\psi\left(\gamma v+(1-\gamma)\frac{u+v}{2}\right)+J_{v^-}^{\alpha}\psi\left(\gamma u+(1-\gamma)\frac{u+v}{2}\right)\right]$$
$$\leq \min\left\{\gamma^{\alpha}\frac{\Gamma(\alpha+1)}{(v-u)^{\alpha}}[J_{u^+}^{\alpha}\psi(v)+J_{v^-}^{\alpha}\psi(u)]\right.$$
$$+(1-\gamma)^{\alpha}\psi\left(\frac{u+v}{2}\right),\left[\frac{1}{2}+\frac{(\Gamma(\alpha+1))^2}{\Gamma(2(\alpha+1))}\right]$$
$$\left.\times\left[\psi\left(\gamma u+(1-\gamma)\frac{u+v}{2}\right)+\psi\left(\gamma v+(1-\gamma)\frac{u+v}{2}\right)\right]\right\}.$$

(ii) Since

$$\tilde{M}=\max\left\{\gamma^{\alpha}\frac{\Gamma(\alpha+1)}{(v-u)^{\alpha}}[J_{u^+}^{\alpha}\psi(v)+J_{v^-}^{\alpha}\psi(u)]+(1-\gamma)^{\alpha}\psi(\frac{u+v}{2}),\right.$$
$$\left.\left[\frac{1}{2}+\frac{(\Gamma(\alpha+1))^2}{\Gamma(2(\alpha+1))}\right]\left[\psi(\gamma u+(1-\gamma)\frac{u+v}{2})+\psi\left(\gamma v+(1-\gamma)\frac{u+v}{2}\right)\right]\right\},$$

we get

$$\tilde{M}=\left[\frac{1}{2}+\frac{\Gamma(\alpha+1)\Gamma(\alpha+1)}{\Gamma(2(\alpha+1))}\right]\left(\gamma^{\alpha}(\psi(u)+\psi(v))+2^{\alpha}(1-\gamma)^{\alpha}\psi\left(\frac{u+v}{2}\right)\right).$$

Theorem 6. *Let $\psi:[u,v]\subset\mathbb{R}_+\to\mathbb{R}^{\alpha}$ be a differentiable function on (u,v) where $0\leq u<v$. For some fixed $q\geq 1$, if $|\psi'|^q$ is generalized s-convex on (u,v), we obtain*

$$\left|\frac{\psi(u)+\psi(v)}{2}-\frac{\Gamma(\alpha+1)}{2(v-u)^{\alpha}}\left[J_{u^+}^{\alpha}\psi(v)+J_{v^-}^{\alpha}\psi(u)\right]\right|\leq\frac{v-u}{2}\left(\frac{2}{\alpha+1}\right)^{1-\frac{1}{q}}$$
$$\times\left(\left[\beta(\alpha+1,\alpha s+1)+\frac{1}{\alpha(s+1)+1}\right]\right.$$
$$\left.\times\left[|\psi'(u)|^q+|\psi'(v)|^q\right]\right)^{\frac{1}{q}}.$$

Proof. Applying Lemma 1, we obtain

$$\left|\frac{\psi(u)+\psi(v)}{2}-\frac{\Gamma(\alpha+1)}{2(v-u)^{\alpha}}\left[J_{u^+}^{\alpha}\psi(v)+J_{v^-}^{\alpha}\psi(u)\right]\right|=\frac{v-u}{2}\left|\int_0^1[(1-\gamma)^{\alpha}-\gamma^{\alpha}]\right.$$
$$\left.\times\,\psi'(\gamma u+(1-\gamma)v)d\gamma\right|. \tag{14}$$

First, suppose $q=1$. Since the function $|\psi'|$ is generalized s-convex on (u,v), we obtain

$$|\psi'(\gamma u+(1-\gamma)v\gamma|\leq\gamma^{\alpha s}|\psi'(u)|+(1-\gamma)^{\alpha s}|\psi'(v)|. \tag{15}$$

Therefore,

$$\begin{aligned}\left|\int_0^1 [(1-\gamma)^\alpha - \gamma^\alpha]\, \psi'(\gamma u + (1-\gamma)v)d\gamma\right| &\leq |\psi'(u)| \int_0^1 [(1-\gamma)^\alpha \gamma^{\alpha s} + \gamma^{\alpha s+\alpha}]d\gamma \\ &+ |\psi'(v)| \int_0^1 [(1-\gamma)^{\alpha s+\alpha} + (1-\gamma^{\alpha s})\gamma^\alpha]d\gamma \\ &= \left(\beta(\alpha+1, \alpha s+1) + \frac{1}{\alpha s+\alpha+1}\right)[\psi'(u) + \psi'(v)]. \quad (16)\end{aligned}$$

Next suppose that $q > 1$. From the power mean inequality and the generalized s-convexity of the function $|\psi'|^q$ we obtain

$$\begin{aligned}\left|\int_0^1 [(1-\gamma)^\alpha - \gamma^\alpha]\, \psi'(\gamma u + (1-\gamma)v)d\gamma\right| &\leq \left|\int_0^1 [(1-\gamma)^\alpha - \gamma^\alpha]^{1-\frac{1}{q}} [(1-\gamma)^\alpha - \gamma^\alpha]^{\frac{1}{q}}\right. \\ &\left.\times \psi'(\gamma u + (1-\gamma)v)d\gamma\right| \\ &\leq \left(\int_0^1 |(1-\gamma)^\alpha - \gamma^\alpha| d\gamma\right)^{1-\frac{1}{q}} \\ &\times \left(\int_0^1 |(1-\gamma)^\alpha - \gamma^\alpha||\psi'(\gamma u + (1-\gamma)v)|^q d\gamma\right)^{\frac{1}{q}} \\ &\leq \left(\int_0^1 (1-\gamma)^\alpha + \gamma^\alpha d\gamma\right)^{1-\frac{1}{q}} \\ &\times \left(\int_0^1 [(1-\gamma)^\alpha + \gamma^\alpha][\gamma^{\alpha s}|\psi'(u)|^q + (1-\gamma)^{\alpha s}|\psi'(v)|^q]d\gamma\right)^{\frac{1}{q}} \\ &= \left(\frac{2}{\alpha+1}\right)^{1-\frac{1}{q}} \left(\left[\beta(\alpha+1, \alpha s+1) + \frac{1}{\alpha s+\alpha+1}\right]\right. \\ &\left.\times [|\psi'(u)|^q + |\psi'(v)|^q]\right)^{\frac{1}{q}}.\end{aligned} \quad (17)$$

In view of inequalities (14), (16) and (17) the proof of Theorem 6 is complete now. □

Corollary 3. *Under the conditions of Theorem 6, we get*

(i) If $q = s = 1$, *then*

$$\left|\frac{\psi(u)+\psi(v)}{2} - \frac{\Gamma(\alpha+1)}{2(v-u)^\alpha}\left[J_{u^+}^\alpha \psi(v) + J_{v^-}^\alpha \psi(u)\right]\right| \leq \frac{v-u}{2}\left[\frac{1}{2\alpha+1} + \beta(\alpha+1, \alpha+1)\right](|\psi'(u)| + |\psi'(v)|).$$

(ii) If $q = \alpha = s = 1$, *then*

$$\left|\frac{\psi(u)+\psi(v)}{2} - \frac{1}{v-u}\int_u^v \psi(x)dx\right| \leq \frac{v-u}{4}(|\psi'(u)| + |\psi'(v)|).$$

(iii) If $q > 1$ *and* $s = 1$

$$\begin{aligned}\left|\frac{\psi(u)+\psi(v)}{2} - \frac{\Gamma(\alpha+1)}{2(v-u)^\alpha}\left[J_{u^+}^\alpha \psi(v) + J_{v^-}^\alpha \psi(u)\right]\right| &\leq \frac{v-u}{2}\left(\frac{2}{\alpha+1}\right)^{1-\frac{1}{q}}\left(\left[\beta(\alpha+1, \alpha s+1) + \frac{1}{\alpha s+\alpha+1}\right]\right. \\ &\left.\times [|\psi'(u)|^q + |\psi'(v)|^q]\right)^{\frac{1}{q}}.\end{aligned}$$

(iv) If $q > 1$ *and* $\alpha = s = 1$ *then*

$$\left|\frac{\psi(u)+\psi(v)}{2} - \frac{1}{v-u}\int_u^v \psi(x)dx\right| \leq \frac{v-u}{2}\left(\frac{|\psi'(u)|^q + |\psi'(v)|^q}{2}\right)^{\frac{1}{q}}.$$

Theorem 7. *Let $\psi : [u,v] \subset \mathbb{R}_+ \to \mathbb{R}^\alpha$ be a differentiable function on (u,v) where $0 \leq u < v$. If $|\psi'|^q$ is generalized s-convex on (u,v) for $q > 1$, we get*

$$\left|\frac{\psi(u)+\psi(v)}{2} - \frac{\Gamma(\alpha+1)}{2(v-u)^\alpha}\left[J^\alpha_{u+}\psi(v) + J^\alpha_{v-}\psi(u)\right]\right| \leq \frac{v-u}{2}\left(\frac{2}{\alpha+1}\left[1-\frac{1}{2^\alpha}\right]\right)^{1-\frac{1}{q}} \times \left(\left[\frac{\Gamma(\alpha s+1)\Gamma(\alpha+1)}{\Gamma(\alpha s+\alpha+2)} + \frac{1}{\alpha s+\alpha+1}\right]\right)^{\frac{1}{q}} \times \left[|\psi'(u)|^q + |\psi'(v)|^q\right]^{\frac{1}{q}}.$$

Proof. Since $|\psi'|$ is generalized s-convex on (u,v), we obtain

$$|\psi'(\gamma u + (1-\gamma)v\gamma| \leq \gamma^{\alpha s}|\psi'(u)| + (1-\gamma)^{\alpha s}|\psi'(v)|.$$

From this fact and applying the Hölder's inequality, we have

$$\begin{aligned}
\left|\int_0^1 [(1-\gamma)^\alpha - \gamma^\alpha]\,\psi'(\gamma u + (1-\gamma)v)d\gamma\right| &\leq \left|\int_0^1 [(1-\gamma)^\alpha - \gamma^\alpha]^{1-\frac{1}{q}}[(1-\gamma)^\alpha - \gamma^\alpha]^{\frac{1}{q}}\,\psi'(\gamma u + (1-\gamma)v)d\gamma\right| \\
&\leq \left(\int_0^{\frac{1}{2}}[(1-\gamma)^\alpha - \gamma^\alpha]d\gamma + \int_{\frac{1}{2}}^1[\gamma^\alpha - (1-\gamma)^\alpha]d\gamma\right)^{1-\frac{1}{q}} \\
&\times\left(\int_0^1 |(1-\gamma)^\alpha - \gamma^\alpha||\psi'(\gamma u + (1-\gamma)v)|^q d\gamma\right)^{\frac{1}{q}} \\
&\leq \left(\int_0^{\frac{1}{2}}[(1-\gamma)^\alpha - \gamma^\alpha]d\gamma + \int_{\frac{1}{2}}^1[\gamma^\alpha - (1-\gamma)^\alpha]d\gamma\right)^{1-\frac{1}{q}} \\
&\times\left(\int_0^1[(1-\gamma)^\alpha + \gamma^\alpha][\gamma^{\alpha s}|\psi'(u)|^q + (1-\gamma)^{\alpha s}|\psi'(v)|^q]d\gamma\right)^{\frac{1}{q}} \\
&= \left(\frac{2}{\alpha+1}\left[1-\frac{1}{2^\alpha}\right]\right)^{1-\frac{1}{q}}\left(\left[\frac{\Gamma(\alpha s+1)\Gamma(\alpha+1)}{\Gamma(\alpha s+\alpha+2)} + \frac{1}{\alpha s+\alpha+1}\right]\right)^{\frac{1}{q}} \\
&\times[|\psi'(u)|^q + |\psi'(v)|^q]^{\frac{1}{q}}.
\end{aligned} \tag{18}$$

Thus, the inequalities (14) and (18) complete the proof of Theorem 7. □

Theorem 8. *Let $\psi : [u,v] \subset \mathbb{R}_+ \to \mathbb{R}^\alpha$ be a differentiable mapping on (u,v) with $0 \leq u < v$. For some fixed $q > 1$, if $|\psi'|^q$ is generalized s-convex on $[u,v]$, then we have:*

$$\left|\frac{\psi(u)+\psi(v)}{2} - \frac{\Gamma(\alpha+1)}{2(v-u)^\alpha}\left[J^\alpha_{u+}\psi(v) + J^\alpha_{v+}\psi(u)\right]\right| \leq \frac{v-u}{2}\left(\frac{2}{\alpha p+1}\left[1-\frac{1}{2^{\alpha p}}\right]\right)^{\frac{1}{p}}\left(\frac{1}{\alpha s+1}\right)^{\frac{1}{q}} \times\left(|\psi'(u)|^q + |\psi'(v)|^q\right)^{\frac{1}{q}}.$$

Proof. By applying Hölder's inequality and (15), we obtain

$$\begin{aligned}
\left|\int_0^1 [(1-\gamma)^\alpha - \gamma^\alpha]\,\psi'(\gamma u + (1-\gamma)v)d\gamma\right| &\leq \left(\int_0^1 |(1-\gamma)^\alpha - \gamma^\alpha|^p d\gamma\right)^{\frac{1}{p}} \\
&\times\left(\int_0^1 |\psi'(\gamma u + (1-\gamma)v)|^q d\gamma\right)^{\frac{1}{q}} \\
&\leq \left(\int_0^{\frac{1}{2}}[(1-\gamma)^\alpha - \gamma^\alpha]^p d\gamma + \int_{\frac{1}{2}}^1[\gamma^\alpha - (1-\gamma)^\alpha]^p d\gamma\right)^{\frac{1}{p}} \\
&\times\left(\int_0^1[\gamma^{\alpha s}|\psi'(u)|^q + (1-\gamma)^{\alpha s}|\psi'(v)|^q]d\gamma\right)^{\frac{1}{q}} \\
&\leq \left(\frac{2}{\alpha p+1}\left[1-\frac{1}{2^{\alpha p}}\right]\right)^{\frac{1}{p}}\left(\frac{|\psi'(u)|^q + |\psi'(v)|^q}{\alpha s+1}\right)^{\frac{1}{q}}.
\end{aligned} \tag{19}$$

Finally, from (14) and (19) we get the desired result. □

Remark 3. *From Theorems 6–8, we obtain the following inequality for $q > 1$*

$$\left|\frac{\psi(u)+\psi(v)}{2} - \frac{\Gamma(\alpha+1)}{2(v-u)^{\alpha}}\left[J^{\alpha}_{u^+}\psi(v) + J^{\alpha}_{v^-}\psi(u)\right]\right| \le \min\{S_1, S_2, S_3\}\frac{(v-u)}{2}\left[|\psi'(u)|^q + |\psi'(v)|^q\right]^{\frac{1}{q}},$$

where

$$S_1 = \left(\frac{2}{\alpha+1}\right)^{1-\frac{1}{q}}\left(\left[\beta(\alpha+1, \alpha s+1) + \frac{1}{\alpha(s+1)+1}\right]\right)^{\frac{1}{q}},$$

$$S_2 = \left(\frac{2}{\alpha+1}\left[1-\frac{1}{2^{\alpha}}\right]\right)^{1-\frac{1}{q}}\left(\left[\frac{\Gamma(\alpha s+1)\Gamma(\alpha+1)}{\Gamma(\alpha(s+1)+2)} + \frac{1}{\alpha(s+1)+1}\right]\right)^{\frac{1}{q}},$$

$$S_3 = \left(\frac{2}{\alpha p+1}\left[1-\frac{1}{2^{\alpha p}}\right]\right)^{\frac{1}{p}}\left(\frac{1}{\alpha s+1}\right)^{\frac{1}{q}}.$$

Theorem 9. *Let $\psi : [u,v] \subset \mathbb{R}_+ \to \mathbb{R}^{\alpha}$ be a twice-differentiable function on (u,v) with $0 \le u < v$. If, for some fixed $q \ge 1$, the function $|\psi''|^q$ is generalized s-convex on the interval $[u,v]$, then we have*

$$\begin{aligned}\left|\frac{\psi(u)+\psi(v)}{2} - \frac{\Gamma(\alpha+1)}{2(v-u)^{\alpha}}\left[J^{\alpha}_{u^+}\psi(v) + J^{\alpha}_{v^-}\psi(u)\right]\right| &\le \frac{(v-u)^2}{2(\alpha+1)}\left(\frac{\alpha}{\alpha+2}\right)^{1-\frac{1}{q}} \\ &\times\left(\frac{1}{\alpha s+1} - \beta(\alpha s+1, \alpha+2) - \frac{1}{\alpha s+\alpha+2}\right)^{\frac{1}{q}} \\ &\times\left(|\psi''(u)|^q + |\psi''(v)|^q\right)^{\frac{1}{q}}.\end{aligned}$$

Proof. Applying Lemma 2, we have

$$\begin{aligned}&\left|\frac{\psi(u)+\psi(v)}{2} - \frac{\Gamma(\alpha+1)}{2(v-u)^{\alpha}}\left[J^{\alpha}_{u^+}\psi(v) + J^{\alpha}_{v^-}\psi(u)\right]\right| \\ =&\frac{(v-u)^2}{2}\int_0^1 \frac{1-(1-\gamma)^{\alpha+1}-\gamma^{\alpha+1}}{\alpha+1}|\psi''(\gamma u + (1-\gamma)v)|d\gamma.\end{aligned} \tag{20}$$

First, suppose that $q = 1$. Since the mapping $|\psi''|$ is generalized s-convex on (u,v), we obtain

$$|\psi''(\gamma u + (1-\gamma)v)| \le \gamma^{\alpha s}|\psi''(u)| + (1-\gamma)^{\alpha s}|\psi''(v)|. \tag{21}$$

Therefore,

$$\begin{aligned}&\left|\frac{\psi(u)+\psi(v)}{2} - \frac{\Gamma(\alpha+1)}{2(v-u)^{\alpha}}\left[J^{\alpha}_{u^+}\psi(v) + J^{\alpha}_{v^-}\psi(u)\right]\right| \\ \le&\frac{(v-u)^2}{2}\int_0^1 \frac{1-(1-\gamma)^{\alpha+1}-\gamma^{\alpha+1}}{\alpha+1}\left(\gamma^{\alpha s}|\psi''(u)| + (1-\gamma)^{\alpha s}|\psi''(v)|\right)d\gamma \\ =&\frac{(v-u)^2}{2(\alpha+1)}\left(\frac{1}{\alpha s+1} - \beta(\alpha s+1, \alpha+2) - \frac{1}{\alpha s+\alpha+2}\right)(|\psi''(u)| + |\psi''(v)|,\end{aligned}$$

where

$$\beta(\alpha s+1, \alpha+2) = \beta(\alpha+2, \alpha s+1).$$

Secondly, for $q > 1$. From Lemma 2 and the power mean inequality, we have

$$\begin{aligned}&\left|\frac{\psi(u)+\psi(v)}{2}-\frac{\Gamma(\alpha+1)}{2(v-u)^{\alpha}}\left[J_{u+}^{\alpha}\psi(v)+J_{v-}^{\alpha}\psi(u)\right]\right|\\&\leq\frac{(v-u)^2}{2(\alpha+1)}\left(\int_0^1(1-(1-\gamma)^{\alpha+1}-\gamma^{\alpha+1})d\gamma\right)^{1-\frac{1}{q}}\\&\times\left(\int_0^1\left(1-(1-\gamma)^{\alpha+1}-\gamma^{\alpha+1}|\psi''(\gamma u+(1-\gamma)v)|^q\right)d\gamma\right)^{\frac{1}{q}}.\end{aligned}\tag{22}$$

Hence, from inequalities (21) and (22), we obtain

$$\begin{aligned}&\left|\frac{\psi(u)+\psi(v)}{2}-\frac{\Gamma(\alpha+1)}{2(v-u)^{\alpha}}\left[J_{u+}^{\alpha}\psi(v)+J_{v-}^{\alpha}\psi(u)\right]\right|\\&\leq\frac{(v-u)^2}{2(\alpha+1)}\left(\int_0^1 1-(1-\gamma)^{\alpha+1}-\gamma^{\alpha+1}\right)^{1-\frac{1}{q}}\left(\int_0^1 1-(1-\gamma)^{\alpha+1}-\gamma^{\alpha+1}|\psi''(\gamma u+(1-\gamma)v)|^q d\gamma\right)^{\frac{1}{q}}\\&\leq\frac{(v-u)^2}{2(\alpha+1)}\left(\int_0^1 1-(1-\gamma)^{\alpha+1}-\gamma^{\alpha+1}\right)^{1-\frac{1}{q}}\\&\times\left(\int_0^1 1-(1-\gamma)^{\alpha+1}-\gamma^{\alpha+1}[\gamma^{\alpha s}|\psi''(u)|^q+(1-\gamma)^{\alpha s}|\psi''(v)|^q]d\gamma\right)^{\frac{1}{q}}\\&\leq\frac{(v-u)^2}{2(\alpha+1)}\left(\frac{\alpha}{\alpha+2}\right)^{1-\frac{1}{q}}\left(\frac{1}{\alpha s+1}-\beta(\alpha s+1,\alpha+2)-\frac{1}{\alpha s+\alpha+2}\right)^{\frac{1}{q}}\left(|\psi''(u)|^q+|\psi''(v)|^q\right)^{\frac{1}{q}}.\end{aligned}$$

This completes the proof of Theorem 9. □

Theorem 10. *Let $0 \leqslant u < v < \infty$ and let the function $\psi : [u, v] \to \mathbb{R}^{\alpha}$ be twice-differentiable on the open interval (u, v), and fix $s \in (0, 1]$ and fix $q > 1$. If, in addition, the function $|\psi''|^q$ is generalized s-convex on $[u, v]$, then*

$$\begin{aligned}\left|\frac{\psi(u)+\psi(v)}{2}-\frac{\Gamma(\alpha+1)}{2(v-u)^{\alpha}}\left[J_{u+}^{\alpha}\psi(v)+J_{v-}^{\alpha}\psi(u)\right]\right|&\leq\frac{(v-u)^2}{2(\alpha+1)}\left(1-\frac{2}{p(\alpha+1)+}\right)^{\frac{1}{p}}\\&\times\left(\frac{|\psi''(u)|^q+|\psi''(v)|^q}{\alpha s+1}\right)^{\frac{1}{q}},\end{aligned}$$

where $\frac{1}{p}+\frac{1}{q}=1$.

Proof. From (20), (21) and the Hölder's inequality, we have

$$\begin{aligned}&\left|\frac{\psi(u)+\psi(v)}{2}-\frac{\Gamma(\alpha+1)}{2(v-u)^{\alpha}}\left[J_{u+}^{\alpha}\psi(v)+J_{b-}^{\alpha}\psi(u)\right]\right|\\&\leq\frac{(v-u)^2}{2(\alpha+1)}\left(\int_0^1(1-(1-\gamma)^{\alpha+1}-\gamma^{\alpha+1})^p d\gamma\right)^{\frac{1}{p}}\left(|\psi''(\gamma u+(1-\gamma)v)|^q d\gamma\right)^{\frac{1}{q}}\\&\leq\frac{(v-u)^2}{2(\alpha+1)}\left(\int_0^1(1-(1-\gamma)^{p(\alpha+1)}-\gamma^{p(\alpha+1)})d\gamma\right)^{\frac{1}{p}}\left(|\psi''(u)|^q\int_0^1\gamma^{\alpha s}d\gamma+|\psi''(v)|^q\int_0^1(1-\gamma)^{\alpha s}d\gamma\right)^{\frac{1}{q}}\\&\leq\frac{(v-u)^2}{2(\alpha+1)}\left(1-\frac{2}{p(\alpha+1)+}\right)^{\frac{1}{p}}\left(\frac{|\psi''(u)|^q+|\psi''(v)|^q}{\alpha s+1}\right)^{\frac{1}{q}}.\end{aligned}$$

We use

$$\left(1-(1-\gamma)^{\alpha+1}-\gamma^{\alpha+1}\right)^{q} \leq 1-(1-\gamma)^{q(\alpha+1)}-t^{q(\alpha+1)},$$

for any $\gamma \in [0,1]$, which follows from

$$(V-N)^q \leq V^q - N^q,$$

where

$$V > N \geq 0 \text{ and } q \geq 1.$$

The proof of Theorem 10 is complete now. □

The following result exhibits another Hermite–Hadamard type inequality in terms of the second derivative of a function.

Theorem 11. *Under the same assumptions of Theorem 10, we have*

$$\begin{aligned}\left|\frac{\psi(u)+\psi(v)}{2}-\frac{\Gamma(\alpha+1)}{2(v-u)^{\alpha}}\left[J_{u^+}^{\alpha}\psi(v)+J_{v^-}^{\alpha}\psi(u)\right]\right| &\leq \frac{(v-u)^2}{2(\alpha+1)} \\ &\times\left(\frac{1}{\alpha s+1}-\beta(\alpha s+1,q(\alpha+s)+1)\right. \\ &\left.-\frac{1}{(\alpha+1)q+\alpha s+1}\right)^{\frac{1}{q}} \\ &\times\left(|\psi''(u)|^q+|\psi''(v)|^q\right)^{\frac{1}{q}}.\end{aligned}$$

Proof. By applying Lemma 2 and the Hölder's inequality, we obtain

$$\begin{aligned}&\left|\frac{\psi(u)+\psi(v)}{2}-\frac{\Gamma(\alpha+1)}{2(v-b)^{\alpha}}\left[J_{u^+}^{\alpha}\psi(v)+J_{v^-}^{\alpha}\psi(u)\right]\right| \\ \leq\ & \frac{(v-u)^2}{2(\alpha+1)}\left(\int_0^1 1d\gamma\right)^{\frac{1}{p}}\left(\int_0^1(1-(1-\gamma)^{\alpha+1}-\gamma^{\alpha+1})^q|\psi''(\gamma u+(1-\gamma)v)|^q d\gamma\right)^{\frac{1}{q}} \\ \leq\ & \frac{(v-u)^2}{2(\alpha+1)}\left(|\psi''(u)|^q\int_0^1(\gamma^{\alpha s}-(1-\gamma)^{q(\alpha+s)}\gamma^{\alpha s}-\gamma^{q(\alpha+1)+\alpha s})d\gamma\right. \\ &\left.+|\psi''(v)|^q\int_0^1((1-\gamma)^{\alpha s}-(1-\gamma)^{q(\alpha+1)+\alpha s}-\gamma^{q(\alpha+1)}(1-\gamma)^{\alpha s})d\gamma\right)^{\frac{1}{q}} \\ \leq\ & \frac{(v-u)^2}{2(\alpha+1)}\times\left(\frac{1}{\alpha s+1}-\beta(\alpha s+1,1+q(\alpha+s))-\frac{1}{(\alpha+1)q+\alpha s+1}\right)^{\frac{1}{q}} \\ &\times\left(|\psi''(u)|^q+|\psi''(v)|^q\right)^{\frac{1}{q}}.\end{aligned}$$

This completes the proof of Theorem 11. □

Remark 4. *From Theorems 9, 10 and 11, we have*

$$\left|\frac{\psi(u)+\psi(v)}{2}-\frac{\Gamma(\alpha+1)}{2(v-u)^{\alpha}}\left[J_{u^+}^{\alpha}\psi(v)+J_{v^-}^{\alpha}\psi(u)\right]\right| \leq \min\{K_1,K_2,K_3\},$$

where

$$K_1 = \frac{(v-u)^2}{2(\alpha+1)}\left(\frac{\alpha}{\alpha+2}\right)^{1-\frac{1}{q}}\left(\frac{1}{\alpha s+1} - \beta(\alpha s+1,\alpha+2) - \frac{1}{\alpha s+\alpha+2}\right)^{\frac{1}{q}}\left(|\psi''(u)|^q + |\psi''(v)|^q\right)^{\frac{1}{q}},$$

$$K_2 = \frac{(v-u)^2}{2(\alpha+1)}\left(1 - \frac{2}{p(\alpha+1)+}\right)^{\frac{1}{p}}\left(\frac{|\psi''(u)|^q + |\psi''(v)|^q}{\alpha s+1}\right)^{\frac{1}{q}},$$

$$K_3 = \frac{(v-u)^2}{2(\alpha+1)}\left(\frac{1}{\alpha s+1} - \beta(\alpha s+1,q(\alpha+s)+1) - \frac{1}{q(\alpha+1)+\alpha s+1}\right)^{\frac{1}{q}}\left(|\psi''(u)|^q + |\psi''(v)|^q\right)^{\frac{1}{q}}.$$

3. Applications to Special Means

Using the obtained results, we examine some applications to special means of non-negative numbers u and v.

1. The arithmetic mean:
   $A = A(u,v) = \frac{u+v}{2}$; $u,v \in \mathbb{R}$, with $u,v > 0$.
2. The logarithmic mean:
   $L(u,v) = \frac{v-u}{\log v - \log u}$; $u,v \in \mathbb{R}$, with $u,v > 0$.
3. The generalized logarithmic mean:
   $L_r(u,v) = \left[\frac{v^{r+1} - u^{r+1}}{(v-u)(r+1)}\right]^{\frac{1}{r}}$; $r \in \mathbb{Z} \setminus \{-1,0\}$ $u,v \in \mathbb{R}$, with $u,v > 0$.

Using the results obtained in Section 2, and the above applications of means, we get the following proposition.

Proposition 1. *Suppose that $r \in \mathbb{Z}$, $|r| \geq 2$ and $u,v \in \mathbb{R}$ such that $0 < u < v$. Then we get the following inequality:*

$$\left|A(u^r,v^r) - L_r^r(u,v)\right| \leq \frac{(v-u)|r|}{2}A(|u|^{r-1},|v|^{r-1}).$$

Proof. This result follows Corollary 3 (ii) applied to the function $\psi(x) = x^r$. □

Proposition 2. *Suppose that $n \in \mathbb{Z}$, $|r| \geq 2$ and $u,v \in \mathbb{R}$ such that $0 < u < v$. Then for $q \geq 1$, we get the following:*

$$\left|A(u^r,v^r) - L_r^r(u,v)\right| \leq \frac{(v-u)|r|}{2}A^{\frac{1}{q}}(|u|^{q(r-1)},|v|^{q(r-1)}).$$

Proof. This result follows from Corollary 3 (iv) applied to the function $\psi(x) = x^r$. □

Proposition 3. *Suppose that $u,v \in \mathbb{R}$ such that $0 < u < v$, then*

$$\left|A(u^{-1},v^{-1}) - L(u,v)\right| \leq \frac{(v-u)|}{2}A(|u|^{-2},|v|^{-2}).$$

Proof. This result follows from Corollary 3 (ii) applied to the function $\psi(x) = x^{-1}$. □

Proposition 4. *Suppose that $u, v \in \mathbb{R}$ such that $0 < u < v$, then*

$$\left| A(u^{-1}, v^{-1}) - L(u,v) \right| \leq \frac{(v-u)|}{2} A^{\frac{1}{q}}(|u|^{-2q}, |v|^{-2q}).$$

Proof. This result follows from Corollary 3 (iv) applied to the function $\psi(x) = x^{-1}$. □

Author Contributions: O.A.: writing—original draft preparation, visualization, A.K.: writing–review and editing, supervision.

Acknowledgments: The authors would like to thank to referees and editors for their very useful and constructive comments and remarks that improved the present manuscript substantially.

References

1. Yang, X.M. On E-convex sets, E-convex functions, and E-convex programming. *J. Optim. Theory Appl.* **2001**, *109*, 699–704. [CrossRef]
2. Hudzik, H.; Maligranda, L. Some remarks ons-convex functions. *Aequ. Math.* **1994**, *48*, 100–111. [CrossRef]
3. Almutairi, O.; Kılıçman, A. New refinements of the Hadamard inequality on coordinated convex function. *J. Inequal. Appl.* **2019**, *1*, 192. [CrossRef]
4. Dragomir, S.S.; Pearce, C. Selected topics on Hermite-Hadamard inequalities and applications. *Math. Prepr. Arch.* **2003**, *3*, 463–817.
5. Agarwal, P.; Jleli, M.; Tomar, M. Certain Hermite-Hadamard type inequalities via generalized k-fractional integrals. *J. Inequal. Appl.* **2017**, *1*, 55. [CrossRef] [PubMed]
6. Jhanthanam, S.; Tariboon, J.; Ntouyas, S.K.; Nonlaopon, K. On *q*-Hermite-Hadamard Inequalities for Differentiable Convex Functions. *Mathematics* **2019**, *7*, 632. [CrossRef]
7. Dragomir, S.S.; Fitzpatrick, S. The Hadamard inequalities for s-convex functions in the second sense. *Demonstr. Math.* **1999**, *32*, 687–696. [CrossRef]
8. Almutairi, O.; Kılıçman, A. New fractional inequalities of midpoint type via s-convexity and their application. *J. Inequal. Appl.* **2019**, *1*, 1–19. [CrossRef]
9. Chen, F. On Hermite-Hadamard type inequalities for Riemann-Liouville fractional integrals via two kinds of convexity. *Chin. J. Math.* **2014**, *92*, 2241–2253. [CrossRef]
10. Sarıkaya, M.Z. On the Hermite–Hadamard-type inequalities for co-ordinated convex function via fractional integrals. *Integr. Transform. Spec. Funct.* **2014**, *25*, 134–147. [CrossRef]
11. Qaisar, S.; Nasir, J.; Butt, S.I.; Hussain, S. On Some Fractional Integral Inequalities of Hermite-Hadamard's Type through Convexity. *Symmetry* **2019**, *11*, 137. [CrossRef]
12. Set, E.; Karaoglan, A. Hermite-Hadamard and Hermite-Hadamard-Fejer Type Inequalities for (*k*, *h*)-Convex Function via Katugampola Fractional Integrals. *Konuralp J. Math.* **2017**, *5*, 181–191.
13. Sun, W. On generalization of some inequalities for generalized harmonically convex functions via local fractional integrals. *Quaest. Math.* **2018**, 1–25. [CrossRef]
14. Abbas, G.; Farid, G. Some integral inequalities for *m*-convex functions via generalized fractional integral operator containing generalized Mittag-Leffler function. *Cogent Math. Stat.* **2016**, *3*, 1269589. [CrossRef]
15. Kılınc, S.; Akkurt, A.; Yildirim, H. Generalization Mittag-Leffer Function Associated with of the Hadamard and Fejer Hadamard Inequalities for (h-m)-Stronly convex Functions via Fractional Integrals. *J. Univ. Math.* **2019**, *2*, 8–15. [CrossRef]
16. Sarikaya, M.Z.; Set, E.; Yaldiz, H.; Başak, N. Hermite–Hadamard's inequalities for fractional integrals and related fractional inequalities. *Math. Comput. Model.* **2013**, *57*, 2403–2407. [CrossRef]
17. Mo, H.; Sui, X. Generalized-convex functions on fractal sets. *Abstr. Appl. Anal.* **2014**, *2014*, 254737. [CrossRef]
18. Kilbas, A.; Srivastava, H.M.; Trujillo, J.J. *Theory and Applications of Fractional Differential Equations*; Elsevier: Amsterdam, The Netherlands, 2014; Volume 204.
19. Wang, J.; Li, X.; Zhou, Y. Hermite-Hadamard Inequalities Involving Riemann-Liouville Fractional Integrals via s-convex Functions and Applications to Special Means. *Filomat* **2016**, *30*, 1143–1150. [CrossRef]

20. Özdemir, M.E.; Merve, A.A.; Kavurmaci-Önalan, H. Hermite-Hadamard type inequalities for s-convex and s-concave functions via fractional integrals. *Turk. J. Sci.* **2016**, *1*, 28–40.
21. Işcan, I. Generalization of different type integral inequalities for s-convex functions via fractional integrals. *Appl. Anal.* **2014**, *93*, 1846–1862. [CrossRef]

13

Fractional q-Difference Inclusions in Banach Spaces

Badr Alqahtani [1], Saïd Abbas [2], Mouffak Benchohra [1,3,*] and Sara Salem Alzaid [1]

1 Department of Mathematics, College of Science, King Saud University, P.O. Box 2455, Riyadh 11451, Saudi Arabia; balqahtani1@ksu.edu.sa (B.A.); sarsalzaid@ksu.edu.sa (S.S.A.)
2 Department of Mathematics, Tahar Moulay University of Saïda, P.O. Box 138, EN-Nasr, 20000 Saïda, Algeria; said.abbas@univ-saida.dz
3 Laboratory of Mathematics, Djillali Liabes University of Sidi Bel-Abbès, P.O. Box 89, 22000 Sidi Bel-Abbès, Algeria
* Correspondence: benchohra@univ-sba.dz

Abstract: In this paper, we study a class of Caputo fractional q-difference inclusions in Banach spaces. We obtain some existence results by using the set-valued analysis, the measure of noncompactness, and the fixed point theory (Darbo and Mönch's fixed point theorems). Finally we give an illustrative example in the last section. We initiate the study of fractional q-difference inclusions on infinite dimensional Banach spaces.

Keywords: fractional q-difference inclusion; measure of noncompactness; solution; fixed point

---

1. Introduction

Fractional differential equations and inclusions have attracted much more interest of mathematicians and physicists which provides an efficiency for the description of many practical dynamical arising in engineering, vulnerability of networks (fractional percolation on random graphs), and other applied sciences [1–8]. Recently, Riemann–Liouville and Caputo fractional differential equations with initial and boundary conditions are studied by many authors; [2,9–14]. In [15–18] the authors present some interesting results for classes of fractional differential inclusions.

q-calculus (quantum calculus) has a rich history and the details of its basic notions, results and methods can be found in [19–21]. The subject of q-difference calculus, initiated in the first quarter of 20th century, has been developed over the years. Some interesting results about initial and boundary value problems of ordinary and fractional q-difference equations can be found in [22–27].

Difference inclusions arise in the mathematical modeling of various problems in economics, optimal control, and stochastic analysis, see for instance [28–30]. However q-difference inclusions are studied in few papers; see for example [31,32]. In this article we consider the Caputo fractional q-difference inclusion

$$({}^cD_q^\alpha u)(t) \in F(t, u(t)),\ t \in I := [0, T], \tag{1}$$

with the initial condition

$$u(0) = u_0 \in E, \tag{2}$$

where $(E, \|\cdot\|)$ is a real or complex Banach space, $q \in (0,1)$, $\alpha \in (0,1]$, $T > 0$, $F : I \times E \to \mathcal{P}(E)$ is a multivalued map, $\mathcal{P}(E) = \{Y \subset E : y \neq \emptyset\}$, and ${}^cD_q^\alpha$ is the Caputo fractional q-difference derivative of order α.

This paper initiates the study of fractional q-difference inclusions on Banach spaces.

2. Preliminaries

Consider the Banach space $C(I) := C(I, E)$ of continuous functions from I into E equipped with the supremum (uniform) norm

$$\|u\|_\infty := \sup_{t\in I} \|u(t)\|.$$

As usual, $L^1(I)$ denotes the space of measurable functions $v : I \to E$ which are Bochner integrable with the norm

$$\|v\|_1 = \int_I \|v(t)\|dt.$$

For $a \in \mathbb{R}$, we set

$$[a]_q = \frac{1-q^a}{1-q}.$$

The q-analogue of the power $(a-b)^n$ is

$$(a-b)^{(0)} = 1,\ (a-b)^{(n)} = \Pi_{k=0}^{n-1}(a-bq^k);\ a, b \in \mathbb{R},\ n \in \mathbb{N}.$$

In general,

$$(a-b)^{(\alpha)} = a^\alpha \Pi_{k=0}^{\infty}\left(\frac{a-bq^k}{a-bq^{k+\alpha}}\right);\ a, b, \alpha \in \mathbb{R}.$$

Definition 1 ([21]). *The q-gamma function is defined by*

$$\Gamma_q(\xi) = \frac{(1-q)^{(\xi-1)}}{(1-q)^{\xi-1}};\ \xi \in \mathbb{R} - \{0, -1, -2, \dots\}$$

Notice that $\Gamma_q(1+\xi) = [\xi]_q \Gamma_q(\xi)$.

Definition 2 ([21]). *The q-derivative of order $n \in \mathbb{N}$ of a function $u : I \to E$ is defined by $(D_q^0 u)(t) = u(t)$,*

$$(D_q u)(t) := (D_q^1 u)(t) = \frac{u(t) - u(qt)}{(1-q)t};\ t \neq 0,\ \ (D_q u)(0) = \lim_{t\to 0}(D_q u)(t),$$

and

$$(D_q^n u)(t) = (D_q D_q^{n-1} u)(t);\ t \in I,\ n \in \{1, 2, \dots\}.$$

Set $I_t := \{tq^n : n \in \mathbb{N}\} \cup \{0\}$.

Definition 3 ([21]). *The q-integral of a function $u : I_t \to E$ is defined by*

$$(I_q u)(t) = \int_0^t u(s)d_q s = \sum_{n=0}^{\infty} t(1-q)q^n f(tq^n),$$

provided that the series converges.

We note that $(D_q I_q u)(t) = u(t)$, while if u is continuous at 0, then

$$(I_q D_q u)(t) = u(t) - u(0).$$

Definition 4 ([33]). *The Riemann–Liouville fractional q-integral of order $\alpha \in \mathbb{R}_+ := [0,\infty)$ of a function $u : I \to E$ is defined by $(I_q^0 u)(t) = u(t)$, and*

$$(I_q^\alpha u)(t) = \int_0^t \frac{(t-qs)^{(\alpha-1)}}{\Gamma_q(\alpha)} u(s) d_q s; \ t \in I.$$

Lemma 1 ([34]). *For $\alpha \in \mathbb{R}_+$ and $\lambda \in (-1,\infty)$ we have*

$$(I_q^\alpha (t-a)^{(\lambda)})(t) = \frac{\Gamma_q(1+\lambda)}{\Gamma(1+\lambda+\alpha)}(t-a)^{(\lambda+\alpha)}; \ 0 < a < t < T.$$

In particular,

$$(I_q^\alpha 1)(t) = \frac{1}{\Gamma_q(1+\alpha)} t^{(\alpha)}.$$

Definition 5 ([35]). *The Riemann–Liouville fractional q-derivative of order $\alpha \in \mathbb{R}_+$ of a function $u : I \to E$ is defined by $(D_q^0 u)(t) = u(t)$, and*

$$(D_q^\alpha u)(t) = (D_q^{[\alpha]} I_q^{[\alpha]-\alpha} u)(t); \ t \in I,$$

where $[\alpha]$ is the integer part of α.

Definition 6 ([35]). *The Caputo fractional q-derivative of order $\alpha \in \mathbb{R}_+$ of a function $u : I \to E$ is defined by $({}^C D_q^0 u)(t) = u(t)$, and*

$$({}^C D_q^\alpha u)(t) = (I_q^{[\alpha]-\alpha} D_q^{[\alpha]} u)(t); \ t \in I.$$

Lemma 2 ([35]). *Let $\alpha \in \mathbb{R}_+$. Then the following equality holds:*

$$(I_q^\alpha \, {}^C D_q^\alpha u)(t) = u(t) - \sum_{k=0}^{[\alpha]-1} \frac{t^k}{\Gamma_q(1+k)} (D_q^k u)(0).$$

In particular, if $\alpha \in (0,1)$, then

$$(I_q^\alpha \, {}^C D_q^\alpha u)(t) = u(t) - u(0).$$

We define the following subsets of $\mathcal{P}(E)$:

$P_{cl}(E) = \{Y \in \mathcal{P}(E) : Y \text{ is closed}\}$,
$P_{bd}(E) = \{Y \in \mathcal{P}(E) : Y \text{ is bounded}\}$,
$P_{cp}(E) = \{Y \in \mathcal{P}(E) : Y \text{ is compact}\}$,
$P_{cv}(E) = \{Y \in \mathcal{P}(E) : Y \text{ is convex}\}$,
$P_{cp,cv}(E) = P_{cp}(E) \cap P_{cv}(E)$.

Definition 7. *A multivalued map $G : E \to \mathcal{P}(E)$ is said to be convex (closed) valued if $G(x)$ is convex (closed) for all $x \in E$. A multivalued map G is bounded on bounded sets if $G(B) = \cup_{x\in B} G(x)$ is bounded in E for all $B \in P_b(E)$ (i.e. $\sup_{x\in B}\{\sup\{|y| : y \in G(x)\}$ exists).*

Definition 8. *A multivalued map $G : E \to \mathcal{P}(E)$ is called upper semi-continuous (u.s.c.) on E if $G(x_0) \in P_{cl}(E)$; for each $x_0 \in E$, and for each open set $N \subset E$ with $G(x_0) \in N$, there exists an open neighborhood N_0 of x_0 such that $G(N_0) \subset N$. G is said to be completely continuous if $G(B)$ is relatively compact for every $B \in P_{bd}(E)$. An element $x \in E$ is a fixed point of G if $x \in G(x)$.*

We denote by $FixG$ the fixed point set of the multivalued operator G.

Lemma 3 ([28]). *Let $G : X \to \mathcal{P}(E)$ be completely continuous with nonempty compact values. Then G is u.s.c. if and only if G has a closed graph, that is,*

$$x_n \to x_*, y_n \to y_*, y_n \in G(x_n) \Longrightarrow y_* \in G(x_*).$$

Definition 9. *A multivalued map $G : J \to P_{cl}(E)$ is said to be measurable if for every $y \in E$, the function*

$$t \to d(y, G(t)) = \inf\{|y - z| : z \in G(t)\}$$

is measurable.

Definition 10. *A multivalued map $F : I \times \mathbb{R} \to \mathcal{P}(E)$ is said to be Carathéodory if:*

(1) $t \to F(t, u)$ is measurable for each $u \in E$;
(2) $u \to F(t, u)$ is upper semicontinuous for almost all $t \in I$.

F is said to be L^1-Carathéodory if Equations (1) and (2) and the following condition holds:

(3) For each $q > 0$, there exists $\varphi_q \in L^1(I, \mathbb{R}_+)$ such that

$$\|F(t, u)\|_{\mathcal{P}} = \sup\{|v| : v \in F(t, u)\} \leq \varphi_q \text{ for all } |u| \leq q \text{ and for a.e. } t \in I.$$

For each $u \in C(I)$, define the set of selections of F by

$$S_{F\circ u} = \{v \in L^1(I) : v(t) \in F(t, u(t)) \text{ a.e. } t \in I\}.$$

Let (E, d) be a metric space induced from the normed space $(E, |\cdot|)$. The function $H_d : \mathcal{P}(E) \times \mathcal{P}(E) \to \mathbb{R}_+ \cup \{\infty\}$ given by:

$$H_d(A, B) = \max\{\sup_{a \in A} d(a, B), \sup_{b \in B} d(A, b)\}.$$

is known as the Hausdorff-Pompeiu metric. For more details on multivalued maps see the books of Hu and Papageorgiou [28].

Let $\mathcal{M}_X$ be the class of all bounded subsets of a metric space X.

Definition 11. *A function $\mu : \mathcal{M}_X \to [0, \infty)$ is said to be a measure of noncompactness on X if the following conditions are verified for all $B, B_1, B_2 \in \mathcal{M}_X$.*

(a) Regularity, i.e., $\mu(B) = 0$ if and only if B is precompact,
(b) invariance under closure, i.e., $\mu(B) = \mu(\overline{B})$,
(c) semi-additivity, i.e., $\mu(B_1 \cup B_2) = \max\{\mu(B_1), \mu(B_2)\}$.

Definition 12 ([36,37]). *Let E be a Banach space and denote by Ω_E the family of bounded subsets of E. the map $\mu : \Omega_E \to [0, \infty)$ defined by*

$$\mu(M) = \inf\{\epsilon > 0 : M \subset \cup_{j=1}^m M_j, \text{diam}(M_j) \leq \epsilon\},\ M \in \Omega_E,$$

is called the Kuratowski measure of noncompactness.

Theorem 1 ([38]). *Let E be a Banach space. Let $C \subset L^1(I)$ be a countable set with $|u(t)| \leq h(t)$ for a.e. $t \in J$ and every $u \in C$, where $h \in L^1(I, \mathbb{R}_+)$. Then $\phi(t) = \mu(C(t)) \in L^1(I, \mathbb{R}_+)$ and verifies*

$$\mu\left(\left\{\int_0^T u(s)\, ds : u \in C\right\}\right) \leq 2 \int_0^T \mu(C(s))\, ds,$$

where μ is the Kuratowski measure of noncompactness on the set E.

Lemma 4 ([39]). *Let F be a Carathéodory multivalued map and $\Theta : L^1(I) \to C(I)$; be a linear continuous map. Then the operator*

$$\Theta \circ S_{F\circ u} : C(I) \to \mathcal{P}_{cv,cp}(C(I)), \quad u \mapsto (\Theta \circ S_{F\circ u})(u) = \Theta(S_{F\circ u})$$

is a closed graph operator in $C(I) \times C(I)$.

Definition 13. *Let E be Banach space. A multivalued mapping $T : E \to \mathcal{P}_{cl,b}(E)$ is called k-set-Lipschitz if there exists a constant $k > 0$, such that $\mu(T(X)) \leq k\mu(X)$ for all $X \in \mathcal{P}_{cl,b}(E)$ with $T(X) \in \mathcal{P}_{cl,b}(E)$. If $k < 1$, then T is called a k-set-contraction on E.*

Now, we recall the set-valued versions of the Darbo and Mönch fixed point theorems.

Theorem 2 ((Darbo fixed point theorem) [40]). *Let X be a bounded, closed, and convex subset of a Banach space E and let $T : X \to \mathcal{P}_{cl,b}(X)$ be a closed and k-set-contraction. Then T has a fixed point.*

Theorem 3 ((Mönch fixed point theorem) [41]). *Let E be a Banach space and $K \subset E$ be a closed and convex set. Also, let U be a relatively open subset of K and $N : \overline{U} \to \mathcal{P}_c(K)$. Suppose that N maps compact sets into relatively compact sets,* graph(N) *is closed and for some $x_0 \in U$, we have*

$$\text{conv}(x_0 \cup N(M)) \supset M \subset \overline{U} \text{ and } \overline{M} = \overline{U} \ (C \subset M \text{ countable}) \text{ imply } \overline{M} \text{ is compact} \tag{3}$$

and

$$x \notin (1-\lambda)x_0 + \lambda N(x) \quad \forall x \in \overline{U} \backslash U, \ \lambda \in (0,1). \tag{4}$$

Then there exists $x \in \overline{U}$ with $x \in N(x)$.

3. Existence Results

First, we state the definition of a solution of the problem found in Equations (1) and (2).

Definition 14. *By a solution of the problem in Equations (1) and (2) we mean a function $u \in C(I)$ that satisfies the initial condition in Equation (2) and the equation $({}^C D_q^\alpha u)(t) = v(t)$ on I, where $v \in S_{F\circ u}$.*

In the sequel, we need the following hypotheses.

Hypothesis 1. *(H_1). The multivalued map $F : I \times E \to \mathcal{P}_{cp,cv}(E)$ is Carathéodory.*

Hypothesis 2. *(H_2). There exists a function $p \in L^\infty(I, \mathbb{R}_+)$ such that*

$$\|F(t,u)\|_{\mathcal{P}} = \sup\{\|v\|_C : v(t) \in F(t,u)\} \leq p(t);$$

for a.e. $t \in I$, and each $u \in E$,

Hypothesis 3. *(H_3). For each bounded set $B \subset C(I)$ and for each $t \in I$, we have*

$$\mu(F(t, B(t)) \leq p(t)\mu(B(t)),$$

where $B(t) = \{u(t) : u \in B\}$,

Hypothesis 4. *(H_4) The function $\phi \equiv 0$ is the unique solution in $C(I)$ of the inequality*

$$\Phi(t) \leq 2p^*(I_q^\alpha \Phi)(t),$$

where p is the function defined in (H_3), *and*

$$p^* = esssup_{t\in I}p(t).$$

Remark 1. *In* (H_3), μ *is the Kuratowski measure of noncompactness on the space E.*

Theorem 4. *If the hypotheses* (H_1)*–*(H_3) *and the condition*

$$L := \frac{p^* T^{(\alpha)}}{\Gamma_q(1+\alpha)} < 1$$

hold, then the problem in Equations (1) and (2) has at least one solution defined on I.

Proof. Consider the multivalued operator $N : C(I) \to \mathcal{P}(C(I))$ defined by:

$$N(u) = \left\{h \in C(I) : h(t) = \mu_0 + \int_0^t \frac{(t-qs)^{(\alpha-1)}}{\Gamma_q(\alpha)} v(s) d_q s;\ v \in S_{F\circ u}\right\}. \tag{5}$$

From Lemma 2, the fixed points of N are solutions of the problem in Equations (1) and (2). Set

$$R := \|u_0\| + \frac{p^* T^{(\alpha)}}{\Gamma_q(1+\alpha)},$$

and let $B_R := \{u \in C(I) : \|u\|_\infty \le R\}$ be the bounded, closed and convex ball of $C(I)$. We shall show in three steps that the multivalued operator $N : B_R \to \mathcal{P}_{cl,b}(C(I))$ satisfies all assumptions of Theorem 2.

Step 1. $N(B_R) \in \mathcal{P}(B_R)$.
Let $u \in B_R$, and $h \in N(u)$. Then for each $t \in I$ we have

$$h(t) = u_0 + \int_0^t \frac{(t-qs)^{(\alpha-1)}}{\Gamma_q(\alpha)} v(s) d_q s,$$

for some $v \in S_{F\circ u}$. On the other hand,

$$\begin{aligned}
\|h(t)\| &\le \|u_0\| + \int_0^t \frac{(t-qs)^{(\alpha-1)}}{\Gamma_q(\alpha)} \|v(s)\| d_q s \\
&\le \|u_0\| + \int_0^t \frac{(t-qs)^{(\alpha-1)}}{\Gamma_q(\alpha)} p(s) d_q s \\
&\le \|u_0\| + esssup_{t\in I}p(t) \int_0^T \frac{(t-qs)^{(\alpha-1)}}{\Gamma_q(\alpha)} d_q s \\
&= \|u_0\| + \frac{p^* T^{(\alpha)}}{\Gamma_q(1+\alpha)}.
\end{aligned}$$

Hence
$\|h\|_\infty \le R$, and so $N(B_R) \in \mathcal{P}(B_R)$.

Step 2. $N(u) \in \mathcal{P}_{cl}(B_R)$ *for each* $u \in B_R$.
Let $\{u_n\}_{n\ge 0} \in N(u)$ such that $u_n \longrightarrow \tilde{u}$ in $C(I)$.Then, $\tilde{u} \in B_R$ and there exists $f_n(\cdot) \in S_{F\circ u}$ be such that, for each $t \in I$, we have

$$u_n(t) = u_0 + \int_0^t \frac{(t-qs)^{(\alpha-1)}}{\Gamma_q(\alpha)} f_n(s) d_q s.$$

From (H_1), and since F has compact values, then we may pass to a subsequence if necessary to get that $f_n(\cdot)$ converges to f in $L^1(I)$, and then $f \in S_{F\circ u}$. Thus, for each $t \in I$, we get

$$u_n(t) \longrightarrow \tilde{u}(t) = u_0 + \int_0^t \frac{(t-qs)^{(\alpha-1)}}{\Gamma_q(\alpha)} f(s) d_q s.$$

Hence $\tilde{u} \in N(u)$.

Step 3. *N satisfies the Darbo condition.*
Let $U \subset B_R$, then for each $t \in I$, we have

$$\mu((NU)(t)) = \mu(\{(Nu)(t) : u \in U\}).$$

Let $h \in N(u)$. Then, there exists $f \in S_{F\circ u}$ such that for each $t \in I$, we have

$$h(t) = u_0 + \int_0^t \frac{(t-qs)^{(\alpha-1)}}{\Gamma_q(\alpha)} f(s) d_q s.$$

From Theorem 1 and since $U \subset B_R \subset C(I)$, then

$$\mu((NU)(t)) \leq 2 \int_0^t \mu\left(\left\{\frac{(t-qs)^{(\alpha-1)}}{\Gamma_q(\alpha)} f(s) : u \in U\right\}\right) d_q s.$$

Now, since $f \in S_{F\circ u}$ and $u(s) \in U(s)$, we have

$$\mu(\{(t-qs)^{(\alpha-1)} f(s)\}) = (t-qs)^{(\alpha-1)} p(s) \mu(U(s)).$$

Then

$$\mu((NU)(t)) \leq 2 \int_0^t \mu\left(\left\{\frac{(t-qs)^{(\alpha-1)}}{\Gamma_q(\alpha)} f(s)\right\}\right) d_q s.$$

Thus

$$\mu((NU)(t)) \leq 2p^* \int_0^t \frac{(t-qs)^{(\alpha-1)}}{\Gamma_q(\alpha)} \mu(U(s)) d_q s.$$

Hence

$$\mu((NU)(t)) \leq \frac{2p^* T^{(\alpha)}}{\Gamma_q(1+\alpha)} \mu(U).$$

Therefore,

$$\mu(N(U)) \leq L\mu(U),$$

which implies the N is a $L-$set-contraction.
As a consequence of Theorem 2, we deduce that N has a fixed point that is a solution of the problem in Equations (1) and (2). □

Now, we prove an other existence result by applying Theorem 3.

Theorem 5. *If the hypotheses $(H_1) - (H_4)$ hold, then there exists at least one solution of our problem in Equations (1) and (2).*

Proof. Consider the multivalued operator $N : C(I) \to \mathcal{P}(C(I))$ defined in Equation (5). We shall show in five steps that the multivalued operator N satisfies all assumptions of Theorem 3.

Step 1. *$N(u)$ is convex for each $u \in C(I)$.*
Let $h_1, h_2 \in N(u)$, then there exist$v_1, v_2 \in S_{F \circ u}$ such that

$$h_i(t) = \mu_0 + \int_0^t \frac{(t-qs)^{(\alpha-1)}}{\Gamma_q(\alpha)} v_i(s) d_q s; \; t \in I, \; i = 1, 2.$$

Let $0 \leq \lambda \leq 1$. Then, for each $t \in I$, we have

$$(\lambda h_1 + (1-\lambda)h_2)(t) = \int_0^t \frac{(t-qs)^{(\alpha-1)}}{\Gamma_q(\alpha)} (\lambda v_1(s) + (1-\lambda)v_2(s)) d_q s.$$

Since $S_{F \circ u}$ is convex (because F has convex values), we have $\lambda h_1 + (1-\lambda)h_2 \in N(u)$.

Step 2. *For each compact $M \subset C(I)$, $N(M)$ is relatively compact.*
Let (h_n) be any sequence in $N(M)$, where $M \subset C(I)$ is compact.We show that (h_n) has a convergent subsequence from Arzéla–Ascoli compactness criterion in $C(I)$. Since $h_n \in N(M)$ there are $u_n \in M$ and $v_n \in S_{F \circ u_n}$ such that

$$h_n(t) = \mu_0 + \int_0^t \frac{(t-qs)^{(\alpha-1)}}{\Gamma_q(\alpha)} v_n(s) d_q s.$$

Using Theorem 1 and the properties of the measure μ, we have

$$\mu(\{h_n(t)\}) \leq 2 \int_0^t \mu\left(\left\{\frac{(t-qs)^{(\alpha-1)}}{\Gamma_q(\alpha)} v_n(s)\right\}\right) d_q s. \tag{6}$$

On the other hand, since M is compact, the set $\{v_n(s) : n \geq 1\}$ is compact. Consequently, $\mu(\{v_n(s) : n \geq 1\}) = 0$ for a.e. $s \in I$. Furthermore

$$\mu(\{(t-qs)^{(\alpha-1)} v_n(s)\}) = (t-qs)^{(\alpha-1)} \mu(\{v_n(s) : n \geq 1\}) = 0.$$

for a.e. $t, s \in I$. Now Equation (6) implies that $\{h_n(t) : n \geq 1\}$ is relatively compact for each $t \in I$. In addition, for each $t_1, t_2 \in I$; with $t_1 < t_2$, we have

$$\begin{aligned}
&\|h_n(t_2) - h_n(t_1)\| \\
&\leq \left\| \int_0^{t_2} \frac{(t_2-qs)^{(\alpha-1)}}{\Gamma_q(\alpha)} p(s) d_q s - \int_0^{t_1} \frac{(t_1-qs)^{(\alpha-1)}}{\Gamma_q(\alpha)} p(s) d_q s \right\| \\
&\leq \int_{t_1}^{t_2} \frac{(t_2-qs)^{(\alpha-1)}}{\Gamma_q(\alpha)} p(s) d_q s \\
&\quad + \int_0^{t_1} \frac{|(t_2-qs)^{(\alpha-1)} - (t_1-qs)^{(\alpha-1)}|}{\Gamma_q(\alpha)} p(s) d_q s \\
&\leq \frac{p^* T^\alpha}{\Gamma_q(1+\alpha)} (t_2 - t_1)^\alpha \\
&\quad + p^* \int_0^{t_1} \frac{|(t_2-qs)^{(\alpha-1)} - (t_1-qs)^{(\alpha-1)}|}{\Gamma_q(\alpha)} d_q s \\
&\quad \to 0 \text{ as } t_1 \longrightarrow t_2.
\end{aligned} \tag{7}$$

This shows that $\{h_n : n \geq 1\}$ is equicontinuous. Consequently, $\{h_n : n \geq 1\}$ is relatively compact in $C(I)$.

Step 3. *The graph of N is closed.*
Let $(u_n, h_n) \in \text{graph}(N)$, $n \geq 1$, with$(\|u_n - u\|, \|h_n - h\|) \to (0.0)$, as $n \to \infty$. We have to show

that $(u,h) \in \text{graph}(N)$. $(u_n,h_n) \in \text{graph}(N)$ means that $h_n \in N(u_n)$, which implies that there exists $v_n \in S_{F\circ u_n}$, such that for each $t \in I$,

$$h_n(t) = u_0 + \int_0^t \frac{(t-qs)^{(\alpha-1)}}{\Gamma_q(\alpha)} v_n(s)d_qs.$$

Consider the continuous linear operator $\Theta : L^1(I) \to C(I)$,

$$\Theta(v)(t) \mapsto h_n(t) = u_0 + \int_0^t \frac{(t-qs)^{(\alpha-1)}}{\Gamma_q(\alpha)} v_n(s)d_qs.$$

Clearly, $\|h_n(t) - h(t)\| \to 0$ as as $n \to \infty$. From Lemma 4 it follows that $\Theta \circ S_F$ is a closed graph operator. Moreover, $h_n(t) \in \Theta(S_{F\circ u_n})$. Since $u_n \to u$, Lemma 4 implies

$$h(t) = u_0 + \int_0^t \frac{(t-qs)^{(\alpha-1)}}{\Gamma_q(\alpha)} v(s)d_qs.$$

for some $v \in S_{F\circ u}$.

Step 4. *M is relatively compact in* $C(I)$.
Let $M \subset \overline{U}$; with $M \subset \text{conv}(\{0\} \cup N(M))$, and let $\overline{M} = \overline{C}$; for some countable set $C \subset M$. the set $N(M)$ is equicontinuous from Equation (7). Therefore,

$$M \subset \text{conv}(\{0\} \cup N(M)) \implies M \textit{ is equicontinuous}.$$

By applying the Arzéla–Ascoli theorem; the set $M(t)$ is relatively compact for each $t \in I$. Since $C \subset M \subset \text{conv}(\{0\} \cup N(M))$, then there exists a countable set $H = \{h_n : n \geq 1\} \subset N(M)$ such that $C \subset \text{conv}(\{0\} \cup H)$. Thus, there exist $u_n \in M$ and $v_n \in S_{F\circ u_n}$ such that

$$h_n(t) = u_0 + \int_0^t \frac{(t-qs)^{(\alpha-1)}}{\Gamma_q(\alpha)} v_n(s)d_qs.$$

From Theorem 1, we get

$$M \subset \overline{C} \subset \overline{conv}(\{0\} \cup H)) \implies \mu(M(t)) \leq \mu(\overline{C}(t)) \leq \mu(H(t)) = \mu(\{h_n(t) : n \geq 1\}).$$

Using now the inequality Equation (6) in step 2, we obtain

$$\mu(M(t)) \leq 2\int_0^t \mu\left(\left\{\frac{(t-qs)^{(\alpha-1)}}{\Gamma_q(\alpha)} v_n(s)\right\}\right) d_qs.$$

Since $v_n \in S_{F\circ u_n}$ and $u_n(s) \in M(s)$, we have

$$\mu(M(t)) \leq 2\int_0^t \mu\left(\left\{\frac{(t-qs)^{(\alpha-1)}}{\Gamma_q(\alpha)} v_n(s) : n \geq 1\right\}\right) d_qs.$$

Also, since $v_n \in S_{F\circ u_n}$ and $u_n(s) \in M(s)$, then from (H_3) we get

$$\mu(\{(t-qs)^{(\alpha-1)} v_n(s);\, n \geq 1\}) = (t-qs)^{(\alpha-1)} p(s)\mu(M(s)).$$

Hence

$$\mu(M(t)) \leq 2p^* \int_0^t \frac{(t-qs)^{(\alpha-1)}}{\Gamma_q(\alpha)} \mu(M(s))d_qs.$$

Consequently, from (H_4), the function Φ given by $\Phi(t) = \mu(M(t))$ satisfies $\Phi \equiv 0$; that is, $\mu(M(t)) = 0$ for all $t \in I$. Finally, the Arzéla–Ascoli theorem implies that M is relatively compact in $C(I)$.

Step 5. *The priori estimate.*
Let $u \in C(I)$ such that $u \in \lambda N(u)$ for some $0 < \lambda < 1$. Then

$$u(t) = \lambda u_0 + \lambda \int_0^t \frac{(t-qs)^{(\alpha-1)}}{\Gamma_q(\alpha)} v(s) d_q s,$$

for each $t \in I$, where $v \in S_{F\circ u}$. On the other hand,

$$\begin{aligned} \|u(t)\| &\leq \|u_0\| + \int_0^t \frac{(t-qs)^{(\alpha-1)}}{\Gamma_q(\alpha)} \|v(s)\| d_q s \\ &\leq \|u_0\| + \int_0^t \frac{(t-qs)^{(\alpha-1)}}{\Gamma_q(\alpha)} p(s) d_q s \\ &\leq \|u_0\| + \frac{p^* T^{(\alpha)}}{\Gamma_q(1+\alpha)}. \end{aligned}$$

Then

$$\|u\| \leq \|u_0\| + \frac{p^* T^{(\alpha)}}{\Gamma_q(1+\alpha)} := d.$$

Set

$$U = \{u \in C_\gamma : \|u\| < 1 + d\}.$$

Hence, the condition in Equation (4) is satisfied. Finally, Theorem 3 implies that N has at least one fixed point $u \in C(I)$ which is a solution of our problem in Equations (1) and (2). □

4. An Example

Let

$$E = l^1 = \left\{ u = (u_1, u_2, \ldots, u_n, \ldots), \sum_{n=1}^{\infty} |u_n| < \infty \right\}$$

be the Banach space with the norm

$$\|u\|_E = \sum_{n=1}^{\infty} |u_n|.$$

Consider now the following problem of fractional $\frac{1}{4}$−difference inclusion

$$\begin{cases} ({}^c D_{\frac{1}{4}}^{\frac{1}{2}} u_n)(t) \in F_n(t, u(t));\ t \in [0, e], \\ u(0) = (1, 0, \ldots, 0, \ldots), \end{cases} \tag{8}$$

where

$$F_n(t, u(t)) = \frac{t^2 e^{-4-t}}{1 + \|u(t)\|_E} [u_n(t) - 1, u_n(t)];\ t \in [0, e],$$

with $u = (u_1, u_2, \ldots, u_n, \ldots)$. Set $\alpha = \frac{1}{2}$, and $F = (F_1, F_2, \ldots, F_n, \ldots)$.
For each $u \in E$ and $t \in [0, e]$, we have

$$\|F(t, u)\|_{\mathcal{P}} \leq p(t),$$

with $p(t) = t^2 e^{-t-4}$. Hence, the hypothesis (H_2) is satisfied with $p^* = e^{-2}$. A simple computation shows that conditions of Theorem 5 are satisfied. Hence, the problem in Equation (8) has at least one solution defined on $[0, e]$.

5. Conclusions

We have provided some sufficient conditions guaranteeing the existence of solutions for some fractional q-difference inclusions involving the Caputo fractional derivative in Banach spaces. The achieved results are obtained using the fixed point theory and the notion of measure of noncompactness. Such notion requires the use of the set-valued analysis conditions on the right-hand side, like the upper semi-continuity. In the forthcoming paper we shall provide sufficient conditions ensuring the existence of weak solutions by using the concept measure of weak noncompactness, the Pettis integration and an appropriate fixed point theorem.

Author Contributions: Funding acquisition, B.A.; Writing—original draft, S.A.; Writing—review & editing, M.B. and S.S.A. All authors have read and agreed to the published version of the manuscript.

Acknowledgments: The authors are grateful to the handling editor and reviewers for their careful reviews and useful comments.

References

1. Abbas, S.; Benchohra, M.; Graef, J.R.; Henderson, J. *Implicit Fractional Differential and Integral Equations: Existence and Stability*; De Gruyter: Berlin, Germany, 2018.
2. Abbas, S.; Benchohra, M.; N'Guérékata, G.M. *Topics in Fractional Differential Equations*; Springer: New York, NY, USA, 2012.
3. Abbas, S.; Benchohra, M.; N'Guérékata, G.M. *Advanced Fractional Differential and Integral Equations*; Nova Science Publishers: New York, NY, USA, 2015.
4. Kilbas, A.A.; Srivastava, H.M.; Juan Trujillo, J. *Theory and Applications of Fractional Differential Equations*; North-Holland Mathematics Studies 204; Elsevier Science B.V.: Amsterdam, The Netherlands, 2006.
5. Tenreiro Machado, J.A.; Kiryakova, V. The chronicles of fractional calculus, *Fract. Calc. Appl. Anal.* **2017**, *20*, 307–336.
6. Samko, S.G.; Kilbas, A.A.; Marichev, O.I. *Fractional Integrals and Derivatives. Theory and Applications*; Gordon and Breach: Amsterdam, The Netherlands, 1987. (Translated from the Russian)
7. Tarasov, V.E. *Fractional Dynamics: Application of Fractional Calculus to Dynamics of Particles, Fields and Media*; Springer: Berlin/Heidelberg, Germany; Higher Education Press: Beijing, China, 2010.
8. Zhou, Y. *Basic Theory of Fractional Differential Equations*; World Scientific: Singapore, 2014.
9. Abdeljawad, T.; Alzabut, J. On Riemann–Liouville fractional q-difference equations and their application to retarded logistic type model. *Math. Methods Appl. Sci.* **2018**, *18*, 8953–8962. [CrossRef]
10. Abbas, S.; Benchohra, M.; Hamani, S.; Henderson, J. Upper and lower solutions method for Caputo–Hadamard fractional differential inclusions. *Math. Moravica* **2019**, *23*, 107–118. [CrossRef]
11. Kilbas, A.A. Hadamard-type fractional calculus. *J. Korean Math. Soc.* **2001**, *38*, 1191–1204.
12. Ren, J.; Zhai, C. Characteristic of unique positive solution for a fractional q-difference equation with multistrip boundary conditions. *Math. Commun.* **2019**, *24*, 181–192.
13. Ren, J.; Zhai, C. A fractional q-difference equation with integral boundary conditions and comparison theorem. *Int. J. Nonlinear Sci. Numer. Simul.* **2017**, *18*, 575–583. [CrossRef]
14. Zhang, T.; Tang, Y. A difference method for solving the q-fractional differential equations. *Appl. Math. Lett.* **2019**, *98*, 292–299. [CrossRef]
15. Abbas, S.; Benchohra, M.; Graef, J.R. Coupled systems of Hilfer fractional differential inclusions in Banach spaces. *Commun. Pure Appl. Anal.* **2018**, *17*, 2479–2493. [CrossRef]
16. Ahmad, B.; Ntouyas, S.K.; Alsaedi, A.; Agarwal, R.P. A study of nonlocal integro-multi-point boundary value problems of sequential fractional integro-differential inclusions. *Dyn. Contin. Disc. Impuls. Syst. Ser. A Math. Anal.* **2018**, *25*, 125–140.
17. Ahmad, B.; Ntouyas, S.K.; Zhou, Y.; Alsaedi, A. A study of fractional differential equations and inclusions with nonlocal Erdélyi-Kober type integral boundary conditions. *Bull. Iran. Math. Soc.* **2018**, *44*, 1315–1328. [CrossRef]

18. Wang, J.; Ibrahim, A.G.; O'Regan, D. Global attracting solutions to Hilfer fractional differential inclusions of Sobolev type with noninstantaneous impulses and nonlocal conditions. *Nonlinear Anal. Model. Control* **2019**, *24*, 775–803. [CrossRef]
19. Adams, C. On the linear ordinary q-difference equation. *Ann. Math.* **1928**, *30*, 195–205. [CrossRef]
20. Carmichael, R.D. The general theory of linear q-difference equations. *Am. J. Math.* **1912**, *34*, 147–168. [CrossRef]
21. Kac, V.; Cheung, P. *Quantum Calculus*; Springer: New York, NY, USA, 2002.
22. Abbas, S.; Benchohra, M.; Laledj, N; Zhou, Y. Existence and Ulam stability for implicit fractional q-difference equation, *Adv. Differ. Equ.* **2019**, *2019*, 480. [CrossRef]
23. Ahmad, B. Boundary value problem for nonlinear third order q-difference equations. *Electron. J. Differ. Equ.* **2011**, *94*, 1–7. [CrossRef]
24. El-Shahed, M.; Hassan, H.A. Positive solutions of q-difference equation. *Proc. Am. Math. Soc.* **2010**, *138*, 1733–1738. [CrossRef]
25. Etemad, S.; Ntouyas, S.K.; Ahmad, B. Existence theory for a fractional q-integro-difference equation with q-integral boundary conditions of different orders. *Mathematics* **2019**, *7*, 659. [CrossRef]
26. Ahmad, B.; Ntouyas, S.K.; Purnaras, L.K. Existence results for nonlocal boundary value problems of nonlinear fractional q-difference equations. *Adv. Differ. Equ.* **2012**, *2012*, 140. [CrossRef]
27. Zuo, M.; Hao, X. Existence results for impulsive fractional q-difference equation with antiperiodic boundary conditions. *J. Funct. Spaces* **2018**, *2018*, 3798342. [CrossRef]
28. Hu, S.; Papageorgiou, N. *Handbook of Multivalued Analysis, Volume I: Theory*; Kluwer: Dordrecht, The Netherlands; Boston, MA, USA; London, UK, 1997.
29. Kisielewicz, M. *Differential Inclusions and Optimal Control*; Kluwer: Dordrecht, The Netherlands, 1991.
30. Smirnov, G.V. *Introduction to the Theory of Differential Inclusions*; Graduate Studies in Mathematics, Volumn 41; American Mathematical Society: Providence, RI, USA, 2002.
31. Ahmad, B.; Ntouyas, S.K. Boundary value problems for q-difference inclusions. *Abstr. Appl. Anal.* **2011**, *2011*, 292860. [CrossRef]
32. Ntouyas, S.K. Existence results for q-difference inclusions with three-point boundary conditions involving different numbers of q. *Discuss. Math. Differ. Incl. Control Optim.* **2014**, *34*, 41–59. [CrossRef]
33. Agarwal, R. Certain fractional q-integrals and q-derivatives. *Proc. Camb. Philos. Soc.* **1969**, *66*, 365–370. [CrossRef]
34. Rajkovic, P.M.; Marinkovic, S.D.; Stankovic, M.S. Fractional integrals and derivatives in q-calculus. *Appl. Anal. Discrete Math.* **2007**, **1**, 311–323.
35. Rajkovic, P.M.; Marinkovic, S.D.; Stankovic, M.S. On q-analogues of Caputo derivative and Mittag-Leffler function. *Fract. Calc. Appl. Anal.* **2007**, *10*, 359–373.
36. Banas, J.; Goebel, K. *Measures of Noncompactness in Banach Spaces*; Marcel Dekker: New York, NY, USA, 1980.
37. Ayerbee Toledano, J.M.; Dominguez Benavides, T.; Lopez Acedo, G. *Measures of Noncompactness in Metric Fixed Point Theory*; Operator Theory, Advances and Applications; Birkhäuser: Basel, Switzerland; Boston, MA, USA; Berlin, Germany, 1997; Volumn 99.
38. Heinz, H.P. On the behaviour of meusure of noncompacteness with respect of differentiation and integration of vector-valued function. *Nonlinear Anal.* **1983**, *7*, 1351–1371. [CrossRef]
39. Losta, A.; Opial, Z. An application of the Kakutani-Ky Fan theorem in the theory of ordinary differential equation. *Bull. Acad. Pol. Sci., Ser. Sci. Math. Astronom. Phys.* **1965**, *13*, 781–786.
40. Dhage, B.C. Some generalization of multi-valued version of Schauder's fixed point theorem with applications. *Cubo* **2010**, *12*, 139–151. [CrossRef]
41. O'Regan, D.; Precup, R. Fixed point theorems for set-valued maps and existence principles for integral inclusions. *J. Math. Anal. Appl.* **2000**, *245*, 594–612. [CrossRef]

14

Nonlinear Integro-Differential Equations Involving Mixed Right and Left Fractional Derivatives and Integrals with Nonlocal Boundary Data

Bashir Ahmad [1,*], Abrar Broom [1], Ahmed Alsaedi [1] and Sotiris K. Ntouyas [1,2]

1 Nonlinear Analysis and Applied Mathematics (NAAM)-Research Group, Department of Mathematics, Faculty of Science, King Abdulaziz University, P.O. Box 80203, Jeddah 21589, Saudi Arabia; abrarbroom1992@gmail.com (A.B.); aalsaedi@hotmail.com (A.A.); sntouyas@uoi.gr (S.K.N.)

2 Department of Mathematics, University of Ioannina, 451 10 Ioannina, Greece

* Correspondence: bashirahmad_qau@yahoo.com or bahmad@kau.edu.sa

Abstract: In this paper, we study the existence of solutions for a new nonlocal boundary value problem of integro-differential equations involving mixed left and right Caputo and Riemann–Liouville fractional derivatives and Riemann–Liouville fractional integrals of different orders. Our results rely on the standard tools of functional analysis. Examples are constructed to demonstrate the application of the derived results.

Keywords: caputo-type fractional derivative; fractional integral; existence; fixed point

---

1. Introduction

In the last few decades, fractional-order single-valued and multivalued boundary value problems containing different fractional derivatives such as Caputo, Riemann–Liouville, Hadamard, etc., and classical, nonlocal, integral boundary conditions have been extensively studied, for example, see the articles [1–12] and the references cited therein.

In the study of variational principles, fractional differential equations involving both left and right fractional derivatives give rise to a special class of Euler–Lagrange equations, for details, see [13] and the references cited therein. Let us consider some works on mixed fractional-order boundary value problems. In [14], the authors discussed the existence of an extremal solution to a nonlinear system involving the right-handed Riemann–Liouville fractional derivative. In [15], a two-point nonlinear higher order fractional boundary value problem involving left Riemann–Liouville and right Caputo fractional derivatives was investigated, while a problem in terms of left Caputo and right Riemann–Liouville fractional derivatives was studied in [16]. A nonlinear fractional oscillator equation containing left Riemann–Liouville and right Caputo fractional derivatives was investigated in [17]. In a recent paper [18], the authors proved some existence results for nonlocal boundary value problems of differential equations and inclusions containing both left Caputo and right Riemann–Liouville fractional derivatives.

Integro-differential equations appear in the mathematical modeling of several real world problems such as, heat transfer phenomena [19,20], forced-convective flow over a heat-conducting plate [21], etc. In [22], the authors studied the steady heat-transfer in fractal media via the local fractional nonlinear Volterra integro-differential equations. Electromagnetic waves in a variety of dielectric media with susceptibility following a fractional power-law are described by the fractional integro-differential equations [23].

Motivated by aforementioned applications of integro-differential equations and [18], we introduce a new kind of integro-differential equation involving right-Caputo and left-Riemann–Liouville fractional derivatives of different orders and right-left Riemann–Liouville fractional integrals and solve it subject to nonlocal boundary conditions. In precise terms, we prove existence and uniqueness of solutions for the problem given by

$$ {}^{C}D^{\alpha}_{1-}\,{}^{RL}D^{\beta}_{0+}y(t) + \lambda I^{p}_{1-}I^{q}_{0+}h(t,y(t)) = f(t,y(t)), \;\; t \in J := [0,1], \tag{1}$$

$$ y(0) = y(\xi) = 0, \quad y(1) = \delta y(\mu), \;\; 0 < \xi < \mu < 1, \tag{2}$$

where ${}^{C}D^{\alpha}_{1-}$ and ${}^{RL}D^{\beta}_{0+}$ denote the right Caputo fractional derivative of order $\alpha \in (1,2]$ and the left Riemann–Liouville fractional derivative of order $\beta \in (0,1]$, I^{p}_{1-} and I^{q}_{0+} denote the right and left Riemann–Liouville fractional integrals of orders $p, q > 0$ respectively, $f, h : [0,1] \times \mathbb{R} \to \mathbb{R}$ are given continuous functions and $\delta, \lambda \in \mathbb{R}$. It is imperative to notice that the integro-differential equation in (1) and (2) contains mixed type (integral and nonintegral) nonlinearities.

We organize the rest of the paper as follows. Section 2 contains some preliminary concepts related to our work. In Section 3, we prove an auxiliary lemma for the linear variant of the problem (1) and (2). Then we derive the existence results for the problem (1) and (2) by applying a fixed point theorem due to Krasnoselski and Leray–Schauder nonlinear alternative, while the uniqueness result is established via Banach contraction mapping principle. Examples illustrating the main results are also presented.

2. Preliminaries

In this section, we recall some related definitions of fractional calculus [1].

Definition 1. *The left and right Riemann–Liouville fractional integrals of order $\beta > 0$ for an integrable function $g : (0,\infty) \to \mathbb{R}$ are respectively defined by*

$$ I^{\beta}_{0+}g(t) = \int_0^t \frac{(t-s)^{\beta-1}}{\Gamma(\beta)} g(s)ds \;\; \text{and} \;\; I^{\beta}_{1-}g(t) = \int_t^1 \frac{(s-t)^{\beta-1}}{\Gamma(\beta)} g(s)ds. $$

Definition 2. *The left Riemann–Liouville fractional derivative and the right Caputo fractional derivative of order $\beta \in (n-1, n], n \in \mathbb{N}$ for a function $g : (0,\infty) \to \mathbb{R}$ with $g \in \mathbb{C}^{n}((0,\infty),\mathbb{R})$ are respectively given by*

$$ D^{\beta}_{0+}g(t) = \frac{d^n}{dt^n}\int_0^t \frac{(t-s)^{n-\beta-1}}{\Gamma(n-\beta)} g(s)ds \;\; \text{and} \;\; {}^{c}D^{\beta}_{1-}g(t) = (-1)^n \int_t^1 \frac{(s-t)^{n-\beta-1}}{\Gamma(n-\beta)} g^{(n)}(s)ds. $$

Lemma 1. *If $p > 0$ and $q > 0$, then the following relations hold almost everywhere on $[a,b]$:*

$$ I^{p}_{1-}I^{q}_{1-}f(x) = I^{p+q}_{1-}f(x), \;\; I^{p}_{0+}I^{q}_{0+}f(x) = I^{p+q}_{0+}f(x). $$

3. Main Results

In the following lemma, we solve a linear variant of the problem (1) and (2).

Lemma 2. *Let $H, F \in C[0,1] \cap L(0,1)$. Then the linear problem*

$$ \begin{cases} {}^{C}D^{\alpha}_{1-}\,{}^{RL}D^{\beta}_{0+}y(t) + \lambda I^{p}_{1-}I^{q}_{0+}H(t) = F(t), \quad t \in J := [0,1], \\ y(0) = y(\xi) = 0, \quad y(1) = \delta y(\mu), \end{cases} \tag{3}$$

is equivalent to the fractional integral equation:

$$\begin{aligned} y(t) &= \int_0^t \frac{(t-s)^{\beta-1}}{\Gamma(\beta)} \Big[I_{1-}^{\alpha} F(s) - \lambda I_{1-}^{\alpha+p} I_{0+}^{q} H(s) \Big] ds \\ &+ a_1(t) \Big\{ \delta \int_0^{\mu} \frac{(\mu-s)^{\beta-1}}{\Gamma(\beta)} \Big[I_{1-}^{\alpha} F(s) - \lambda I_{1-}^{\alpha+p} I_{0+}^{q} H(s) \Big] ds \\ &- \int_0^{1} \frac{(1-s)^{\beta-1}}{\Gamma(\beta)} \Big[I_{1-}^{\alpha} F(s) - \lambda I_{1-}^{\alpha+p} I_{0+}^{q} H(s) \Big] ds \Big\} \\ &+ a_2(t) \int_0^{\xi} \frac{(\xi-s)^{\beta-1}}{\Gamma(\beta)} \Big[I_{1-}^{\alpha} F(s) - \lambda I_{1-}^{\alpha+p} I_{0+}^{q} H(s) \Big] ds, \end{aligned} \tag{4}$$

where

$$a_1(t) = \frac{1}{\Lambda}\Big[\xi^{\beta+1}t^{\beta} - \xi^{\beta}t^{\beta+1}\Big], \; a_2(t) = \frac{1}{\Lambda}\Big[t^{\beta}(1-\delta\mu^{\beta+1}) - t^{\beta+1}(1-\delta\mu^{\beta})\Big], \tag{5}$$

and it is assumed that

$$\Lambda = \xi^{\beta+1}(1-\delta\mu^{\beta}) - \xi^{\beta}(1-\delta\mu^{\beta+1}) \neq 0. \tag{6}$$

Proof. Applying the left and right fractional integrals I_{1-}^{α} and I_{0+}^{β} successively to the integro-differential equation in (3), and then using Lemma 1, we get

$$y(t) = I_{0+}^{\beta} I_{1-}^{\alpha} F(t) - \lambda I_{0+}^{\beta} I_{1-}^{\alpha+p} I_{0+}^{q} H(t) + c_0 \frac{t^{\beta}}{\Gamma(\beta+1)} + c_1 \frac{t^{\beta+1}}{\Gamma(\beta+2)} + c_2 t^{\beta-1}, \tag{7}$$

where c_0, c_1 and c_2 are unknown arbitrary constants.

In view of the condition $y(0) = 0$, it follows from (7) that $c_2 = 0$. Inserting $c_2 = 0$ in (7) and then using the nonlocal boundary conditions $y(\xi) = 0, y(1) = \delta y(\mu)$ in the resulting equation, we obtain a system of equations in c_0 and c_1 given by

$$\begin{aligned} c_0\Big(\frac{1-\delta\mu^{\beta}}{\Gamma(\beta+1)}\Big) + c_1\Big(\frac{1-\delta\mu^{\beta+1}}{\Gamma(\beta+2)}\Big) &= \delta A_1 - A_2, \\ c_0\Big(\frac{\xi^{\beta}}{\Gamma(\beta+1)}\Big) + c_1\Big(\frac{\xi^{\beta+1}}{\Gamma(\beta+2)}\Big) &= -A_3, \end{aligned} \tag{8}$$

where

$$A_1 = I_{0+}^{\beta} I_{1-}^{\alpha} F(\mu) - \lambda I_{0+}^{\beta} I_{1-}^{\alpha+p} I_{0+}^{q} H(\mu), \; A_2 = I_{0+}^{\beta} I_{1-}^{\alpha} F(1) - \lambda I_{0+}^{\beta} I_{1-}^{\alpha+p} I_{0+}^{q} H(1),$$
$$A_3 = I_{0+}^{\beta} I_{1-}^{\alpha} F(\xi) - \lambda I_{0+}^{\beta} I_{1-}^{\alpha+p} I_{0+}^{q} H(\xi).$$

Solving the system (8), we find that

$$\begin{aligned} c_0 &= \frac{\Gamma(\beta+1)}{\Lambda}\Big[\xi^{\beta+1}(\delta A_1 - A_2) + (1-\delta\mu^{\beta+1})A_3\Big], \\ c_1 &= \frac{-\Gamma(\beta+2)}{\Lambda}\Big[\xi^{\beta}(\delta A_1 - A_2) + (1-\delta\mu^{\beta})A_3\Big], \end{aligned}$$

where Λ is defined by (6). Substituting the values of c_0 and c_1 together with the notations (5) in (7), we obtain the solution (4). The converse follows by direct computation. This completes the proof. □

Let $\mathcal{X} = C([0,1],\mathbb{R})$ denote the Banach space of all continuous functions from $[0,1] \to \mathbb{R}$ equipped with the norm $\|y\| = \sup\{|y(t)| : t \in [0,1]\}$. By Lemma 2, we define an operator $\mathcal{G} : \mathcal{X} \to \mathcal{X}$ associated with the problem (1) and (2) as

$$\begin{aligned}
\mathcal{G}y(t) &= \int_0^t \frac{(t-s)^{\beta-1}}{\Gamma(\beta)}\Big[I_{1-}^{\alpha} f(s,y(s)) - \lambda I_{1-}^{\alpha+p} I_{0+}^{q} h(s,y(s))\Big] ds \\
&+a_1(t)\Bigg[\delta \int_0^{\mu} \frac{(\mu-s)^{\beta-1}}{\Gamma(\beta)}\Big[I_{1-}^{\alpha} f(s,y(s)) - \lambda I_{1-}^{\alpha+p} I_{0+}^{q} h(s,y(s))\Big] ds \\
&- \int_0^1 \frac{(1-s)^{\beta-1}}{\Gamma(\beta)}\Big[I_{1-}^{\alpha} f(s,y(s)) - \lambda I_{1-}^{\alpha+p} I_{0+}^{q} h(s,y(s))\Big] ds\Bigg] \\
&+a_2(t)\int_0^{\xi} \frac{(\xi-s)^{\beta-1}}{\Gamma(\beta)}\Big[I_{1-}^{\alpha} f(s,y(s)) - \lambda I_{1-}^{\alpha+p} I_{0+}^{q} h(s,y(s))\Big] ds.
\end{aligned}$$

Notice that the fixed points of the operator $\mathcal{G}$ are solutions of the problem (1) and (2). In the forthcoming analysis, we use the following estimates:

$$\begin{aligned}
\int_0^t \frac{(t-s)^{\beta-1}}{\Gamma(\beta)} I_{1-}^{\alpha+p} I_{0+}^{q} ds &= \int_0^t \frac{(t-s)^{\beta-1}}{\Gamma(\beta)} \int_s^1 \frac{(u-s)^{\alpha+p-1}}{\Gamma(\alpha+p)} \int_0^u \frac{(u-r)^{q-1}}{\Gamma(q)} dr\, du\, ds \\
&\le \frac{t^{\beta}}{\Gamma(\beta+1)\Gamma(\alpha+p+1)\Gamma(q+1)}, \\
\int_0^t \frac{(t-s)^{\beta-1}}{\Gamma(\beta)} I_{1-}^{\alpha} ds &= \int_0^t \frac{(t-s)^{\beta-1}}{\Gamma(\beta)} \int_s^1 \frac{(u-s)^{\alpha-1}}{\Gamma(\alpha)} du\, ds \le \frac{t^{\beta}}{\Gamma(\alpha+1)\Gamma(\beta+1)},
\end{aligned}$$

where we have used $u^q \le 1$, $(1-s)^{\alpha+p} < 1$; $p, q > 0, 1 < \alpha \le 2$.

In the sequel, we set

$$\Omega_1 = \frac{\Delta}{\Gamma(\alpha+1)},\quad \Omega_2 = \frac{|\lambda|\Delta}{\Gamma(\alpha+p+1)\Gamma(q+1)}, \tag{9}$$

where

$$\begin{aligned}
\Delta &= \frac{1}{\Gamma(\beta+1)}\Big[1 + \bar{a}_1(|\delta|\mu^{\beta}+1) + \bar{a}_2\xi^{\beta}\Big], \\
\bar{a}_1 &= \max_{t\in[0,1]} |a_1(t)|,\quad \bar{a}_2 = \max_{t\in[0,1]} |a_2(t)|.
\end{aligned}$$

3.1. Existence Results

In the following, we prove our first existence result for the problem (1) and (2), which relies on Krasnoselskii's fixed point theorem [24].

Theorem 1. *Assumed that:*

(B_1) *There exist* $L > 0$ *such that* $|f(t,x) - f(t,y)| \le L|x-y|$, $\forall t \in [0,1]$, $x, y \in \mathbb{R}$;
(B_2) *There exist* $K > 0$ *such that* $|h(t,x) - h(t,y)| \le K|x-y|$, $\forall t \in [0,1]$, $x, y \in \mathbb{R}$;
(B_3) $|f(t,y)| \le \sigma(t)$ *and* $|h(t,y)| \le \rho(t)$, *where* $\sigma, \rho \in C([0,1],\mathbb{R}^+)$.

Then the problem (1) and (2) has at least one solution on $[0,1]$ *if* $L\gamma_1 + K\gamma_2 < 1$, *where*

$$\gamma_1 = \frac{1}{\Gamma(\beta+1)\Gamma(\alpha+1)},\quad \gamma_2 = \frac{|\lambda|}{\Gamma(\beta+1)\Gamma(\alpha+p+1)\Gamma(q+1)}. \tag{10}$$

Proof. Introduce the ball $B_\theta = \{y \in \mathcal{X} : \|y\| \leq \theta\}$, where $\|\sigma\| = \sup_{t\in[0,1]} |\sigma(t)|$, $\|\rho\| = \sup_{t\in[0,1]} |\rho(t)|$ and

$$\theta \geq \|\sigma\|\Omega_1 + \|\rho\|\Omega_2. \tag{11}$$

Let us split the operator $\mathcal{G} : \mathcal{X} \to \mathcal{X}$ on B_θ as $\mathcal{G} = \mathcal{G}_1 + \mathcal{G}_2$, where

$$\begin{aligned}
\mathcal{G}_1 y(t) &= \int_0^t \frac{(t-s)^{\beta-1}}{\Gamma(\beta)} I_{1-}^{\alpha} f(s,y(s))ds - \lambda \int_0^t \frac{(t-s)^{\beta-1}}{\Gamma(\beta)} I_{1-}^{\alpha+p} I_{0+}^{q} h(s,y(s))ds, \\
\mathcal{G}_2 y(t) &= a_1(t)\Bigg[\delta\Big(\int_0^{\mu} \frac{(\mu-s)^{\beta-1}}{\Gamma(\beta)} I_{1-}^{\alpha} f(s,y(s))ds - \lambda \int_0^{\mu} \frac{(\mu-s)^{\beta-1}}{\Gamma(\beta)} I_{1-}^{\alpha+p} I_{0+}^{q} h(s,y(s))ds\Big) \\
&\quad -\Big(\int_0^{1} \frac{(1-s)^{\beta-1}}{\Gamma(\beta)} I_{1-}^{\alpha} f(s,y(s))ds - \lambda \int_0^{1} \frac{(1-s)^{\beta-1}}{\Gamma(\beta)} I_{1-}^{\alpha+p} I_{0+}^{q} h(s,y(s))ds\Big)\Bigg] \\
&\quad +a_2(t)\Bigg[\int_0^{\xi} \frac{(\xi-s)^{\beta-1}}{\Gamma(\beta)} I_{1-}^{\alpha} f(s,y(s))ds - \lambda \int_0^{\xi} \frac{(\xi-s)^{\beta-1}}{\Gamma(\beta)} I_{1-}^{\alpha+p} I_{0+}^{q} h(s,y(s))ds\Bigg].
\end{aligned}$$

Now, we verify that the operators $\mathcal{G}_1$ and $\mathcal{G}_2$ satisfy the hypothesis of Krasnoselskii's theorem [24] in three steps.

(i) For $y, x \in B_\theta$, we have

$$\begin{aligned}
&\|\mathcal{G}_1 y + \mathcal{G}_2 x\| \\
\leq\ & \sup_{t\in[0,1]} \Bigg\{ \int_0^t \frac{(t-s)^{\beta-1}}{\Gamma(\beta)} I_{1-}^{\alpha} |f(s,y(s))| ds + |\lambda| \int_0^t \frac{(t-s)^{\beta-1}}{\Gamma(\beta)} I_{1-}^{\alpha+p} I_{0+}^{q} |h(s,y(s))| ds \\
&+|a_1(t)|\Bigg\{|\delta|\Big(\int_0^{\mu} \frac{(\mu-s)^{\beta-1}}{\Gamma(\beta)} I_{1-}^{\alpha} |f(s,x(s))| ds + |\lambda| \int_0^{\mu} \frac{(\mu-s)^{\beta-1}}{\Gamma(\beta)} I_{1-}^{\alpha+p} I_{0+}^{q} |h(s,x(s))| ds\Big) \\
&+\Big(\int_0^{1} \frac{(1-s)^{\beta-1}}{\Gamma(\beta)} I_{1-}^{\alpha} |f(s,x(s))| ds + |\lambda| \int_0^{1} \frac{(1-s)^{\beta-1}}{\Gamma(\beta)} I_{1-}^{\alpha+p} I_{0+}^{q} |h(s,x(s))| ds\Big)\Bigg\} \\
&+|a_2(t)|\Bigg\{\int_0^{\xi} \frac{(\xi-s)^{\beta-1}}{\Gamma(\beta)} I_{1-}^{\alpha} |f(s,x(s))| ds + |\lambda| \int_0^{\xi} \frac{(\xi-s)^{\beta-1}}{\Gamma(\beta)} I_{1-}^{\alpha+p} I_{0+}^{q} |h(s,x(s))| ds\Bigg\}\Bigg\} \\
\leq\ & \|\sigma\| \sup_{t\in[0,1]} \Bigg\{ \int_0^t \frac{(t-s)^{\beta-1}}{\Gamma(\beta)} I_{1-}^{\alpha} ds + |a_1(t)|\Bigg[|\delta| \int_0^{\mu} \frac{(\mu-s)^{\beta-1}}{\Gamma(\beta)} I_{1-}^{\alpha} ds \\
&+ \int_0^{1} \frac{(1-s)^{\beta-1}}{\Gamma(\beta)} I_{1-}^{\alpha} ds\Bigg] + |a_2(t)| \int_0^{\xi} \frac{(\xi-s)^{\beta-1}}{\Gamma(\beta)} I_{1-}^{\alpha} ds\Bigg\} \\
&+\|\rho\||\lambda| \sup_{t\in[0,1]} \Bigg\{ \int_0^t \frac{(t-s)^{\beta-1}}{\Gamma(\beta)} I_{1-}^{\alpha+p} I_{0+}^{q} ds + |a_1(t)|\Bigg[|\delta| \int_0^{\mu} \frac{(\mu-s)^{\beta-1}}{\Gamma(\beta)} I_{1-}^{\alpha+p} I_{0+}^{q} ds \\
&+ \int_0^{1} \frac{(1-s)^{\beta-1}}{\Gamma(\beta)} I_{1-}^{\alpha+p} I_{0+}^{q} ds\Bigg] + |a_2(t)| \int_0^{\xi} \frac{(\xi-s)^{\beta-1}}{\Gamma(\beta)} I_{1-}^{\alpha+p} I_{0+}^{q} ds\Bigg\} \\
\leq\ & \Bigg\{\frac{\|\sigma\|}{\Gamma(\alpha+1)} + \frac{\|\rho\||\lambda|}{\Gamma(\alpha+p+1)\Gamma(q+1)}\Bigg\}\Delta \\
=\ & \|\sigma\|\Omega_1 + \|\rho\|\Omega_2 < \theta,
\end{aligned}$$

where we used (11). Thus $\mathcal{G}_1 y + \mathcal{G}_2 x \in B_\theta$.

(ii) Using (B_1) and (B_2), it is easy to show that

$$\begin{aligned}\|\mathcal{G}_1y-\mathcal{G}_1x\| &\le \sup_{t\in[0,1]}\Big\{\int_0^t \frac{(t-s)^{\beta-1}}{\Gamma(\beta)} I_{1-}^{\alpha}|f(s,y(s))-f(s,x(s))|ds \\ &\quad +|\lambda|\int_0^t \frac{(t-s)^{\beta-1}}{\Gamma(\beta)} I_{1-}^{\alpha+p}I_{0+}^{q}|h(s,y(s))-h(s,x(s))|ds\Big\} \\ &\le (L\gamma_1+K\gamma_2)\|y-x\|,\end{aligned}$$

which, in view of the condition: $L\gamma_1+K\gamma_2<1$, implies that the operator $\mathcal{G}_1$ is a contraction.

(iii) Continuity of the functions f, h implies that the operator $\mathcal{G}_2$ is continuous. In addition, $\mathcal{G}_2$ is uniformly bounded on B_θ as

$$\begin{aligned}\|\mathcal{G}_2y\| &\le \sup_{t\in[0,1]}\Big\{|a_1(t)|\Big\{|\delta|\int_0^\mu \frac{(\mu-s)^{\beta-1}}{\Gamma(\beta)}\Big[I_{1-}^{\alpha}|f(s,y(s))|+|\lambda|I_{1-}^{\alpha+p}I_{0+}^{q}|h(s,y(s))|\Big]ds \\ &\quad +\int_0^1 \frac{(1-s)^{\beta-1}}{\Gamma(\beta)}\Big[I_{1-}^{\alpha}|f(s,y(s))|+|\lambda|I_{1-}^{\alpha+p}I_{0+}^{q}|h(s,y(s))|\Big]ds\Big\} \\ &\quad +|a_2(t)|\int_0^\xi \frac{(\xi-s)^{\beta-1}}{\Gamma(\beta)}\Big[I_{1-}^{\alpha}|f(s,y(s))|+|\lambda|I_{1-}^{\alpha+p}I_{0+}^{q}|h(s,y(s))|\Big]ds\Big\} \\ &\le \|\sigma\|\sup_{t\in[0,1]}\Big\{|a_1(t)|\Big[|\delta|\int_0^\mu \frac{(\mu-s)^{\beta-1}}{\Gamma(\beta)}I_{1-}^{\alpha}ds+\int_0^1 \frac{(1-s)^{\beta-1}}{\Gamma(\beta)}I_{1-}^{\alpha}ds\Big] \\ &\quad +|a_2(t)|\int_0^\xi \frac{(\xi-s)^{\beta-1}}{\Gamma(\beta)}I_{1-}^{\alpha}ds\Big\} \\ &\quad +\|\rho\||\lambda|\sup_{t\in[0,1]}\Big\{|a_1(t)|\Big[|\delta|\int_0^\mu \frac{(\mu-s)^{\beta-1}}{\Gamma(\beta)}I_{1-}^{\alpha+p}I_{0+}^{q}ds+\int_0^1 \frac{(1-s)^{\beta-1}}{\Gamma(\beta)}I_{1-}^{\alpha+p}I_{0+}^{q}ds\Big] \\ &\quad +|a_2(t)|\int_0^\xi \frac{(\xi-s)^{\beta-1}}{\Gamma(\beta)}I_{1-}^{\alpha+p}I_{0+}^{q}ds\Big\} \\ &\le \|\sigma\|(\Omega_1-\gamma_1)+\|\rho\|(\Omega_2-\gamma_2),\end{aligned}$$

where Ω_i, and γ_i $(i=1,2)$ are defined by (9) and (10) respectively.

To show the compactness of $\mathcal{G}_2$, we fix $\sup_{(t,y)\in[0,1]\times B_\theta}|f(t,y)|=\overline{f}$, $\sup_{(t,y)\in[0,1]\times B_\theta}|h(t,y)|=\overline{h}$. Then, for $0<t_1<t_2<1$, we have

$$\begin{aligned}&|(\mathcal{G}_2y)(t_2)-(\mathcal{G}_2y)(t_1)| \\ &\le |a_1(t_2)-a_1(t_1)|\Big\{|\delta|\int_0^\mu \frac{(\mu-s)^{\beta-1}}{\Gamma(\beta)}\Big[I_{1-}^{\alpha}|f(s,y(s))|+|\lambda|I_{1-}^{\alpha+p}I_{0+}^{q}|h(s,y(s))|\Big]ds \\ &\quad +\int_0^1 \frac{(1-s)^{\beta-1}}{\Gamma(\beta)}\Big[I_{1-}^{\alpha}|f(s,y(s))|+|\lambda|I_{1-}^{\alpha+p}I_{0+}^{q}|h(s,y(s))|\Big]ds\Big\} \\ &\quad +|a_2(t_2)-a_2(t_1)|\Big\{\int_0^\xi \frac{(\xi-s)^{\beta-1}}{\Gamma(\beta)}\Big[I_{1-}^{\alpha}|f(s,y(s))|+|\lambda|I_{1-}^{\alpha+p}I_{0+}^{q}|h(s,y(s))|\Big]ds\Big\} \\ &\le (\gamma_1\overline{f}+\gamma_2\overline{h})\Big\{(\xi^{\beta+1}|t_2^\beta-t_1^\beta|+\xi^\beta|t_1^{\beta+1}-t_2^{\beta+1}|)\frac{(|\delta|\mu^\beta+1)}{|\Lambda|} \\ &\quad +(|1-\delta\mu^{\beta+1}||t_2^\beta-t_1^\beta|+|1-\delta\mu^\beta||t_1^{\beta+1}-t_2^{\beta+1}|)\frac{\xi^\beta}{|\Lambda|}\Big\},\end{aligned}$$

which tends to zero independent of y as $t_2\to t_1$. This shows that $\mathcal{G}_2$ is equicontinuous. It is clear from the foregoing arguments that the operator $\mathcal{G}_2$ is relatively compact on B_θ. Hence, by the Arzelá-Ascoli theorem, $\mathcal{G}_2$ is compact on B_θ.

In view of the foregoing arguments (*i*)-(*iii*), the hypothesis of the Krasnoselskii's fixed point theorem [24] holds true. Thus, the operator $\mathcal{G}_1 + \mathcal{G}_2 = \mathcal{G}$ has a fixed point, which implies that the problem (1) and (2) has at least one solution on $[0,1]$. The proof is finished. □

Remark 1. *If we interchange the roles of the operators $\mathcal{G}_1$ and $\mathcal{G}_2$ in the previous result, the condition $L\gamma_1 + K\gamma_2 < 1$, is replaced with the following one:*

$$L(\Omega_1 - \gamma_1) + K(\Omega_2 - \gamma_2) < 1,$$

where Ω_1, Ω_2 and γ_1, γ_2 are defined by (9), (10) respectively.

The following existence result relies on Leray–Schauder nonlinear alternative [25].

Theorem 2. *Suppose that the following conditions hold:*

(B_4) *There exist continuous nondecreasing functions $\phi, \psi : [0,\infty) \to (0,\infty)$ such that $\forall (t,y) \in [0,1] \times \mathbb{R}$, $|f(t,y)| \leq \omega_1(t)\phi(\|y\|)$ and $|h(t,y)| \leq \omega_2(t)\psi(\|y\|)$, where $\omega_1, \omega_2 \in C([0,T], \mathbb{R}^+)$.*

(B_5) *There exist a constant $M > 0$ such that*

$$\frac{M}{\|\omega_1\|\phi(M)\Omega_1 + \|\omega_2\|\psi(M)\Omega_2} > 1,$$

Then, the problem (1) and (2) has at least one solution on $[0,1]$.

Proof. First we show that the operator $\mathcal{G}$ is completely continuous. This will be established in several steps.

(*i*) $\mathcal{G}$ maps bounded sets into bounded sets in $\mathcal{X}$.

Let $y \in \mathcal{B}_r = \{y \in \mathcal{X} : \|y\| \leq r\}$, where r is a fixed number. Then, using the strategy employed in the proof of Theorem 1, we obtain

$$\begin{aligned} \|\mathcal{G}y(t)\| &\leq \left\{ \frac{\|\omega_1\|\phi(r)}{\Gamma(\alpha+1)} + \frac{\|\omega_2\|\psi(r)|\lambda|}{\Gamma(\alpha+p+1)\Gamma(q+1)} \right\}\Delta \\ &= \|\omega_1\|\phi(r)\Omega_1 + \|\omega_2\|\psi(r)\Omega_2 < \infty. \end{aligned}$$

(ii) $\mathcal{G}$ maps bounded sets into equicontinuous sets.

Let $0 < t_1 < t_2 < 1$ and $y \in \mathcal{B}_r$, where $\mathcal{B}_r$ is bounded set in $\mathcal{X}$. Then we obtain

$$
\begin{aligned}
&|\mathcal{G}y(t_2)-\mathcal{G}y(t_1)| \\
\le\;& \left|\int_0^{t_1}\frac{(t_2-s)^{\beta-1}-(t_1-s)^{\beta-1}}{\Gamma(\beta)}\Big[I_{1-}^{\alpha}|f(s,y(s))|+|\lambda|I_{1-}^{\alpha+p}I_{0+}^{q}|h(s,y(s))|\Big]ds\right| \\
&+\left|\int_{t_1}^{t_2}\frac{(t_2-s)^{\beta-1}}{\Gamma(\beta)}\Big[I_{1-}^{\alpha}|f(s,y(s))|+|\lambda|I_{1-}^{\alpha+p}I_{0+}^{q}|h(s,y(s))|\Big]ds\right| \\
&+|a_1(t_2)-a_1(t_1)|\bigg\{|\delta|\bigg(\int_0^{\mu}\frac{(\mu-s)^{\beta-1}}{\Gamma(\beta)}\Big[I_{1-}^{\alpha}|f(s,y(s))|+|\lambda|I_{1-}^{\alpha+p}I_{0+}^{q}|h(s,y(s))|\Big]ds\bigg) \\
&+\bigg(\int_0^{1}\frac{(1-s)^{\beta-1}}{\Gamma(\beta)}\Big[I_{1-}^{\alpha}|f(s,y(s))|+|\lambda|I_{1-}^{\alpha+p}I_{0+}^{q}|h(s,y(s))|\Big]ds\bigg)\bigg\} \\
&+|a_2(t_2)-a_2(t_1)|\bigg\{\int_0^{\xi}\frac{(\xi-s)^{\beta-1}}{\Gamma(\beta)}\Big[I_{1-}^{\alpha}|f(s,y(s))|+|\lambda|I_{1-}^{\alpha+p}I_{0+}^{q}|h(s,y(s))|\Big]ds\bigg\} \\
\le\;& \left[\frac{\|\omega_1\|\phi(r)}{\Gamma(\beta+1)\Gamma(\alpha+1)}+\frac{\|\omega_2\|\psi(r)|\lambda|}{\Gamma(\beta+1)\Gamma(\alpha+p+1)\Gamma(q+1)}\right] \\
&\times\bigg\{2(t_2-t_1)^{\beta}+|t_2{}^{\beta}-t_1{}^{\beta}|+\frac{(|\delta|\mu^{\beta}+1)}{|\Lambda|}\Big(\xi^{\beta+1}|t_2^{\beta}-t_1^{\beta}|+\xi^{\beta}|t_2^{\beta+1}-t_1^{\beta+1}|\Big) \\
&+\frac{\xi^{\beta}}{|\Lambda|}\Big(|1-\delta\mu^{\beta+1}||t_2^{\beta}-t_1^{\beta}|+|1-\delta\mu^{\beta}||t_2^{\beta+1}-t_1^{\beta+1}|\Big)\bigg\}.
\end{aligned}
$$

Notice that the right-hand side of the above inequality tends to 0 as $t_2 \to t_1$, independent of $y \in \mathcal{B}_r$. In view of the foregoing arguments, it follows by the Arzelá–Ascoli theorem that $\mathcal{G} : \mathcal{X} \to \mathcal{X}$ is completely continuous.

The conclusion of the Leray–Schauder nonlinear alternative [25] will be applicable once it is shown that there exists an open set $\mathcal{U} \subset C([0,1],\mathbb{R})$ with $y \neq \nu\mathcal{G}y$ for $\nu \in (0,1)$ and $y \in \partial\mathcal{U}$. Let $y \in C([0,1],\mathbb{R})$ such that $y = \nu\mathcal{G}y$ for $\nu \in (0,1)$. As argued in proving that the operator $\mathcal{G}$ is bounded, one can obtain that

$$|y(t)| = |\nu\mathcal{G}y(t)| \;\; \le |\omega_1(t)|\phi(\|y\|)\Omega_1 + |\omega_2(t)|\psi(\|y\|)\Omega_2,$$

which can be written as

$$\frac{\|y\|}{\|\omega_1\|\phi(\|y\|)\Omega_1 + \|\omega_2\|\psi(\|y\|)\Omega_2} \le 1.$$

On the other hand, we can find a positive number M such that $\|y\| \neq M$ by assumption (B_5). Let us set

$$\mathcal{U} = \{y \in \mathcal{X} : \|y\| < M\}.$$

Clearly, $\partial\mathcal{U}$ contains a solution only when $\|y\| = M$. In other words, there is no solution $y \in \partial\mathcal{U}$ such that $y = \nu\mathcal{G}y$ for some $\nu \in (0,1)$. Therefore, $\mathcal{G}$ has a fixed point $y \in \overline{\mathcal{U}}$ which is a solution of the problem (1) and (2). The proof is finished. □

3.2. Uniqueness Result

Here we prove a uniqueness result for the problem (1) and (2) with the aid of Banach contraction mapping principle.

Theorem 3. *If the conditions (B_1) and (B_2) hold, then the problem (1) and (2) has a unique solution on $[0,1]$ if*

$$L\Omega_1 + K\Omega_2 < 1, \tag{12}$$

where Ω_1 and Ω_2 are defined by (9).

Proof. In the first step, we show that $\mathcal{G}\mathcal{B}_r \subset \mathcal{B}_r$, where $\mathcal{B}_r = \{y \in \mathcal{X} : \|y\| \leq r\}$ with

$$r \geq \frac{f_0\Omega_1 + h_0\Omega_2}{1-(L\Omega_1 + K\Omega_2)}, \; f_0 = \sup_{t\in[0,1]} |f(t,0)|, \; h_0 = \sup_{t\in[0,1]} |h(t,0)|.$$

For $y \in \mathcal{B}_r$ and using the condition (B_1), we have

$$\begin{aligned} |f(t,y)| &= |f(t,y) - f(t,0) + f(t,0)| \leq |f(t,y) - f(t,0)| + |f(t,0)| \\ &\leq L\|y\| + f_0 \leq Lr + f_0. \end{aligned} \tag{13}$$

Similarly, using (B_2), we get

$$|h(t,y)| \leq Kr + h_0. \tag{14}$$

In view of (13) and (14), we obtain

$$\begin{aligned} \|\mathcal{G}y\| &= \sup_{t\in[0,1]} |\mathcal{G}y(t)| \\ &\leq \sup_{t\in[0,1]} \Bigg\{ \int_0^t \frac{(t-s)^{\beta-1}}{\Gamma(\beta)} \left[I_{1-}^{\alpha} |f(s,y(s))| + |\lambda| I_{1-}^{\alpha+p} I_{0+}^{q} |h(s,y(s))| \right] ds \\ &\quad + |a_1(t)| \Bigg\{ |\delta| \int_0^{\mu} \frac{(\mu-s)^{\beta-1}}{\Gamma(\beta)} \left[I_{1-}^{\alpha} |f(s,y(s))| + |\lambda| I_{1-}^{\alpha+p} I_{0+}^{q} |h(s,y(s))| \right] ds \\ &\quad + \int_0^1 \frac{(1-s)^{\beta-1}}{\Gamma(\beta)} \left[I_{1-}^{\alpha} |f(s,y(s))| + |\lambda| I_{1-}^{\alpha+p} I_{0+}^{q} |h(s,y(s))| \right] ds \Bigg\} \\ &\quad + |a_2(t)| \int_0^{\xi} \frac{(\xi-s)^{\beta-1}}{\Gamma(\beta)} \left[I_{1-}^{\alpha} |f(s,y(s))| + |\lambda| I_{1-}^{\alpha+p} I_{0+}^{q} |h(s,y(s))| \right] ds \Bigg\} \\ &\leq (Lr + f_0) \sup_{t\in[0,1]} \Bigg\{ \int_0^t \frac{(t-s)^{\beta-1}}{\Gamma(\beta)} I_{1-}^{\alpha} ds + |a_1(t)| \Bigg[|\delta| \int_0^{\mu} \frac{(\mu-s)^{\beta-1}}{\Gamma(\beta)} I_{1-}^{\alpha} ds \\ &\quad + \int_0^1 \frac{(1-s)^{\beta-1}}{\Gamma(\beta)} I_{1-}^{\alpha} ds \Bigg] + |a_2(t)| \int_0^{\xi} \frac{(\xi-s)^{\beta-1}}{\Gamma(\beta)} I_{1-}^{\alpha} ds \Bigg\} \\ &\quad + (Kr + h_0)|\lambda| \sup_{t\in[0,1]} \Bigg\{ \int_0^t \frac{(t-s)^{\beta-1}}{\Gamma(\beta)} I_{1-}^{\alpha+p} I_{0+}^{q} ds \\ &\quad + |a_1(t)| \Bigg[|\delta| \int_0^{\mu} \frac{(\mu-s)^{\beta-1}}{\Gamma(\beta)} I_{1-}^{\alpha+p} I_{0+}^{q} ds + \int_0^1 \frac{(1-s)^{\beta-1}}{\Gamma(\beta)} I_{1-}^{\alpha+p} I_{0+}^{q} ds \Bigg] \\ &\quad + |a_2(t)| \int_0^{\xi} \frac{(\xi-s)^{\beta-1}}{\Gamma(\beta)} I_{1-}^{\alpha+p} I_{0+}^{q} ds \Bigg\} \\ &\leq \left\{ \frac{(Lr + f_0)}{\Gamma(\alpha+1)} + \frac{(Kr + h_0)|\lambda|}{\Gamma(\alpha+p+1)\Gamma(q+1)} \right\} \Delta \\ &= (Lr + f_0)\Omega_1 + (Kr + h_0)\Omega_2 < r, \end{aligned}$$

which implies that $\mathcal{G}y \in \mathcal{B}_r$, for any $y \in \mathcal{B}_r$. Therefore, $\mathcal{G}\mathcal{B}_r \subset \mathcal{B}_r$. Next, we prove that $\mathcal{G}$ is a contraction. For that, let $x, y \in \mathcal{X}$ and $t \in [0,1]$. Then, by the conditions (B_1) and (B_2), we obtain

$$\begin{aligned}
\|\mathcal{G}y-\mathcal{G}x\| &= \sup_{t\in[0,1]} \left|(\mathcal{G}y)(t)-(\mathcal{G}x)(t)\right| \\
&\leq \sup_{t\in[0,1]} \Bigg\{ \int_0^t \frac{(t-s)^{\beta-1}}{\Gamma(\beta)} I_{1-}^{\alpha}|f(s,y(s))-f(s,x(s))|ds \\
&\quad +|\lambda| \int_0^t \frac{(t-s)^{\beta-1}}{\Gamma(\beta)} I_{1-}^{\alpha+p} I_{0+}^{q}|h(s,y(s))-h(s,x(s))|ds \\
&\quad +|a_1(t)| \Bigg[|\delta| \bigg(\int_0^{\mu} \frac{(\mu-s)^{\beta-1}}{\Gamma(\beta)} I_{1-}^{\alpha}|f(s,y(s))-f(s,x(s))|ds \\
&\quad +|\lambda| \int_0^{\mu} \frac{(\mu-s)^{\beta-1}}{\Gamma(\beta)} I_{1-}^{\alpha+p} I_{0+}^{q}|h(s,y(s))-h(s,x(s))|ds \bigg) \\
&\quad +\bigg(\int_0^{1} \frac{(1-s)^{\beta-1}}{\Gamma(\beta)} I_{1-}^{\alpha}|f(s,y(s))-f(s,x(s))|ds \\
&\quad +|\lambda| \int_0^{1} \frac{(1-s)^{\beta-1}}{\Gamma(\beta)} I_{1-}^{\alpha+p} I_{0+}^{q}|h(s,y(s))-h(s,x(s))|ds \bigg) \Bigg] \\
&\quad +|a_2(t)| \Bigg[\int_0^{\xi} \frac{(\xi-s)^{\beta-1}}{\Gamma(\beta)} I_{1-}^{\alpha}|f(s,y(s))-f(s,x(s))|ds \\
&\quad +|\lambda| \int_0^{\xi} \frac{(\xi-s)^{\beta-1}}{\Gamma(\beta)} I_{1-}^{\alpha+p} I_{0+}^{q}|h(s,y(s))-h(s,x(s))|dr\,du\,ds \Bigg] \Bigg\} \\
&\leq L\|y-x\| \sup_{t\in[0,1]} \Bigg\{ \int_0^t \frac{(t-s)^{\beta-1}}{\Gamma(\beta)} I_{1-}^{\alpha} ds \\
&\quad +|a_1(t)| \Bigg[|\delta| \int_0^{\mu} \frac{(\mu-s)^{\beta-1}}{\Gamma(\beta)} I_{1-}^{\alpha} ds + \int_0^{1} \frac{(1-s)^{\beta-1}}{\Gamma(\beta)} I_{1-}^{\alpha} ds \Bigg] \\
&\quad +|a_2(t)| \int_0^{\xi} \frac{(\xi-s)^{\beta-1}}{\Gamma(\beta)} I_{1-}^{\alpha} ds \Bigg\} \\
&\quad +K\|y-x\||\lambda| \sup_{t\in[0,1]} \Bigg\{ \int_0^t \frac{(t-s)^{\beta-1}}{\Gamma(\beta)} I_{1-}^{\alpha+p} I_{0+}^{q} ds \\
&\quad +|a_1(t)| \Bigg[|\delta| \int_0^{\mu} \frac{(\mu-s)^{\beta-1}}{\Gamma(\beta)} I_{1-}^{\alpha+p} I_{0+}^{q} ds + \int_0^{1} \frac{(1-s)^{\beta-1}}{\Gamma(\beta)} I_{1-}^{\alpha+p} I_{0+}^{q} ds \Bigg] \\
&\quad +|a_2(t)| \int_0^{\xi} \frac{(\xi-s)^{\beta-1}}{\Gamma(\beta)} I_{1-}^{\alpha+p} I_{0+}^{q} ds \Bigg\} \\
&\leq \left\{ \frac{L\Delta}{\Gamma(\alpha+1)} + \frac{K|\lambda|\Delta}{\Gamma(\alpha+p+1)\Gamma(q+1)} \right\} \|y-x\| \\
&= (L\Omega_1+K\Omega_2)\|y-x\|.
\end{aligned}$$

From the above inequality, it follows by the assumption $(L\Omega_1+K\Omega_2)<1$ that $\mathcal{G}$ is a contraction. Therefore, we deduce by Banach contraction mapping principle that there exists a unique fixed point for the operator $\mathcal{G}$, which corresponds to a unique solution for the problem (1) and (2) on $[0,1]$. The proof is completed. □

3.3. Examples

In this subsection, we construct examples to illustrate the existence and uniqueness results obtained in the last two subsections. Let us consider the following problem:

$$\begin{cases} D_{1-}^{3/2} D_{0+}^{1/2} y(t) + 2 I_{1-}^{4/3} I_{0+}^{5/4} h(t,y(t)) = f(t,y(t)), \ t \in J := [0,1], \\ y(0) = y(2/3) = 0, \ \ y(1) = \frac{1}{2} y(3/4). \end{cases} \tag{15}$$

Here $\alpha = 3/2, \beta = 1/2, \lambda = 2, p = 4/3, q = 5/4, \mu = 3/4, \delta = 1/2, \xi = 2/3$, and

$$f(t,y) = \frac{1}{(t^2+8)}\big(\tan^{-1} y + e^{-t}\big), \quad h(t,y) = \frac{1}{2\sqrt{t^2+9}}\Big(\frac{|y|}{1+|y|} + e^{-t}\Big). \tag{16}$$

Using the given data, it is found that $L = 1/8, K = 1/6$,

$$\begin{aligned}\bar{a}_1 &= \max_{t\in[0,1]} |a_1(t)| = |a_1(t)|_{t=1} \approx 1.121394517474712,\\ \bar{a}_2 &= \max_{t\in[0,1]} |a_2(t)| = |a_2(t)|_{t=t_{a_2}} \approx 1.168623082364286,\end{aligned}$$

where

$$t_{a_2} = \frac{\beta(1-\delta\mu^{\beta+1})}{(1-\delta\mu^{\beta})(\beta+1)} \approx 0.396975661732535.$$

In consequence, we get

$$\Omega_1 \approx 3.022797441671726, \Omega_2 \approx 1.451691300771574, |\Lambda| \approx 0.242702744426469,$$

where Ω_1,Ω_2 are defined by (9) and Λ is given by (6).
(i) For illustrating Theorem 1, we have

$$|f(t,y)| \leq \sigma(t) = \frac{e^{-t} + (\pi/2)}{t^2+8}, \qquad |h(t,y)| \leq \rho(t) = \frac{e^{-t}+1}{2\sqrt{t^2+9}},$$

and that

$$L\gamma_1 + K\gamma_2 \approx 0.174044436618777 < 1,$$

where $\gamma_1 \approx 0.848826363156775$ and $\gamma_2 \approx 0.407646847345084$. Clearly, the hypothesis of Theorem 1 is satisfied and consequently its conclusion applies to the problem (15).
(ii) In order to explain Theorem 2, we take the following values (instead of (16)) in the problem (15):

$$f(t,y) = \frac{1}{\sqrt{t^2+25}}(y\cos y + \pi/2), \quad h(t,y) = \frac{1}{5\sqrt{t^2+4}}(\sin y + 1/4), \tag{17}$$

and note that $\omega_1(t) = \frac{1}{\sqrt{t^2+25}}, \|\omega_1\| = 1/5, \omega_2(t) = \frac{1}{5\sqrt{t^2+4}}, \|\omega_2\| = 1/10, \phi(\|y\|) = \|y\| + \pi/2$ and $\psi(\|y\|) = \|y\| + 1/4$. By the condition (B_5), we find that $M > 3.939452045479877$. Thus, all the conditions of Theorem 2 are satisfied and, hence the problem (15) with $f(t,y)$ and $h(t,y)$ given by (17) has at least one solution on $[0,1]$.
(iii) It is easy to show that $f(t,y)$ and $h(t,y)$ satisfy the conditions (B_1) and (B_2) respectively with $L = 1/8$ and $K = 1/6$ and that $L\Omega_1 + K\Omega_2 \approx 0.619798230337561 < 1$. Thus, all the assumptions of Theorem 3 hold true and hence the problem (15) has a unique solution on $[0,1]$.

4. Conclusions

We considered a fractional differential equation involving left Caputo and right Riemann–Liouville fractional derivatives of different orders and a pair of nonlinearities: $I^p_{1-} I^q_{0+} h(t,y(t)) = \int_t^1 \frac{(s-t)^{p-1}}{\Gamma(p)} \int_0^s \frac{(s-v)^{q-1}}{\Gamma(q)} h(v,y(v))dvds$ (integral type) and $f(t,y(t))$, equipped with four-point nonlocal boundary conditions. Different criteria ensuring the existence of solutions for the given problem are presented in Theorems 1 and 2, while the uniqueness of solutions is shown in Theorem 3. An interesting and scientific feature of the fractional differential Equation (1) is that the integral type of nonlinearity can describe composition of a physical quantity (like density) over two different arbitrary subsegments of the given domain. In the case of $p = q = 1$, this composition takes the form

$\int_t^1 \int_0^s h(v, y(v))dvds$. As pointed out in the introduction, fractional differential equations containing mixed (left Caputo and right Riemann–Liouville) fractional derivatives appear as Euler–Lagrange equations in the study of variational principles. So, such equations in the presence of the integral type of nonlinearity of the form introduced in (1) enhances the scope of Euler–Lagrange equations studied in [26]. Moreover, the fractional integro-differential Equation (1) can improve the description of the electromagnetic waves in dielectric media considered in [23]. As a special case, our results correspond to a three-point nonlocal mixed fractional order boundary value problem by letting $\delta = 0$, which is indeed new in the given configuration.

Author Contributions: Conceptualization, B.A.; Formal analysis, B.A., A.B., A.A. and S.K.N.; Funding acquisition, A.A.; Methodology, B.A., A.B., A.A. and S.K.N. All authors have read and agreed to the published version of the manuscript.

Acknowledgments: This project was funded by the Deanship of Scientific Research (DSR), King Abdulaziz University, Jeddah, Saudi Arabia under grant no. (KEP-MSc-23-130-40). The authors, therefore, acknowledge with thanks DSR technical and financial support. We also thank the reviewers for their useful remarks on our work.

References

1. Kilbas, A.A.; Srivastava, H.M.; Trujillo, J.J. *Theory and Applications of Fractional Differential Equations*; North-Holland Mathematics Studies, 204; Elsevier Science B.V.: Amsterdam, The Netherlands, 2006.
2. Henderson, J.; Luca, R.; Tudorache, A. On a system of fractional differential equations with coupled integral boundary conditions. *Fract. Calc. Appl. Anal.* **2015**, *18*, 361–386. [CrossRef]
3. Peng, L.; Zhou, Y. Bifurcation from interval and positive solutions of the three-point boundary value problem for fractional differential equations. *Appl. Math. Comput.* **2015**, *257*, 458–466. [CrossRef]
4. Ahmad, B.; Alsaedi, A.; Ntouyas, S.K.; Tariboon, J. *Hadamard-Type Fractional Differential Equations, Inclusions and Inequalities*; Springer: Cham, Switzerland, 2017.
5. Ahmad, B.; Ntouyas, S.K. Nonlocal initial value problems for Hadamard-type fractional differential equations and inclusions. *Rocky Mt. J. Math.* **2018**, *48*, 1043–1068. [CrossRef]
6. Cui, Y.; Ma, W.; Sun, Q.; Su, X. New uniqueness results for boundary value problem of fractional differential equation. *Nonlinear Anal. Model. Control* **2018**, *23*, 31–39. [CrossRef]
7. Ahmad, B.; Luca, R. Existence of solutions for sequential fractional integro-differential equations and inclusions with nonlocal boundary conditions. *Appl. Math. Comput.* **2018**, *339*, 516–534. [CrossRef]
8. Alsaedi, A.; Ahmad, B.; Alghanmi, M.; Ntouyas, S.K. On a generalized Langevin type nonlocal fractional integral multivalued problem. *Mathematics* **2019**, *7*, 1015. [CrossRef]
9. Ahmad, B.; Alghamdi, N.; Alsaedi, A.; Ntouyas, S.K. A system of coupled multi-term fractional differential equations with three-point coupled boundary conditions. *Fract. Calc. Appl. Anal.* **2019**, *22*, 601–618. [CrossRef]
10. Alsaedi, A.; Ahmad, B.; Alghanmi, M. Extremal solutions for generalized Caputo fractional differential equations with Steiltjes-type fractional integro-initial conditions. *Appl. Math. Lett.* **2019**, *91*, 113120. [CrossRef]
11. Ahmad, B.; Alsaedi, A.; Alruwaily, Y.; Ntouyas, S.K. Nonlinear multi-term fractional differential equations with Riemann-Stieltjes integro-multipoint boundary conditions. *AIMS Math.* **2020**, *5*, 1446–1461. [CrossRef]
12. Liang, S.; Wang, L.; Yin, G. Fractional differential equation approach for convex optimization with convergence rate analysis. *Optim. Lett.* **2020**, *14*, 145–155. [CrossRef]
13. Atanackovic, T.M.; Stankovic, B. On a differential equation with left and right fractional derivatives. *Fract. Calc. Appl. Anal.* **2007**, *10*, 139–150. [CrossRef]
14. Zhang, L.; Ahmad, B.; Wang, G. The existence of an extremal solution to a nonlinear system with the right-handed Riemann-Liouville fractional derivative. *Appl. Math. Lett.* **2014**, *31*, 1–6. [CrossRef]
15. Khaldi, R.; Guezane-Lakoud, A. Higher order fractional boundary value problems for mixed type derivatives. *J. Nonlinear Funct. Anal.* **2017**, *30*, 9.

16. Lakoud, A.G.; Khaldi, R.; Kilicman, A. Existence of solutions for a mixed fractional boundary value problem. *Adv. Differ. Equ.* **2017**, *2017*, 164. [CrossRef]
17. Guezane-Lakoud, A.; Khaldi, R.; Torres, D.F.M. On a fractional oscillator equation with natural boundary conditions. *Progr. Fract. Differ. Appl.* **2017**, *3*, 191–197. [CrossRef]
18. Ahmad, B.; Ntouyas, S.K.; Alsaedi, A. Existence theory for nonlocal boundary value problems involving mixed fractional derivatives. *Nonlinear Anal. Model. Control* **2019**, *24*, 937–957. [CrossRef]
19. Laitinen, M.; Tiihonen, T. Heat transfer in conducting and radiating bodies. *Appl. Math. Lett.* **1997**, *10*, 5–8. [CrossRef]
20. Laitinen, M.; Tiihonen, T. Integro-differential equation modelling heat transfer in conducting, radiating and semitransparent materials. *Math. Methods Appl. Sci.* **1998**, *21*, 375–392. [CrossRef]
21. Hajmohammadi, M.R.; Nourazar, S.S.; Manesh, A.H. Semi-analytical treatments of conjugate heat transfer. *J. Mech. Eng. Sci.* **2013**, *227*, 492–503. [CrossRef]
22. Yang, A.M.; Han, Y.; Zhang, Y.Z.; Wang, L.T.; Zhang, D.; Yang, X.J. On nonlocal fractional Volterra integro-differential equations in fractional steady heat transfer. *Therm. Sci.* **2016**, *20*, S789–S793. [CrossRef]
23. Tarasov, V.E. Fractional integro-differential equations for electromagnetic waves in dielectric media. *Theor. Math. Phys.* **2009**, *158*, 355–359. [CrossRef]
24. Krasnoselskii, M.A. Two remarks on the method of successive approximations. *Uspekhi Mat. Nauk.* **1955**, *10* , 123–127.
25. Granas, A.; Dugundji, J. *Fixed Point Theory*; Springer: New York, NY, USA, 2005.
26. Agrawal, O.P. Formulation of Euler-Lagrange equations for fractional variational problems. *J. Math. Anal. Appl.* **2002**, *272*, 368–379. [CrossRef]

15

Some Fractional Dynamic Inequalities of Hardy's type via Conformable Calculus

Samir Saker [1,*], Mohammed Kenawy [2], Ghada AlNemer [3] and Mohammed Zakarya [4,5]

1 Department of Mathematics, Faculty of Science, Mansoura University, Mansoura 35516, Egypt
2 Department of Mathematics, Faculty of Science, Fayoum University, Fayoum 63514, Egypt; mrz00@fayoum.edu.eg
3 Department of Mathematical Science, College of Science, Princess Nourah bint Abdulrahman University, P.O. Box 105862, Riyadh, Saudi 11656, Arabia; gnnemer@pnu.edu.sa
4 Department of Mathematics, College of Science, King Khalid University, P.O. Box 9004, Abha 61413, Saudi Arabia; Mohammed_Zakaria1983@yahoo.com
5 Department of Mathematics, Faculty of Science, Al-Azhar University, Assiut 71524, Egypt
* Correspondence: shsaker@mans.edu.eg

Abstract: In this article, we prove some new fractional dynamic inequalities on time scales via conformable calculus. By using chain rule and Hölder's inequality on timescales we establish the main results. When $\alpha = 1$ we obtain some well-known time-scale inequalities due to Hardy, Copson, Bennett and Leindler inequalities.

Keywords: fractional hardy's inequality; fractional bennett's inequality; fractional copson's inequality; fractional leindler's inequality; timescales; conformable fractional calculus; fractional hölder inequality

---

In 1920, Hardy [1] established the inequality

$$\sum_{n=1}^{\infty}\left(\frac{1}{n}\sum_{i=1}^{n}w(i)\right)^{k} \leq \left(\frac{k}{k-1}\right)^{k}\sum_{n=1}^{\infty}w^{k}(n), \quad k>1. \tag{1}$$

where $w(n)$ is a positive sequence defined for all $n \geq 1$. After that, Hardy [2], by using the calculus of variations, proved the continuous inequality of (1) which has the form

$$\int_{0}^{\infty}\left(\frac{1}{x}\int_{0}^{x}g(s)ds\right)^{k}dx \leq \left(\frac{k}{k-1}\right)^{k}\int_{0}^{\infty}g^{k}(x)dx, \tag{2}$$

for a given positive function g, which is integrable over $(0,x)$, and g^k is convergent and integrable over $(0,\infty)$ and $k>1$. In (1) and (2), $(k/\,(k-1))^k$ is a sharp constant. As a generalization of (2), Hardy [3] showed that when $k>1$, then

$$\int_{0}^{\infty}x^{-h}\left(\int_{0}^{x}g(s)ds\right)^{k}dx \leq \left(\frac{k}{h-1}\right)^{k}\int_{0}^{\infty}x^{k-h}g^{k}(x)dx, \text{ for } h>1, \tag{3}$$

and

$$\int_{0}^{\infty}x^{-h}\left(\int_{x}^{\infty}g(s)ds\right)^{k}dx \leq \left(\frac{k}{1-h}\right)^{k}\int_{0}^{\infty}x^{k-h}g^{k}(x)dx, \text{for } h<1. \tag{4}$$

The constants $(k/(h-1))^k$ and $(k/(1-h))^k$ in (3) and (4) are the best possible. Copson [4] demonstrated that if $g(x) > 0$, $k > 1$ and $g^k(x)$ is integrable on the interval $(0, \infty)$, then

$$\int_x^\infty \left(\frac{g(s)}{s}\right) ds,$$

converges for $x > 0$ and

$$\int_0^\infty \left(\int_x^\infty \frac{g(s)}{s} ds\right)^k dx \leq k^k \int_0^\infty g^k(x) dx, \tag{5}$$

where k^k is the best possible constant. Some of the generalizations of the discrete Hardy inequality (1) and the discrete version of (5) and its extensions are due to Leindler, we refer to the papers the papers [5–8]. For example, Leindler in [5] proved that if $p > 1$, $\lambda(n), g(n) > 0$, then

$$\sum_{n=1}^\infty \lambda(n) \left(\sum_{s=1}^n g(s)\right)^p \leq p^p \sum_{n=1}^\infty \lambda^{1-p}(n) \left(\sum_{s=n}^\infty \lambda(s)\right)^p g^p(n), \tag{6}$$

and

$$\sum_{n=1}^\infty \lambda(n) \left(\sum_{k=n}^\infty g(k)\right)^p \leq p^p \sum_{n=1}^\infty \lambda^{1-p}(n) \left(\sum_{k=1}^n \lambda(k)\right)^p g^p(n). \tag{7}$$

The converses of (6) and (7) are proved by Leindler in [6]. He proved that if $0 < p \leq 1$, then

$$\sum_{n=1}^\infty \lambda(n) \left(\sum_{k=1}^n g(k)\right)^p \geq p^p \sum_{n=1}^\infty \lambda^{1-p}(n) \left(\sum_{k=n}^\infty \lambda(k)\right)^p g^p(n), \tag{8}$$

and

$$\sum_{n=1}^\infty \lambda(n) \left(\sum_{k=n}^\infty g(k)\right)^p \geq p^p \sum_{n=1}^\infty \lambda^{1-p}(n) \left(\sum_{k=1}^n \lambda(p)\right)^p g^p(n). \tag{9}$$

For more generalization Copson in [9] showed that if $k > 1$, $\lambda(j) \geq 0$, $w(j) \geq 0$, $\forall j \geq 1$, $\Omega(m) = \sum_{j=1}^m \lambda(j)$, and $h > 1$, then

$$\sum_{m=1}^\infty \frac{\lambda(m)}{\Omega^h(m)} \left(\sum_{j=1}^m w(j)\lambda(j)\right)^k \leq \left(\frac{k}{h-1}\right)^k \sum_{m=1}^\infty \lambda(m)\Omega^{k-h}(m) w^k(m), \tag{10}$$

and if $0 \leq h < 1$and $k > 1$, then

$$\sum_{m=1}^\infty \frac{\lambda(m)}{\Omega^h(m)} \left(\sum_{j=m}^\infty w(j)\lambda(j)\right)^k \leq \left(\frac{k}{1-h}\right)^k \sum_{m=1}^\infty \lambda(m)\Omega^{k-h}(m) w^k(m). \tag{11}$$

The integral versions of the inequalities (10) and (11) was proved by Copson in [10] (Theorems 1 and 3). In particular, he proved that if $k \geq 1$, $h > 1$, and $\Omega(s) = \int_0^s \lambda(t)dt$, then

$$\int_0^\infty \frac{\lambda(s)}{\Omega^h(s)} \Phi^k(s) ds \leq \left(\frac{k}{h-1}\right)^k \int_0^\infty \frac{\lambda(s)}{\Omega^{h-k}(s)} g^k(s) ds, \tag{12}$$

where $\Phi(s) = \int_0^s \lambda(t) g(t) dt$, and if $k > 1$, $0 \leq h < 1$, then

$$\int_0^\infty \frac{\lambda(s)}{\Omega^h(s)} \Phi^k(s) ds \leq \left(\frac{k}{1-h}\right)^k \int_0^\infty \frac{\lambda(s)}{\Omega^{h-k}(s)} g^k(s) ds, \tag{13}$$

where $\Phi(s) = \int_s^\infty \lambda(t)g(t)dt$. Leindler in [5] and Bennett in [11] presented interesting different inequalities. Leindler established that if $k > 1$, $\Omega^*(m) = \sum_{j=m}^\infty \lambda(j) < \infty$, and $0 \le h < 1$, then

$$\sum_{m=1}^{\infty} \frac{\lambda(m)}{(\Omega^*(m))^h} \left(\sum_{j=1}^{m} w(j)\lambda(j) \right)^k \le \left(\frac{k}{1-h} \right)^k \sum_{m=1}^{\infty} \lambda(m)(\Omega^*(m))^{k-h} w^k(m), \tag{14}$$

and Bennett in [11] showed that if $1 < h \le k$, then

$$\sum_{m=1}^{\infty} \frac{\lambda(m)}{(\Omega^*(m))^h} \left(\sum_{j=m}^{\infty} w(j)\lambda(j) \right)^k \le \left(\frac{k}{h-1} \right)^k \sum_{m=1}^{\infty} \lambda(m)(\Omega^*(m))^{k-h} w^k(m). \tag{15}$$

In last decades, studying the dynamic equations and inequalities on time scales become a main field in applied and pure mathematics, we refer to [12–14] and the references they are cited. In fact, the book [13] includes forms of the above inequalities on time-scale and their extensions. The timescales idea is returned to Stefan Hilger [15], who investigated the research of dynamic equations on timescales. The books by Bohner and Peterson in [16,17] summarized and organized most timescales calculus. The three most common timescales calculuses are difference, differential, and quantum calculus (see [18]), i.e., at $\mathbb{T} = \mathbb{N}$, $\mathbb{T} = \mathbb{R}$, and $\mathbb{T} = q^{\mathbb{N}_0} = \{q^s : s \in \mathbb{N}_0\}$ where $q > 1$.

In recent years, a lot of work has been published for fractional inequalities and the subject becomes an active field of research and several authors were interested in proving inequalities of fractional type by using the Riemann-Liouville and Caputo derivative (see [19–21]).

On the other hand, the authors in [22,23] introduced a new fractional calculus called the conformable calculus and gave a new definition of the derivative with the base properties of the calculus based on the new definition of derivative and integrals. By using conformable calculus, some authors have studied classical inequalities like Chebyshev's inequality [24], Hermite-Hadamard's inequality [25–27], Opial's inequality [28,29] and Steffensen's inequality [30].

The main question that arises now is: Is it possible to prove new fractional inequalities on timescales and give a unified approach of such studies? This in fact needs a new fractional calculus on timescales. Very recently Torres and others, in [31,32], combined a time scale calculus and conformable calculus and obtained the new fractional calculus on timescales. So, it is natural to look on new fractional inequalities on timescales and give an affirmative answer to the above question.

In particular, in this paper, we will prove the fractional forms of the classical Hardy, Bennett, Copson and Leindler inequalities. The paper is divided into two sections. Section 2 is an introduction of basics of fractional calculus on timescales and Section 3 contains the main results.

1. Preliminaries and Basic Lemmas

We present the fundamental results about the fractional timescales calculus. The results are adapted from [16,17,31,32]. A time-scale $\mathbb{T}$ is non-empty closed subset of $\mathbb{R}$ ($\mathbb{R}$ is the real numbers). The operators of backward jump and forward jump express of the closest point $t \in \mathbb{T}$ on the right and left of t is defined by, respectively:

$$\rho(t) := \sup\{s \in \mathbb{T} : s < t\}, \tag{16}$$

$$\sigma(t) := \inf\{s \in \mathbb{T} : s > t\}, \tag{17}$$

where $\sup \phi = \inf \mathbb{T}$ and $\inf \phi = \sup \mathbb{T}$ (ϕ denotes the empty set), for any $t \in \mathbb{T}$ the notation $f^\sigma(t)$ refer to $f(\sigma(t))$, i.e., $f^\sigma = f \circ \sigma$. The graininess function $\mu : \mathbb{T} \to [0, \infty)$, defined by $\mu(t) := \sigma(t) - t$.

Definition 1. *The number $T_\alpha^\Delta(f)(t)$ (provided it exists) of the function $f : \mathbb{T} \to \mathbb{R}$, for $t > 0$ and $\alpha \in (0,1]$ is the number which has the property that for any $\epsilon > 0$, there exists a neighborhood U of t S. T.*

$$\left|[f^\sigma(t) - f(s)]t^{1-\alpha} - T_\alpha^\Delta(f(t))(\sigma(t) - s)\right| \leq \epsilon\,|\sigma(t) - s|\,, \quad \textit{for all } t \in U$$

$T_\alpha^\Delta(f(t))$ is called the conformable $\alpha-$fractional derivative of function f of order α at t , for conformable fractional derivative on $\mathbb{T}$ at 0, we define it with $T_\alpha^\Delta(f(0) = \lim_{t\to 0^+} T_\alpha^\Delta(f(t))$.

The conformable fractional derivative has the following properties

Theorem 1. *Let $v, u : \mathbb{T} \to \mathbb{R}$ are conformable fractional derivative from order $\alpha \in (0,1]$, then the following properties are hold:*

(i) *The $v + u : \mathbb{T} \to \mathbb{R}$ is conformable fractional derivative and*

$$T_\alpha^\Delta(v + u) = T_\alpha^\Delta(v) + T_\alpha^\Delta(u).$$

(ii) *Fora all $k \in \mathbb{R}$, then $kv : \mathbb{T} \to \mathbb{R}$ is $\alpha-$fractional differentiable and*

$$T_\alpha^\Delta(kv) = kT_\alpha^\Delta(v).$$

(iii) *If v and u are $\alpha-$fractional differentiable, we have $vu : \mathbb{T} \to \mathbb{R}$ is $\alpha-$fractional differentiable and*

$$T_\alpha^\Delta(vu) = T_\alpha^\Delta(v)\,u + (v \circ \sigma)\,T_\alpha^\Delta(u) = T_\alpha^\Delta(v)\,(u \circ \sigma) + v\,T_\alpha^\Delta(u).$$

(iv) *If v is $\alpha-$fractional differentiable, then $1/v$ is $\alpha-$fractional differentiable with*

$$T\left(\frac{1}{v}\right) = -\frac{T_\alpha^\Delta(v)}{v\,(v \circ \sigma)}.$$

(v) *If v and u are $\alpha-$fractional differentiable, then v/u is $\alpha-$fractional differentiable with*

$$T_\alpha^\Delta(v/u) = \frac{T_\alpha^\Delta(v)u - vT_\alpha^\Delta(u)}{u(u \circ \sigma)},$$

valid $\forall\ t \in \mathbb{T}^k$,where $u(t)(u(\sigma(t)) \neq 0$.

Lemma 1. *Let $v : \mathbb{T} \to \mathbb{R}$ is continuous and $\alpha-$fractional differentiable at $t \in \mathbb{T}$ for $\alpha \in (0,1]$, and $u : \mathbb{R} \to \mathbb{R}$ is continuous and differentiable. Then there exists $d \in [t, \sigma(t)]$ with*

$$T_\alpha^\Delta(u \circ v)(t) = u'(v(d))T_\alpha^\Delta(v(t)). \tag{18}$$

Lemma 2. *Let $u : \mathbb{R} \to \mathbb{R}$ is continuously differentiable, $\alpha \in (0,1]$, and $v : \mathbb{T} \to \mathbb{R}$ be $\alpha-$fractional differentiable. Then $(u \circ v) : \mathbb{T} \to \mathbb{R}$ is $\alpha-$fractional differentiable and we have*

$$T_\alpha^\Delta(u \circ v)(s) = \left(\int_0^1 u'\left(v(s) + h\mu(s)s^{\alpha-1}T_\alpha^\Delta(v(s))\right)dh\right)T_\alpha^\Delta(v(s)). \tag{19}$$

Definition 2. *Let $0 < \alpha \leq 1$, the $\alpha-$fractional integral of f, is defined as*

$$\int f(s)\Delta_\alpha s = \int f(s)s^{\alpha-1}\Delta s.$$

The conformable fractional integral satisfying the next properties

Theorem 2. *Assume $a, b, c \in \mathbb{T}$, $\lambda \in \mathbb{R}$. Let $u, v : \mathbb{T} \to \mathbb{R}$. Then*

(i) $\int_a^b [v(s) + u(s)]\Delta_\alpha s = \int_a^b v(s)\Delta_\alpha s + \int_a^b u(s)\Delta_\alpha s.$

(ii) $\int_a^b \lambda v(s)\Delta_\alpha s = \lambda \int_a^b v(s)\Delta_\alpha s.$

(iii) $\int_a^b v(s)\Delta_\alpha s = -\int_b^a v(s)\Delta_\alpha s.$

(iv) $\int_a^b v(s)\Delta_\alpha s = \int_a^c v(s)\Delta_\alpha s + \int_c^b v(s)\Delta_\alpha s.$

(v) $\int_a^a v(s)\Delta_\alpha s = 0.$

Lemma 3. *Assume $\mathbb{T}$ be a time-scale, $a, b \in \mathbb{T}$ where $b > a$. Let u, v are conformable α—fractional differentiable, $\alpha \in (0, 1]$. Then the formula of integration by parts is given by*

$$\int_a^b v(s)T_\alpha^\Delta u(s)\Delta_\alpha s = [v(s)u(s)]_a^b - \int_a^b u^\sigma(s)T_\alpha^\Delta v(s)\Delta_\alpha s. \tag{20}$$

Lemma 4. *Assume $\mathbb{T}$ be a time-scale, $a, b \in \mathbb{T}$ and $\alpha \in (0, 1]$. Let $u, v : \mathbb{T} \to \mathbb{R}$. Then*

$$\int_a^b |v(s)u(s)|\Delta_\alpha s \le \left[\int_a^b |v(s)|^k \Delta_\alpha s\right]^{\frac{1}{k}} \left[\int_a^b |u(s)|^l \Delta_\alpha s\right]^{\frac{1}{l}}, \tag{21}$$

where $k > 1$ and $1/k + 1/l = 1$.

2. Main Results

Throughout the paper, we will assume that the functions are nonnegative on $[a, \infty)_\mathbb{T}$ and its integrals exist and are finite. We start with the fractional time-scale inequality of Copson's type.

Theorem 3. *Assume $1 < c < k$, define*

$$\Phi(x) := \int_a^x \lambda(s)\Delta_\alpha s \text{ and } \Omega(x) := \int_a^x \lambda(s)g(s)\Delta_\alpha s.$$

If

$$\Omega(\infty) < \infty, \text{ and } \int_a^\infty \frac{\lambda(s)}{(\Phi^\sigma(s))^{c-\alpha+1}}\Delta_\alpha s < \infty,$$

then

$$\int_a^\infty \frac{\lambda(x)}{(\Phi^\sigma(x))^{c-\alpha+1}} (\Omega^\sigma(x))^k \Delta_\alpha x \le \left(\frac{k}{c-\alpha}\right)^k \int_a^\infty \frac{\lambda(x)\Phi^{k(\alpha-c)}(x)}{(\Phi^\sigma(x))^{(c-\alpha+1)(1-k)}} g^k(x)\Delta_\alpha x. \tag{22}$$

Proof. By employing the formula of integration by parts (20) on the term

$$\int_a^\infty \frac{\lambda(x)}{(\Phi^\sigma(x))^{c-\alpha+1}} (\Omega^\sigma(x))^k \Delta_\alpha x,$$

with $u^\sigma(x) = (\Omega^\sigma(x))^k$ and $x_\alpha^\Delta v(x) = \frac{\lambda(x)}{(\Phi^\sigma(x))^{c-\alpha+1}}$, we have that

$$\int_a^\infty \frac{\lambda(x)}{(\Phi^\sigma(x))^{c-\alpha+1}} (\Omega^\sigma(x))^k \Delta_\alpha x = v(x)\Omega^k(x)\Big|_a^\infty + \int_a^\infty -v(x)x_\alpha^\Delta\left(\Omega^k(x)\right)\Delta_\alpha x, \tag{23}$$

where

$$-v(x) = \int_x^\infty \frac{\lambda(s)}{(\Phi^\sigma(s))^{c-\alpha+1}}\Delta_\alpha s = \int_x^\infty x_\alpha^\Delta\Phi(s)\left(\Phi^\sigma(s)\right)^{\alpha-c-1}\Delta_\alpha s.$$

By using the chain rule (18), we obtain that

$$\begin{aligned} -x_{\alpha}^{\Delta}\left(\Phi^{\alpha-c}(x)\right) &= -(\alpha-c)\Phi^{\alpha-c-1}(d)x_{\alpha}^{\Delta}\Phi(x), \text{ where } d\in[x,\sigma(x)] \\ &= \frac{(c-\alpha)\,x_{\alpha}^{\Delta}\Phi(x)}{\Phi^{c-\alpha+1}(d)} \\ &\geq \frac{(c-\alpha)\,x_{\alpha}^{\Delta}\left(\Phi(x)\right)}{\left(\Phi^{\sigma}(x)\right)^{c-\alpha+1}}. \end{aligned}$$

Then we have

$$x_{\alpha}^{\Delta}\left(\Phi(x)\right)\left(\Phi^{\sigma}(x)\right)^{\alpha-c-1} \leq \frac{-1}{c-\alpha}x_{\alpha}^{\Delta}\Phi^{\alpha-c}(x),$$

and thus

$$-v(x) = \int_{x}^{\infty}\frac{\lambda(s)}{\left(\Phi^{\sigma}(s)\right)^{c-\alpha+1}}\Delta_{\alpha}s \leq \frac{-1}{c-\alpha}\int_{x}^{\infty}x_{\alpha}^{\Delta}\Phi^{\alpha-c-1}(s)\Delta_{\alpha}s \leq \frac{\Phi^{\alpha-c}(x)}{c-\alpha}. \tag{24}$$

Again, by using the chain rule (18) to calculate

$$x_{\alpha}^{\Delta}\left(\Omega^{k}(x)\right) = k\Omega^{k-1}(d)x_{\alpha}^{\Delta}\left(\Omega(x)\right), \text{ where } d\in[x,\sigma(x)],$$

at $x_{\alpha}^{\Delta}\left(\Omega(x)\right) = \lambda(x)g(x) \geq 0$ and $d \leq \sigma(x)$, we get that

$$x_{\alpha}^{\Delta}\left(\Omega^{k}(x)\right) \leq k\lambda(x)g(x)\left(\Omega^{\sigma}(x)\right)^{k-1}. \tag{25}$$

Since $\Omega(a) = 0$, $v(\infty) = 0$ and from (24), (25) and (23) we have

$$\int_{a}^{\infty}\frac{\lambda(x)}{\left(\Phi^{\sigma}(x)\right)^{c-\alpha+1}}\left(\Omega^{\sigma}(x)\right)^{k}\Delta_{\alpha}x \leq \frac{k}{c-\alpha}\int_{a}^{\infty}\Phi^{\alpha-c}(x)\lambda(x)g(x)\left(\Omega^{\sigma}(x)\right)^{k-1}\Delta_{\alpha}x,$$

which reformulated as

$$\begin{aligned} &\int_{a}^{\infty}\frac{\lambda(x)}{\left(\Phi^{\sigma}(x)\right)^{c-\alpha+1}}\left(\Omega^{\sigma}(x)\right)^{k}\Delta_{\alpha}x \\ &= \frac{k}{c-\alpha}\int_{a}^{\infty}\frac{\lambda(x)\Phi^{\alpha-c}(x)g(x)}{\left(\lambda(x)\left(\Phi^{\sigma}(x)\right)^{\alpha-c-1}\right)^{\frac{k-1}{k}}}\left(\frac{\lambda(x)\left(\Omega^{\sigma}(x)\right)^{k}}{\left(\Phi^{\sigma}(x)\right)^{c-\alpha+1}}\right)^{\frac{k-1}{k}}\Delta_{\alpha}x. \end{aligned}$$

Using Hölder's inequality (21) on

$$\int_{a}^{\infty}\frac{\lambda(x)\Phi^{\alpha-c}(x)g(x)}{\left(\lambda(x)\left(\Phi^{\sigma}(x)\right)^{\alpha-c-1}\right)^{\frac{k-1}{k}}}\left(\frac{\lambda(x)\left(\Omega^{\sigma}(x)\right)^{k}}{\left(\Phi^{\sigma}(x)\right)^{c-\alpha+1}}\right)^{\frac{k-1}{k}}\Delta_{\alpha}x$$

with indices k and $k/(k-1)$, we have

$$\int_a^\infty \frac{\lambda(x)}{(\Phi^\sigma(x))^{c-\alpha+1}} (\Omega^\sigma(x))^k \Delta_\alpha x \leq$$

$$\frac{k}{c-\alpha}\left[\int_a^\infty \left[\frac{\lambda(x)\Phi^{\alpha-c}(x)g(x)}{\left[\lambda(x)\left(\Phi^\sigma(x)\right)^{\alpha-c-1}\right]^{\frac{k-1}{k}}}\right]^k \Delta_\alpha x\right]^{\frac{1}{k}}$$

$$\times\left[\int_a^\infty \left[\left[\frac{\lambda(x)\left(\Omega^\sigma(x)\right)^k}{(\Phi^\sigma(x))^{c-\alpha+1}}\right]^{\frac{k-1}{k}}\right]^{\frac{k}{k-1}} \Delta_\alpha x\right]^{\frac{k-1}{k}},$$

then

$$\left[\int_a^\infty \frac{\lambda(x)}{(\Phi^\sigma(x))^{c-\alpha+1}} (\Omega^\sigma(x))^k \Delta_\alpha x\right]^{\frac{1}{k}} \leq \frac{k}{c-\alpha}\left[\int_a^\infty \frac{\lambda(x)\Phi^{k(\alpha-c)}(x)g^k(x)}{(\Phi^\sigma(x))^{(c-\alpha+1)(1-k)}}\Delta_\alpha x\right]^{\frac{1}{k}}.$$

This leads to

$$\int_a^\infty \frac{\lambda(x)}{(\Phi^\sigma(x))^{c-\alpha+1}} (\Omega^\sigma(x))^k \Delta_\alpha x \leq \left(\frac{k}{c-\alpha}\right)^k \int_a^\infty \frac{\lambda(x)\Phi^{k(\alpha-c)}(x)g^k(x)}{(\Phi^\sigma(x))^{(c-\alpha+1)(1-k)}}\Delta_\alpha x,$$

that is the desired inequality (22). The proof is complete. □

Corollary 1. *At $\alpha = 1$ in Theorem 3, we obtain the inequality*

$$\int_a^\infty \frac{\lambda(x)}{(\Phi^\sigma(x))^c} (\Omega^\sigma(x))^k \Delta x \leq \left(\frac{k}{c-1}\right)^k \int_a^\infty \frac{\Phi^{k(1-c)}(x)}{(\Phi^\sigma(x))^{c(1-k)}}\lambda(x)g^k(x)\Delta x.$$

that is the timescales version of inequality (2.8) in [33].

Corollary 2. *At $\alpha = 1$, and $\mathbb{T} = \mathbb{R}$ $(\Phi^\sigma(x) = \Phi(x))$ in Theorem 3, we obtain the integral inequality*

$$\int_a^\infty \frac{\lambda(x)}{\Phi^c(x)}\left(\int_a^x \lambda(s)g(s)ds\right)^k dx \leq \left(\frac{k}{c-1}\right)^k \int_a^\infty \Phi^{k-c}(x)\lambda(x)g^k(x)dx,$$

which is of Copson type.

Corollary 3. *At $\alpha = 1$, $T = \mathbb{R}$, $\lambda(x) = 1$ and $a = 0$, $(\Phi(x) = \int_0^x \lambda(s)ds = x)$ in Theorem 3, we have Hardy-Littlewood integral inequality (3)*

$$\int_0^\infty \frac{1}{x^c}\left(\int_0^x g(s)ds\right)^k dx \leq \left(\frac{k}{c-1}\right)^k \int_0^\infty \frac{1}{x^{c-k}}g^k(x)dx.$$

Also, if $c = k$, we obtain the standard Hardy inequality (2)

$$\int_0^\infty \frac{1}{x^k}\left(\int_0^x g(s)ds\right)^k dx \leq \left(\frac{k}{k-1}\right)^k \int_0^\infty g^k(x)dx.$$

Theorem 4. *Let $0 \leq c < 1$ and $k > 1$. Define*

$$\Phi(x) := \int_a^x \lambda(s)\Delta_\alpha s \text{ and } \Omega(x) := \int_x^\infty \lambda(s)g(s)\Delta_\alpha s.$$

If

$$\Omega(a) < \infty, \text{ and } \int_a^\infty \frac{\lambda(s)}{(\Phi^\sigma(s))^{c-\alpha+1}} \Delta_\alpha s < \infty,$$

then

$$\int_a^\infty \frac{\lambda(x)}{(\Phi^\sigma(x))^{c-\alpha+1}} \Omega^k(x) \Delta_\alpha x \leq \left(\frac{k}{\alpha-c}\right)^k \int_a^\infty (\Phi^\sigma(x))^{k-c+\alpha-1} \lambda(x) g^k(x) \Delta_\alpha x. \tag{26}$$

Proof. By using the formula of integration by parts (20) on

$$\int_a^\infty \frac{\lambda(x)}{(\Phi^\sigma(x))^{c-\alpha+1}} \Omega^k(x) \Delta_\alpha x,$$

with $v(x) = \Omega^k(x)$ and $x_\alpha^\Delta u(x) = \frac{\lambda(x)}{(\Phi^\sigma(x))^{c-\alpha+1}}$, we have

$$\int_a^\infty \frac{\lambda(x)}{(\Phi^\sigma(x))^{c-\alpha+1}} \Omega^k(x) \Delta_\alpha x = u(x)\Omega^k(x)\Big|_a^\infty + \int_a^\infty u^\sigma(x) x_\alpha^\Delta \left(-\Omega^k(x)\right) \Delta_\alpha x, \tag{27}$$

where

$$u(x) = \int_a^x \frac{\lambda(s)}{(\Phi^\sigma(s))^{c-\alpha+1}} \Delta_\alpha s = \int_a^x x_\alpha^\Delta \Phi(s) \left(\Phi^\sigma(s)\right)^{\alpha-c-1} \Delta_\alpha s.$$

By using chain rule (18), then for $d \in [x, \sigma(x)]$, we get that

$$\begin{aligned} x_\alpha^\Delta \left(\Phi^{\alpha-c}(x)\right) &= (\alpha-c)\Phi^{\alpha-c-1}(d) x_\alpha^\Delta \left(\Phi(x)\right) = \frac{(\alpha-c)\, x_\alpha^\Delta \left(\Phi(x)\right)}{\Phi^{c-\alpha+1}(d)} \\ &\geq \frac{(\alpha-c)\, x_\alpha^\Delta \left(\Phi(x)\right)}{(\Phi^\sigma(x))^{c-\alpha+1}}. \end{aligned}$$

So

$$x_\alpha^\Delta \left(\Phi(x)\right) \left(\Phi^\sigma(x)\right)^{\alpha-c-1} \leq \frac{1}{\alpha-c} x_\alpha^\Delta \left(\Phi^{\alpha-c}(x)\right),$$

and then,

$$\begin{aligned} u^\sigma(x) &= \int_a^{\sigma(x)} x_\alpha^\Delta \left(\Phi(s)\right) \left(\Phi^\sigma(s)\right)^{\alpha-c-1} \Delta_\alpha s \\ &\leq \frac{1}{\alpha-c} \int_a^{\sigma(x)} x_\alpha^\Delta \left(\Phi^{\alpha-c}(s)\right) \Delta_\alpha s \leq \frac{(\Phi^\sigma(x))^{\alpha-c}}{\alpha-c}. \end{aligned} \tag{28}$$

Again, by using chain rule (18), we obtain

$$-x_\alpha^\Delta \left(\Omega^k(x)\right) = -k\Omega^{k-1}(d) x_\alpha^\Delta \left(\Omega(x)\right), \text{where } d \in [x, \sigma(x)],$$

since $x_\alpha^\Delta (\Omega(x)) = -\lambda(x)g(x) \geq 0$ and $d \geq x$, then

$$-\, x_\alpha^\Delta \left(\Omega^k(x)\right) \leq k\lambda(x)g(x)\Omega^{k-1}(x). \tag{29}$$

Using $\Phi(a) = 0, \Omega(\infty) = 0$ and (28), (29) and (27), we have that

$$\int_a^\infty \frac{\lambda(x)}{(\Phi^\sigma(x))^{c-\alpha+1}} \Omega^k(x) \Delta_\alpha x \leq \frac{k}{\alpha-c} \int_a^\infty (\Phi^\sigma(x))^{\alpha-c} \lambda(x) g(x) \Omega^{k-1}(x) \Delta_\alpha x,$$

which reformulated as

$$\int_a^\infty \frac{\lambda(x)}{(\Phi^\sigma(x))^{c-\alpha+1}} \Omega^k(x) \Delta_\alpha x =$$

$$\frac{k}{\alpha - c} \int_a^\infty \frac{(\Phi^\sigma(x))^{\alpha-c} \lambda(x) g(x)}{\left(\lambda(x) (\Phi^\sigma(x))^{\alpha-c-1}\right)^{\frac{k-1}{k}}} \left(\frac{\lambda(x)\Omega^k(x)}{(\Phi^\sigma(x))^{c-\alpha+1}} \right)^{\frac{k-1}{k}} \Delta_\alpha x.$$

By employing Hölder's inequality (21) on

$$\int_a^\infty \frac{(\Phi^\sigma(x))^{\alpha-c} \lambda(x) g(x)}{\left(\lambda(x) (\Phi^\sigma(x))^{\alpha-c-1}\right)^{\frac{k-1}{k}}} \left(\frac{\lambda(x)\Omega^k(x)}{(\Phi^\sigma(x))^{c-\alpha+1}} \right)^{\frac{k-1}{k}} \Delta_\alpha x$$

with indices k and $k/(k-1)$, we have

$$\int_a^\infty \frac{\lambda(x)}{(\Phi^\sigma(x))^{c-\alpha+1}} \Omega^k(x) \Delta_\alpha x \leq$$

$$\frac{k}{\alpha - c} \left[\int_a^\infty \left[\frac{(\Phi^\sigma(x))^{1-c} \lambda(x) g(x)}{\left[\lambda(x) (\Phi^\sigma(x))^{\alpha-c-1}\right]^{\frac{k-1}{k}}} \right]^k \Delta_\alpha x \right]^{\frac{1}{k}}$$

$$\times \left[\int_a^\infty \left[\left[\frac{\lambda(x)\Omega^k(x)}{(\Phi^\sigma(x))^{c-\alpha+1}} \right]^{\frac{k-1}{k}} \right]^{\frac{k}{k-1}} \Delta_\alpha x \right]^{\frac{k-1}{k}},$$

then we have

$$\left[\int_a^\infty \frac{\lambda(x)}{(\Phi^\sigma(x))^{c-\alpha+1}} \Omega^k(x) \Delta_\alpha x \right]^{\frac{1}{k}} \leq \frac{k}{\alpha - c} \left[\int_a^\infty \frac{\lambda(x) g^k(x)}{(\Phi^\sigma(x))^{c-k-\alpha+1}} \Delta_\alpha x \right]^{\frac{1}{k}}.$$

This leads to

$$\int_a^\infty \frac{\lambda(x)}{(\Phi^\sigma(x))^c} \Omega^k(x) \Delta_\alpha x \leq \left(\frac{k}{\alpha - c} \right)^k \int_a^\infty (\Phi^\sigma(x))^{k-c+\alpha-1} \lambda(x) g^k(x) \Delta_\alpha x,$$

that is the desired inequality (26). The proof is complete. □

Corollary 4. *At $\alpha = 1$ in Theorem 4, then*

$$\int_a^\infty \frac{\lambda(x)}{(\Phi^\sigma(x))^c} \Omega^k(x) \Delta x \leq \left(\frac{k}{1 - c} \right)^k \int_a^\infty (\Phi^\sigma(x))^{k-c} \lambda(x) g^k(x) \Delta x.$$

which is the timescales version inequality (2.22) in [33].

Corollary 5. *At $\alpha = 1$, and $T= \mathbb{R}$ in Theorem 4, we obtain the next integral inequality*

$$\int_a^\infty \frac{\lambda(x)}{\Phi^c(x)} \left(\int_x^\infty \lambda(s) g(s) ds \right)^k dx \leq \left(\frac{k}{1 - c} \right)^k \int_a^\infty \Phi^{k-c}(x) \lambda(x) g^k(x) dx,$$

which considered an extension of Hardy's inequality (4) as in the following corollary.

Corollary 6. *At $\alpha = 1$, $T= \mathbb{R}$, $\lambda(x) = 1$ and $a = 0$ in Theorem 4, we have Hardy-Littlewood integral inequality (4)*

$$\int_0^\infty \frac{1}{x^c}\left(\int_x^\infty g(s)ds\right)^k dx \leq \left(\frac{k}{1-c}\right)^k \int_0^\infty \frac{1}{x^{c-k}} g^k(x)dx.$$

A generalization of Leindler's inequality (14) on fractional time scales will be proved in the next theorem.

Theorem 5. *Assume $0 \leq c < 1 < k$, define*

$$\Phi(x) := \int_x^\infty \lambda(s)\Delta_\alpha s \text{ and } \Omega(x) := \int_a^x \lambda(s)g(s)\Delta_\alpha s.$$

If

$$\Omega(\infty) < \infty, \text{ and } \int_a^\infty \frac{\lambda(s)}{\Phi^{c-\alpha+1}(s)}\Delta_\alpha s < \infty,$$

then

$$\int_a^\infty \frac{\lambda(x)}{\Phi^{c-\alpha+1}(x)}\left(\Omega^\sigma(x)\right)^k \Delta_\alpha x \leq \left(\frac{k}{\alpha-c}\right)^k \int_a^\infty \Phi^{k-c+\alpha-1}(x)\lambda(x)g^k(x)\Delta_\alpha x. \tag{30}$$

Proof. Using the formula of integration by parts (20) on

$$\int_a^\infty \frac{\lambda(x)}{\Phi^{c-\alpha+1}(x)}\Omega^k(x)\Delta_\alpha x,$$

with $u^\sigma(x) = (\Omega^\sigma(x))^k$ and $x_\alpha^\Delta v(x) = \frac{\lambda(x)}{\Phi^{c-\alpha+1}(x)}$, we have

$$\int_a^\infty \frac{\lambda(x)}{\Phi^{c-\alpha+1}(x)}\left(\Omega^\sigma(x)\right)^k \Delta_\alpha x = v(x)\Omega^k(x)\Big|_a^\infty + \int_a^\infty -v(x)x_\alpha^\Delta\left(\Omega^k(x)\right)\Delta_\alpha x, \tag{31}$$

where

$$v(x) = -\int_x^\infty \frac{\lambda(s)}{\Phi^{c-\alpha+1}(s)}\Delta_\alpha s.$$

By chain rule (18), we see for $d \in [x, \sigma(x)]$ that

$$\begin{aligned} -x_\alpha^\Delta \Phi^{\alpha-c}(x) &= -(\alpha-c)\Phi^{\alpha-c-1}(d)x_\alpha^\Delta\left(\Phi(x)\right) = \frac{-(\alpha-c)(-\lambda(x))}{\Phi^{c-\alpha+1}(d)} \\ &\geq \frac{(\alpha-c)\,\lambda(x)}{\Phi^{c-\alpha+1}(x)}. \end{aligned}$$

Hence

$$-v(x) = \int_x^\infty \frac{\lambda(s)}{\Phi^{c-\alpha+1}(s)}\Delta_\alpha s \leq \frac{-1}{\alpha-c}\int_x^\infty x_\alpha^\Delta \Phi^{\alpha-c}(s)\Delta_\alpha s \leq \frac{\Phi^{\alpha-c}(x)}{\alpha-c}, \tag{32}$$

from chain rule (18), we obtain

$$x_\alpha^\Delta\left(\Omega^k(x)\right) = k\Omega^{k-1}(d)x_\alpha^\Delta\left(\Omega(x)\right), \text{ where } d \in [x, \sigma(x)],$$

since

$$x_\alpha^\Delta\left(\Omega(x)\right) = \lambda(x)g(x) \geq 0 \text{ and } d \leq \sigma(x),$$

we get

$$x_\alpha^\Delta\left(\Omega^k(x)\right) \leq k\lambda(x)g(x)\left(\Omega^\sigma(x)\right)^{k-1}. \tag{33}$$

Using $\Omega(a) = 0$, $v(\infty) = 0$ and (32), (33) and (31), we get that

$$\int_a^\infty \frac{\lambda(x)}{\Phi^{c-\alpha+1}(x)} \left(\Omega^\sigma(x)\right)^k \Delta_\alpha x \leq \frac{k}{\alpha - c} \int_a^\infty \Phi^{\alpha-c}(x)\lambda(x)g(x) \left(\Omega^\sigma(x)\right)^{k-1} \Delta_\alpha x,$$

which reformulated as

$$\int_a^\infty \frac{\lambda(x)}{\Phi^{c-\alpha+1}(x)} \left(\Omega^\sigma(x)\right)^k \Delta_\alpha x \leq$$
$$\frac{k}{\alpha - c} \int_a^\infty \frac{\Phi^{(c-\alpha+1)\left(\frac{k-1}{k}\right)}(x)}{\lambda^{\frac{k-1}{k}}(x)\Phi^{c-\alpha}(x)} \lambda(x)g(x) \frac{\lambda^{\frac{k-1}{k}}(x)\left(\Omega^\sigma(x)\right)^{k-1}}{\Phi^{(c-\alpha+1)\left(\frac{k-1}{k}\right)}(x)} \Delta_\alpha x.$$

Using Hölder's inequality (21) on

$$\int_a^\infty \frac{\Phi^{(c-\alpha+1)\left(\frac{k-1}{k}\right)}(x)}{\lambda^{\frac{k-1}{k}}(x)\Phi^{c-\alpha}(x)} \lambda(x)g(x) \frac{\lambda^{\frac{k-1}{k}}(x)\left(\Omega^\sigma(x)\right)^{k-1}}{\Phi^{(c-\alpha+1)\left(\frac{k-1}{k}\right)}(x)} \Delta_\alpha x$$

with indices k and $k/(k-1)$, we have

$$\int_a^\infty \frac{\lambda(x)}{\Phi^{c-\alpha+1}(x)} \left(\Omega^\sigma(x)\right)^k \Delta_\alpha x \leq$$
$$\frac{k}{\alpha - c} \left[\int_a^\infty \left[\frac{\Phi^{(c-\alpha+1)\left(\frac{k-1}{k}\right)}(x)}{\lambda^{\frac{k-1}{k}}(x)\Phi^{c-\alpha}(x)} \lambda(x)g(x)\right]^k \Delta_\alpha x\right]^{\frac{1}{k}} \times$$
$$\left[\int_a^\infty \left[\lambda^{\frac{k-1}{k}}(x) \frac{\left(\Omega^\sigma(x)\right)^{k-1}}{\Phi^{(c-\alpha+1)\left(\frac{k-1}{k}\right)}(x)}\right]^{\frac{k}{k-1}} \Delta_\alpha x\right]^{\frac{k-1}{k}},$$

then

$$\left[\int_a^\infty \frac{\lambda(x)}{\Phi^{c-\alpha+1}(x)} \left(\Omega^\sigma(x)\right)^k \Delta_\alpha x\right]^{\frac{1}{k}} \leq \frac{k}{\alpha - c} \left[\int_a^\infty \Phi^{k-c+\alpha-1}(x)\lambda(x)g^k(x)\Delta_\alpha x\right]^{\frac{1}{k}}.$$

This leads to

$$\int_a^\infty \frac{\lambda(x)}{\Phi^{c-\alpha+1}(x)} \left(\Omega^\sigma(x)\right)^k \Delta_\alpha x \leq \left(\frac{k}{\alpha - c}\right)^k \int_a^\infty \Phi^{k-c+\alpha-1}(x)\lambda(x)g^k(x)\Delta_\alpha x,$$

that is the desired inequality (30). The proof is complete. □

Corollary 7. *At $\alpha = 1$ in Theorem 5, we get*

$$\int_a^\infty \frac{\lambda(x)}{\Phi^c(x)} \left(\Omega^\sigma(x)\right)^k \Delta x \leq \left(\frac{k}{1 - c}\right)^k \int_a^\infty \Phi^{k-c}(x)\lambda(x)g^k(x)\Delta x,$$

which inequality (2.36) in [33].

A generalization of Bennett's inequality (15) on fractional timescales will be proved in the next theorem.

Theorem 6. *Assume $0 < \alpha \leq 1$, $1 < c \leq k$, and define*

$$\Phi(x) := \int_x^\infty \lambda(s)\Delta_\alpha s \text{ and } \Omega(x) := \int_x^\infty \lambda(s)g(s)\Delta_\alpha s.$$

If

$$\Omega(a) < \infty \text{ and } \int_a^\infty \frac{\lambda(s)}{\Phi^{c-\alpha+1}(s)} \Delta_\alpha s < \infty,$$

then

$$\int_a^\infty \frac{\lambda(x)}{\Phi^{c-\alpha+1}(x)} \Omega^k(x) \Delta_\alpha x \leq \left(\frac{k}{c-\alpha}\right)^k \int_a^\infty \Phi^{k-c+\alpha-1}(x) \lambda(x) g^k(x) \Delta_\alpha x. \tag{34}$$

Proof. Using the formula of integration by parts (20) on

$$\int_a^\infty \frac{\lambda(x)}{\Phi^{c-\alpha+1}(x)} \Omega^k(x) \Delta_\alpha x,$$

with $v(x) = \Omega^k(x)$ and $x_\alpha^\Delta u(x) = \frac{\lambda(x)}{\Phi^{c-\alpha+1}(x)}$, then

$$\int_a^\infty \frac{\lambda(x)}{\Phi^{c-\alpha+1}(x)} \Omega^k(x) \Delta_\alpha x = u(x) \Omega^k(x) \Big|_a^\infty + \int_a^\infty u^\sigma(x) x_\alpha^\Delta \left(-\Omega^k(x)\right) \Delta_\alpha x, \tag{35}$$

where

$$u(x) = \int_a^x \frac{\lambda(s)}{\Phi^{c-\alpha+1}(s)} \Delta_\alpha s.$$

By using chain rule (18), we see for $d \in [x, \sigma(x)]$ that

$$\begin{aligned} x_\alpha^\Delta \left(\Phi^{\alpha-c}(x)\right) &= (\alpha - c) \Phi^{\alpha-c-1}(d) x_\alpha^\Delta \left(\Phi(x)\right) = \frac{(\alpha-c)(-\lambda(x))}{\Phi^{c-\alpha+1}(d)} \\ &\geq \frac{(c-\alpha)\lambda(x)}{\Phi^{c-\alpha+1}(x)}. \end{aligned}$$

we get,

$$\begin{aligned} u^\sigma(x) &= \int_a^{\sigma(x)} \frac{\lambda(s)}{\Phi^{c-\alpha+1}(s)} \Delta_\alpha s \leq \frac{1}{c-\alpha} \int_a^{\sigma(x)} x_\alpha^\Delta \Phi^{\alpha-c}(s) \Delta_\alpha s \\ &= \frac{(\Phi^\sigma(x))^{\alpha-c}}{c-\alpha} - \frac{\Phi^{\alpha-c}(a)}{c-\alpha} \leq \frac{\Phi^{\alpha-c}(x)}{c-\alpha}, \end{aligned} \tag{36}$$

from chain rule (18), we find that

$$x_\alpha^\Delta \left(\Omega^k(x)\right) = k \Omega^{k-1}(d) x_\alpha^\Delta \Omega(x), \text{ where } d \in [x, \sigma(x)],$$

since

$$x_\alpha^\Delta \left(\Omega(x)\right) = -\lambda(x) g(x) \leq 0 \text{ and } x \leq d,$$

then

$$-x_\alpha^\Delta \left(\Omega^k(x)\right) \leq k \lambda(x) g(x) \Omega^{k-1}(x). \tag{37}$$

Using $v(a) = 0$, $\Omega(\infty) = 0$ and (36), (37) and (35), we get that

$$\int_a^\infty \frac{\lambda(x)}{\Phi^{c-\alpha+1}(x)} \left(\Omega^\sigma(x)\right)^k \Delta_\alpha x \leq \frac{k}{c-\alpha} \int_a^\infty \Phi^{\alpha-c}(x) \lambda(x) g(x) \Omega^{k-1}(x) \Delta_\alpha x,$$

which reformulated as

$$\int_a^\infty \frac{\lambda(x)}{\Phi^{c-\alpha+1}(x)}\Omega^k(x)\Delta_\alpha x \leq$$

$$\frac{k}{c-\alpha}\int_a^\infty \frac{\Phi^{(c-\alpha+1)(\frac{k-1}{k})}(x)}{\lambda^{\frac{k-1}{k}}(x)\Phi^{c-\alpha}(x)}\lambda(x)g(x)\lambda^{\frac{k-1}{k}}(x)\frac{\Omega^{k-1}(x)}{\Phi^{(c-\alpha+1)(\frac{k-1}{k})}(x)}\Delta_\alpha x,$$

Using Hölder's inequality (21) on

$$\int_a^\infty \lambda(x)g(x)\frac{\Phi^{(c-\alpha+1)(\frac{k-1}{k})}(x)}{\lambda^{\frac{k-1}{k}}(x)\Phi^{c-\alpha}(x)}\frac{\lambda^{\frac{k-1}{k}}(x)\Omega^{k-1}(x)}{\Phi^{(c-\alpha+1)(\frac{k-1}{k})}(x)}\Delta_\alpha x$$

with indices k and $k/(k-1)$, we have

$$\int_a^\infty \frac{\lambda(x)}{\Phi^{c-\alpha+1}(x)}\Omega^k(x)\Delta_\alpha x \leq$$

$$\frac{k}{c-\alpha}\left[\int_a^\infty \left[\lambda(x)g(x)\frac{\Phi^{(c-\alpha+1)(\frac{k-1}{k})}(x)}{\lambda^{\frac{k-1}{k}}(x)\Phi^{c-\alpha}(x)}\right]^k \Delta_\alpha x\right]^{\frac{1}{k}} \times$$

$$\left[\int_a^\infty \left[\lambda^{\frac{k-1}{k}}(x)\frac{\Omega^{k-1}(x)}{\Phi^{(c-\alpha+1)(\frac{k-1}{k})}(x)}\right]^{\frac{k}{k-1}} \Delta_\alpha x\right]^{\frac{k-1}{k}},$$

then

$$\left[\int_a^\infty \frac{\lambda(x)}{\Phi^{c-\alpha+1}(x)}(\Omega(x))^k \Delta_\alpha x\right]^{\frac{1}{k}} \leq \frac{k}{c-\alpha}\left[\int_a^\infty \Phi^{k-c+\alpha-1}(x)\lambda(x)g^k(x)\Delta_\alpha x\right]^{\frac{1}{k}}.$$

This leads to

$$\int_a^\infty \frac{\lambda(x)}{\Phi^{c-\alpha+1}(x)}(\Omega^\sigma(x))^k \Delta_\alpha x \leq \left(\frac{k}{c-\alpha}\right)^k \int_a^\infty \Phi^{k-c+\alpha-1}(x)\lambda(x)g^k(x)\Delta_\alpha x,$$

that is the desired inequality (34). The proof is complete. □

Corollary 8. *At $\alpha = 1$ in Theorem 5, we have the inequality*

$$\int_a^\infty \frac{\lambda(x)}{\Phi^c(x)}(\Omega^\sigma(x))^k \Delta x \leq \left(\frac{k}{c-1}\right)^k \int_a^\infty \Phi^{k-c}(x)\lambda(x)g^k(x)\Delta x,$$

which the inequality (2.49) in [33].

3. Conclusions

The new fractional calculus on timescales is presented with applications to some new fractional inequalities on timescales like Hardy, Bennett, Copson and Leindler types. Inequalities are considered in rather general forms and contain several special integral and discrete inequalities. The technique is based on the applications of well-known inequalities and new tools from fractional calculus.

Author Contributions: S.S. contributed in preparing the introduction, preliminaries and formulate theorem 3, its proof and its corollaries (cor.1, cor. 2 and cor.3). M.K. contributed in preparing the introduction, preliminaries and formulate theorem 4, its proof and its corollaries (cor.4, cor. 5 and cor.6). G.A. contributed in preparing the introduction, preliminaries and formulate theorem 5, its proof and its corollaries (cor.7). M.Z. contributed in preparing the introduction, preliminaries and formulate theorem 6, its proof and its corollaries (cor.8). All authors contributed equally to the writing of this manuscript. All authors have read and agreed to the published version of the manuscript.

Acknowledgments: This research was funded by the Deanship of Scientific Research at Princess Nourah bint Abdulrahman University through the Fast-track Research Funding Program. The authors thank the referees for helpful comments that lead to the improvement of the presentation of the results in this paper.

References

1. Hardy, G.H. Notes on a theorem of Hilbert. *Math. Z.* **1920**, *6*, 314–317. [CrossRef]
2. Hardy, G.H. Notes on some points in the integral calculus, LX. An inequality between integrals. *Mess. Math.* **1925**, *54*, 150–156.
3. Hardy, G.H. Notes on some points in the integral calculus, LXIV. Further inequalities between integrals. *Mess. Math.* **1928**, *57*, 12–16.
4. Copson, E.T. Note on series of positive terms. *J. Lond. Math. Soc.* **1927**, *2*, 9–12. [CrossRef]
5. Leindler, L. Generalization of inequalities of Hardy and Littlewood. *Acta Sci. Math.* **1970**, *31*, 297–285.
6. Leindler, L. Further sharpining of inequalities of Hardy and Littlewood. *Acta Sci. Math.* **1990**, *54*, 285–289.
7. Leindler, L. A theorem of Hardy-Bennett-type. *Acta Math. Hungar.* **1998**, *87*, 315–325. [CrossRef]
8. Leindler, L. Two Hardy-Bennett-type theorems. *Acta Math. Hungar.* **1999**, *85*, 265–276. [CrossRef]
9. Copson, E.T. Note on series of positive terms. *J. Lond. Math. Soc.* **1928**, *3*, 49–51. [CrossRef]
10. Copson, E.T. Some integral inequalities. *Prof. R. Soc. Edinburg. Sect. A* **1976**, *75*, 157–164. [CrossRef]
11. Bennett, G. Some elementary inequalities. *Quart. J. Math. Oxford* **1987**, *38*, 401–425. [CrossRef]
12. Agarwal, R.P.; Bohner, M.; O'Regan, D.; Saker, S.H. Some dynamic wirtinger-type inequalities and their applications. *Pacific J. Math.* **2011**, *252*, 1–18. [CrossRef]
13. Agarwal, R.P.; O'Regan, D.; Saker, S.H. *Hardy Type Ineqaualities on Time Scales*; Springer: Basel, Switzerland, 2016.
14. Saker, S.H. Some nonlinear dynamic inequalities on time scales and applications. *J. Math. Inequal.* **2010**, *4*, 561–579. [CrossRef]
15. Hilger, S. Analysis on measure chains—A unified approach to continuous and discrete calculus. *Results Math.* **1990**, *18*, 18–56. [CrossRef]
16. Bohner, M.; Peterson, A. *Dynamic Equations on Time Scales: An Introduction with Applications*; Birkhäuser: Boston, MA, USA, 2001.
17. Bohner, M.; Peterson, A. *Advances in Dynamic Equations on Time Scales*; Birkhäuser: Boston, MA, USA, 2003.
18. Kac, V.; Cheung, P. *Quantum Calculus*; Springer: New York, NY, USA, 2001.
19. Bogdan, K.; Dyda, B. The best constant in a fractional Hardy inequality. *Math. Nach.* **2011**, *284*, 629–638. [CrossRef]
20. Jleli, M.; Samet, B. Lyapunov-type inequalities for a fractional differential equation with mixed boundary conditions. *Math. Inequal. Appl.* **2015**, *18*, 443–451. [CrossRef]
21. Yildiz, C.; Ozdemir, M.E.; Onelan, H.K. Fractional integral inequalities for different functions. *New Trends Math. Sci.* **2015**, *3*, 110–117.
22. Abdeljawad, T. On conformable fractional calculus. *J. Comp. Appl. Math.* **2015**, *279*, 57–66. [CrossRef]
23. Khalil, R.; Horani, M.A.; Yousef, A.; Sababheh, M. A new definition of fractional derivative. *J. Comp. Appl. Math.* **2014**, *264*, 65–70. [CrossRef]
24. Akkurt, A.; Yildirim, M.E.; Yildirim, H. On some integral inequalities for conformable fractional integrals. *RGMIA Res. Rep. Collect.* **2016**, *19*, 107.
25. Chu, Y.M.; Khan, M.A.; Ali, T.; Dragomir, S.S. Inequalities for α-fractional differentiable functions. *J. Inequal. Appl.* **2017**, *1*, 93. [CrossRef]
26. Khan, M.A.; Ali, T.; Dragomir, S.S.; Sarikaya, M.Z. Hermite-Hadamard type inequalities for conformable fractional integrals. *Rev. Real Acad. Cienc. Exactas* **2018**, *112*, 1033–1048. [CrossRef]
27. Set, E.; Gözpınar, A.; Ekinci, A. Hermite-Hadamard type inequalities via conformable fractional integrals. *Acta Math. Univ. Comenian.* **2017**, *86*, 309–320. [CrossRef]
28. Sarikaya, M.Z.; Budak, H. New inequalities of Opial type for conformable fractional integrals. *Turkish J. Math.* **2017**, *41*, 1164–1173. [CrossRef]
29. Sarikaya, M.Z.; Budak, H. Opial type inequalities for conformable fractional integrals. *Rgmia Res. Rep. Collect.* **2016**, *19*, 93.

30. Sarikaya, M.Z.; Yaldiz, H.; Budak, H. Steffensen's integral inequality for conformable fractional integrals. *Int. J. Anal. Appl.* **2017**, *15*, 23–30.
31. Benkhettou, N.; Salima, H.; Torres, D.F.M. A conformable fractional calculus on arbitrary time scales. *J. King Saud Univ. Science* **2016**, *1*, 93–98. [CrossRef]
32. Nwaeze, E.R.; Torres, D.F.M. Chain rules and inequalities for the BHT fractional calculus on arbitrary timescales. *Arab J. Math.* **2017**, *6*, 13–20. [CrossRef]
33. Saker, S.H.; O'Regan, D.; Agarwal, R.P. Generalized Hardy, Copson, Leindler and Bennett inequalities on time scales. *Math. Nachr.* **2014**, *287*, 686–698. [CrossRef]

16

Certain Hadamard Proportional Fractional Integral Inequalities

Gauhar Rahman [1], **Kottakkaran Sooppy Nisar** [2,*] **and Thabet Abdeljawad** [3,4,5,*]

1 Department of Mathematics, Shaheed Benazir Bhutto University, Sheringal, Upper Dir 18000, Pakistan; gauhar55uom@gmail.com
2 Department of Mathematics, College of Arts and Sciences, Prince Sattam bin Abdulaziz University, Wadi Aldawaser 11991, Saudi Arabia
3 Department of Mathematics and General Sciences, Prince Sultan University, Riyadh 11586, Saudi Arabia
4 Department of Medical Research, China Medical University, Taichung 40402, Taiwan
5 Department of Computer Science and Information Engineering, Asia University, Taichung 40402, Taiwan
* Correspondence: n.sooppy@psau.edu.sa or ksnisar1@gmail.com (K.S.N.); tabdeljawad@psu.edu.sa (T.A.)

Abstract: In this present paper we study the non-local Hadmard proportional integrals recently proposed by Rahman et al. (Advances in Difference Equations, (2019) 2019:454) which containing exponential functions in their kernels. Then we establish certain new weighted fractional integral inequalities involving a family of n $(n \in \mathbb{N})$ positive functions by utilizing Hadamard proportional fractional integral operator. The inequalities presented in this paper are more general than the inequalities existing in the literature.

Keywords: fractional integrals; hadamard proportional fractional integrals; fractional integral inequalities

---

1. Introduction

The field of fractional integral inequalities play an important role in the field of differential equations and applied mathematics. These inequalities have many applications in applied sciences such as probability, statistical problems, numerical quadrature and transform theory. In the last few decades, many mathematicians have paid their valuable considerations to this field and obtained a bulk of various fractional integral inequalities and their applications. The interested readers are referred to the work of [1–5] and the references cited therein. A variety of different kinds of certain classical integral inequalities and their extensions have been investigated by considering the classical Riemann–Liouville (RL) fractional integrals, fractional derivatives and their various extensions. In [6], the authors presented integral inequalities via generalized (k,s)-fractional integrals. Dahmani and Tabharit [7] proposed weighted Grüss-type inequalities by utilizing Riemann–Liouville fractional integrals. Dahmani [8] presented several new inequalities in fractional integrals. Nisar et al. [9] investigated several inequalities for extended gamma function and confluent hypergeometric k-function. Gronwall type inequalities associated with Riemann–Liouville k and Hadamard k-fractional derivative with applications can be found in the work of Nisar et al. [10]. Rahman et al. [11] presented certain inequalities for generalized fractional integrals. Sarikaya and Budak [12] investigated Ostrowski type inequalities by employing local fractional integrals. The generalized (k,s)-fractional integrals and their applications can be found in the work of Sarikaya et al. [13]. In [14], Set et al. proposed the generalized Grüss-type inequalities by employing generalized k-fractional integrals. Set et al. [15] have introduced the generalized version of Hermite–Hadamard type inequalities via fractional integrals. Agarwal et al. [16] studied the Hermite–Hadamard type inequalities by considering the Riemann–Liouville k-fractional integrals. Fractional integral inequalities via Hadamard fractional integral, Saigo fractional integral and fractional q-integral operators are found in [17–19]. In [20], Huang et al. investigated Hermite–Hadamard type inequalities for k-fractional conformable integrals. Mubeen et al. [21] proposed the Minkowski's inequalities involving the generalized k-fractional conformable integrals. The chebyshev type inequalities involving generalized k-fractional conformable integrals can be found in the work of Qi et al. [22]. Rahman et al. investigated Chebyshev type inequalities by utilizing fractional conformable integrals [23,24]. In [25,26], Nisar et al. proposed generalized Chebyshev-type inequalities and certain Minkowski's type inequalities by employing generalized conformable integrals. Recently, Tassaddiq et al. [27] investigated certain inequalities for the weighted and the extended Chebyshev functionals by using fractional conformable integrals. Nisar et al. [28] established certain new inequalities for a class of $n(n \in \mathbb{N})$ positive continuous and decreasing functions by employing generalized conformable fractional integrals. Certain generalized fractional integral inequalities for Marichev–Saigo–Maeda (MSM) fractional integral operators were recently established by Nisar et al. [29]. Rahman et al. [30] recently investigated Grüss type inequalities for generalized k-fractional conformable integrals. In [31–33], Rahman et al. recently investigated Minkowski's inequalities, fractional proportional integral inequalities and fractional proportional inequalities for convex functions by employing fractional proportional integrals.

Moreover, the recent research focuses to the development of the theory of fractional calculus and its applications in multiple disciplines of sciences. In last three centuries, the field of fractional calculus has earned more recognition due to its wide applications in diverse domains. Recently, several kinds of various fractional integral and derivative operators have been investigated. The idea of fractional conformable derivative operators with some drawbacks was proposed by Khalil et al. [34]. The properties of the fractional conformable derivative operators was investigated by Abdeljawad [35]. Abdeljawad and Baleanu [36] proposed several monotonicity results for fractional difference operators with discrete exponential kernels. Abdeljawad and Baleanu [37] have presented fractional derivative operator with exponential kernel and their discrete version. A new fractional derivative operator with the non-local and non-singular kernel was proposed by Atangana and Baleanu [38]. Jarad et al. [39] proposed the fractional conformable integral and derivative operators. The idea of conformable derivative by employing local proportional derivatives was proposed by Anderson and Unless [40]. In [41], Caputo and Fabrizio proposed fractional derivative without a singular kernel. Later on, Losada and Nieto [42] gave certain properties of fractional derivative without a singular kernel.

In [43], Liu et al. investigated several integral inequalities. Later on, Dahmani [44] proposed certain classes of weighted fractional integral inequalities for a family of n positive increasing and decreasing functions by utilizing Riemann–Liouville fractional integrals. Houas [45] utilized Hadamard fractional integrals and established several weighted type integral inequalities. Recently, Jarad et al. [46] and Rahman et al. [47] proposed the idea of non-local fractional proportional and Hadamard proportional integrals which concerning exponential functions in their kernels. The aim of this paper is to establish weighted type inequalities by using the non-local Hdamard proportional integrals. The paper is organized as follows. In Section 2, we present some basic definitions and mathematical formulas. In Section 3, we establish certain weighted Hadamard proportional fractional integral inequalities. Section 4 containing concluding remarks.

2. Preliminaries

This section is devoted to some well-known definitions and mathematical preliminaries of fractional calculus which will be used in this article. Jarad et al. [46] presented the left and right proportional integral operators.

Definition 1. *The left and right sided proportional fractional integrals are respectively defined by*

$$\left({}_{a}\mathcal{J}^{\tau,\nu}\mathcal{U}\right)(\theta)=\frac{1}{\nu^{\tau}\Gamma(\tau)}\int_{a}^{\theta}\exp[\frac{\nu-1}{\nu}(\theta-t)](\theta-t)^{\tau-1}\mathcal{U}(t)dt, a<\theta \tag{1}$$

and

$$\left(\mathcal{J}_{b}^{\tau,\nu}\mathcal{U}\right)(x)=\frac{1}{\nu^{\tau}\Gamma(\tau)}\int_{\theta}^{b}\exp[\frac{\nu-1}{\nu}(t-\theta)](t-\theta)^{\tau-1}\mathcal{U}(t)dt, \theta<b. \tag{2}$$

where the proportionality index $\nu\in(0,1]$ *and* $\tau\in\mathbb{C}$ *with* $\Re(\tau)>0$.

Remark 1. *If we consider* $\nu=1$ *in (1) and (2), then we get the well-known left and right Riemann–Liouville integrals which are respectively defined by*

$$\left({}_{a}\mathcal{J}^{\tau}\mathcal{U}\right)(x)=\frac{1}{\Gamma(\tau)}\int_{a}^{\theta}(\theta-t)^{\tau-1}\mathcal{U}(t)dt, a<\theta \tag{3}$$

and

$$\left(\mathcal{J}_{b}^{\tau}\mathcal{U}\right)(\theta)=\frac{1}{\Gamma(\tau)}\int_{\theta}^{b}(t-\theta)^{\tau-1}\mathcal{U}(t)dt, \theta<b \tag{4}$$

where $\tau\in\mathbb{C}$ *with* $\Re(\tau)>0$.

Recently, Rahman et al. [47] proposed the following generalized Hadamard proportional fractional integrals.

Definition 2. *The left sided generalized Hadamard proportional fractional integral of order* $\tau>0$ *and proportional index* $\nu\in(0,1]$ *is defined by*

$$\left({}_{a}\mathcal{H}^{\tau,\nu}\mathcal{U}\right)(\theta)=\frac{1}{\nu^{\tau}\Gamma(\tau)}\int_{a}^{\theta}\exp[\frac{\nu-1}{\nu}(\ln\theta-\ln t)](\ln\theta-\ln t)^{\tau-1}\frac{\mathcal{U}(t)}{t}dt, a<\theta. \tag{5}$$

Definition 3. *The right sided generalized Hadamard proportional fractional integral of order* $\tau>0$ *and proportional index* $\nu\in(0,1]$ *is defined by*

$$\left(\mathcal{H}_{b}^{\tau,\nu}\mathcal{U}\right)(\theta)=\frac{1}{\nu^{\tau}\Gamma(\tau)}\int_{\theta}^{b}\exp[\frac{\nu-1}{\nu}(\ln t-\ln\theta)](\ln t-\ln\theta)^{\tau-1}\frac{\mathcal{U}(t)}{t}dt, \theta<b. \tag{6}$$

Definition 4. *The one sided generalized Hadamard proportional fractional integral of order* $\tau>0$ *and proportional index* $\nu\in(0,1]$ *is defined by*

$$\left(\mathcal{H}_{1,\theta}^{\tau,\nu}\mathcal{U}\right)(\theta)=\frac{1}{\nu^{\tau}\Gamma(\tau)}\int_{1}^{\theta}\exp[\frac{\nu-1}{\nu}(\ln\theta-\ln t)](\ln\theta-\ln t)^{\tau-1}\frac{\mathcal{U}(t)}{t}dt, \theta>1, \tag{7}$$

where $\Gamma(\tau)$ *is the classical well-known gamma function.*

Remark 2. *If we consider* $\nu=1$, *then (5)–(7) will led to the following well-known Hadamard fractional integrals*

$$\left({}_{a}\mathcal{H}^{\tau}\mathcal{U}\right)(\theta)=\frac{1}{\Gamma(\tau)}\int_{a}^{\theta}(\ln\theta-\ln t)^{\tau-1}\frac{\mathcal{U}(t)}{t}dt, a<\theta, \tag{8}$$

$$\left(\mathcal{H}_{b}^{\tau}\mathcal{U}\right)(\theta)=\frac{1}{\Gamma(\tau)}\int_{\theta}^{b}(\ln t-\ln\theta)^{\tau-1}\frac{\mathcal{U}(t)}{t}dt, \theta<b \tag{9}$$

and

$$\left(\mathcal{H}_{1,\theta}^{\tau}\mathcal{U}\right)(\theta) = \frac{1}{\Gamma(\tau)}\int_{1}^{\theta}(\ln\theta - \ln t)^{\tau-1}\frac{\mathcal{U}(t)}{t}dt, \theta > 1. \tag{10}$$

One can easily prove the following results

Lemma 1.

$$\left(\mathcal{H}_{1,\theta}^{\tau,\nu}\exp[\frac{\nu-1}{\nu}(\ln\theta)](\ln\theta)^{\lambda-1}\right)(\theta) = \frac{\Gamma(\lambda)}{\nu^{\tau}\Gamma(\tau+\lambda)}\exp[\frac{\nu-1}{\nu}(\ln\theta)](\ln\theta)^{\tau+\lambda-1} \tag{11}$$

and the semi group property

$$\left(\mathcal{H}_{1,\theta}^{\tau,\nu}\right)\left(\mathcal{H}_{1,\theta}^{\lambda,\nu}\right)\mathcal{U}(\theta) = \left(\mathcal{H}_{1,\theta}^{\tau+\lambda,\nu}\right)\mathcal{U}(\theta). \tag{12}$$

Remark 3. *If $\nu = 1$, then* (11) *will reduce to the result of [48] as defined by*

$$\left(\mathcal{H}_{1,\theta}^{\tau}(\ln\theta)^{\lambda-1}\right)(\theta) = \frac{\Gamma(\lambda)}{\Gamma(\tau+\lambda)}(\ln\theta)^{\tau+\lambda-1}. \tag{13}$$

3. Main Results

In this section, we present certain new proportional fractional integral inequalities by utilizing Hadamard proportional fractional integral. Employ the left generalized proportional fractional integral operator to establish the generalization of some classical inequalities.

Theorem 1. *Let the two functions $\mathcal{U}$ and $\mathcal{V}$ be positive and continuous on the interval $[1,\infty)$ and satisfy*

$$\left(\mathcal{V}^{\sigma}(\zeta)\mathcal{U}^{\sigma}(\rho) - \mathcal{V}^{\sigma}(\rho)\mathcal{U}^{\sigma}(\zeta)\right)\left(\mathcal{U}^{\delta-\xi}(\rho) - \mathcal{U}^{\delta-\xi}(\zeta)\right) \geq 0, \tag{14}$$

where $\rho, \zeta \in (1,\theta), \theta > 1$ and for any $\sigma > 0, \delta \geq \xi > 0$. Assume that the function $\mathcal{W} : [1,\infty) \to \mathbb{R}^{+}$ is positive and continuous. Then for all $\theta > 1$, the following inequality for Hadamard proportional fractional integral operator (7) *holds;*

$$\begin{aligned} &\mathcal{H}_{1,\theta}^{\tau,\nu}\left[\mathcal{W}(\theta)\mathcal{U}^{\delta+\sigma}(\theta)\right]\mathcal{H}_{1,\theta}^{\tau,\nu}\left[\mathcal{W}(\theta)\mathcal{V}^{\sigma}(\theta)\mathcal{U}^{\xi}(\theta)\right] \\ \geq &\mathcal{H}_{1,\theta}^{\tau,\nu}\left[\mathcal{W}(\theta)\mathcal{U}^{\sigma+\xi}(\theta)\right]\mathcal{H}_{1,\theta}^{\tau,\nu}\left[\mathcal{W}(\theta)\mathcal{V}^{\sigma}(\theta)\mathcal{U}^{\delta}(\theta)\right], \end{aligned} \tag{15}$$

where $\delta \geq \xi > 0$, $\tau, \nu, \sigma > 0$.

Proof. Consider the function

$$\mathcal{F}(\theta,\rho) = \frac{1}{\nu^{\tau}\Gamma(\tau)}e^{\frac{\nu-1}{\nu}(\ln\theta - \ln\rho)}(\ln\theta - \ln\rho)^{\tau-1}\frac{\mathcal{W}(\rho)\mathcal{U}^{\xi}(\rho)}{\rho}, \tag{16}$$

where $\tau > 0, \xi > 0$ and $\rho \in (1,\theta), \theta > 1$.

Since the functions $\mathcal{U}$ and $\mathcal{V}$ satisfy (14) for all $\rho, \zeta \in (1,\theta), \theta > 1$ and for any $\sigma > 0, \delta \geq \xi > 0$. Therefore, from (14), we have

$$\mathcal{V}^{\sigma}(\zeta)\mathcal{U}^{\sigma+\delta-\xi}(\rho) + \mathcal{V}^{\sigma}(\rho)\mathcal{U}^{\sigma+\delta-\xi}(\zeta) \geq \mathcal{V}^{\sigma}(\zeta)\mathcal{U}^{\sigma}(\rho)\mathcal{U}^{\delta-\xi}(\zeta) + \mathcal{V}^{\sigma}(\rho)\mathcal{U}^{\sigma}(\zeta)\mathcal{U}^{\delta-\xi}(\rho). \tag{17}$$

We observe that the function $\mathcal{F}(\theta,\rho)$ remains positive for all $\rho \in (1,\theta)$, $\theta > 1$. Therefore multiplying (17) by $\mathcal{F}(\theta,\rho)$ (where $\mathcal{F}(\theta,\rho)$ is defined in (16)) and integrating the resultant estimates with respect to ρ over $(1,\theta)$, we get

$$\begin{aligned}
&\mathcal{V}^{\sigma}(\zeta)\frac{1}{\nu^{\tau}\Gamma(\tau)}\int_1^{\theta} e^{\frac{\nu-1}{\nu}(\ln\theta-\ln\rho)}(\ln\theta-\ln\rho)^{\tau-1}\frac{\mathcal{W}(\rho)\mathcal{U}^{\xi}(\rho)}{\rho}\mathcal{U}^{\sigma+\delta-\xi}(\rho)d\rho\\
&+\mathcal{U}^{\sigma+\delta-\xi}(\zeta)\frac{1}{\nu^{\tau}\Gamma(\tau)}\int_1^{\theta} e^{\frac{\nu-1}{\nu}(\ln\theta-\ln\rho)}(\ln\theta-\ln\rho)^{\tau-1}\frac{\mathcal{W}(\rho)\mathcal{U}^{\xi}(\rho)}{\rho}\mathcal{V}^{\sigma}(\rho)d\rho\\
&\geq\mathcal{V}^{\sigma}(\zeta)\mathcal{U}^{\delta-\xi}(\zeta)\frac{1}{\nu^{\tau}\Gamma(\tau)}\int_1^{\theta} e^{\frac{\nu-1}{\nu}(\ln\theta-\ln\rho)}(\ln\theta-\ln\rho)^{\tau-1}\frac{\mathcal{W}(\rho)\mathcal{U}^{\xi}(\rho)}{\rho}\mathcal{U}^{\sigma}(\rho)d\rho\\
&+\mathcal{U}^{\sigma}(\zeta)\frac{1}{\nu^{\tau}\Gamma(\tau)}\int_1^{\theta} e^{\frac{\nu-1}{\nu}(\ln\theta-\ln\rho)}(\ln\theta-\ln\rho)^{\tau-1}\frac{\mathcal{W}(\rho)\mathcal{U}^{\xi}(\rho)}{\rho}\mathcal{V}^{\sigma}(\rho)\mathcal{U}^{\delta-\xi}(\rho)d\rho,
\end{aligned}$$

which in view of (7) becomes,

$$\begin{aligned}
&\mathcal{V}^{\sigma}(\zeta)\mathcal{H}_{1,\theta}^{\tau,\nu}\left[\mathcal{W}(\theta)\mathcal{U}^{\sigma+\delta}(\theta)\right]+\mathcal{U}^{\sigma+\delta-\xi}(\zeta)\mathcal{H}_{1,\theta}^{\tau,\nu}\left[\mathcal{W}(\theta)\mathcal{V}^{\sigma}(\theta)\mathcal{U}^{\xi}(\theta)\right]\\
&\geq\mathcal{V}^{\sigma}(\zeta)\mathcal{U}^{\delta-\xi}(\zeta)\mathcal{H}_{1,\theta}^{\tau,\nu}\left[\mathcal{W}(\theta)\mathcal{U}^{\sigma+\xi}(\rho)\right]+\mathcal{U}^{\sigma}(\zeta)\mathcal{H}_{1,\theta}^{\tau,\nu}\left[\mathcal{W}(\theta)\mathcal{V}^{\sigma}(\theta)\mathcal{U}^{\delta}(\theta)\right].
\end{aligned}\tag{18}$$

Now, multiplying (18) by $\mathcal{F}(\theta,\zeta)$ (where $\mathcal{F}(\theta,\zeta)$ can be obtain from (16) by replacing ρ by ζ) and integrating the resultant estimates with respect to ζ over $(1,\theta)$ and then by using (7), we obtain

$$\begin{aligned}
&\mathcal{H}_{1,\theta}^{\tau,\nu}\left[\mathcal{W}(\theta)\mathcal{V}^{\sigma}(\theta)\mathcal{U}^{\xi}(\theta)\right]\mathcal{H}_{1,\theta}^{\tau,\nu}\left[\mathcal{W}(\theta)\mathcal{U}^{\sigma+\delta}(\theta)\right]+\mathcal{H}_{1,\theta}^{\tau,\nu}\left[\mathcal{W}(\theta)\mathcal{U}^{\sigma+\delta}(\theta)\right]\mathcal{H}_{1,\theta}^{\tau,\nu}\left[\mathcal{W}(\theta)\mathcal{V}^{\sigma}(\theta)\mathcal{U}^{\xi}(\theta)\right]\\
&\geq\mathcal{H}_{1,\theta}^{\tau,\nu}\left[\mathcal{W}(\theta)\mathcal{V}^{\sigma}(\theta)\mathcal{U}^{\delta}(\theta)\right]\mathcal{H}_{1,\theta}^{\tau,\nu}\left[\mathcal{W}(\theta)\mathcal{U}^{\sigma+\xi}(\rho)\right]+\mathcal{H}_{1,\theta}^{\tau,\nu}\left[\mathcal{W}(\theta)\mathcal{U}^{\sigma+\xi}(\theta)\right]\mathcal{H}_{1,\theta}^{\tau,\nu}\left[\mathcal{W}(\theta)\mathcal{V}^{\sigma}(\theta)\mathcal{U}^{\delta}(\theta)\right],
\end{aligned}$$

which gives the desired assertion (15). □

Remark 4. *If the inequality* (14) *reverses, then the inequality* (15) *will also reverse.*

Theorem 2. *Let the two functions $\mathcal{U}$ and $\mathcal{V}$ be positive and continuous on the interval $[1,\infty)$ and satisfy* (14). *Assume that the function $\mathcal{W}:[1,\infty)\to\mathbb{R}^+$ is positive and continuous. Then for all $\theta>1$, the following inequality for Hadamard proportional fractional integral operator* (7) *holds;*

$$\begin{aligned}
&\mathcal{H}_{1,\theta}^{\tau,\nu}\left[\mathcal{W}(\theta)\mathcal{U}^{\sigma+\delta}(\theta)\right]\mathcal{H}_{1,\theta}^{\lambda,\nu}\left[\mathcal{W}(\theta)\mathcal{V}^{\sigma}(\theta)\mathcal{U}^{\xi}(\theta)\right]+\mathcal{H}_{1,\theta}^{\tau,\nu}\left[\mathcal{W}(\theta)\mathcal{V}^{\sigma}(\theta)\mathcal{U}^{\xi}(\theta)\right]\mathcal{H}_{1,\theta}^{\lambda,\nu}\left[\mathcal{W}(\theta)\mathcal{U}^{\sigma+\delta}(\theta)\right]\\
\geq\ &\mathcal{H}_{1,\theta}^{\tau,\nu}\left[\mathcal{W}(\theta)\mathcal{U}^{\sigma+\xi}(\rho)\right]\mathcal{H}_{1,\theta}^{\lambda,\nu}\left[\mathcal{W}(\theta)\mathcal{V}^{\sigma}(\theta)\mathcal{U}^{\delta}(\theta)\right]+\mathcal{H}_{1,\theta}^{\tau,\nu}\left[\mathcal{W}(\theta)\mathcal{V}^{\sigma}(\theta)\mathcal{U}^{\delta}(\theta)\right]\mathcal{H}_{1,\theta}^{\lambda,\nu}\left[\mathcal{W}(\theta)\mathcal{U}^{\sigma+\xi}(\theta)\right],
\end{aligned}\tag{19}$$

where $\delta\geq\xi>0$, $\tau,\lambda,\nu,\sigma>0$.

Proof. Taking product on both sides of (18) by $\mathcal{G}(\theta,\zeta)=\frac{1}{\nu^{\lambda}\Gamma(\lambda)}e^{\frac{\nu-1}{\nu}(\ln\theta-\ln\zeta)}(\ln\theta-\ln\zeta)^{\lambda-1}\frac{\mathcal{W}(\zeta)\mathcal{U}^{\xi}(\zeta)}{\zeta}$, where $\lambda>0,\xi>0$ and $\zeta\in(1,\theta),\theta>1$. and integrating the resultant estimates with respect to ζ over $(1,\theta)$, we have

$$\begin{aligned}
&\mathcal{H}_{1,\theta}^{\tau,\nu}\left[\mathcal{W}(\theta)\mathcal{U}^{\sigma+\delta}(\theta)\right]\frac{1}{\nu^{\lambda}\Gamma(\lambda)}\int_1^{\theta} e^{\frac{\nu-1}{\nu}(\ln\theta-\ln\zeta)}(\ln\theta-\ln\zeta)^{\lambda-1}\frac{\mathcal{W}(\zeta)\mathcal{U}^{\xi}(\zeta)}{\zeta}\mathcal{V}^{\sigma}(\zeta)d\zeta\\
&+\mathcal{H}_{1,\theta}^{\tau,\nu}\left[\mathcal{W}(\theta)\mathcal{V}^{\sigma}(\theta)\mathcal{U}^{\xi}(\theta)\right]\frac{1}{\nu^{\lambda}\Gamma(\lambda)}\int_1^{\theta} e^{\frac{\nu-1}{\nu}(\ln\theta-\ln\zeta)}(\ln\theta-\ln\zeta)^{\lambda-1}\frac{\mathcal{W}(\zeta)\mathcal{U}^{\xi}(\zeta)}{\zeta}\mathcal{U}^{\sigma+\delta-\xi}(\zeta)d\zeta\\
&\geq\mathcal{H}_{1,\theta}^{\tau,\nu}\left[\mathcal{W}(\theta)\mathcal{U}^{\sigma+\xi}(\rho)\right]\frac{1}{\nu^{\lambda}\Gamma(\lambda)}\int_1^{\theta} e^{\frac{\nu-1}{\nu}(\ln\theta-\ln\zeta)}(\ln\theta-\ln\zeta)^{\lambda-1}\frac{\mathcal{W}(\zeta)\mathcal{U}^{\xi}(\zeta)}{\zeta}\mathcal{V}^{\sigma}(\zeta)\mathcal{U}^{\delta-\xi}(\zeta)d\zeta\\
&+\mathcal{H}_{1,\theta}^{\tau,\nu}\left[\mathcal{W}(\theta)\mathcal{V}^{\sigma}(\theta)\mathcal{U}^{\delta}(\theta)\right]\frac{1}{\nu^{\lambda}\Gamma(\lambda)}e^{\frac{\nu-1}{\nu}(\ln\theta-\ln\zeta)}(\ln\theta-\ln\zeta)^{\lambda-1}\frac{\mathcal{W}(\zeta)\mathcal{U}^{\xi}(\zeta)}{\zeta}\mathcal{U}^{\sigma}(\zeta)d\zeta,
\end{aligned}$$

which in view of (7) yields,

$$\begin{aligned}&\mathcal{H}^{\tau,\nu}_{1,\theta}\left[\mathcal{W}(\theta)\mathcal{U}^{\sigma+\delta}(\theta)\right]\mathcal{H}^{\lambda,\nu}_{1,\theta}\left[\mathcal{W}(\theta)\mathcal{V}^{\sigma}(\theta)\mathcal{U}^{\xi}(\theta)\right]+\mathcal{H}^{\tau,\nu}_{1,\theta}\left[\mathcal{W}(\theta)\mathcal{V}^{\sigma}(\theta)\mathcal{U}^{\xi}(\theta)\right]\mathcal{H}^{\lambda,\nu}_{1,\theta}\left[\mathcal{W}(\theta)\mathcal{U}^{\sigma+\delta}(\theta)\right]\\ \geq&\mathcal{H}^{\tau,\nu}_{1,\theta}\left[\mathcal{W}(\theta)\mathcal{U}^{\sigma+\xi}(\rho)\right]\mathcal{H}^{\lambda,\nu}_{1,\theta}\left[\mathcal{W}(\theta)\mathcal{V}^{\sigma}(\theta)\mathcal{U}^{\delta}(\theta)\right]+\mathcal{H}^{\tau,\nu}_{1,\theta}\left[\mathcal{W}(\theta)\mathcal{V}^{\sigma}(\theta)\mathcal{U}^{\delta}(\theta)\right]\mathcal{H}^{\lambda,\nu}_{1,\theta}\left[\mathcal{W}(\theta)\mathcal{U}^{\sigma+\xi}(\theta)\right],\end{aligned}$$

which completes the proof of (19). □

Remark 5. *Applying Theorem 2 for $\tau=\lambda$, we get Theorem 1.*

Theorem 3. *Let the two functions $\mathcal{U}$ and $\mathcal{V}$ be positive and continuous on the interval $[1,\infty)$ such that the function $\mathcal{U}$ is decreasing and the function $\mathcal{V}$ is increasing on $[1,\infty)$ and assume that the function $\mathcal{W}:[1,\infty)\to\mathbb{R}^+$ is positive and continuous. Then for all $\theta>1$, the following inequality for Hadamard proportional fractional integral operator (7) holds;*

$$\begin{aligned}&\mathcal{H}^{\tau,\nu}_{1,\theta}\left[\mathcal{W}(\theta)\mathcal{U}^{\delta}(\theta)\right]\mathcal{H}^{\tau,\nu}_{1,\theta}\left[\mathcal{W}(\theta)\mathcal{V}^{\sigma}(\theta)\mathcal{U}^{\xi}(\theta)\right]\\ \geq&\mathcal{H}^{\tau,\nu}_{1,\theta}\left[\mathcal{W}(\theta)\mathcal{U}^{\xi}(\theta)\right]\mathcal{H}^{\tau,\nu}_{1,\theta}\left[\mathcal{W}(\theta)\mathcal{V}^{\sigma}(\theta)\mathcal{U}^{\delta}(\theta)\right],\end{aligned}\tag{20}$$

where $\delta\geq\xi>0$, $\tau,\nu,\sigma>0$.

Proof. Since the two functions $\mathcal{U}$ and $\mathcal{V}$ are positive and continuous on $[1,\infty)$ such that $\mathcal{U}$ is decreasing and $\mathcal{V}$ is increasing on $[1,\infty)$, then for all $\sigma>0$, $\delta\geq\xi>0$, $\rho,\zeta\in(1,\theta)$, $\theta>1$, we have

$$(\mathcal{V}^{\sigma}(\zeta)-\mathcal{V}^{\sigma}(\rho))\left(\mathcal{U}^{\delta-\xi}(\rho)-\mathcal{U}^{\delta-\xi}(\zeta)\right)\geq 0,$$

which follows that

$$\mathcal{V}^{\sigma}(\zeta)\mathcal{U}^{\delta-\xi}(\rho)+\mathcal{V}^{\sigma}(\rho)\mathcal{U}^{\delta-\xi}(\zeta)\geq\mathcal{V}^{\sigma}(\zeta)\mathcal{U}^{\delta-\xi}(\zeta)+\mathcal{V}^{\sigma}(\rho)\mathcal{U}^{\delta-\xi}(\rho).\tag{21}$$

Multiplying (21) by $\mathcal{F}(\theta,\rho)$ (where $\mathcal{F}(\theta,\rho)$ is defined in (16)) and integrating the resultant estimates with respect to ρ over $(1,\theta)$, we get

$$\begin{aligned}&\mathcal{V}^{\sigma}(\zeta)\frac{1}{\nu^{\tau}\Gamma(\tau)}\int_1^{\theta}e^{\frac{\nu-1}{\nu}(\ln\theta-\ln\rho)}(\ln\theta-\ln\rho)^{\tau-1}\frac{\mathcal{W}(\rho)\mathcal{U}^{\xi}(\rho)}{\rho}\mathcal{U}^{\delta-\xi}(\rho)d\rho\\ &+\mathcal{U}^{\delta-\xi}(\zeta)\frac{1}{\nu^{\tau}\Gamma(\tau)}\int_1^{\theta}e^{\frac{\nu-1}{\nu}(\ln\theta-\ln\rho)}(\ln\theta-\ln\rho)^{\tau-1}\frac{\mathcal{W}(\rho)\mathcal{U}^{\xi}(\rho)}{\rho}\mathcal{V}^{\sigma}(\rho)d\rho\\ \geq&\mathcal{V}^{\sigma}(\zeta)\mathcal{U}^{\delta-\xi}(\zeta)\frac{1}{\nu^{\tau}\Gamma(\tau)}\int_1^{\theta}e^{\frac{\nu-1}{\nu}(\ln\theta-\ln\rho)}(\ln\theta-\ln\rho)^{\tau-1}\frac{\mathcal{W}(\rho)\mathcal{U}^{\xi}(\rho)}{\rho}d\rho\\ &+\frac{1}{\nu^{\tau}\Gamma(\tau)}\int_1^{\theta}e^{\frac{\nu-1}{\nu}(\ln\theta-\ln\rho)}(\ln\theta-\ln\rho)^{\tau-1}\frac{\mathcal{W}(\rho)\mathcal{U}^{\xi}(\rho)}{\rho}\mathcal{V}^{\sigma}(\rho)\mathcal{U}^{\delta-\xi}(\rho)d\rho,\end{aligned}$$

which in view of (7) becomes,

$$\begin{aligned}&\mathcal{V}^{\sigma}(\zeta)\mathcal{H}^{\tau,\nu}_{1,\theta}\left[\mathcal{W}(\theta)\mathcal{U}^{\delta}(\theta)\right]+\mathcal{U}^{\delta-\xi}(\zeta)\mathcal{H}^{\tau,\nu}_{1,\theta}\left[\mathcal{W}(\theta)\mathcal{V}^{\sigma}(\theta)\mathcal{U}^{\xi}(\theta)\right]\\ \geq&\mathcal{V}^{\sigma}(\zeta)\mathcal{U}^{\delta-\xi}(\zeta)\mathcal{H}^{\tau,\nu}_{1,\theta}\left[\mathcal{W}(\theta)\mathcal{U}^{\xi}(\theta)\right]+\mathcal{H}^{\tau,\nu}_{1,\theta}\left[\mathcal{W}(\theta)\mathcal{V}^{\sigma}(\theta)\mathcal{U}^{\delta}(\theta)\right].\end{aligned}\tag{22}$$

Again, multiplying (22) by $\mathcal{F}(\theta,\zeta)$ (where $\mathcal{F}(\theta,\zeta)$ can be obtained from (16)) and integrating the resultant estimates with respect to ζ over $(1,\theta)$ and then employing (7), we get

$$\mathcal{H}_{1,\theta}^{\tau,\nu}\left[\mathcal{W}(\theta)\mathcal{V}^{\sigma}(\theta)\mathcal{U}^{\xi}(\theta)\right]\mathcal{H}_{1,\theta}^{\tau,\nu}\left[\mathcal{W}(\theta)\mathcal{U}^{\delta}(\theta)\right]+\mathcal{H}_{1,\theta}^{\tau,\nu}\left[\mathcal{W}(\theta)\mathcal{U}^{\delta}(\theta)\right]\mathcal{H}_{1,\theta}^{\tau,\nu}\left[\mathcal{W}(\theta)\mathcal{V}^{\sigma}(\theta)\mathcal{U}^{\xi}(\theta)\right]$$
$$\geq\mathcal{H}_{1,\theta}^{\tau,\nu}\left[\mathcal{W}(\theta)\mathcal{V}^{\sigma}(\theta)\mathcal{U}^{\delta}(\theta)\right]\mathcal{H}_{1,\theta}^{\tau,\nu}\left[\mathcal{W}(\theta)\mathcal{U}^{\xi}(\theta)\right]+\mathcal{H}_{1,\theta}^{\tau,\nu}\left[\mathcal{W}(\theta)\mathcal{U}^{\xi}(\theta)\right]\mathcal{H}_{1,\theta}^{\tau,\nu}\left[\mathcal{W}(\theta)\mathcal{V}^{\sigma}(\theta)\mathcal{U}^{\delta}(\theta)\right],$$

which gives the desired assertion (20). □

Theorem 4. *Let the two functions $\mathcal{U}$ and $\mathcal{V}$ be positive and continuous on the interval $[1,\infty)$ such that the function $\mathcal{U}$ is decreasing and the function $\mathcal{V}$ is increasing on $[1,\infty)$ and assume that the function $\mathcal{W}:[1,\infty)\to\mathbb{R}^{+}$ is positive and continuous. Then for all $\theta>1$, the following inequality for Hadamard proportional fractional integral operator (7) holds;*

$$\mathcal{H}_{1,\theta}^{\tau,\nu}\left[\mathcal{W}(\theta)\mathcal{U}^{\delta}(\theta)\right]\mathcal{H}_{1,\theta}^{\lambda,\nu}\left[\mathcal{W}(\theta)\mathcal{V}^{\sigma}(\theta)\mathcal{U}^{\xi}(\theta)\right]+\mathcal{H}_{1,\theta}^{\lambda,\nu}\left[\mathcal{W}(\theta)\mathcal{U}^{\delta}(\theta)\right]\mathcal{H}_{1,\theta}^{\tau,\nu}\left[\mathcal{W}(\theta)\mathcal{V}^{\sigma}(\theta)\mathcal{U}^{\xi}(\theta)\right]$$
$$\geq\mathcal{H}_{1,\theta}^{\tau,\nu}\left[\mathcal{W}(\theta)\mathcal{U}^{\xi}(\theta)\right]\mathcal{H}_{1,\theta}^{\lambda,\nu}\left[\mathcal{W}(\theta)\mathcal{V}^{\sigma}(\theta)\mathcal{U}^{\delta}(\theta)\right]+\mathcal{H}_{1,\theta}^{\lambda,\nu}\left[\mathcal{W}(\theta)\mathcal{U}^{\xi}(\theta)\right]\mathcal{H}_{1,\theta}^{\tau,\nu}\left[\mathcal{W}(\theta)\mathcal{V}^{\sigma}(\theta)\mathcal{U}^{\delta}(\theta)\right],\quad(23)$$

where $\delta\geq\xi>0$, $\tau,\lambda,\nu,\sigma>0$.

Proof. Taking product on both sides of (22) by $\mathcal{G}(\theta,\zeta)=\frac{1}{\nu^{\lambda}\Gamma(\lambda)}e^{\frac{\nu-1}{\nu}(\ln\theta-\ln\zeta)}(\ln\theta-\ln\zeta)^{\lambda-1}\frac{\mathcal{W}(\zeta)\mathcal{U}^{\xi}(\zeta)}{\zeta}$, where $\lambda>0,\xi>0$ and $\zeta\in(1,\theta),\theta>1.$ and integrating the resultant estimates with respect to ζ over $(1,\theta)$, we have

$$\mathcal{H}_{1,\theta}^{\tau,\nu}\left[\mathcal{W}(\theta)\mathcal{U}^{\delta}(\theta)\right]\frac{1}{\nu^{\lambda}\Gamma(\lambda)}\int_{1}^{\theta}e^{\frac{\nu-1}{\nu}(\ln\theta-\ln\zeta)}(\ln\theta-\ln\zeta)^{\lambda-1}\frac{\mathcal{W}(\zeta)\mathcal{U}^{\xi}(\zeta)}{\zeta}\mathcal{V}^{\sigma}(\zeta)d\zeta$$
$$+\mathcal{H}_{1,\theta}^{\tau,\nu}\left[\mathcal{W}(\theta)\mathcal{V}^{\sigma}(\theta)\mathcal{U}^{\xi}(\theta)\right]\frac{1}{\nu^{\lambda}\Gamma(\lambda)}\int_{1}^{\theta}e^{\frac{\nu-1}{\nu}(\ln\theta-\ln\zeta)}(\ln\theta-\ln\zeta)^{\lambda-1}\frac{\mathcal{W}(\zeta)\mathcal{U}^{\xi}(\zeta)}{\zeta}\mathcal{U}^{\delta-\xi}(\zeta)d\zeta$$
$$\geq\mathcal{H}_{1,\theta}^{\tau,\nu}\left[\mathcal{W}(\theta)\mathcal{U}^{\xi}(\theta)\right]\frac{1}{\nu^{\lambda}\Gamma(\lambda)}\int_{1}^{\theta}e^{\frac{\nu-1}{\nu}(\ln\theta-\ln\zeta)}(\ln\theta-\ln\zeta)^{\lambda-1}\frac{\mathcal{W}(\zeta)\mathcal{U}^{\xi}(\zeta)}{\zeta}\mathcal{V}^{\sigma}(\zeta)\mathcal{U}^{\delta-\xi}(\zeta)d\zeta$$
$$+\mathcal{H}_{1,\theta}^{\tau,\nu}\left[\mathcal{W}(\theta)\mathcal{V}^{\sigma}(\theta)\mathcal{U}^{\delta}(\theta)\right]\frac{1}{\nu^{\lambda}\Gamma(\lambda)}\int_{1}^{\theta}e^{\frac{\nu-1}{\nu}(\ln\theta-\ln\zeta)}(\ln\theta-\ln\zeta)^{\lambda-1}\frac{\mathcal{W}(\zeta)\mathcal{U}^{\xi}(\zeta)}{\zeta}d\zeta.$$

Consequently, in view of (7) it can be written as,

$$\mathcal{H}_{1,\theta}^{\tau,\nu}\left[\mathcal{W}(\theta)\mathcal{U}^{\delta}(\theta)\right]\mathcal{H}_{1,\theta}^{\lambda,\nu}\left[\mathcal{W}(\theta)\mathcal{V}^{\sigma}(\theta)\mathcal{U}^{\xi}(\theta)\right]+\mathcal{H}_{1,\theta}^{\tau,\nu}\left[\mathcal{W}(\theta)\mathcal{V}^{\sigma}(\theta)\mathcal{U}^{\xi}(\theta)\right]\mathcal{H}_{1,\theta}^{\lambda,\nu}\left[\mathcal{W}(\theta)\mathcal{U}^{\delta}(\theta)\right]$$
$$\geq\mathcal{H}_{1,\theta}^{\tau,\nu}\left[\mathcal{W}(\theta)\mathcal{U}^{\xi}(\theta)\right]\mathcal{H}_{1,\theta}^{\lambda,\nu}\left[\mathcal{W}(\theta)\mathcal{V}^{\sigma}(\theta)\mathcal{U}^{\delta}(\theta)\right]+\mathcal{H}_{1,\theta}^{\tau,\nu}\left[\mathcal{W}(\theta)\mathcal{V}^{\sigma}(\theta)\mathcal{U}^{\delta}(\theta)\right]\mathcal{H}_{1,\theta}^{\lambda,\nu}\left[\mathcal{W}(\theta)\mathcal{U}^{\xi}(\theta)\right],$$

which gives the desired inequality (23). □

Remark 6. *Applying Theorem 4 for $\tau=\lambda$, we get Theorem 3.*

Next, we present some fractional proportional inequalities for a family of n positive functions defined on $[1,\infty)$ by utilizing Hadamard proportional fractional integral (7).

Theorem 5. *Let the functions $\mathcal{U}_j$, $(j=1,2,\cdots,n)$ and $\mathcal{V}$ be positive and continuous on the interval $[1,\infty)$. Suppose that for any fixed $k=1,2,\cdots,n$,*

$$\left(\mathcal{V}^{\sigma}(\zeta)\mathcal{U}_k^{\sigma}(\rho)-\mathcal{V}^{\sigma}(\rho)\mathcal{U}_k^{\sigma}(\zeta)\right)\left(\mathcal{U}_k^{\delta-\xi_k}(\rho)-\mathcal{U}_k^{\delta-\xi_k}(\zeta)\right)\geq0,\quad(24)$$

where $\rho,\zeta\in(1,\theta)$, $\theta>1$, $\sigma>0$, $\delta>\xi_k>0$. Assume that the function $\mathcal{W}:[1,\infty)\to\mathbb{R}^{+}$ is positive and continuous. Then for all $\theta>1$, the following inequality for Hadamard proportional fractional integral operator (7) holds;

$$\mathcal{H}_{1,\theta}^{\tau,\nu}\left[\mathcal{W}(\theta)\mathcal{U}_k^{\delta+\sigma}(\theta)\prod_{j\neq k}^{n}\mathcal{U}_j^{\xi_j}(\theta)\right]\mathcal{H}_{1,\theta}^{\tau,\nu}\left[\mathcal{W}(\theta)\mathcal{V}^{\sigma}(\theta)\prod_{j=1}^{n}\mathcal{U}_j^{\xi_j}(\theta)\right]$$
$$\geq\mathcal{H}_{1,\theta}^{\tau,\nu}\left[\mathcal{W}(\theta)\mathcal{U}_k^{\sigma}(\theta)\prod_{j=1}^{n}\mathcal{U}_j^{\xi_j}(\theta)\right]\mathcal{H}_{1,\theta}^{\tau,\nu}\left[\mathcal{W}(\theta)\mathcal{V}^{\sigma}(\theta)\mathcal{U}_k^{\delta}(\theta)\prod_{j\neq k}^{n}\mathcal{U}_j^{\xi_j}(\theta)\right], \tag{25}$$

where $\delta\geq\xi_k>0,\ \tau,\nu,\sigma>0$ *and* $k=1,2,\cdots,n$.

Proof. Consider the function

$$\mathcal{F}_1(\theta,\rho)=\frac{1}{\nu^{\tau}\Gamma(\tau)}e^{\frac{\nu-1}{\nu}(\ln\theta-\ln\rho)}(\ln\theta-\ln\rho)^{\tau-1}\frac{\mathcal{W}(\rho)\prod_{j=1}^{n}\mathcal{U}_j^{\xi_j}(\rho)}{\rho}, \tag{26}$$
$$\xi_j>0, j=1,2,\cdots,n,\ \rho\in(1,\theta),\theta>1.$$

Since the functions $\mathcal{U}_j$, $(j=1,2,\cdots,n)$ and $\mathcal{V}$ satisfy (24) for any fixed $k=1,2,\cdots,n$. Therefore, we can write

$$\mathcal{V}^{\sigma}(\zeta)\mathcal{U}_k^{\delta+\sigma-\xi_k}(\rho)+\mathcal{V}^{\sigma}(\rho)\mathcal{U}_k^{\delta+\sigma-\xi_k}(\zeta)\geq\mathcal{V}^{\sigma}(\zeta)\mathcal{U}_k^{\sigma}(\rho)\mathcal{U}_k^{\delta-\xi_k}(\zeta)+\mathcal{V}^{\sigma}(\rho)\mathcal{U}_k^{\sigma}(\zeta)\mathcal{U}_k^{\delta-\xi_k}(\rho). \tag{27}$$

Multiplying (27) by $\mathcal{F}_1(\theta,\rho)$ (where $\mathcal{F}_1(\theta,\rho)$ is defined in (26)) and integrating the resultant estimates with respect to ρ over $(1,\theta),\theta>1$, we have

$$\mathcal{V}^{\sigma}(\zeta)\frac{1}{\nu^{\tau}\Gamma(\tau)}\int_1^{\infty}e^{\frac{\nu-1}{\nu}(\ln\theta-\ln\rho)}(\ln\theta-\ln\rho)^{\tau-1}\frac{\mathcal{W}(\rho)\prod_{j=1}^{n}\mathcal{U}_j^{\xi_j}(\rho)}{\rho}\mathcal{U}_k^{\delta+\sigma-\xi_k}(\rho)d\rho$$
$$+\mathcal{U}_k^{\delta+\sigma-\xi_k}(\zeta)\frac{1}{\nu^{\tau}\Gamma(\tau)}\int_1^{\infty}e^{\frac{\nu-1}{\nu}(\ln\theta-\ln\rho)}(\ln\theta-\ln\rho)^{\tau-1}\frac{\mathcal{W}(\rho)\prod_{j=1}^{n}\mathcal{U}_j^{\xi_j}(\rho)}{\rho}\mathcal{V}^{\sigma}(\rho)d\rho$$
$$\geq\mathcal{V}^{\sigma}(\zeta)\mathcal{U}_k^{\delta-\xi_k}(\zeta)\frac{1}{\nu^{\tau}\Gamma(\tau)}\int_1^{\infty}e^{\frac{\nu-1}{\nu}(\ln\theta-\ln\rho)}(\ln\theta-\ln\rho)^{\tau-1}\frac{\mathcal{W}(\rho)\prod_{j=1}^{n}\mathcal{U}_j^{\xi_j}(\rho)}{\rho}\mathcal{U}_k^{\sigma}(\rho)d\rho$$
$$+\mathcal{U}_k^{\sigma}(\zeta)\frac{1}{\nu^{\tau}\Gamma(\tau)}\int_1^{\infty}e^{\frac{\nu-1}{\nu}(\ln\theta-\ln\rho)}(\ln\theta-\ln\rho)^{\tau-1}\frac{\mathcal{W}(\rho)\prod_{j=1}^{n}\mathcal{U}_j^{\xi_j}(\rho)}{\rho}\mathcal{V}^{\sigma}(\rho)\mathcal{U}_k^{\delta-\xi_k}(\rho)d\rho,$$

which with aid of (7) gives

$$\mathcal{V}^{\sigma}(\zeta)\mathcal{H}_{1,\theta}^{\tau,\nu}\left[\mathcal{W}(\theta)\mathcal{U}_k^{\delta+\sigma}(\theta)\prod_{j\neq k}^{n}\mathcal{U}_j^{\xi_j}(\theta)\right]+\mathcal{U}_k^{\delta+\sigma-\xi_k}(\zeta)\mathcal{H}_{1,\theta}^{\tau,\nu}\left[\mathcal{W}(\theta)\mathcal{V}^{\sigma}(\theta)\prod_{j=1}^{n}\mathcal{U}_j^{\xi_j}(\theta)\right] \tag{28}$$

$$\geq\mathcal{V}^{\sigma}(\zeta)\mathcal{U}_k^{\delta-\xi_k}(\zeta)\mathcal{H}_{1,\theta}^{\tau,\nu}\left[\mathcal{W}(\theta)\mathcal{U}_k^{\sigma}(\theta)\prod_{j=1}^{n}\mathcal{U}_j^{\xi_j}(\theta)\right]+\mathcal{U}_k^{\sigma}(\zeta)\mathcal{H}_{1,\theta}^{\tau,\nu}\left[\mathcal{W}(\theta)\mathcal{V}^{\sigma}(\theta)\mathcal{U}_k^{\delta+\sigma}(\theta)\prod_{j\neq k}^{n}\mathcal{U}_j^{\xi_j}(\theta)\right]. \tag{29}$$

Now, multiplying (28) by $\mathcal{F}_1(\theta,\zeta)$ (where $\mathcal{F}_1(\theta,\zeta)$ can be obtained from (26) by replacing ρ by ζ) and integrating the resultant estimates with respect to ζ over $(1,\theta),\theta>1$ and then by applying (7), we obtain

$$\begin{aligned}
&\mathcal{H}_{1,\theta}^{\tau,\nu}\left[\mathcal{W}(\theta)\mathcal{U}_k^{\delta+\sigma}(\theta)\prod_{j\neq k}^{n}\mathcal{U}_j^{\xi_j}(\theta)\right]\mathcal{H}_{1,\theta}^{\tau,\nu}\left[\mathcal{W}(\theta)\mathcal{V}^{\sigma}(\theta)\prod_{j=1}^{n}\mathcal{U}_j^{\xi_j}(\theta)\right]\\
&+\mathcal{H}_{1,\theta}^{\tau,\nu}\left[\mathcal{W}(\theta)\mathcal{V}^{\sigma}(\theta)\prod_{j=1}^{n}\mathcal{U}_j^{\xi_j}(\theta)\right]\mathcal{H}_{1,\theta}^{\tau,\nu}\left[\mathcal{W}(\theta)\mathcal{U}_k^{\delta+\sigma}(\theta)\prod_{j\neq k}^{n}\mathcal{U}_j^{\xi_j}(\theta)\right]\\
&\geq\mathcal{H}_{1,\theta}^{\tau,\nu}\left[\mathcal{W}(\theta)\mathcal{U}_k^{\sigma}(\theta)\prod_{j=1}^{n}\mathcal{U}_j^{\xi_j}(\theta)\right]\mathcal{H}_{1,\theta}^{\tau,\nu}\left[\mathcal{W}(\theta)\mathcal{V}^{\sigma}(\theta)\mathcal{U}_k^{\delta}(\theta)\prod_{j\neq k}^{n}\mathcal{U}_j^{\xi_j}(\theta)\right]\\
&+\mathcal{H}_{1,\theta}^{\tau,\nu}\left[\mathcal{W}(\theta)\mathcal{V}^{\sigma}(\theta)\mathcal{U}_k^{\delta}(\theta)\prod_{j\neq k}^{n}\mathcal{U}_j^{\xi_j}(\theta)\right]\mathcal{H}_{1,\theta}^{\tau,\nu}\left[\mathcal{W}(\theta)\mathcal{U}_k^{\sigma}(\theta)\prod_{j=1}^{n}\mathcal{U}_j^{\xi_j}(\theta)\right],
\end{aligned}$$

which completes the desired assertion (25). □

Theorem 6. *Let the functions $\mathcal{U}_j$, $(j=1,2,\cdots,n)$ and $\mathcal{V}$ be positive and continuous on the interval $[1,\infty)$ and satisfy (24) for any fixed $k=1,2,\cdots,n$. Assume that the function $\mathcal{W}:[1,\infty)\to\mathbb{R}^+$ is positive and continuous. Then for all $\theta>1$, the following inequality for Hadamard proportional fractional integral operator (7) holds;*

$$\begin{aligned}
&\mathcal{H}_{1,\theta}^{\tau,\nu}\left[\mathcal{W}(\theta)\mathcal{U}_k^{\delta+\sigma}(\theta)\prod_{j\neq k}^{n}\mathcal{U}_j^{\xi_j}(\theta)\right]\mathcal{H}_{1,\theta}^{\lambda,\nu}\left[\mathcal{W}(\theta)\mathcal{V}^{\sigma}(\theta)\prod_{j=1}^{n}\mathcal{U}_j^{\xi_j}(\theta)\right]\\
&+\mathcal{H}_{1,\theta}^{\lambda,\nu}\left[\mathcal{W}(\theta)\mathcal{U}_k^{\delta+\sigma}(\theta)\prod_{j\neq k}^{n}\mathcal{U}_j^{\xi_j}(\theta)\right]\mathcal{H}_{1,\theta}^{\tau,\nu}\left[\mathcal{W}(\theta)\mathcal{V}^{\sigma}(\theta)\prod_{j=1}^{n}\mathcal{U}_j^{\xi_j}(\theta)\right]\\
&\geq\mathcal{H}_{1,\theta}^{\tau,\nu}\left[\mathcal{W}(\theta)\mathcal{U}_k^{\sigma}(\theta)\prod_{j=1}^{n}\mathcal{U}_j^{\xi_j}(\theta)\right]\mathcal{H}_{1,\theta}^{\lambda,\nu}\left[\mathcal{W}(\theta)\mathcal{V}^{\sigma}(\theta)\mathcal{U}_k^{\delta}(\theta)\prod_{j\neq k}^{n}\mathcal{U}_j^{\xi_j}(\theta)\right]\\
&+\mathcal{H}_{1,\theta}^{\lambda,\nu}\left[\mathcal{W}(\theta)\mathcal{U}_k^{\sigma}(\theta)\prod_{j=1}^{n}\mathcal{U}_j^{\xi_j}(\theta)\right]\mathcal{H}_{1,\theta}^{\tau,\nu}\left[\mathcal{W}(\theta)\mathcal{V}^{\sigma}(\theta)\mathcal{U}_k^{\delta}(\theta)\prod_{j\neq k}^{n}\mathcal{U}_j^{\xi_j}(\theta)\right],
\end{aligned}\tag{30}$$

where $\delta\geq\xi_k>0$, $\tau,\lambda,\nu,\sigma>0$ and $k=1,2,\cdots,n$.

Proof. Taking product on both sides of (28) by $\mathcal{G}_1(\theta,\zeta)=\frac{1}{\nu^{\lambda}\Gamma(\lambda)}e^{\frac{\nu-1}{\nu}(\ln\theta-\ln\zeta)}(\ln\theta-\ln\zeta)^{\lambda-1}\frac{\mathcal{W}(\zeta)\prod_{j=1}^{n}\mathcal{U}_j^{\xi_j}(\zeta)}{\zeta}$, where $\lambda>0,\xi_j>0$ and $\zeta\in(1,\theta),\theta>1$, $j=1,2,\cdots,n$ and integrating the resultant estimates with respect to ζ over $(1,\theta)$, we have

$$\begin{aligned}
&\mathcal{H}_{1,\theta}^{\tau,\nu}\left[\mathcal{W}(\theta)\mathcal{U}_k^{\delta+\sigma}(\theta)\prod_{j\neq k}^{n}\mathcal{U}_j^{\xi_j}(\theta)\right]\frac{1}{\nu^{\lambda}\Gamma(\lambda)}\int_1^{\theta}e^{\frac{\nu-1}{\nu}(\ln\theta-\ln\zeta)}(\ln\theta-\ln\zeta)^{\lambda-1}\frac{\mathcal{W}(\zeta)\prod_{j=1}^{n}\mathcal{U}_j^{\xi_j}(\zeta)}{\zeta}\mathcal{V}^{\sigma}(\zeta)d\zeta\\
&+\mathcal{H}_{1,\theta}^{\tau,\nu}\left[\mathcal{W}(\theta)\mathcal{V}^{\sigma}(\theta)\prod_{j=1}^{n}\mathcal{U}_j^{\xi_j}(\theta)\right]\frac{1}{\nu^{\lambda}\Gamma(\lambda)}\int_1^{\theta}e^{\frac{\nu-1}{\nu}(\ln\theta-\ln\zeta)}(\ln\theta-\ln\zeta)^{\lambda-1}\frac{\mathcal{W}(\zeta)\prod_{j=1}^{n}\mathcal{U}_j^{\xi_j}(\zeta)}{\zeta}\mathcal{U}_k^{\delta+\sigma-\xi_k}(\zeta)d\zeta\\
&\geq\mathcal{H}_{1,\theta}^{\tau,\nu}\left[\mathcal{W}(\theta)\mathcal{U}_k^{\sigma}(\theta)\prod_{j=1}^{n}\mathcal{U}_j^{\xi_j}(\theta)\right]\frac{1}{\nu^{\lambda}\Gamma(\lambda)}\int_1^{\theta}e^{\frac{\nu-1}{\nu}(\ln\theta-\ln\zeta)}(\ln\theta-\ln\zeta)^{\lambda-1}\frac{\mathcal{W}(\zeta)\prod_{j=1}^{n}\mathcal{U}_j^{\xi_j}(\zeta)}{\zeta}\mathcal{V}^{\sigma}(\zeta)\mathcal{U}_k^{\delta-\xi_k}(\zeta)d\zeta\\
&+\mathcal{H}_{1,\theta}^{\tau,\nu}\left[\mathcal{W}(\theta)\mathcal{V}^{\sigma}(\theta)\mathcal{U}_k^{\delta+\sigma}(\theta)\prod_{j\neq k}^{n}\mathcal{U}_j^{\xi_j}(\theta)\right]\frac{1}{\nu^{\lambda}\Gamma(\lambda)}\int_1^{\theta}e^{\frac{\nu-1}{\nu}(\ln\theta-\ln\zeta)}(\ln\theta-\ln\zeta)^{\lambda-1}\frac{\mathcal{W}(\zeta)\prod_{j=1}^{n}\mathcal{U}_j^{\xi_j}(\zeta)}{\zeta}\mathcal{U}_k^{\sigma}(\zeta)d\zeta,
\end{aligned}$$

which in view of (7) yields the desired assertion

$$\mathcal{H}_{1,\theta}^{\tau,\nu}\left[\mathcal{W}(\theta)\mathcal{U}_k^{\delta+\sigma}(\theta)\prod_{j\neq k}^{n}\mathcal{U}_j^{\xi_j}(\theta)\right]\mathcal{H}_{1,\theta}^{\lambda,\nu}\left[\mathcal{W}(\theta)\mathcal{V}^{\sigma}(\theta)\prod_{j=1}^{n}\mathcal{U}_j^{\xi_j}(\theta)\right]$$
$$+\mathcal{H}_{1,\theta}^{\lambda,\nu}\left[\mathcal{W}(\theta)\mathcal{U}_k^{\delta+\sigma}(\theta)\prod_{j\neq k}^{n}\mathcal{U}_j^{\xi_j}(\theta)\right]\mathcal{H}_{1,\theta}^{\tau,\nu}\left[\mathcal{W}(\theta)\mathcal{V}^{\sigma}(\theta)\prod_{j=1}^{n}\mathcal{U}_j^{\xi_j}(\theta)\right]$$
$$\geq\mathcal{H}_{1,\theta}^{\tau,\nu}\left[\mathcal{W}(\theta)\mathcal{U}_k^{\sigma}(\theta)\prod_{j=1}^{n}\mathcal{U}_j^{\xi_j}(\theta)\right]\mathcal{H}_{1,\theta}^{\lambda,\nu}\left[\mathcal{W}(\theta)\mathcal{V}^{\sigma}(\theta)\mathcal{U}_k^{\delta}(\theta)\prod_{j\neq k}^{n}\mathcal{U}_j^{\xi_j}(\theta)\right]$$
$$+\mathcal{H}_{1,\theta}^{\lambda,\nu}\left[\mathcal{W}(\theta)\mathcal{U}_k^{\sigma}(\theta)\prod_{j=1}^{n}\mathcal{U}_j^{\xi_j}(\theta)\right]\mathcal{H}_{1,\theta}^{\tau,\nu}\left[\mathcal{W}(\theta)\mathcal{V}^{\sigma}(\theta)\mathcal{U}_k^{\delta}(\theta)\prod_{j\neq k}^{n}\mathcal{U}_j^{\xi_j}(\theta)\right].$$

□

Remark 7. *Applying Theorem 6 for $\tau=\lambda$, we get Theorem 5.*

Theorem 7. *Let the functions $\mathcal{U}_j$, $(j=1,2,\cdots,n)$ and $\mathcal{V}$ be positive and continuous on the interval $[1,\infty)$ such that the function $\mathcal{V}$ is increasing and the function $\mathcal{U}_j$ for $j=1,2,\cdots,n$ are decreasing on $[1,\infty)$. Assume that the function $\mathcal{W}:[1,\infty)\to\mathbb{R}^+$ is positive and continuous. Then for all $\theta>1$, the following inequality for Hadamard proportional fractional integral operator (7) holds;*

$$\mathcal{H}_{1,\theta}^{\tau,\nu}\left[\mathcal{W}(\theta)\mathcal{U}_k^{\delta}(\theta)\prod_{j\neq k}^{n}\mathcal{U}_j^{\xi_j}(\theta)\right]\mathcal{H}_{1,\theta}^{\tau,\nu}\left[\mathcal{W}(\theta)\mathcal{V}^{\sigma}(\theta)\prod_{j=1}^{n}\mathcal{U}_j^{\xi_j}(\theta)\right]$$
$$\geq\mathcal{H}_{1,\theta}^{\tau,\nu}\left[\mathcal{W}(\theta)\mathcal{V}^{\sigma}(\theta)\mathcal{U}_k^{\delta}(\theta)\prod_{j\neq k}^{n}\mathcal{U}_j^{\xi_j}(\theta)\right]\mathcal{H}_{1,\theta}^{\tau,\nu}\left[\mathcal{W}(\theta)\prod_{j=1}^{n}\mathcal{U}_j^{\xi_j}(\theta)\right],\tag{31}$$

where $\delta\geq\xi_k>0$, $\tau,\nu,\sigma>0$ and $k=1,2,\cdots,n$.

Proof. Under the conditions stated in Theorem 7, we can write

$$(\mathcal{V}^{\sigma}(\zeta)-\mathcal{V}^{\sigma}(\rho))\left(\mathcal{U}_k^{\delta-\xi_k}(\rho)-\mathcal{U}_k^{\delta-\xi_k}(\zeta)\right)\geq 0\tag{32}$$

for any $\rho,\zeta\in(1,\theta),\theta>1,\sigma>0,\delta>\xi_k>0,k=1,2,3,\cdots,n$.

From (32), we have

$$\mathcal{V}^{\sigma}(\zeta)\mathcal{U}_k^{\delta-\xi_k}(\rho)+\mathcal{V}^{\sigma}(\rho)\mathcal{U}_k^{\delta-\xi_k}(\zeta)\geq\mathcal{V}^{\sigma}(\zeta)\mathcal{U}_k^{\delta-\xi_k}(\zeta)+\mathcal{V}^{\sigma}(\rho)\mathcal{U}_k^{\delta-\xi_k}(\rho).\tag{33}$$

Multiplying (27) by $\mathcal{F}_1(\theta,\rho)$ (where $\mathcal{F}_1(\theta,\rho)$ is defined in (26)) and integrating the resultant estimates with respect to ρ over $(1,\theta),\theta>1$, we have

$$\mathcal{V}^{\sigma}(\zeta)\frac{1}{\nu^{\tau}\Gamma(\tau)}\int_1^{\theta}e^{\frac{\nu-1}{\nu}(\ln\theta-\ln\rho)}(\ln\theta-\ln\rho)^{\tau-1}\frac{\mathcal{W}(\rho)\prod_{j=1}^{n}\mathcal{U}_j^{\xi_j}(\rho)}{\rho}\mathcal{U}_k^{\delta-\xi_k}(\rho)d\rho$$
$$+\mathcal{U}_k^{\delta-\xi_k}(\zeta)\frac{1}{\nu^{\tau}\Gamma(\tau)}\int_1^{\theta}e^{\frac{\nu-1}{\nu}(\ln\theta-\ln\rho)}(\ln\theta-\ln\rho)^{\tau-1}\frac{\mathcal{W}(\rho)\prod_{j=1}^{n}\mathcal{U}_j^{\xi_j}(\rho)}{\rho}\mathcal{V}^{\sigma}(\rho)d\rho$$
$$\geq\mathcal{V}^{\sigma}(\zeta)\mathcal{U}_k^{\delta-\xi_k}(\zeta)\frac{1}{\nu^{\tau}\Gamma(\tau)}\int_1^{\theta}e^{\frac{\nu-1}{\nu}(\ln\theta-\ln\rho)}(\ln\theta-\ln\rho)^{\tau-1}\frac{\mathcal{W}(\rho)\prod_{j=1}^{n}\mathcal{U}_j^{\xi_j}(\rho)}{\rho}d\rho$$
$$+\frac{1}{\nu^{\tau}\Gamma(\tau)}\int_1^{\theta}e^{\frac{\nu-1}{\nu}(\ln\theta-\ln\rho)}(\ln\theta-\ln\rho)^{\tau-1}\frac{\mathcal{W}(\rho)\prod_{j=1}^{n}\mathcal{U}_j^{\xi_j}(\rho)}{\rho}\mathcal{V}^{\sigma}(\rho)\mathcal{U}_k^{\delta-\xi_k}(\rho)d\rho,$$

which in view of Hadamard proportional fractional integral (7) becomes

$$\mathcal{V}^{\sigma}(\zeta)\mathcal{H}_{1,\theta}^{\tau,\nu}\left[\mathcal{W}(\theta)\mathcal{U}_k^{\delta}(\theta)\prod_{j\neq k}^{n}\mathcal{U}_j^{\xi_j}(\theta)\right]+\mathcal{U}_k^{\delta-\xi_k}(\zeta)\mathcal{H}_{1,\theta}^{\tau,\nu}\left[\mathcal{W}(\theta)\mathcal{V}^{\sigma}(\theta)\prod_{j=1}^{n}\mathcal{U}_j^{\xi_j}(\theta)\right]$$
$$\geq\mathcal{V}^{\sigma}(\zeta)\mathcal{U}_k^{\delta-\xi_k}(\zeta)\mathcal{H}_{1,\theta}^{\tau,\nu}\left[\mathcal{W}(\theta)\prod_{j=1}^{n}\mathcal{U}_j^{\xi_j}(\theta)\right]+\mathcal{H}_{1,\theta}^{\tau,\nu}\left[\mathcal{W}(\theta)\mathcal{V}^{\sigma}(\theta)\mathcal{U}_k^{\delta}(\theta)\prod_{j\neq k}^{n}\mathcal{U}_j^{\xi_j}(\theta)\right]. \tag{34}$$

Now, multiplying (34) by $\mathcal{F}_1(\theta,\zeta)$ (where $\mathcal{F}_1(\theta,\zeta)$ can be obtained from (26) by replacing ρ by ζ) and integrating the resultant estimates with respect to ζ over $(1,\theta),\theta>1$ and then by applying (7), we obtain

$$\mathcal{H}_{1,\theta}^{\tau,\nu}\left[\mathcal{W}(\theta)\mathcal{U}_k^{\delta}(\theta)\prod_{j\neq k}^{n}\mathcal{U}_j^{\xi_j}(\theta)\right]\mathcal{H}_{1,\theta}^{\tau,\nu}\left[\mathcal{W}(\theta)\mathcal{V}^{\sigma}(\theta)\prod_{j=1}^{n}\mathcal{U}_j^{\xi_j}(\theta)\right]$$
$$\geq\mathcal{H}_{1,\theta}^{\tau,\nu}\left[\mathcal{W}(\theta)\mathcal{V}^{\sigma}(\theta)\mathcal{U}_k^{\delta}(\theta)\prod_{j\neq k}^{n}\mathcal{U}_j^{\xi_j}(\theta)\right]\mathcal{H}_{1,\theta}^{\tau,\nu}\left[\mathcal{W}(\theta)\prod_{j=1}^{n}\mathcal{U}_j^{\xi_j}(\theta)\right],$$

which is the desired assertion (31). □

Theorem 8. *Let the functions $\mathcal{U}_j$, $(j=1,2,\cdots,n)$ and $\mathcal{V}$ be positive and continuous on the interval $[1,\infty)$ such that the function $\mathcal{V}$ is increasing and the function $\mathcal{U}_j$ for $j=1,2,\cdots,n$ are decreasing on $[1,\infty)$. Assume that the function $\mathcal{W}:[1,\infty)\to\mathbb{R}^+$ is positive and continuous. Then for all $\theta>1$, the following inequality for Hadamard proportional fractional integral operator (7) holds;*

$$\mathcal{H}_{1,\theta}^{\tau,\nu}\left[\mathcal{W}(\theta)\mathcal{U}_k^{\delta}(\theta)\prod_{j\neq k}^{n}\mathcal{U}_j^{\xi_j}(\theta)\right]\mathcal{H}_{1,\theta}^{\lambda,\nu}\left[\mathcal{W}(\theta)\mathcal{V}^{\sigma}(\theta)\prod_{j=1}^{n}\mathcal{U}_j^{\xi_j}(\theta)\right]$$
$$+\mathcal{H}_{1,\theta}^{\lambda,\nu}\left[\mathcal{W}(\theta)\mathcal{U}_k^{\delta}(\theta)\prod_{j\neq k}^{n}\mathcal{U}_j^{\xi_j}(\theta)\right]\mathcal{H}_{1,\theta}^{\tau,\nu}\left[\mathcal{W}(\theta)\mathcal{V}^{\sigma}(\theta)\prod_{j=1}^{n}\mathcal{U}_j^{\xi_j}(\theta)\right]$$
$$\geq\mathcal{H}_{1,\theta}^{\tau,\nu}\left[\mathcal{W}(\theta)\mathcal{V}^{\sigma}(\theta)\mathcal{U}_k^{\delta}(\theta)\prod_{j\neq k}^{n}\mathcal{U}_j^{\xi_j}(\theta)\right]\mathcal{H}_{1,\theta}^{\lambda,\nu}\left[\mathcal{W}(\theta)\prod_{j=1}^{n}\mathcal{U}_j^{\xi_j}(\theta)\right] \tag{35}$$
$$+\mathcal{H}_{1,\theta}^{\tau,\nu}\left[\mathcal{W}(\theta)\mathcal{V}^{\sigma}(\theta)\mathcal{U}_k^{\delta}(\theta)\prod_{j\neq k}^{n}\mathcal{U}_j^{\xi_j}(\theta)\right]\mathcal{H}_{1,\theta}^{\lambda,\nu}\left[\mathcal{W}(\theta)\prod_{j=1}^{n}\mathcal{U}_j^{\xi_j}(\theta)\right],$$

where $\delta\geq\xi_k>0$, $\tau,\lambda,\nu,\sigma>0$ and $k=1,2,\cdots,n$.

Proof. To obtain the desire assertion (35), we multiply (34) by $\mathcal{G}_1(\theta,\zeta)=\dfrac{1}{\nu^{\lambda}\Gamma\binom{\lambda}{}}e^{\frac{\nu-1}{\nu}(\ln\theta-\ln\zeta)}(\ln\theta-\ln\zeta)^{\lambda-1}\dfrac{\mathcal{W}(\zeta)\prod_{j=1}^{n}\mathcal{U}_j^{\xi_j}(\zeta)}{\zeta}$, where $\lambda>0,\xi_j>0$ and $\zeta\in(1,\theta),\theta>1$, $j=1,2,\cdots,n$ and integrating the resultant estimates with respect to ζ over $(1,\theta)$, we have

$$\frac{1}{\nu^{\lambda}\Gamma(\lambda)}\int_1^{\theta} e^{\frac{\nu-1}{\nu}(\ln\theta-\ln\zeta)}(\ln\theta-\ln\zeta)^{\lambda-1}\frac{\mathcal{W}(\zeta)\prod_{j=1}^{n}\mathcal{U}_j^{\xi_j}(\zeta)}{\zeta}\mathcal{V}^{\sigma}(\zeta)d\zeta\mathcal{H}_{1,\theta}^{\tau,\nu}\left[\mathcal{W}(\theta)\mathcal{U}_k^{\delta}(\theta)\prod_{j\neq k}^{n}\mathcal{U}_j^{\xi_j}(\theta)\right]$$
$$+\frac{1}{\nu^{\lambda}\Gamma(\lambda)}\int_1^{\theta} e^{\frac{\nu-1}{\nu}(\ln\theta-\ln\zeta)}(\ln\theta-\ln\zeta)^{\lambda-1}\frac{\mathcal{W}(\zeta)\prod_{j=1}^{n}\mathcal{U}_j^{\xi_j}(\zeta)}{\zeta}\mathcal{U}_k^{\delta-\xi_k}(\zeta)d\zeta\mathcal{H}_{1,\theta}^{\tau,\nu}\left[\mathcal{W}(\theta)\mathcal{V}^{\sigma}(\theta)\prod_{j=1}^{n}\mathcal{U}_j^{\xi_j}(\theta)\right]$$
$$\geq\frac{1}{\nu^{\lambda}\Gamma(\lambda)}\int_1^{\theta} e^{\frac{\nu-1}{\nu}(\ln\theta-\ln\zeta)}(\ln\theta-\ln\zeta)^{\lambda-1}\frac{\mathcal{W}(\zeta)\prod_{j=1}^{n}\mathcal{U}_j^{\xi_j}(\zeta)}{\zeta}\mathcal{V}^{\sigma}(\zeta)\mathcal{U}_k^{\delta-\xi_k}(\zeta)d\zeta\mathcal{H}_{1,\theta}^{\tau,\nu}\left[\mathcal{W}(\theta)\prod_{j=1}^{n}\mathcal{U}_j^{\xi_j}(\theta)\right]$$
$$+\mathcal{H}_{1,\theta}^{\tau,\nu}\left[\mathcal{W}(\theta)\mathcal{V}^{\sigma}(\theta)\mathcal{U}_k^{\delta}(\theta)\prod_{j\neq k}^{n}\mathcal{U}_j^{\xi_j}(\theta)\right]\frac{1}{\nu^{\lambda}\Gamma(\lambda)}\int_1^{\theta} e^{\frac{\nu-1}{\nu}(\ln\theta-\ln\zeta)}(\ln\theta-\ln\zeta)^{\lambda-1}\frac{\mathcal{W}(\zeta)\prod_{j=1}^{n}\mathcal{U}_j^{\xi_j}(\zeta)}{\zeta}.$$

which in view of Hadamard proportional fractional integral (7) gives

$$\mathcal{H}_{1,\theta}^{\tau,\nu}\left[\mathcal{W}(\theta)\mathcal{U}_k^{\delta}(\theta)\prod_{j\neq k}^{n}\mathcal{U}_j^{\xi_j}(\theta)\right]\mathcal{H}_{1,\theta}^{\lambda,\nu}\left[\mathcal{W}(\theta)\mathcal{V}^{\sigma}(\theta)\prod_{j=1}^{n}\mathcal{U}_j^{\xi_j}(\theta)\right]$$
$$+\mathcal{H}_{1,\theta}^{\lambda,\nu}\left[\mathcal{W}(\theta)\mathcal{U}_k^{\delta}(\theta)\prod_{j\neq k}^{n}\mathcal{U}_j^{\xi_j}(\theta)\right]\mathcal{H}_{1,\theta}^{\tau,\nu}\left[\mathcal{W}(\theta)\mathcal{V}^{\sigma}(\theta)\prod_{j=1}^{n}\mathcal{U}_j^{\xi_j}(\theta)\right]$$
$$\geq\mathcal{H}_{1,\theta}^{\tau,\nu}\left[\mathcal{W}(\theta)\mathcal{V}^{\sigma}(\theta)\mathcal{U}_k^{\delta}(\theta)\prod_{j\neq k}^{n}\mathcal{U}_j^{\xi_j}(\theta)\right]\mathcal{H}_{1,\theta}^{\lambda,\nu}\left[\mathcal{W}(\theta)\prod_{j=1}^{n}\mathcal{U}_j^{\xi_j}(\theta)\right]$$
$$+\mathcal{H}_{1,\theta}^{\tau,\nu}\left[\mathcal{W}(\theta)\mathcal{V}^{\sigma}(\theta)\mathcal{U}_k^{\delta}(\theta)\prod_{j\neq k}^{n}\mathcal{U}_j^{\xi_j}(\theta)\right]\mathcal{H}_{1,\theta}^{\lambda,\nu}\left[\mathcal{W}(\theta)\prod_{j=1}^{n}\mathcal{U}_j^{\xi_j}(\theta)\right],$$

which completes the proof of (35). □

Remark 8. *Applying Theorem 8 for $\tau=\lambda$, we get Theorem 7.*

4. Concluding Remarks

Recently Jarad et al. [46] introduced the idea of generalized proportional fractional integral operators which comprises exponential in their kernels. Later on, Rahman et al. [47] studied these operators and defined Hadamard proportional fractional integrals. They established certain inequalities for convex functions by employing Hadamard proportional fractional integrals. In [49], the authors defined bounds of generalized proportional fractional integral operators for convex functions and their applications. Motivated by the above, here we presented certain inequalities by employing Hadamard proportional fractional integrals. The inequalities obtained in this paper generalized the inequalities presented earlier by Houas [45].

Author Contributions: Conceptualization, K.S.N.; Formal analysis, T.A.; Writing—original draft, G.R. and K.S.N.; Writing—review & editing, T.A. All authors have read and agreed to the published version of the manuscript.

References

1. Anber, A.; Dahmani, Z.; Bendoukha, B. New integral inequalities of Feng Qi type via Riemann–Liouville fractional integration. *Facta Univ. Ser. Math. Inform.* **2012**, *27*, 13–22.
2. Belarbi, S.; Dahmani, Z. On some new fractional integral inequalities. *J. Inequal. Pure Appl. Math.* **2009**, *10*, 1–12.

3. Houas, M. Certain weighted integral inequalities involving the fractional hypergeometric operators. *Sci. Ser. A Math. Sci.* **2016**, *27*, 87–97.
4. Kilbas, A.A.; Srivastava, H.M.; Trujillo, J.J. *Theory and Application of Fractional Differential Equations*; Elsevier: Amersterdam, The Netherlands, 2006.
5. Podlubny, I. *Fractional Differential Equations*; Academic Press: London, UK, 1999.
6. Aldhaifallah, M.; Tomar, M.; Nisar, K.S.; Purohit, S.D. Some new inequalities for (k,s)-fractional integrals. *J. Nonlinear Sci. Appl.* **2016**, *9*, 5374–5381. [CrossRef]
7. Dahmani, Z.; Tabharit, L. On weighted Gruss type inequalities via fractional integration. *J. Adv. Res. Pure Math.* **2010**, *2*, 31–38. [CrossRef]
8. Dahmani, Z. New inequalities in fractional integrals. *Int. J. Nonlinear Sci.* **2010**, *9*, 493–497.
9. Nisar, K.S.; Qi, F.; Rahman, G.; Mubeen, S.; Arshad, M. Some inequalities involving the extended gamma function and the Kummer confluent hypergeometric k-function. *J. Inequal. Appl.* **2018**, *2018*, 135. [CrossRef]
10. Nisar, K.S.; Rahman, G.; Choi, J.; Mubeen, S.; Arshad, M. Certain Gronwall type inequalities associated with Riemann–Liouville k- and Hadamard k-fractional derivatives and their applications. *East Asian Math. J.* **2018**, *34*, 249–263.
11. Rahman, G.; Nisar, K.S.; Mubeen, S.; Choi, J. Certain Inequalities involving the (k,ρ)-fractional integral operator. *Far East J. Math. Sci. (FJMS)* **2018**, *103*, 1879–1888. [CrossRef]
12. Sarikaya, M.Z.; Budak, H. Generalized Ostrowski type inequalities for local fractional integrals. *Proc. Am. Math. Soc.* **2017**, *145*, 1527–1538. [CrossRef]
13. Sarikaya, M.Z.; Dahmani, Z.; Kiris, M.E.; Ahmad, F. (k,s)-Riemann–Liouville fractional integral and applications. *Hacet. J. Math. Stat.* **2016**, *45*, 77–89. [CrossRef]
14. Set, E.; Tomar, M.; Sarikaya, M.Z. On generalized Grüss type inequalities for k-fractional integrals. *Appl. Math. Comput.* **2015**, *269*, 29–34. [CrossRef]
15. Set, E.; Noor, M.A.; Awan, M.U.; Gözpinar, A. Generalized Hermite–Hadamard type inequalities involving fractional integral operators. *J. Inequal. Appl.* **2017**, *2017*, 169. [CrossRef]
16. Agarwal, P.; Jleli, M.; Tomar, M. Certain Hermite–Hadamard type inequalities via generalized k-fractional integrals. *J. Inequal. Appl.* **2017**, *2017*, 55. [CrossRef] [PubMed]
17. Chinchane, V.L.; Pachpatte, D.B. Some new integral inequalities using Hadamard fractional integral operator. *Adv. Inequal. Appl.* **2014**, *2014*, 12.
18. Dahmani, Z.; Benzidane, A. New inequalities using Q-fractional theory. *Bull. Math. Anal. Appl.* **2012**, *4*, 190–196.
19. Yang, W. Some new Chebyshev and Gruss-type integral inequalities for Saigo fractional integral operators and Their q-analogues. *Filomat* **2015**, *29*, 1269–1289. [CrossRef]
20. Huang, C.J.; Rahman, G.; Nisar, K.S.; Ghaffar, A.; Qi, F. Some Inequalities of Hermite–Hadamard type for k-fractional conformable integrals. *Aust. J. Math. Anal. Appl.* **2019**, *16*, 1–9.
21. Mubeen, S.; Habib, S.; Naeem, M.N. The Minkowski inequality involving generalized k-fractional conformable integral. *J. Inequalities Appl.* **2019**, *2019*, 81. [CrossRef]
22. Qi, F.; Rahman, G.; Hussain, S.M.; Du, W.S.; Nisar, K.S. Some inequalities of Čebyšev type for conformable k-fractional integral operators. *Symmetry* **2018**, *10*, 614. [CrossRef]
23. Rahman, G.; Nisar, K.S.; Qi, F. Some new inequalities of the Gruss type for conformable fractional integrals. *AIMS Math.* **2018**, *3*, 575–583. [CrossRef]
24. Rahman, G.; Ullah, Z.; Khan, A.; Set, E.; Nisar, K.S. Certain Chebyshev type inequalities involving fractional conformable integral operators. *Mathematics* **2019**, *7*, 364. [CrossRef]
25. Nisar, K.S.; Rahman, G.; Mehrez, K. Chebyshev type inequalities via generalized fractional conformable integrals. *J. Inequal. Appl.* **2019**, *2019*, 245. [CrossRef]
26. Niasr, K.S.; Tassadiq, A.; Rahman, G.; Khan, A. Some inequalities via fractional conformable integral operators. *J. Inequal. Appl.* **2019**, *2019*, 217. [CrossRef]
27. Tassaddiq, A.; Rahman, G.; Nisar, K.S.; Samraiz, M. Certain fractional conformable inequalities for the weighted and the extended Chebyshev functionals. *Adv. Differ. Equ.* **2020**, *2020*, 96. [CrossRef]
28. Nisar, K.S.; Rahman, G.; Khan, A. Some new inequalities for generalized fractional conformable integral operators. *Adv. Differ. Equ.* **2019**, *2019*, 427. [CrossRef]
29. Nisar, K.S.; Rahman, G.; Tassaddiq, A.; Khan, A.; Abouzaid, M.S. Certain generalized fractional integral inequalities. *AIMS Math.* **2020**, *5*, 1588–1602. [CrossRef]

30. Rahman, G.; Nisar, K.S.; Ghaffar, A.; Qi, F. Some inequalities of the Grüss type for conformable *k*-fractional integral operators. *RACSAM* **2020**, *114*, 9. [CrossRef]
31. Rahman, G.; Abdeljawad, T.; Khan, A.; Nisar, K.S. Some fractional proportional integral inequalities. *J. Inequal. Appl.* **2019**, *2019*, 244. [CrossRef]
32. Rahman, G.; Khan, A.; Abdeljawad, T.; Nisar, K.S. The Minkowski inequalities via generalized proportional fractional integral operators. *Adv. Differ. Equ.* **2019**, *2019*, 287. [CrossRef]
33. Rahman, G.; Nisar, K.S.; Abdeljawad, T.; Ullah, S. Certain Fractional Proportional Integral Inequalities via Convex Functions. *Mathematics* **2020**, *8*, 222. [CrossRef]
34. Khalil, R.; Horani, M.A.; Yousef, A.; Sababheh, M. A new definition of fractional derivative. *J. Comput. Appl. Math.* **2014**, *264*, 65–70. [CrossRef]
35. Abdeljawad, T. On Conformable Fractional Calculus. *J. Comput. Appl. Math.* **2015**, *279*, 57–66. [CrossRef]
36. Abdeljawad, T.; Baleanu, D. Monotonicity results for fractional difference operators with discrete exponential kernels. *Adv. Differ. Equ.* **2017**, *2017*, 78. [CrossRef]
37. Abdeljawad, T.; Baleanu, D. On Fractional Derivatives with Exponential Kernel and their Discrete Versions. *Rep. Math. Phys.* **2017**, *80*, 11–27. [CrossRef]
38. Atangana, A.; Baleanu, D. New fractional derivatives with nonlocal and non-singular kernel: Theory and application to heat transfer model. *Thermal Sci.* **2016**, *20*, 763–769. [CrossRef]
39. Jarad, F.; Ugurlu, E.; Abdeljawad, T.; Baleanu, D. On a new class of fractional operators. *Adv. Differ. Equ.* **2017**, *2017*, 247. [CrossRef]
40. Anderson, D.R.; Ulness, D.J. Newly defined conformable derivatives. *Adv. Dyn. Syst. Appl.* **2015**, *10*, 109–137.
41. Caputo, M.; Fabrizio, M. A new Definition of Fractional Derivative without Singular Kernel. *Progr. Fract. Differ. Appl.* **2015**, *1*, 73–85.
42. Losada, J.; Nieto, J.J. Properties of a New Fractional Derivative without Singular Kernel. *Progr. Fract. Differ. Appl.* **2015**, *1*, 87–92.
43. Liu, W.; Ngô, Q.A.; Huy, V.N. Several interesting integral inequalities. *J. Math. Inequ.* **2009**, *3*, 201–212. [CrossRef]
44. Dahmani, Z. New classes of integral inequalities of fractional order. *Le Mat.* **2014**, *69*, 237–247. [CrossRef]
45. Houas, M. On some generalized integral inequalities for Hadamard fractional integrals. *Med. J. Model. Simul.* **2018**, *9*, 43–52.
46. Jarad, F.; Abdeljawad, T.; Alzabut, J. Generalized fractional derivatives generated by a class of local proportional derivatives. *Eur. Phys. J. Spec. Top.* **2017**, *226*, 3457–3471. [CrossRef]
47. Rahman, G.; Abdeljawad, T.; Jarad, F.; Khan, A.; Nisar, K.S. Certain inequalities via generalized proportional Hadamard fractional integral operators. *Adv. Differ. Equ.* **2019**, *2019*, 454. [CrossRef]
48. Samko, S.G.; Kilbas, A.A.; Marichev, O.I. *Fractional Integrals and Derivatives, Theory and Applications*; Edited and with a foreword by S. M. Nikol'skiĭ, Translated from the 1987 Russian original, Revised by the authors; Gordon and Breach Science Publishers: Yverdon, Switzerland, 1993.
49. Rahmnan, G.; Abdeljawad, T.; Jarad, F.; Nisar, K.S. Bounds of Generalized Proportional Fractional Integrals in General Form via Convex Functions and their Applications. *Mathematics* **2020**, *8*, 113. [CrossRef]

Permissions

All chapters in this book were first published by MDPI; hereby published with permission under the Creative Commons Attribution License or equivalent. Every chapter published in this book has been scrutinized by our experts. Their significance has been extensively debated. The topics covered herein carry significant findings which will fuel the growth of the discipline. They may even be implemented as practical applications or may be referred to as a beginning point for another development.

The contributors of this book come from diverse backgrounds, making this book a truly international effort. This book will bring forth new frontiers with its revolutionizing research information and detailed analysis of the nascent developments around the world.

We would like to thank all the contributing authors for lending their expertise to make the book truly unique. They have played a crucial role in the development of this book. Without their invaluable contributions this book wouldn't have been possible. They have made vital efforts to compile up to date information on the varied aspects of this subject to make this book a valuable addition to the collection of many professionals and students.

This book was conceptualized with the vision of imparting up-to-date information and advanced data in this field. To ensure the same, a matchless editorial board was set up. Every individual on the board went through rigorous rounds of assessment to prove their worth. After which they invested a large part of their time researching and compiling the most relevant data for our readers.

The editorial board has been involved in producing this book since its inception. They have spent rigorous hours researching and exploring the diverse topics which have resulted in the successful publishing of this book. They have passed on their knowledge of decades through this book. To expedite this challenging task, the publisher supported the team at every step. A small team of assistant editors was also appointed to further simplify the editing procedure and attain best results for the readers.

Apart from the editorial board, the designing team has also invested a significant amount of their time in understanding the subject and creating the most relevant covers. They scrutinized every image to scout for the most suitable representation of the subject and create an appropriate cover for the book.

The publishing team has been an ardent support to the editorial, designing and production team. Their endless efforts to recruit the best for this project, has resulted in the accomplishment of this book. They are a veteran in the field of academics and their pool of knowledge is as vast as their experience in printing. Their expertise and guidance has proved useful at every step. Their uncompromising quality standards have made this book an exceptional effort. Their encouragement from time to time has been an inspiration for everyone.

The publisher and the editorial board hope that this book will prove to be a valuable piece of knowledge for researchers, students, practitioners and scholars across the globe.

List of Contributors

Bashir Ahmad, Ymnah Alruwaily and Ahmed Alsaedi
Nonlinear Analysis and Applied Mathematics (NAAM)-Research Group, Department of Mathematics, Faculty of Science, King Abdulaziz University, Jeddah 21589, Saudi Arabia

Sotiris K. Ntouyas
Nonlinear Analysis and Applied Mathematics (NAAM)-Research Group, Department of Mathematics, Faculty of Science, King Abdulaziz University, Jeddah 21589, Saudi Arabia
Department of Mathematics, University of Ioannina, Ioannina 45110, Greece

Yu Chen
Department of Mathematics, Guizhou University, Guiyang 550025, China

Jin Rong Wang
Department of Mathematics, Guizhou University, Guiyang 550025, China
School of Mathematical Sciences, Qufu Normal University, Qufu 273165, China

Nazim I Mahmudov and Areen Al-Khateeb
Eastern Mediterranean University, Gazimagusa 99628, T.R. North Cyprus, Mersin 10, Turkey

Abdeljabbar Ghanmi
Faculté des Sciences de Tunis, Université de Tunis El Manar, Tunis 1060, Tunisia

Samah Horrigue
Department of Mathematics, Higher Institute of Applied Science and Technology, University of Monastir, Monastir 5000, Tunisia

Samar Madian
Basic Sciences Department, Higher Institute for Engineering and Technology, New Damietta 34517, Egypt

Guotao Wang
School of Mathematics and Computer Science, Shanxi Normal University, Linfen, Shanxi 041004, China
College of Mathematics and System Science, Shandong University of Science and Technology, Qingdao, Shandong 266590, China
Nonlinear Analysis and Applied Mathematics (NAAM) Research Group, Department of Mathematics, Faculty of Science, King Abdulaziz University, Jeddah 21589, Saudi Arabia

Madeaha Alghanmi
Nonlinear Analysis and Applied Mathematics (NAAM)-Research Group, Department of Mathematics, Faculty of Science, King Abdulaziz University, Jeddah 21589, Saudi Arabia

Hari M. Srivastava
Department of Mathematics and Statistics, University of Victoria, Victoria, BC V8W 3R4, Canada
Department of Medical Research, China Medical University Hospital, China Medical University, Taichung 40402, Taiwan

Jia Wei He
Faculty of Mathematics and Computational Science, Xiangtan University, Xiangtan 411105, China

Yong Liang
Faculty of Information Technology, Macau University of Science and Technology, Macau 999078, China

Yong Zhou
Faculty of Mathematics and Computational Science, Xiangtan University, Xiangtan 411105, China
Nonlinear Analysis and Applied Mathematics (NAAM) Research Group, Faculty of Science, King Abdulaziz University, Jeddah 21589, Saudi Arabia

Sina Etemad
Department of Mathematics, Azarbaijan Shahid Madani University, Azarshahr, Tabriz, Iran

Saima Rashid and Muhammad Aslam Noor
Department of Mathematics, COMSATS University Islamabad, Islamabad 45550, Pakistan

Fahd Jarad
Department of Mathematics, Çankaya University, Etimesgut, 06790 Ankara, Turkey

Akbar Zada, Shaheen Fatima and Zeeshan Ali
Department of Mathematics, University of Peshawar, Khyber Pakhtunkhwa 25000, Pakistan

Jiafa Xu
School of Mathematical Sciences, Chongqing Normal University, Chongqing 401331, China

Yujun Cui
State Key Laboratory of Mining Disaster Prevention and Control Co-founded by Shandong Province and the Ministry of Science and Technology, Shandong University of Science and Technology, Qingdao 266590, China

Imed Bachar and Hassan Eltayeb
Mathematics Department, College of Science, King Saud University, Riyadh 11451, Saudi Arabia

Habib Mâagli
Department of Mathematics, College of Sciences and Arts, King Abdulaziz University, Rabigh Campus, Rabigh 21911, Saudi Arabia
Analyse Harmonique et Théorie du Potentiel, LR10ES09 Modélisation Mathématique, Faculté des Sciences de Tunis, Université de Tunis El Manar, Tunis 2092, Tunisie

Ohud Almutairi
Department of Mathematics, University of Hafr Al-Batin, Hafr Al-Batin 31991, Saudi Arabia

Adem Kılıçman
Department of Mathematics and Institute for Mathematical Research, University Putra Malaysia, Serdang 43400, Malaysia

Badr Alqahtani and Sara Salem Alzaid
Department of Mathematics, College of Science, King Saud University, Riyadh 11451, Saudi Arabia

Saïd Abbas
Department of Mathematics, Tahar Moulay University of Saïda, EN-Nasr, 20000 Saïda, Algeria

Mouffak Benchohra
Department of Mathematics, College of Science, King Saud University, Riyadh 11451, Saudi Arabia
Laboratory of Mathematics, Djillali Liabes University of Sidi Bel-Abbès, 22000 Sidi Bel-Abbès, Algeria

Abrar Broom
Nonlinear Analysis and Applied Mathematics (NAAM)-Research Group, Department of Mathematics, Faculty of Science, King Abdulaziz University, Jeddah 21589, Saudi Arabia

Samir Saker
Department of Mathematics, Faculty of Science, Mansoura University, Mansoura 35516, Egypt

Mohammed Kenawy
Department of Mathematics, Faculty of Science, Fayoum University, Fayoum 63514, Egypt

Ghada AlNemer
Department of Mathematical Science, College of Science, Princess Nourah bint Abdulrahman University, Riyadh, Saudi 11656, Arabia

Mohammed Zakarya
Department of Mathematics, College of Science, King Khalid University, Abha 61413, Saudi Arabia
Department of Mathematics, Faculty of Science, Al-Azhar University, Assiut 71524, Egypt

Gauhar Rahman
Department of Mathematics, Shaheed Benazir Bhutto University, Sheringal, Upper Dir 18000, Pakistan

Kottakkaran Sooppy Nisar
Department of Mathematics, College of Arts and Sciences, Prince Sattam bin Abdulaziz University, Wadi Aldawaser 11991, Saudi Arabia

Thabet Abdeljawad
Department of Mathematics and General Sciences, Prince Sultan University, Riyadh 11586, Saudi Arabia
Department of Medical Research, China Medical University, Taichung 40402, Taiwan
Department of Computer Science and Information Engineering, Asia University, Taichung 40402, Taiwan

Index

www.ingramcontent.com/pod-product-compliance
Lightning Source LLC
LaVergne TN
LVHW081017210926
848485LV00004B/887